ORGANIC COATING TECHNOLOGY

BOOKS BY PAYNE, H. F.

Organic Coating Technology

Volume 1, First Edition; Volume 2

ORGANIC COATING TECHNOLOGY

VOLUME I

Oils, Resins, Varnishes, and Polymers

HENRY FLEMING PAYNE

TECHNICAL EDITOR, AMERICAN CYANAMID CO.
ADJUNCT PROFESSOR, POLYTECHNIC INSTITUTE OF BROOKLYN

NEW YORK · JOHN WILEY & SONS, INC.
London · Sydney

SECOND PRINTING, JULY, 1964

Library of Congress Catalog Card Number: 54–5971

PRINTED IN THE UNITED STATES OF AMERICA

Dedicated to

NOBILITY of THOUGHT

as exemplified by

my wife

LEE FLEMING PAYNE

"As a man thinketh . . . so is he."

Count each day lost wherein no new knowledge is gained.

Knowledge is the power with which men make history.

Tyrants abuse the power of knowledge; their selfish gains soon crumble to dust, and are scattered by the winds.

Honest men use the power of knowledge; through devotion to truth, love of beauty, and compassion for weakness, they build civilization on these foundations:

the development of language
the cultivation of earth
the building of communities
the wonder of writing
the logic of mathematics
the challenge of philosophy
the beauty of art
the Golden Rule
the confines of liberty
the marvel of printing
the harnessing of wind and water, steam and electricity, the atom, and radiant energy
the spread of industry
the elimination of disease
the research of science
the conquest of space and time
the triumph of the mind
the strength of righteousness
the guidance of the Spirit

Man is more than the sum of his physical parts.

PREFACE

This book, together with Volume II, which is in preparation, is intended for students in paint courses, and for new employees in the oil, resin, pigment, and paint industries. The two volumes are designed to outline in compact form the current theory and practice regarding organic coatings. It is with some regret that much interesting material has been omitted in the interest of brevity; practically no historical background has been included. It is hoped that the general form and arrangement will lend themselves to periodic revision without bulky increase in size. The two volumes are based on two lecture courses which I give at the Polytechnic Institute of Brooklyn.

A work of this kind is made possible only by reviewing and compiling the brilliant ideas and ingenious experimental techniques which have been published in the scientific literature. We of today should raise our voices in heartfelt thanks for this treasure house of knowledge which was not available to those mental giants who pioneered our scientific age. I have given a credit line to authors, scientific societies, and manufacturers for direct citations from their literature. It would be almost impossible to give full credit to my many friends in research and industry who have reviewed various sections of this work and contributed valuable ideas; even to mention their names would require much space. However, despite the best of human efforts there will be inevitable errors of omission and commission. It is my earnest desire that these be brought to my attention so that future revisions may be more satisfactory.

Many specific formulations are given which have been taken from the trade literature. It should be clearly understood that these are suggestions only, intended to provide the reader with a starting point for his own investigations. Although I made an effort to select the most representative materials and formulations, my choice was influenced somewhat by personal familiarity. The reader should check with other manufacturers for comparable materials and formulations. The quality of trade literature is now such that it is a credit to most manufacturers, and of real value to consumers. The reader should always check the patent situation with respect to materials and processes to avoid unintentional violation of patent rights.

I repeat, it is with reluctance that I refrain from citing the names of those who have assisted by helpful comments during the preparation of this work. They have had my personal appreciation, and will derive satisfaction from realizing that their efforts may be of value to seekers after knowledge. I thank the Polytechnic Institute of Brooklyn for providing facilities whereby men of industry can contribute of their experience to the scientists and engineers of tomorrow. I also thank the American Cyanamid Company for its generous support of the present work. To Mae Conway and Anna von Plees my sincere appreciation is expressed for their untiring efforts in converting penciled notes into typewritten form. Finally, I acknowledge my deep gratitude for the vital support and understanding which could be given only by one as closely associated with me as my wife.

The random thoughts in this book are intended for those who do not live by bread alone. Life becomes barren when the pursuit of material things is its sole object. To live successfully it is necessary to maintain equilibrium among material, mental, and spiritual activities. Also, there is the need to give and to receive the stimulus of noble thoughts expressed by illustrious predecessors or by those with whom we associate.

HENRY FLEMING PAYNE

Old Greenwich, Conn.
January 1, 1954

CONTENTS

There is more to physical phenomena than meets the eye; the unseen is revealed through diligent thought and steadfast purpose.

Chapter 1. FUNDAMENTALS OF FILM FORMATION

INTRODUCTION

Some organic coating materials produce clear, transparent coatings, and others produce black, white, or colored opaque coatings. This volume describes the general chemistry and manufacture of the oils, resins, varnishes, and high polymers which are used for the clear coatings, and which are also used for the pigment binder in the opaque coatings. Volume II will describe the general chemistry and manufacture of black, white, and colored pigments, and also the manufacture and application of paints, enamels, and lacquers used for architectural and industrial coatings. The volatile solvents, and metallic driers which are used in many clear and opaque coatings, are described in this volume.

A variety of film-forming materials may be used for the clear coatings. These include drying oils, varnishes, synthetic resins, and high-polymeric materials such as the cellulosic and vinyl polymers. Usually, the film-forming material is dissolved in a volatile solvent so that it will have the correct consistency for brushing, spraying, dipping, or other method of application. Metallic driers are added to many oils and resins to shorten the time required for them to change from the liquid state to the dry film. Coatings are available which change from the liquid state to the dry film at normal room temperatures; others require baking at elevated temperatures.

The opaque coatings require pigments in addition to the film-former, solvent, driers, etc. The pigments are black, white, or colored powders which are insoluble in the film-former. In the manufacture of paint the pigment is mixed and finely dispersed in the combination of film-former, solvent, etc. This combination is known in the trade as the vehicle, since it carries the pigment in suspension. When the paint is applied the solvent evaporates, and the film-former binds the discrete particles of pigment into a continuous film or coating. For this reason the film-former is referred to as the pigment binder.

Film-formers change from the liquid state to the solid state by a variety of methods. These methods and their mechanisms constitute the main subject of this chapter. The various film-formers produce films which include a wide range of physical properties. Some films are very flexible, others are strong and tough, and still others are hard and rigid. Combinations may be made to meet the wide range of requirements for architectural and industrial coatings. The useful life of a coating is determined chiefly by the changes which take place in the film-former under the service requirements of the coating. In pigmented coatings the pigment provides color and opacity, and it also contributes to the hardness. In many cases it provides some protection against degradation of the film-former by the ultraviolet component of sunshine. In coatings which contain basic pigments, such as white lead or zinc oxide, there is a chemical reaction between such pigments and any acidic components of the film-former. This reaction produces metallic soaps which have some influence on the physical properties of the coating. However, in the majority of coatings, there is relatively little change in the pigment during the life of the coating. In contrast, the film-former or pigment-binder is in a state of more or less continual change from the time the paint is applied to the time when the surface must be repainted because of the wearing away of the coating. Thus the rate of change of the binder under the service conditions of the coating determines the useful life of the coating. The elements in the life cycle of a paint are illustrated in Fig. 1.

Figure 1 is intended to illustrate in a very general way the elements which affect the useful life of a paint. The pigment and the vehicle must be chosen carefully with the particular service requirements in mind. To produce a satisfactory paint the pigment must be mixed with the vehicle, then ground or dispersed thoroughly. Each batch of paint should be tested to be sure that it meets the required specifications before it is packaged in a suitable container. Faulty manufacturing procedure can spoil the performance of a coating. Also, faulty application of the coating can ruin its performance. The importance of the film-forming process, and also of the aging process, will be discussed later in this chapter. A paint which has been properly formulated, manufactured, and applied will wear away slowly to a condition which is suitable for repainting. The surface should be repainted before the coating has deteriorated to the extent that it must be removed before repainting. Since any system is no stronger than its weakest link, satisfactory paint performance will not be obtained unless each element (formulation, manufacture, application, etc.) is thoroughly understood and properly executed. Further details regarding these elements will be given in Volume II.

Fig. 1. Life Cycle of a Paint

PIGMENT VEHICLE

Mix → grind → test → package.

↓

PIGMENT-VEHICLE as a liquid dispersion → PAINT

Apply → solvent evaporates → film forms.

↓

PIGMENT-BINDER as a solid dispersion → COATING

Ages under service conditions → coating fails.

↓

PIGMENT-DETERIORATED BINDER

↓

REPAINT before deterioration becomes excessive.

METHODS OF FILM FORMATION

There are three general methods by which film-formers change from the liquid to the solid state:

1. Evaporation of the solvent.
2. Oxidation.
3. Polymerization.

Method 1 may be illustrated by the drying of shellac. A solution of shellac in alcohol, when applied in a thin layer, dries by simple evaporation of the alcohol, leaving a continuous film of shellac as the coating. This method applies to spirit varnishes, nitrocellulose lacquers, polyvinyl coatings, and many others. The solvent may be allowed to evaporate from these coatings under normal room conditions, or evaporation may be accelerated by heating the coated object in an oven. The latter procedure is generally used for industrial coating operations.

Method 2 may be illustrated by the drying of linseed oil. When linseed oil is spread as a thin film, it absorbs oxygen from the air, and changes from the liquid oil to the solid film of linoxyn. This method applies to drying oils, oleoresinous varnishes, oxidizing alkyd resins, and many other film-forming materials. As a result of the oxidation of these materials, linkages are formed between two or more molecules, thereby producing a polymeric structure. The general process is referred to as autoxidative polymerization. The known mechanism of this process will be explained later.

Fig. 2. Fundamentals of Film Formation and Film Properties

CHEMICAL COMPOSITION

↓

FUNCTIONALITY

↓

POLYMERIZATION

↙ ↘

CONVERTIBLE

Thermosetting
Cross-linked
Type of cross-linking
Extent of cross-linking

Autoxidative
Condensation
Addition
Copolymerization
Heteropolymerization

NON-CONVERTIBLE

Thermoplastic
Linear
Molecular size and shape
Polar or non-polar

Type and extent of pigmentation
Type of solvent

↘ ↙

FILM PROPERTIES

Resistance to:
- acid, alkali, grease, solvents
- water immersion and permeation
- discoloration with heat or aging
- oxidation decomposition

Adhesion, hardness
Flexibility, impact resistance
Abrasion resistance
Tensile strength, elongation

Flow and leveling
Color, opacity, gloss

Method 3 is designated as polymerization, but we have seen that one type of polymerization occurs in method 2. There are two other general types, namely, condensation polymerization and addition polymerization. Condensation polymerization occurs when coatings containing urea- or melamine-formaldehyde resins, heat-reactive phenolic resins, etc., are dried by baking at temperatures ranging from 200°F to 400°F. In this process a small molecule, usually water, is split off between two film-forming molecules, thereby producing a chemical bond between them. Addition polymerization occurs when unsaturated molecules, such as styrene or vinyl chloride, are activated by a suitable catalyst so that they add together to form large molecules. Both condensation and addition polymerization will be discussed in detail later.

The particular process which is used to form a film of coating material depends on the chemical composition of the film-former and on a characteristic usually referred to as its functionality. We shall see later that, if a film-former has a functionality no greater than 2, it is capable of forming linear polymers only. If the film-former has a functionality of 3, it is capable of forming three-dimensional polymeric structures. The materials made up of linear polymers are known as thermoplastic, and also as non-convertible materials, since they soften when heat is applied, and remain soluble instead of being converted into insoluble materials. The materials consisting of three-dimensional polymeric structures are known as thermosetting, and also as convertible materials, since they do not soften appreciably when heated, and they become insoluble in the solvents from which they were applied. Film formation is much more complex than is indicated by this brief outline of the general processes. It is an extremely interesting phenomenon because of the many variables which are available, and consequently the many types of films or coatings which can be produced. The fundamental factors in film formation, and their connection with film properties, are indicated in Fig. 2. In view of the complexity of film formation it will be necessary to consider some of the chemical and physical principles involved in the various factors. This will be done in a general way in this chapter, and again in more detail in later chapters dealing with specific film-formers.

FUNDAMENTALS OF FILM FORMATION

The three fundamentals of film formation in Fig. 2 are:

1. Chemical composition.
2. Functionality of the film-former.
3. Type and degree of polymerization of the film-former.

These fundamentals are related in that the degree of functionality depends on the number of certain reactive groups in the chemical composition, and the ability to polymerize is related to the type and arrangement of the reactive groups. However, each fundamental has its own particular significance. Since the majority of film-formers are organic compounds, they are composed of carbon and hydrogen atoms, and many also contain oxygen, nitrogen, or chlorine. The type and arrangement of these elements determine whether the compound is a hydrocarbon, ester, ether, amide, amino resin, chlorinated hydrocarbon, etc. Paint and resin chemists have learned to associate specific properties with these chemical compounds when they are used as film-formers. These properties include relative degree of chemical reactivity, sensitivity to water and alkalies, and type of solvent required. For example, films of hydrocarbons, such as paraffin wax or polyethylene, are resistant to attack by acids and alkalies, but films of esters, such as linseed oil, are destroyed by saponification with alkalies. Alkyd resins also contain the ester group; therefore alkyd films are somewhat sensitive to alkali. However, the sensitivity may be reduced by blending alkyd resins with phenolic resins, or amino resins, because these do not contain ester groups. Linseed oil is saponified more readily than dried films of the oil because the linkages between molecules formed in the drying process are more resistant to alkali than the ester group is. We shall see that the solubility of film-formers is related to their chemical composition. Hydrocarbon film-formers usually are soluble in hydrocarbon solvents, and esters in ester solvents, and so on. However, the degree of polymerization is also an important factor in solubility.

The functionality of the film-former indicates whether it may or may not be polymerized, and also the type of polymer which may be obtained, i.e. convertible or non-convertible. For example, a monobasic acid such as acetic acid may be reacted with a monohydric alcohol such as ethyl alcohol to produce the ester, ethyl acetate. In this case both acid and alcohol are monofunctional, and the product is a monomeric ester which cannot be polymerized. However, if we react the dibasic acid, adipic acid, with the dihydric alcohol, glycol, we can obtain glycol adipate as a linear polymer because each reactant is difunctional. It does not require much mental effort to visualize these results, and to see that if trifunctional materials are used we should obtain cross-linked, three-dimensional polymers. Of course, the reactive points in the materials need not be acid groups or alcohol groups only. Any point of reactivity in a molecule through which it can be connected to another molecule may be considered a potential functional point or functional group. The unsaturated system in a drying oil, or in styrene and ethylene, is a functional group because polymers of these materials are formed by reaction through the unsaturation. We shall discuss functionality in more detail later in this chapter, and again in other chapters, because it is a very important fundamental in film formation.

Polymerization is another very important fundamental in film formation because the type and degree of polymerization directly affect the physical properties of films. Strong, continuous films cannot be made with small molecules or monomers. Monomeric styrene is a volatile liquid, whereas polystyrene is a tough, solid material. Linseed oil is a mobile liquid, but a dried film of linseed oil is a rather weak, rubbery solid. The conversion of the oil is due to autoxidative polymerization which produces chemical linkages between the individual oil molecules. A very important discovery was made when it was found that, under proper conditions, two different monomers may be polymerized simultaneously to produce a copolymer. In many cases the copolymer has better film properties than either of the polymers which could be produced from the two monomers separately. For example, the styrenated oil produced by copolymerizing styrene and linseed oil gives a stronger film and dries faster than straight linseed oil. Perhaps you are wondering whether the same result could be obtained by mixing a solution of polystyrene with linseed oil. The answer is no, because linseed is not compatible with a solution of polystyrene. The two liquids form a cloudy, turbid mixture which separates on standing and which would be useless as a film-former. However, these liquids may be combined through copolymerization, and we shall see later that there are many very useful copolymers in film-forming materials.

The extent to which a monomer has been polymerized determines the molecular size or molecular weight of the polymer. The degree of polymerization also affects the physical properties of the polymer; a high polymer is usually stronger and tougher than a low polymer of the same monomer. Unfortunately high polymers are more difficult to dissolve, and their solutions are much more viscous than low-polymer solutions. This means that high polymers require more solvent than low polymers to bring them to the viscosity required for application as coatings. Since more solvent produces a lower percentage of solids in solution, it will be seen that more coats would be required of the high-polymer solution to produce a film of specified thickness. This would increase the cost of application, and there would also be an added cost due to the use of more solvent. In view of this situation a compromise is usually made in commercial high-polymeric materials between maximum film properties and cost of application.

This preliminary discussion of the three fundamentals of film formation has indicated that the structure of films is composed of molecular aggregates of various shapes and sizes. We must now consider the various forces which bind atoms into molecules, and which bind molecules into aggregates, or micelles, and also those forces which bind micelles together to produce the finished film or coating. A complete study of these forces would involve a great deal of theoretical physical chemistry. Since this book is directed more along technological lines, only the minimum of theory will be given to meet the requirements for an understanding of the subject. If more of the theoretical aspects are desired, the reader should consult the many excellent books which are available on the physical chemistry of organic materials.

STATES OF MATTER

We are all familiar with the three general states of matter: gaseous, liquid, and solid. We know that dry steam can be condensed to liquid water and cooled further to produce solid ice. This is an example of the same chemical composition existing in different physical states at different energy levels. However, under conditions of normal temperature and pressure the physical state of many materials depends largely on the size of their molecules and the forces operating between the molecules. If a small quantity of a gaseous material is placed in a relatively large container, the molecules of gas rapidly diffuse through the entire space available. This indicates that the forces operating between the molecules are quite weak in comparison with the kinetic energy of the molecules. Since some of the gas may be found in all parts of the container, it appears that the molecules of the gas must be relatively far apart.

In the liquid state the molecules are much closer together, and the forces operating between them are considerably stronger. However, the forces are not strong enough to prevent fairly free movement of the molecules past one another. This permits liquids to flow so that they may be poured from a container, but there is a great difference in the rates of flow of liquids such as water, molasses, and very viscous oils. The rate of flow is a measure of the strength of the forces operating between the molecules of the liquid; it is referred to as the viscosity of the liquid. We shall see later that the viscosity is also influenced by the size and shape of the molecules and by temperature and pressure.

In the solid state the forces operating between molecules, or other fundamental units, are so great that the material exists in a relatively rigid condition. Again, we can recognize great differences in the degree of rigidity of various solid materials. Putty is a plastic solid which may be deformed easily, but rubber is an elastic solid which recovers its original shape when deformed within certain limits. However, the elasticity of rubber may be varied considerably by changing the number of bonds between rubber molecules through variation in vulcanization. Rosin is a weak, brittle resin which softens readily at about 80°C, but ethyl cellulose is a strong, tough resin which requires heating to 160–180°C before it softens appreciably. Many other comparisons could be made among coating materials which would show wide variations in physical properties in the solid state.

From observation of these variations in solids, liquids, and gases two assumptions might be made: first, that there are many types of forces operating between molecules; second, that there are many degrees of the same force in operation. Eventually it may be proved that the second assumption is correct, but for the present it is convenient to consider the forces as being of two types: primary valence forces and secondary valence forces. Primary valence forces are direct chemical bonds operating between specific atoms in a molecule, or between two atoms in different molecules. Secondary valence forces are the residual forces in the surface of molecules. Both forces hold molecules together in bundles known as fibrils, micelles, etc., but the primary valence forces are stronger than the secondary valence forces.

PRIMARY VALENCE FORCES

We stated above that a primary valence force is a direct chemical bond between two atoms. We find that chemical bonds may be classified into three types:

1. The metallic bond between similarly charged particles.
2. The ionic bond between oppositely charged ions.
3. The covalent bond between neutral atoms.

Since the metallic bond is not a factor in organic coatings, it will not be discussed here. Again, the ionic bond is important for inorganic reactions but has little application in organic coatings. The ionic bond results from the attraction between oppositely charged ions. The ions are produced by the transfer of an electron from one atom to another atom. The covalent bond results from the sharing of an electron between two atoms. The sharing of the electron completes the outer electron shells of the atoms, thereby producing a more stable configuration. The electrons attached to atoms are often indicated by dots near the symbol for the element. The electrons attached to hydrogen, carbon, and oxygen, and the sharing of the electrons by these elements to form the single covalent bonds in methyl alcohol, are shown below:

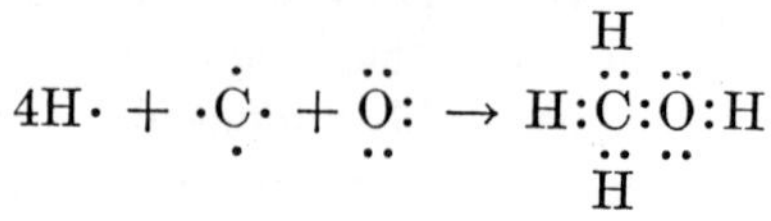

Methyl alcohol

Since it would be quite cumbersome to express chemical formulas in this manner, the single covalent bond is indicated by a single line in structural formulas, and omitted or inferred in general formulas.

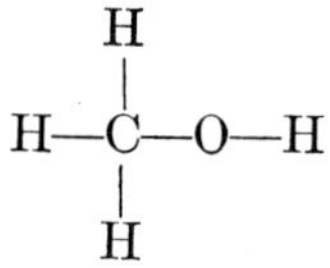

Structural formula

CH_3OH

General formula

The double and triple covalent bonds which occur in compounds such as ethylene and acetylene respectively may be indicated as follows:

Ethylene: C_2H_4, $CH_2{=}CH_2$, 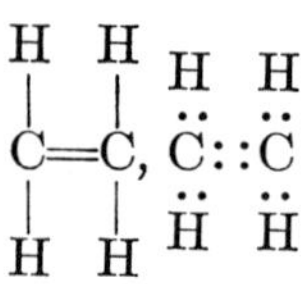

Acetylene: C_2H_2, $CH{\equiv}CH$, $H{-}C{\equiv}C{-}H$, $H{:}C{:}{:}{:}C{:}H$

The single, double, and triple covalent bonds in the above compounds are primary valence bonds because they operate between two atoms in a compound in accordance with our previous definition. These bonds bind the elements together to form a definite chemical compound. The primary valence bonds are strong enough to maintain the identity of the compound under conditions of heat, distillation, etc.

Our definition for primary valence bonds included their operating between two atoms in different molecules as well as between two atoms in the same molecule. We see at once that primary valence bonds can be used to connect monomeric molecules so that they form polymeric molecules. Here we remember the importance of functionality, so we add the proviso that polymers can be formed only if the monomers have the correct functionality. We recall that polymers could not be formed from the reaction between ethyl alcohol and acetic acid because these materials are monofunctional. However, a primary valence bond was formed between them when they reacted to produce ethyl acetate. Also primary valence bonds were formed between the monomers to produce polymerized linseed oil, glycol adipate, polystyrene, and other compounds referred to previously. The covalent type of primary valence bonds is the force which operates between the atoms in the vast number of organic compounds which are available today. Further details regarding these bonds are given in treatises on the physical chemistry of organic compounds by Wheland (7), Waters (8), and many others.

SECONDARY VALENCE FORCES

It was stated previously that secondary valence forces are the residual forces in the surface of molecules. They are often referred to as van der Waals' forces because of van der Waals' work on the forces of association between molecules. These forces are the result of attraction between electrostatically active areas in the molecular structure. A molecule contains both positive and negative charges, but since the charges are balanced electrically, the molecule is neutral electrically. When the arrangement of the positive and negative charges in a molecule is symmetrical, the molecule is known as a non-polar molecule. When the arrangement of the charges is unsymmetrical, the molecule is called a polar molecule. The difference in polarity of molecules affects the strength of the secondary valence forces between them. In general, the secondary valence forces between non-polar molecules are quite weak but the secondary valence forces between polar molecules are relatively strong. Since there is a wide range in the degree of polarity of molecules from zero upwards, there is a corresponding range in the strength of the secondary valence forces between molecules. It will be realized that these forces increase as the size of the molecules increases since the number of polar or non-polar units will also increase.

The dipole moment is an indication of the strength of the secondary valence forces since it is related to the strength of the electrical charges and the distance between them. The values for the dipole moments given in Table 1 are in Debye units (10^{-18} esu) and are the sum of the electrical charges multiplied by the distance between them. It will be noticed in Table 1 that the non-polar molecules have symmetrical structures, and their dipole moments are zero. The polar molecules have unsymmetrical structures, and the dipole moments vary from 0.37 to 2.71 for those illustrated. It should be remembered that a dipole moment is produced when the centers of gravity of the positive and negative charges do not coincide. The strength of the dipole moment is related to the strength of the charges as well as to their

Table 1 Polar and Non-Polar Molecules

			Dipole Moment, Debye units (10^{-18} esu)
Saturated hydrocarbon	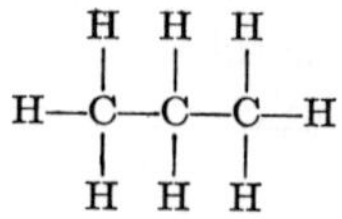	Symmetrical	0
Unsaturated hydrocarbon	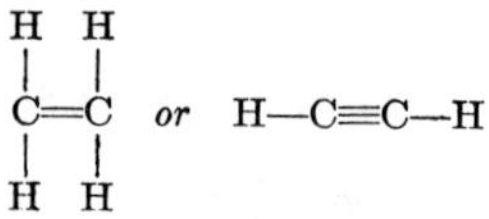	Symmetrical	0
Unsaturated hydrocarbon	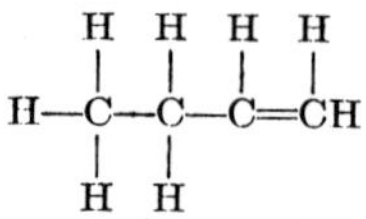	Unsymmetrical	0.37
Alcohol	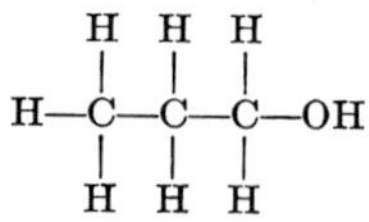	Unsymmetrical	1.66
Ketone	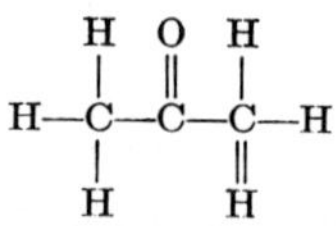	Unsymmetrical	2.71
Carboxylic acid	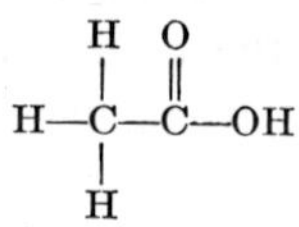	Unsymmetrical	1.40
Ring compounds	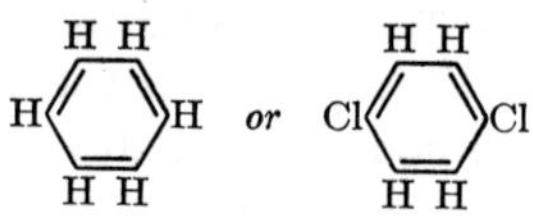	Symmetrical	0
Ring compound	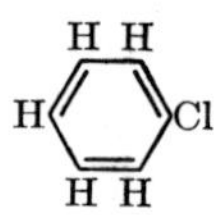	Unsymmetrical	2.25

unsymmetrical arrangements so that structural dissymmetry is not the only criterion for the strength of the secondary valence forces. This point will be illustrated later in Fig. 3.

POLAR AND NON-POLAR MOLECULES

A good understanding of the characteristics of polar and non-polar molecules is distinctly advantageous when formulating organic coating materials. We have seen that the degree of polarity of the molecules of thermoplastic materials is an indication of the strength of the attractive forces between the molecules, and consequently a measure of the strength of their films. We shall see that the solubility of coating materials is also related to their relative polarity. In general, non-polar materials are soluble in non-polar solvents, and polar ma-

Fig. 3. Polarity

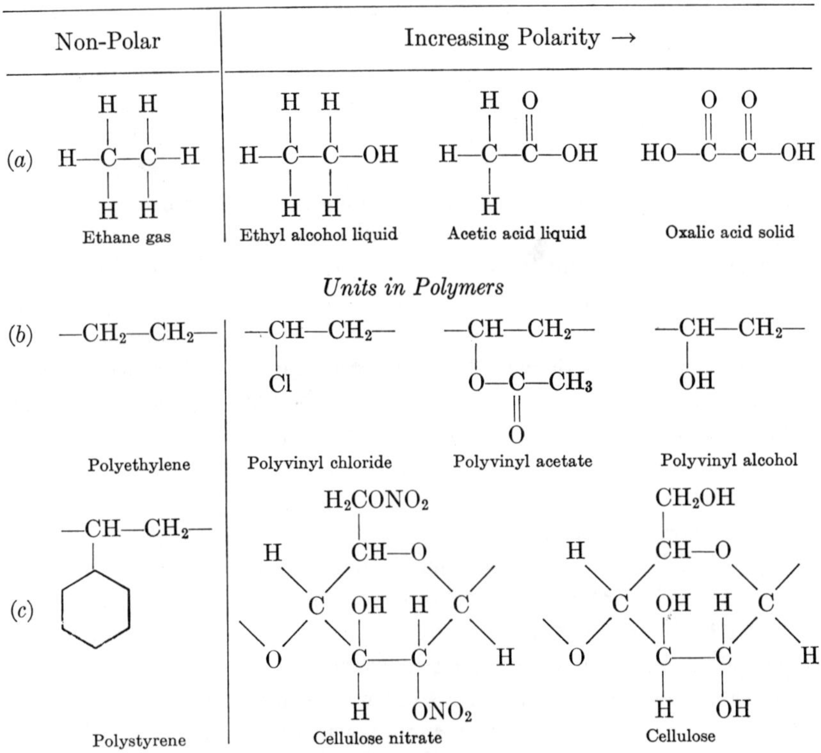

terials are soluble in polar solvents. The ability of coating materials to mix with one another to form homogeneous compositions is also related to their relative polarity. However, we stated previously that the degree of polymerization, or molecular weight, is a contributing factor in the three properties: film strength, solubility, and compatibility. An indication of the effect of relative polarity on solubility and molecular cohesion is given in Fig. 3.

If we consider the four small molecules in group (*a*), Fig. 3, we shall see that ethane has a symmetrical structure, and the other three have unsymmetrical structures. This indicates that ethane is a non-polar material, whereas ethyl alcohol, acetic acid, and oxalic acid are polar materials. Ethane is a simple hydrocarbon which is soluble in other hydrocarbons but insoluble in water. Ethyl alcohol is miscible with many hydrocarbons and also with water. Acetic and oxalic acids are practically insoluble in hydrocarbons but are soluble in water. Hydrocarbons are non-polar materials, and water is a very polar material. Therefore difference in solubility of the materials in group (*a*) will be seen to be related to the fact that they increase in polarity from the non-polar ethane to the very polar oxalic acid.

A striking indication of the effect of this increasing polarity on molecular cohesion is given by the consistencies of these four materials. Although each is based on a structure of two carbon atoms, we see that they vary in consistency from gas, through liquid, to a solid material. The secondary valence forces between the molecules of the non-polar ethane are so small compared to their kinetic energy that the molecules are free to move about readily, as is typical of the gaseous state. The replacement of one hydrogen atom in ethane with a hydroxyl group produces ethyl alcohol. The hydroxyl group is a strongly polar group, and the molecule is unsymmetrical; therefore the secondary valence forces between the alcohol molecules are quite strong. There are no primary valence bonds between the alcohol molecules, but the secondary valence forces are strong enough to associate the molecules sufficiently so that the material is a liquid.

Acetic acid is a more strongly polar liquid than ethyl alcohol because the carboxyl group is more polar than the hydroxyl group. An indication of this fact may be seen in their boiling points; the boiling point for ethyl alcohol is 78.4°C, but the temperature must be raised to 118.1°C before the more strongly associated acetic acid boils. However, in oxalic acid, having two carboxyl groups, the association is so great that the material is a solid with a melting point of 187°C.

Group (*b*) of Fig. 3 shows a monomeric unit of each of the four polymers, polyethylene, polyvinyl chloride, polyvinyl acetate, and polyvinyl alcohol. The unit of polyethylene is symmetrical; therefore this material is non-polar. The units of the other three polymers are unsymmetrical. If structural dissymmetry were the only criterion of polarity, polyvinyl acetate would be the most polar. However, the electrical characteristics of the hydroxyl group make it more polar than the acetyl group; therefore polyvinyl alcohol is the most polar material in this series. As may be expected from the discussion of group (*a*), polyethylene is soluble in hydrocarbons, and polyvinyl alcohol is soluble in water. The two intermediate polymers are soluble in solvents having intermediate polarity, such as ketones and esters.

Since the commercial polymers have high molecular weights, their degree of solubility is low, as discussed previously.

Another series of monomeric units is shown in group (*c*) of Fig. 3. The unit for polystyrene is the same as that for polyethylene except that a hydrogen has been replaced with a phenyl group. The unit is structurally unsymmetrical, but since the phenyl group is non-polar, polystyrene is practically non-polar. It is softened by aromatic hydrocarbons and is unaffected by water, but its molecular weight is so high that it is practically insoluble. The same situation exists for cellulose, the most polar material in this series. We might expect cellulose to be soluble in water because of its hydroxyl groups. However, its molecular weight is so high that it simply swells in water, but does not dissolve. The cellulose nitrate is an ester of cellulose, and is soluble in the ester solvents, ethyl acetate and butyl acetate, and in other polar solvents.

MOLECULAR COHESION

An illustration of the effect produced by an increase in molecular weight on the cohesion between molecules is the change in physical properties of the homologous hydrocarbons listed below:

CH_4: methane, a gas.
C_5H_{12}: pentane, a liquid.
$C_{18-22}H_{38-46}$: paraffin wax, a weak, low-melting solid.
$C_{2000+}H_{4002+}$: polyethylene, a strong, high-melting solid.

The elements (carbon and hydrogen) in these hydrocarbons are held together by primary valence forces, but the physical state of the compounds is due to the secondary valence forces between their molecules. Polyethylene is a strong rubber-like solid because of the strong secondary valence forces which hold these large molecules together. The forces are much less in the smaller molecules of paraffin wax, as indicated by its relatively weak physical condition. The secondary valence forces continue to decrease as the molecular weight decreases until the materials become gases under normal conditions.

Polyethylene has rubber-like qualities for two reasons: (1) the long-chain polyethylene molecules are coiled or kinked instead of being arranged in parallel order; (2) a certain amount of slippage is possible between these molecules because of the absence of primary bonds between them. This situation is typical of natural and synthetic rubbers and other elastic high polymers. An interesting effect was observed when these polymers were stretched; it was found that their

Table 2 Non-Polar to Polar Polymeric Materials

Material	Repeating Unit in Polymer
1. Paraffin wax	$-CH_2-$
2. Polyethylene	$-CH_2-CH_2-$
3. Polyvinyl chloride	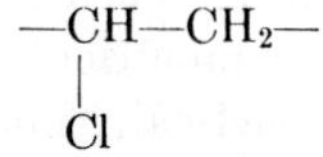
4. Polystyrene	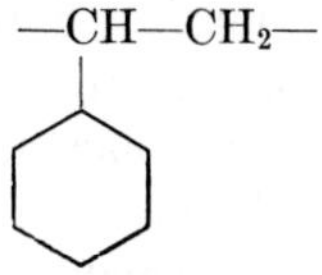
5. Polyamide	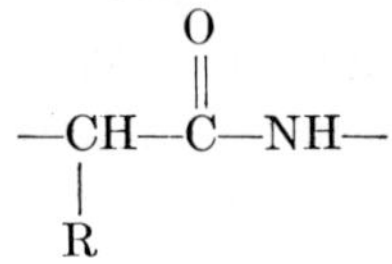
6. Nitrocellulose *	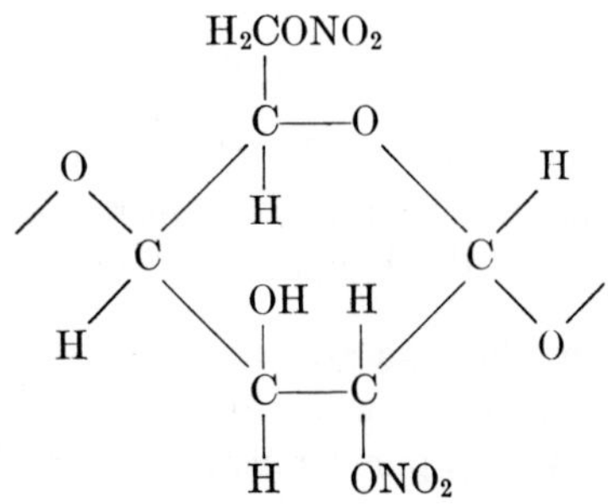
7. Polyvinyl acetate	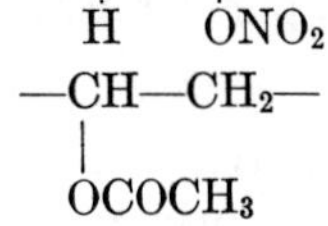
8. Ethyl cellulose *	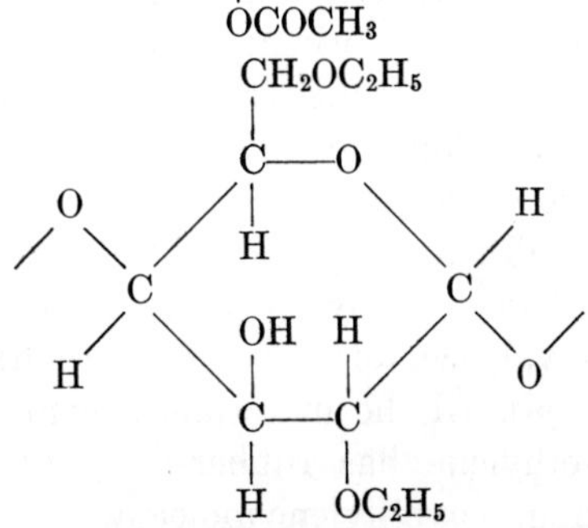
9. Cellulose acetate	$-OCOCH_3$ instead of $-NO_3$ in #6
10. Methyl cellulose	$-OCH_3$ instead of $-OC_2H_5$ in #8
11. Polyvinyl alcohol	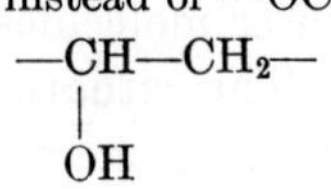

* These structures are shown as the diester and diether; commercial materials may vary from di compounds.

tensile strength increased. X-ray analysis showed that the stretching had uncoiled many of the molecular chains so that they became arranged in more nearly parallel order. In some cases the molecules were packed so closely, and in such an orderly manner, that they appeared crystalline. Increased strength was obtained as this crystalline condition developed.

There is a double reason for this increase in strength with the orderly alignment, or orientation, of the molecules. Since the secondary valence forces are "in the surface of the molecules," their strength will be greatest when the maximum surface is available for contact. This condition exists when the arrangement of the molecules is orderly and parallel. The other factor is that the rate of decrease in strength of the secondary valence forces is proportional to the $1/6$ power of the distance between the molecules. This means that the forces have relatively little strength if the molecules are far apart. Indications are that the forces are not particularly effective unless the molecules are within 5 Å of each other. This is a very important idea which has practical applications in the coating industry. For example, nitrocellulose lacquer and other high-polymer films have greater tensile strength when the correct solvent, rather than a poor solvent, is used. A good solvent, with the correct rate of evaporation, will uncoil the high-polymer molecules and permit them to align themselves in the most orderly arrangement when the film is drying.

It is well known that high-polymer coatings have poor adhesion to smooth metal surfaces. This is due to the extremely high cohesion between the large molecules of the coatings; they have a greater attraction among themselves than they have for the metal. The usual addition of plasticizers, resins, etc., which have much smaller molecules, improves the adhesion; but, as may be expected, these materials lower the strength of the film. A series of typical high polymers is shown in Table 2 with the repeating units of their structures. These materials range in polarity from non-polar to polar.

ORIENTATION

We are all familiar with the classical experiment in which a dilute solution of fatty acid is placed on water; after the solvent evaporates, the fatty acid molecules are oriented with the carboxyl group in the water, and the hydrocarbon chain in the air. This orientation occurs because there is a strong attraction of the polar carboxyl group for the highly polar water, and also because the fatty acid molecules are free to move about to orient themselves in this position. A similar situation exists in many coating materials. When coating materials are applied as solutions, orientation of the molecules occurs provided the solvent does not evaporate so rapidly that the molecules are immobilized before they can orient to the best advantage. It has also been shown that the addition of some highly polar groups will improve the adhesion of certain coatings to smooth metals because of the affinity of the polar groups for the metal, and the fact that the polar groups are oriented to the metal at the coating-metal interface. An indication of the relative strength of the attraction between polar groups, as compared with non-polar groups, is given by Mark (1) in the values for molar cohesion in calories per group.

	Non-polar		Polar
Group	$—CH_3$	$—CH_2—$	—COOH
Calories per group	1.8	1.0	8.97

These values indicate that the attraction between carboxyl groups in a number of fatty acid molecules will be much greater than that between the hydrocarbon groups. Therefore, when fatty acids are free to orient, they arrange themselves in a head-to-head manner, that is, with the carboxyl in one fatty acid oriented to the carboxyl in another fatty acid. However, when hydrocarbons having the same chain length as fatty acids are free to orient, they simply assume a parallel arrangement because of the relative equality of attraction of the groups along the chain and those at either end.

We shall see that this idea may be extended to the sensitivity of polar and non-polar films to moisture. Water molecules are strongly attracted to polar groups on molecules in films, but there is no such attraction for water in non-polar films. We have seen that when relatively small molecules, such as those of plasticizers or resins in lacquers, come between the large polymer molecules, they reduce the attractive forces between the polymer molecules, and the film strength is reduced correspondingly. A similar situation exists when the small water molecules permeate a film; they reduce the attractive forces between the micelles and molecules in the film structure, and the film is weakened. Of course, when the permeation is carried to the limit, the molecules are separated to such an extent that we say the film is dissolved. It will be realized that, although non-polar films are not affected by water, they are permeated, and may be dissolved, by non-polar materials having small molecules.

One important qualification should be applied to the above discussion of permeation and dissolution; the statements refer to non-convertible or thermoplastic films much more than they do to convertible or thermosetting films.

CONVERTIBLE AND NON-CONVERTIBLE FILMS

It will be seen from Fig. 2 that two types of films are formed from polymers: convertible and non-convertible. Figure 2 also indicates that convertible films are thermosetting, and the molecules in the film structure are cross-linked by primary valence bonds. In contrast, non-convertible films are thermoplastic, and their structure is composed of linear molecules held together by secondary valence forces. From the discussion in the previous section we realize that the size, shape, orientation, and degree of polarity of the linear molecules affect the physical properties of non-convertible films. The factors which affect the physical properties of convertible films are the type of cross-linkages and the extent to which they are distributed throughout the film structure. Since these cross-linkages are primary forces or direct chemical bonds, they are not so readily ruptured by heat, water, solvents, or mechanical stresses. Also, if there are a great many of them between the molecules, the structure will be rigid. Conversely, a small number of linkages between large molecules will produce an open and more flexible structure.

We shall apply this idea later on to the relative hardness and flexibility of oil-modified alkyd resins and many other film-formers. The alkyd resins which have a small percentage of oil are quite hard because the cross-linkages in the glyceryl phthalate part of these resins are relatively close together, whereas those in the oil part are far apart. Therefore the alkyds which have a high proportion of oil will be softer and much more flexible. Varnishes may be considered as another illustration of this phenomenon. Varnish resins are hard materials because of frequent cross-linkages. However, they may be cooked with oils in a wide range of proportions to make varnishes which produce films ranging from brittle to flexible, depending on the ratio of resin to oil. Rubber may be considered as an extreme case. Before vulcanization the very large rubber molecules are held together by secondary valence forces, and the structure is soft, tacky, and extremely extensible. Vulcanization introduces sulfur linkages between the molecules which strengthen the structure because they are direct chemical bonds. Vulcanized rubber is non-tacky, stronger, and non-plastic, and the extent of these properties is influenced by the degree of vulcanization or cross-linking.

It will be appreciated that many fascinating possibilities exist when planning the architecture of film structures. There are many sizes and types of molecular building blocks, and a variety of forces to bind them together. Oil and resin chemists have worked with these theories and raw materials and have provided the truly amazing range of organic coatings in use today. It is even more remarkable to realize that considerably greater development is possible from the continued application of theory to practice.

FUNCTIONALITY

The concept of functionality is essential when planning the architecture of molecular structures. The functionality of the molecules determines the type of structure which can be built; therefore molecules of definite functionality must be selected for specific structures. This concept was developed by Carothers (2, 3), Kienle (4), Bradley (5), and others whose names are synonymous with high polymers, coating resins, and the resinous state.

It was stated previously that the functionality of a molecule is the number of reactive groups which it contains through which it can react with other molecules. Figure 4 shows the development of this idea, using linoleic acid and a variety of alcohols. Reference to Table 3, Chapter 2, indicates that linoleic acid has the following formula:

$$CH_3(CH_2)_4CH{=}CH{-}CH_2{-}CH{=}CH(CH_2)_7COOH$$

This shows that linoleic is a monobasic acid, having two non-conjugated double bonds. This molecule has a functionality of 2 despite the fact that there are three potentially reactive groups: the carboxyl group and the two unsaturated ethylenic groups. Bradley (6) has shown that this type of unsaturation, referred to as non-terminal unsaturation, must be assigned a functionality of 1 only, irrespective of whether the unsaturated system contains one, two, or three double bonds. We shall see the reason for this when we discuss oxidation later in this chapter. The four alcohols shown in Fig. 4 have functionalities of 1, 2, 3, and 4, indicated by the number of hydroxyl groups they contain. These are as follows: methyl alcohol, $f = 1$; ethylene glycol, $f = 2$; glycerol, $f = 3$; pentaerythritol, $f = 4$.

The first line in Fig. 4 shows the reaction between linoleic acid and methyl alcohol to produce methyl linoleate. This is referred to as a 1:1 reaction because it involves two monofunctional molecules. It will be noticed that this reaction does not involve the double bonds. The next reaction in Fig. 4 shows the polymerization of methyl linoleate by heat or oxidation through the double bonds to produce the dimer of methyl linoleate. This is also a 1:1 reaction because methyl linoleate is monofunctional. What is the functionality of the dimer of methyl linoleate? Since there are no further groups through which reaction may take place, its functionality is zero, $f = 0$. Therefore polymers no higher than the dimer can be formed from this type of reaction. This is characteristic of 1:1 reactions; only simple molecules or dimers can be formed. The 1:1 reaction between ethyl alcohol and acetic acid to produce ethyl acetate was referred to earlier in this chapter.

Fig. 4. The Concept of Functionality

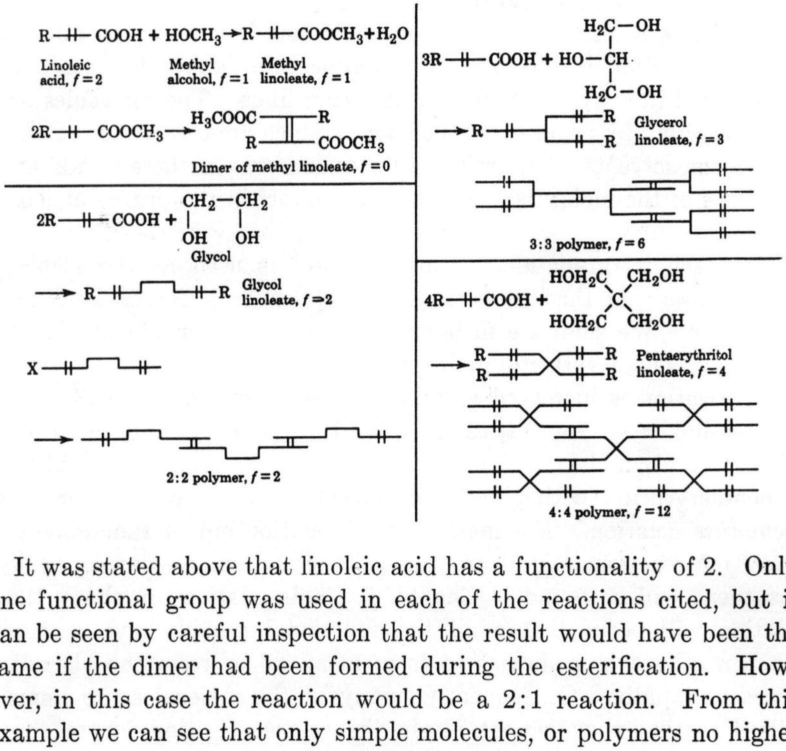

It was stated above that linoleic acid has a functionality of 2. Only one functional group was used in each of the reactions cited, but it can be seen by careful inspection that the result would have been the same if the dimer had been formed during the esterification. However, in this case the reaction would be a 2:1 reaction. From this example we can see that only simple molecules, or polymers no higher than the dimer, can be produced from either 1:1 reactions or 2:1 reactions.

Now let us consider a 2:2 reaction. This is shown in Fig. 4 in the reaction between two molecules of linoleic acid and one molecule of ethylene glycol, to produce glycol linoleate. This is only a 2:1 reaction because the double bonds in the linoleate radicals have not been used. However, if we polymerize the glycol linoleate through the unsaturation as shown, the 2:2 reaction will be complete. Now we see that the polymer always has a functionality of 2 regardless of whether it is terminated with a linoleate radical or a glycol. Therefore it can continue to react with difunctional molecules to produce long, chain-like polymers. From this we can state that 2:2 reactions produce linear polymers.

Now consider the reaction in Fig. 4 between linoleic acid and the trifunctional alcohol, glycerol. The first reaction produces glycerol linoleate, which is a molecule similar to linseed oil. It has a functionality of 3 because of the unsaturation in the three linoleic radicals. If we polymerize this type of molecule the reaction is 3:3. The diagram indicates that, if four of these molecules are reacted through the unsaturation, a tetramer will be formed which has a functionality of 6. This reaction is the cross-linked type referred to previously as being found in convertible or thermosetting films. The molecules are bound together by primary valence bonds which are not broken readily by heat or solvents. The relative distance between these bonds and the shape of the molecular structure are factors in the degree of flexibility of the film.

When the tetrafunctional pentaerythritol is used as the alcohol, Fig. 4 shows that there is even greater opportunity for cross-linking in the film. We shall see in later chapters that when glycerol is replaced with pentaerythritol in drying oils and resins the rate of thermal polymerization is increased greatly because of the ready availability of cross-linkages. In fact, the rate is so fast that gelation or insolubility may occur during the heating process. For this reason blends of pentaerythritol with glycerol or glycol are often used to prevent premature gelation. The mathematical relationship of functionality, extent of reaction, and degree of polymerization, as developed by Carothers, will be found in Chapter 7. This relationship shows that gelation will occur before all the functional groups have reacted. When a gel structure begins to form in an oil or resin the viscosity increases rapidly. Consequently, many of the molecules are immobilized so that they cannot attain the correct position for reaction with other molecules.

The pentaerythritol-modified oils and resins also air-dry faster than those containing alcohols of lower functionality. This may be expected from their greater potentialities for cross-linking by oxidation and subsequent reactions.

Another interesting illustration of the effect of increasing functionality is the series of products obtained by reacting rosin with the alcohols indicated in Fig. 4. Rosin is a monobasic acid, chiefly abietic acid, but its structure is quite different from that of linoleic acid. It is composed of three rings with the side groups as shown:

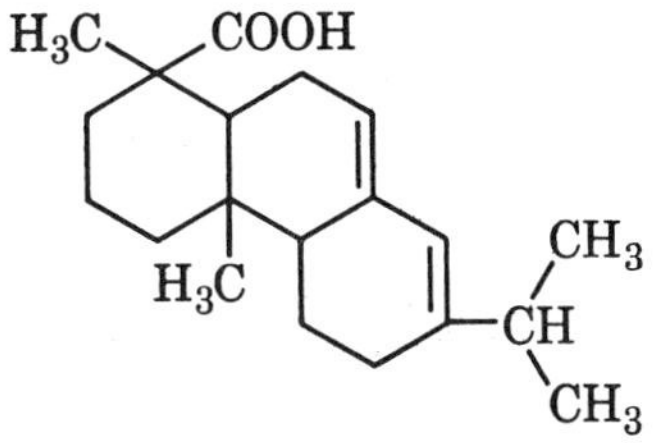

This structure is compact and rigid instead of long and flexible like that of linoleic acid, given previously. We know that rosin is a brittle solid, whereas linoleic acid is a liquid. Rosin has considerable polarity from the carboxyl group, but the large hydrocarbon group is non-polar. Therefore, it is not surprising to find that rosin is soluble in both ethyl alcohol and mineral spirits. It is not polar enough to be soluble in water. However, if its polarity is increased by replacing the hydrogen of the carboxyl group with a sodium atom, it is then soluble in water. Sodium abietate may be used in soap, and as a paper sizing to be added to the water slurry of paper pulp.

Rosin may be esterified with methyl alcohol to produce methyl abietate. Since this is a 1:1 reaction we know that only monomers will be obtained. We find that methyl abietate is a viscous liquid used as a plasticizer for high-polymeric materials. A great quantity of rosin is esterified with glycerol to produce the resin known as ester gum. Rosin is also esterified with pentaerythritol to produce the "penta" ester. However, if we compare the softening points of rosin, ester gum, and the penta ester, we find that they are in the following order: 80°C, 91°C, and 105°C. The increase in functionality of the alcohol produces an increase in hardness of the resins.

We can use these three resins to illustrate the relationship between solubility, polarity, and molecular size. The ester group in the esterified resins is less polar than the carboxyl group in rosin; therefore the esterified resins are not soluble in alcohol. In addition, their molecular weights are greater than that of rosin so that ester gum is not as soluble in mineral spirits, and the penta ester requires the stronger aromatic hydrocarbon solvents.

How can the functionality of rosin be increased so that harder ester resins may be made from it? We shall see in Chapter 3 that this question was answered by resin chemists when they reacted maleic anhydride with rosin. Maleic anhydride adds to rosin through the well-known Diels-Alder reaction. The resulting rosin-maleic adduct has three carboxyl groups: one from the rosin and two from the maleic anhydride. Therefore the adduct is trifunctional. If the adduct were esterified with glycerol the reaction would be 3:3, and we would expect rapid gelation of the product. This is precisely what happens; therefore only small percentages of the maleic adduct are used with rosin for esterification with glycerol. However, we find that the softening point of the mixed ester is raised considerably:

	Softening Point, °C
Rosin	80
Ester gum	91
Maleic-type ester	140–155

The maleic-type ester is used in varnishes and nitrocellulose lacquers, and its higher softening point produces harder films. It also requires less time than ester gum to cook with drying oils in making varnishes. Both these economic advantages are derived from our ability to increase the functionality of rosin.

POLYMERIZATION

The term "polymer" is generally understood to refer to a compound structurally composed of a number of units known as monomers. Dimers, trimers, and tetramers are composed of two, three, and four monomers respectively. High polymers may contain thousands of monomeric units. Polymerization is indicated in Fig. 2 to be one of the important fundamentals in film formation. Five types of polymerization are shown: autoxidative, condensation, and addition polymerization, copolymerization, and heteropolymerization. The last two types are modifications of addition polymerization. The three main types of polymerization, autoxidative, condensation, and addition, were described briefly in a previous section. We shall now consider them in more detail so that their general significance will be understood when we apply them to specific oils and resins in later chapters.

One general feature that should be kept in mind regarding polymerization is the large difference in reactivity which may be found among the various monomers. Some monomers are so reactive that stabilizers are added to prevent polymerization during storage and shipment. Despite this situation, catalysts are used in practically all polymerization processes to speed the reaction, influence the polymer size, and make the polymerization reaction as complete as possible. These catalysts include heat, light, peroxides, and metallo-organic compounds.

Autoxidative Polymerization. Autoxidative polymerization occurs when polymers are formed from unsaturated monomers by absorption of oxygen, and the subsequent formation of cross-linkages between the monomers. The best-known example of autoxidative polymerization is the "drying" of linseed oil referred to previously. The drying action is accelerated by addition of metallic driers described in Chapter 5. We shall see in Chapter 2 that oils are classified as drying, semi-drying, and non-drying. The degree of drying is related to the type and amount of unsaturation in the oil molecules. Oils with little or no unsaturation are the non-drying type. Oils such as soybean, which have only one or two double bonds in their fatty acid radicals, are the semi-drying type. The good-drying-type oils have two or three double bonds in each fatty acid radical.

There is a difference in the position of the unsaturation in oils, and that which is found in reactive monomers such as styrene and vinyl chloride. The unsaturation in oils is toward the middle of the 18-carbon chain of the fatty acid radical and is referred to as non-terminal unsaturation. The unsaturation in styrene, etc., may be considered as being at the ends of the ethylene group and is called terminal unsaturation. Autoxidative polymerization is usually associated with non-terminal unsaturation, and addition polymerization with terminal unsaturation. However, addition polymerization may take place between molecules having non-terminal unsaturation. Ring structures are formed in such cases, whereas linear structures are produced when addition polymerization takes place between molecules having terminal unsaturation.

Some of the reactions involved in autoxidative polymerization are shown in Fig. 5, using the non-terminal unsaturation of the linoleic acid radical as the illustration.

It has been shown that, when oxygen is absorbed by non-terminal unsaturation, a hydroperoxide is formed on a carbon atom adjacent to a double bond. This is indicated in Fig. 5 by the addition of oxygen to 5(*a*) to form the hydroperoxide on 5(*b*). It has also been shown that isomerization occurs with a shift of a double bond to the conjugated position, and the transfer of the hydroperoxide to another carbon atom. These moves result in the configuration shown in 5(*c*). The next step in autoxidative polymerization has not been elucidated completely despite the efforts of the best oil chemists in the world. However, several possibilities have been deduced on the basis of tests made during the drying process. One of these possibilities is the formation of an ether linkage between two molecules as indicated in 5(*d*); another is the ring structure connecting two molecules, shown in 5(*e*). We shall consider these two possibilities separately.

The hydroperoxide content of the oil as it is drying is indicated by determination of the peroxide value. It has been shown that the peroxide value increases to a certain maximum, then decreases, but

Fig. 5. Autoxidative Polymerization

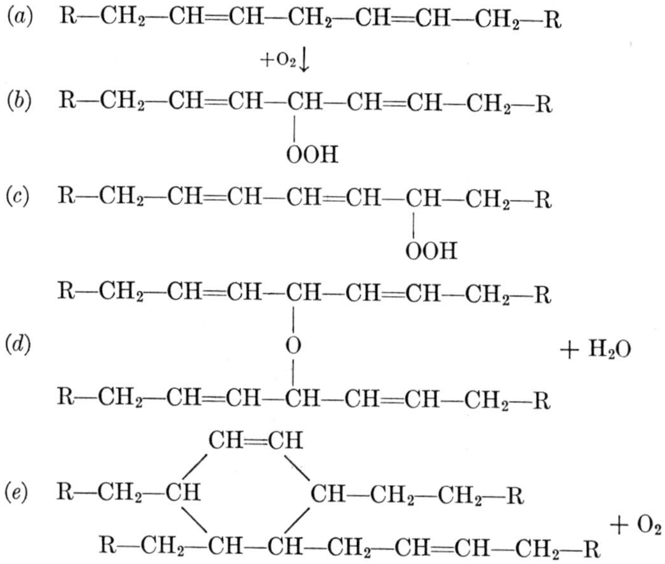

it does not become entirely zero. When the peroxide value reaches the maximum the oil film is observed to change from the liquid to a soft solid condition, or to become "set." Thereafter the peroxide value decreases, indicating a connection between the decomposition of the hydroperoxides and the formation of the solid film. Hydroperoxides are not very stable compounds and are quite easily decomposed. It has been postulated that the ether linkage in 5(*d*) may be formed between two molecules by decomposition of the hydroperoxide with the elimination of a molecule of water. This postulation is strengthened by the fact that water is one of the products of autoxidative polymerization, and that ether groups have been shown to exist in polymerized oil films. It was stated previously that oil films are more resistant to saponification than the oil itself is. One reason for this may be traced to the fact that ether linkages are more resistant than the ester linkages in the oil molecule.

The formation of the ring structure in 5(*e*) would require a reaction between 5(*a*) and 5(*c*) with the elimination of a molecule of oxygen. This would necessitate the decomposition of the hydroperoxide, probably by the drier, and the absorption of the oxygen, also by the drier. The subsequent addition reaction between the conjugated unsaturation in 5(*c*) and the double bond in 5(*a*) has been shown to be possible, although it is usually associated with thermal polymerization. However, it is conceivable that the removal of the oxygen by the drier may catalyze this reaction so that it occurs at normal temperatures. This may explain the fact that oil films have lower oxygen content when air-dried with driers present than when air-dried without driers. It may also account for the low oxygen content of highly conjugated oil films and their fast-drying properties. Since the ring structure is composed entirely of carbon-carbon linkages it would be highly resistant to saponification. It is well known that films of the highly conjugated tung oil are more alkali resistant than those of linseed oil. It is possible that the continued development of ultraviolet and infrared methods of analysis may reveal more about the mechanism of autoxidative polymerization. Further details regarding this type of polymerization are given in Chapter 2, and the mechanism of action of driers is discussed more fully in Chapter 5.

Condensation Polymerization. It was stated previously that water or some other small molecule is given off in the condensation polymerization reaction. Therefore the esterification of polybasic acids with polyhydric alcohols to form resinous polymers is one form of condensation polymerization. This is illustrated in Fig. 6(*a*) by the partial reaction of phthalic anhydride and glycerol. From our discussion of functionality it can be seen that this reaction could produce highly cross-linked polymers. These two materials form the basis of the widely used alkyd resins described in Chapter 7. Usually fatty acids or drying oils are reacted with the phthalic anhydride and glycerol to produce the oil-modified alkyd resins.

Figures 6(*b*) and (*c*) show another type of condensation polymerization with the elimination of a molecule of water. The methylolphenol in (*c*) condenses under heat and catalyst to form the phenolic resins used in varnish making. The different phenols available for these resins are described in Chapter 3, and the use of the resins in varnishes is covered in Chapter 4. The partial condensation of dimethylol urea in 6(*c*) indicates the type of reaction which takes place to form the urea-formaldehyde resins and plastics. Many of the modifications of this reaction are discussed in Chapter 8 on urea- and melamine-formaldehyde resins.

The resin manufacturer usually carries the condensation reaction only to the point at which the resin is still soluble in solvents or oils. In the case of certain phenolic resins the condensation is continued to some extent during the cooking of the resin with oil to produce

Fig. 6. Condensation Polymerization

(a) (b) (c)

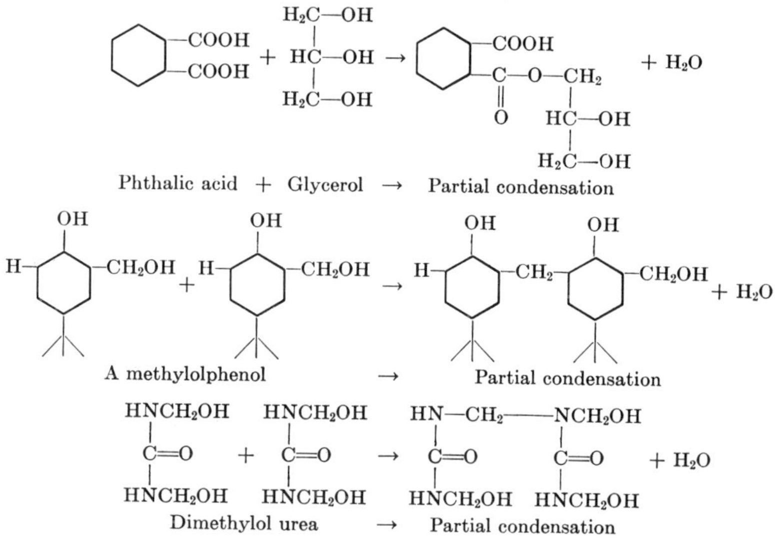

varnishes. The condensation is carried further in the baking type of phenolic resins which are used as coatings without oil modification, the coating being baked after application.

A comparison of the three reactions in Fig. 6 shows that the first contains ester linkages whereas the other two contain methylene linkages between the monomeric units. From our previous discussion of polarity we can see that the first (alkyd resin) is quite polar because of the ester group, but the others are non-polar because of the methylene linkages. Therefore we would expect that alkyd resins would be more sensitive to water and alkali than phenolic or urea-formaldehyde resins. This expectation is confirmed in the use of these resins, and it is also known that, in general, phenolic resins are more resistant than urea-formaldehyde resins. This may be explained in part by the greater degree of polarity of the urea-formaldehyde resins. It will also be realized that films produced from baking materials will be more resistant than those produced by air-drying or autoxidative polymerization because of their lower degree of polarity.

Addition Polymerization. As the name implies, addition polymers are produced by the simple addition of monomers to each other. Since nothing is eliminated in this reaction the composition of the polymer will be identical with that of the monomer, but the polymer will have a higher molecular weight. However, the two end groups on a linear polymer may not be identical with the monomer, as we shall see in later chapters. Usually they are not significant since they constitute only a very small percentage of a high polymer.

There are three types of addition polymers: homopolymers, copolymers, and heteropolymers. Homopolymers are those in which all the monomeric units are identical. Copolymers contain different monomeric units, but, by definition, each of the different monomeric types must be capable of forming polymers alone. Heteropolymers also contain different monomeric units, but, by definition, one or more of them are not capable of forming polymers alone. Specific examples of these addition polymers are given in Chapter 12, Table 1. The general reaction of addition polymerization is illustrated in Fig. 7.

Figure 7(*a*) shows monomeric styrene in both the normal and activated forms, and polystyrene produced by addition of x number of monomers. The mechanism of activation and polymerization will be discussed in a moment. Figure 7(*b*) shows the inactive and activated forms of butadiene, and a part of a copolymer of styrene and butadiene. It will be seen in later chapters that this type of copolymer is widely used in synthetic rubber and in rubber resins. It will be recalled that this type of unsaturation is the terminal type and produces long-chain linear polymers.

Figure 7(*c*) shows the addition polymerization between two monomers having non-terminal unsaturation, but close examination reveals a difference from the reaction shown in Fig. 5(*e*). In 7(*c*) both monomers have conjugated unsaturation, whereas in 5(*e*) one has conjugated and the other has non-conjugated unsaturation. Figure 7(*c*) illustrates the classical Diels-Alder 1:4,1:2 addition reaction between two molecules each having conjugated unsaturation. If the carbon atoms in the diene unsaturation are numbered from 1 to 4 as shown in 7(*c*), the addition takes place between the 1:4 carbon atoms in one molecule and the 1:2 carbon atoms in the other. There is a simultaneous shift of the unsaturation to the 2:3 position between the 1:4 carbon atoms, as shown in 7(*d*). We shall see in subsequent chapters that this important reaction is used in the production of improved drying oils and certain synthetic resins.

Mechanism of Addition Polymerization. The mechanism of addition polymerization is discussed in later chapters; therefore only a brief outline is given here. Mark (1) has shown that addition polymerization takes place in three general stages: initiation, propaga-

Fig. 7. Addition Polymerization

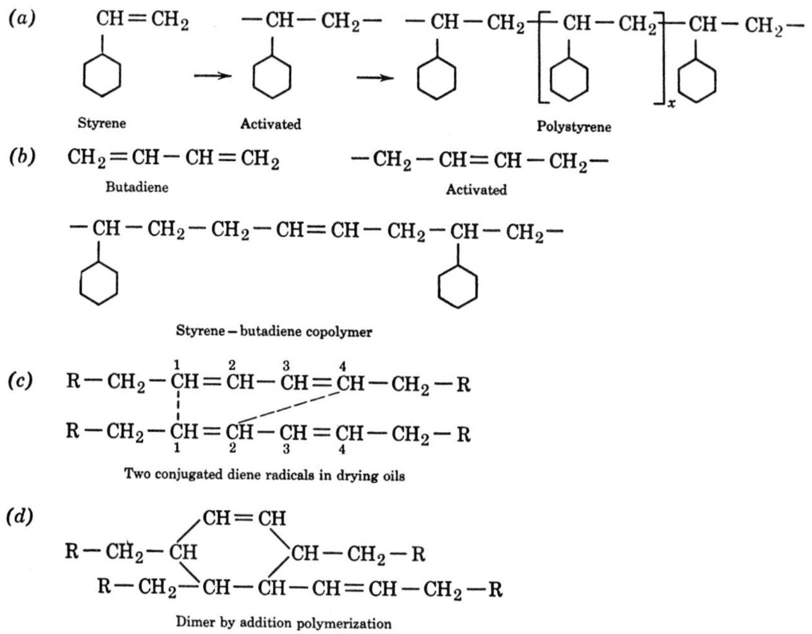

tion, and termination. Polymerization is initiated when the monomer becomes activated by heat, light, or catalyst. The polymer grows, or is propagated, by successive additions of monomers. The rate of growth depends on the polymerizing conditions, the type and amount of catalyst, and the type of monomer. The polymerization is terminated by a collision between the active ends of two growing chains, or by addition of a terminator such as a free radical from the catalyst or solvent, or by a chain-transfer mechanism.

Peroxides are frequently used as catalysts for addition polymerization. Peroxides are relatively unstable compounds which may decompose to produce free radicals. The free radicals provide the energy to activate the monomers, thereby initiating the polymerization reaction. Benzoyl peroxide is widely used as a catalyst, and its decomposition to produce two free phenyl radicals and carbon dioxide is indicated below. The free-radical condition is usually indicated by the large dot at the side of the atom.

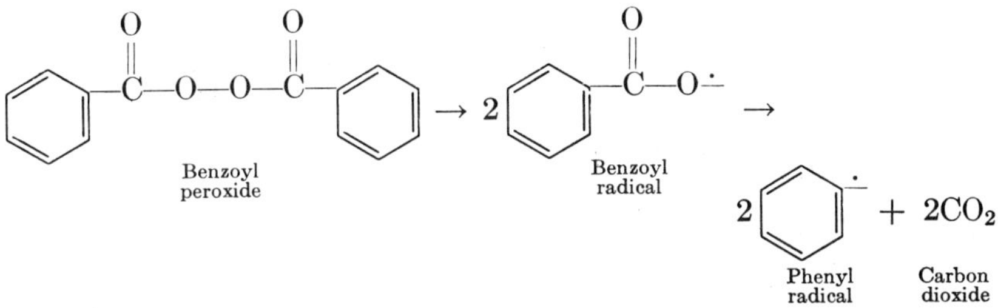

The activation of ethylene by a free phenyl radical may be indicated as follows:

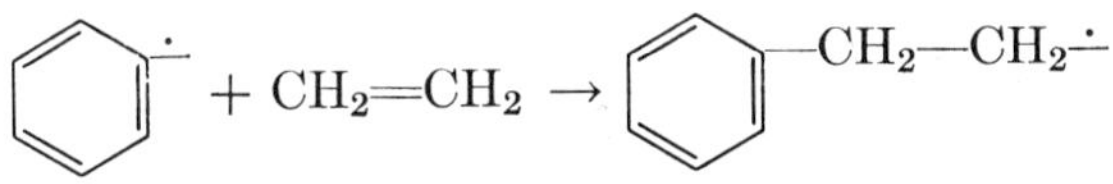

The chain grows by successive additions of ethylene until it is terminated as indicated previously. When initiated and terminated by free phenyl radicals the polymerization of ethylene may be indicated as follows:

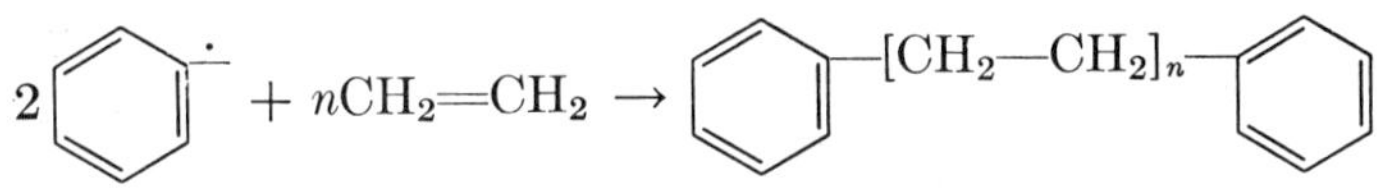

The polymer size is determined largely by the conditions under which the polymerization takes place. There are four general methods used for addition polymerization: mass, solution, emulsion, and dispersion methods. The reaction may be quite rapid in the mass method and difficult to control. When the monomers are dispersed in a solvent (solvent method) better control of the reaction is obtained. The effects produced by different solvents are discussed in later chapters. Excellent control is obtained when the monomers are emulsified with soap or other emulsifying agent to form an oil-in-water emulsion before polymerization. When this type of polymerization has been carried to the required extent, the product may be used as a latex, or the emulsion may be broken and the polymer coagulated, washed, and dried. Good control is also obtained in the dispersion method, in which the monomers are dispersed in a non-solvent, usually water. The polymer is obtained as a very fine bead, which is not contaminated with emulsifying agent.

Mechanism of Emulsion Polymerization. The mechanism of polymerization in emulsion systems is quite complicated and has not been elucidated completely despite the extensive research which has been done on it and on related phenomena. Harkins (9) proposed a general theory of the mechanism, and D'Alelio (10) gives a comprehensive review of the subject in his recent book. At present the generally

Fig. 8. Emulsion Polymerization Process

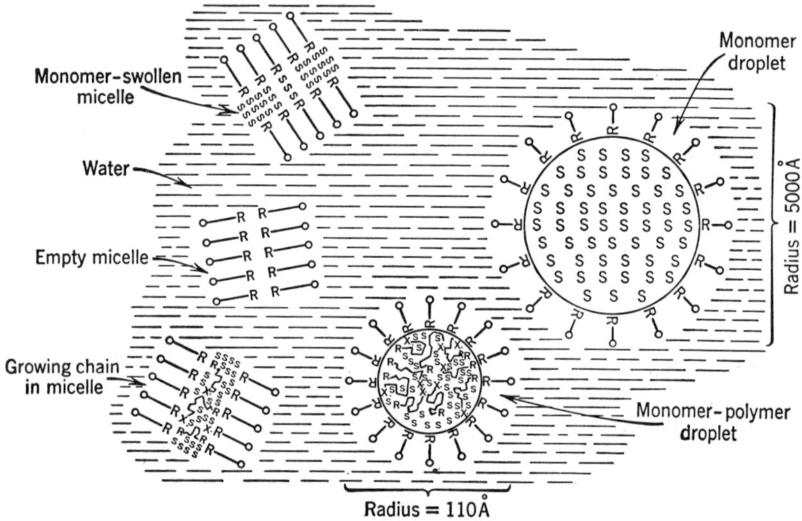

Monomers are solubilized in the micelles, then activated by catalyst from the water phase. Activated polymer chains grow to large polymers by diffusion of monomers from monomer droplets into the micelle. The swollen micelle changes to a monomer-polymer droplet at a certain size and becomes the polymer particle when polymerization is complete. The monomer droplets disappear before polymerization is complete. Courtesy of G. F. D'Alelio, *Fundamental Principles of Polymerization,* page 377, John Wiley & Sons, 1952.

accepted theory proposes that polymerization is initiated in a monomer-swollen soap micelle which grows to a monomer-polymer droplet by migration of monomer from a monomer droplet. The various stages in the proposed process are shown diagrammatically in Fig. 8 in an idealized manner.

The process requires that the concentration of soap in the water phase be greater than the critical micelle concentration so that aggregations of soap molecules in double layers will be present to form the empty micelles as shown. The micelles become swollen by absorption or "solubilization" of the monomers. The monomer-swollen micelles are believed to be widely separated from each other in dilute solutions, but they may form groups in more concentrated solutions. The swollen micelles are loosely organized; therefore free diffusion of the water-soluble catalyst into the micelle may occur readily to produce active nuclei for the initiation of polymerization.

It has been shown (11) that the energy of activation of styrene in emulsion polymerization is about 8000 calories less than in bulk or solution polymerization. The lower energy requirement, and consequent faster rate of polymerization, are believed to be due to the orientation of the monomer molecules in the micelle. Thus the monomer-containing micelle becomes the primary locus of initiation of the polymerization reaction. Since the monomer has an equilibrium solubility in the water phase, the dissolved monomer may react with catalyst radicals to produce a minor locus of initiation, but this would be very small compared with the micelle initiation. It is assumed that the micelles contain from 200 to 1000 monomer molecules; therefore the degree of polymerization would be low if it were restricted to the micelle. In order to obtain high polymers it is necessary to supply monomers to the active points of the growing polymer chains in the micelle. The normal emulsion polymerization system contains a high monomer-soap ratio so that more monomer is present than can be solubilized in the micelles. The excess monomer is present as droplets surrounded by a very thin layer of soap molecules which is probably monomolecular. The monomer droplets serve as reservoirs from which monomers migrate to the micelle to increase the size of the growing polymer. When the polymer particle becomes sufficiently large the soap which formed the micelle becomes a protective layer around the polymer particle. The monomer droplets which served as reservoirs for the monomer disappear entirely before polymerization is complete (12).

From the above results it appears that the principal locus of initiation and propagation for emulsion polymerization is the solubilized monomers in the micellar soap. The monomers in the micelle are activated by absorption of catalyst from the water phase, and the energy of activation is low because of orientation of the monomers in the micelle. Additional monomers are supplied to the growing polymer chain by diffusion from monomer droplets which are thousands of times larger than the finished polymer particle. The monomer droplets decrease in size and disappear before the polymerization is complete.

The general process may be expected to vary somewhat with different monomers, particularly with those which have a significant solubility in water. Modifications may be made by varying the concentration of soap, type of soap, and monomer-soap ratio. The importance of the soap is evident from the data of Harkins (13) that the rate of polymerization of styrene in water without soap is from 100 to 200 times slower than in water with soap.

A glance at the various monomeric units in Table 1, Chapter 12, will indicate the many possibilities which are available for the linear type of addition polymer. The possibilities may be enlarged greatly by blending many of the monomers to produce copolymers. Variation in the components of copolymers may also be used to produce different products. For example, Fig. 7(*b*) shows the possibility of producing styrene-butadiene copolymers. When the percentage of butadiene is high the products are soft and rubbery and are known as GR-S rubbers. When the ratio is reversed (high styrene content) the products are much harder and are known as rubber resins, as described in Chapter 9. The high-styrene type of copolymer is polymerized by the emulsion method, and the resulting latex used as the vehicle for interior architectural paints.

AIR-DRYING AND BAKING COATINGS

Our brief outline of film formation has indicated that we may produce a great variety of films some of which are formed under air-drying conditions and others under baking conditions. Air-drying coatings are used chiefly for architectural work, and on objects made of wood and other materials which would be adversely affected by the elevated temperatures required in baking ovens. Baking coatings are used chiefly for industrial finishing of metal objects. Coatings are formulated to meet the durability and service requirements of the various applications. However, it is possible to formulate baking coatings which have greater durability than could be obtained from the oxidizing type of air-drying coatings. There are three broad reasons for this situation: (1) oxidizing coatings do not dry through as uniformly as baking coatings; (2) oxidation continues at a reduced rate after the coating is "dry"; (3) oxidation-decomposition products are formed which detract from durability.

The surface of oxidizing coatings dries hard faster than the subsurface, thereby setting up mechanical stresses in the film as it ages. Coatings which have been air-dried for 1 or 2 weeks have been found to be quite tacky at the metal-coating interface although they are dry and hard at the surface in contact with the air. Such coatings have good adhesion to the metal at this stage, but the adhesion may be reduced considerably when the subsurface has dried completely. We have seen that a film-former may become immobilized, or set to a dry condition, before all the possible cross-linkages have been formed. Therefore oxidation will continue for a long time at a greatly reduced rate. Since more cross-linkages increase the hardness of the oxidizing coatings, these coatings will become progressively harder with time. This condition need not exist in baking coatings since they can be made with practically non-oxidizing film-formers which are less susceptible to change with age. The aging of oxidizing films is discussed in detail in Chapter 2, and it will be shown that a certain amount of film deterioration is connected with oxidation. The oxidation-decomposition products include water and carbon dioxide as mentioned previously, and also short-chain fatty acids, such as formic acid. Such acids would react with basic pigments to form soaps. If basic pigments were not present the acids might incite corrosion on steel surfaces. It will be recalled that Fig. 1 indicates that the film-former is continually changing during the life cycle of a paint. Some of the reasons for these changes will be apparent from this brief discussion of oxidizing films.

It has been well established that oxidation produces chromophoric or colored compounds. If a white enamel were made with linseed oil, the film would lose its whiteness as it aged because of the formation of yellow products of the oxidation reaction. Since baking coatings are cured chiefly by thermal polymerization instead of oxidation, there is little or no "yellowing" as the films age.

LACQUER FILMS

The term "lacquer" is used chiefly for coatings based on nitrocellulose. However, an increasing number of high-polymeric materials are being used as film-formers in air-drying coatings. We have seen that these materials dry chiefly by evaporation of the solvent and do not require either oxidation or thermal polymerization to form films. We shall see in later chapters that some of the high-polymeric materials are heated after application, but this is done to eliminate the solvent completely and to develop better adhesion, rather than as a film-forming process. This applies particularly to the polyvinyl resins.

The polymeric units in lacquer films are formed before the coating is applied, and they are bound together by the secondary valence forces discussed previously. We saw that the strength of the film increases as the molecular weight of the polymers increases, and that there is also a related increase in viscosity of the coating. An indication of the increase in strength of nitrocellulose films with increase in molecular weight is given in Table 3. The values in the table are comparative only for this particular group since they vary somewhat with film thickness and other conditions of the test. However, they show that, as the viscosity or size of the molecules increases, the cohesive force between them increases as shown by greater tensile strength. This complies with the statement made earlier that the secondary valence forces which bind these molecules into coherent films increase as the size of the molecules increases. These forces may be partially neutralized by the addition of resin or plasticizer molecules which lie between the high-polymer molecules in the film. Obviously, the ratio of resin or plasticizer to high polymer will be a factor in the extent of change of force, but the type of resin or plasticizer is also important. This is illustrated in Table 4 by the comparison of three plasticizers used in different ratios in nitrocellulose films.

Table 3 Variation of Tensile Strength and Per Cent Elongation with Increasing Viscosity of Nitrocellulose

Viscosity *	Tensile Strength, psi	% Elongation
23 cp	9,000	2
0.25 sec	12,000	5
0.5 sec	12,800	7.5
5 sec	14,200	12.5
40 sec	14,600	13.0
1000 sec	15,000	13.0

* See Chapter 11 on cellulosic polymers for interpretation.

Table 4 Variation of Tensile Strength and Per Cent Elongation with Type and Ratio of Plasticizer to 0.5-Sec Nitrocellulose

	Dibutyl Phthalate		Paraplex * C20		Paraplex * C25	
% Plasticizer	TS, psi	% Elong.	TS, psi	% Elong.	TS, psi	% Elong.
0	8000	2.5	8000	2.5	8000	2.5
25	5400	3.5	6400	3.0	7100	3.0
35	2700	5.0	3900	6.0	6100	5.0
45	900	11.5	1800	28.0	2900	17.0

* Trade mark, Rohm and Haas Co.

In all cases the strength of the film is lowered with the addition of plasticizer, and the flexibility is increased. The plasticizer molecules neutralize some of the secondary valence force between the high-polymer molecules and so permit greater freedom of movment between them, resulting in the increased flexibility. However, there is less sacrifice of tensile strength, at equal flexibility, with the resinous plasticizers than with the dibutyl phthalate. With some of the newer polymeric materials the plasticizer may be "built into" the molecule by the use of branched structures or by copolymerizing a softer material with a harder material. Widely used examples of the copolymers are the vinyl chloride-acetate copolymers and the styrene-butadiene copolymers.

EFFECT OF VOLATILE SOLVENTS

Since a large proportion of high-polymeric materials are applied from solution it is necessary to consider the effect of the volatile solvent on the film properties. The two characteristics of the solvent which are important in this respect are its solvent power and its rate of evaporation. If the evaporation rate is very fast, the consistency of the film increases rapidly after application and the polymer molecules have less opportunity to align themselves in the best position for maximum film strength, as indicated previously. In addition, if the solvent is a relatively poor one for the polymer, the consistency of the film increases even more rapidly with solvent evaporation, giving the molecules still less opportunity for the most favorable alignment. The situation is complicated further by the fact that lacquer solvents are usually mixtures of true solvents, latent solvents, and diluents. These must be properly balanced in order to obtain both the lowest cost and satisfactory performance. Many examples are given in subsequent chapters, but it will be realized from this discussion that film properties can be affected considerably by the composition of such solvent combinations. The lacquer-formulator must study carefully the relationship between solvent and film properties if he is to obtain the maximum performance from his coatings.

It should be apparent that the solvent is not so important to the film properties of oils, varnishes, alkyds, and other relatively low-polymeric film-forming materials. Frequently the film-forming process does not begin with these materials until most of the solvent has evaporated. However, even with these materials some improvement in film properties may be obtained from the use of the best possible solvent. Further discussion of the effects of the various types of solvents is given in Chapter 6.

EFFECT OF PIGMENTS

Pigments are used primarily to impart color and opacity, but they also affect other physical properties of coatings. Pigments vary widely in particle size and shape, wettability by oils and resins, chemical reactivity, effect on drying rate of coatings, ultraviolet absorption, etc., so that their effects on the physical properties of coatings may be expected to vary considerably. This subject will be covered in detail in Volume II, but a brief outline of these effects may be of assistance in this study of oils, resins, and other film-formers.

The ratio of pigment to binder varies with the type of pigment and coating. An indication of the general range used may be seen in Table 5, which gives the pounds of pigment used with 100 lb of binder, or vehicle solids.

Table 5 Pigment-Binder Ratios

Coating	Pounds of Pigment per 100 Lb of Binder
Black	6–8
Bright red	20–30
Blue	20–30
Green	35–45
White	90–125
Exterior house paint	
White	175–225
Flat wall finish	
White	200–400

In general the inorganic pigments contribute to the hardness and firmness of coatings. In gloss coatings the pigment reduces the permeability to moisture, but the high pigment content of flat wall coatings tends to make them more permeable than clear coatings. Laminar or plate-like pigments such as mica or aluminum powder decrease permeability, but bulky pigments such as diatomaceous silica tend to make the coating spongy or porous. Many pigments are nodular in shape, but some are acicular or needle-like, and these tend to physically reinforce the film structure.

Some pigments are organophilic or easily wetted by oils and resins, whereas others are organophobic or difficult to wet. If the pigment is poorly wetted by the binder, the coating will be less compact, and therefore more permeable and not as strong. Many binders contain a small percentage of free fatty acids, and other acids may develop as a result of oxidation during the drying process. Some pigments are inert to the action of these acids, but basic pigments such as zinc oxide, white lead, and red lead react with the acids to form metallic soaps. Zinc soaps tend to make coatings hard, whereas lead soaps tend to soften coatings, but both appear to improve water resistance slightly.

One of the most destructive agents for exterior coatings is the ultraviolet component of sunshine. Ultraviolet accelerates the oxidation decomposition of drying oils, and it degrades nitrocellulose. Pigments such as zinc oxide absorb ultraviolet, thereby protecting the film-former from degradation.

These few comments are intended to point out some of the effects of pigments on film properties other than color and opacity. It will be apparent that consideration must be given to all the characteristics of pigments when they are selected for use with the various vehicles.

SUMMARY

This survey of the physical and chemical factors which affect the formation and properties of organic coatings has indicated something of the complexity of film formation. The factors are referred to again in subsequent chapters in connection with specific film-forming materials. The successful formulator of organic coatings must be fully aware of all these factors in order to produce the most satisfactory coating for any particular service requirement. He should also balance cost with performance, since performance greater than that required is not justified if obtained at higher cost. A brief review of the pertinent factors may be helpful.

Primary valence, or the covalent bond, is based on the sharing of electrons between two atoms in contrast with the transfer of electrons from one atom to another in the electrovalent bond. The covalent bond is the general type of direct chemical linkage between the atoms of organic materials. It is the force which binds a group of atoms into a molecule, and also connects the molecules in the heat- or element-convertible type of film.

Secondary valence, or van der Waals' forces, are weaker than the primary valance. These forces operate between molecules in the non-convertible types of film. Their strength is affected by the molecular size, shape, and degree of polarity and the distance between molecules.

Polarity is a residual activity of molecules determined by the arrangement of the electropositive and negative parts of a molecule. When the positive and negative parts are arranged symmetrically the molecule is non-polar. An unsymmetrical arrangement produces a polar molecule. The distance between the electrical charges multiplied by the strength of the charge is the dipole moment of a polar molecule.

Functionality is the number of reactive points in a molecule at which primary valence bonds may be formed with another molecule. Reactions between monofunctional molecules produce simple molecules. Linear polymers may be formed from reactions between difunctional molecules, and cross-linked polymers will result when one of the reactants has a functionality of at least 2 and the other of 3 or more.

Thermoplastic materials soften when heated below their decomposition temperature, and return to their normal condition when cooled to room temperature. They are also known as non-convertible materials and are usually composed of linear polymers.

Thermosetting materials are converted to an infusible and relatively insoluble condition by heat and do not soften appreciably when reheated. They are referred to as heat-convertible materials, and the film structure is composed of cross-linked molecules.

Oxygen-convertible materials are changed to a cross-linked, insoluble condition by absorption of oxygen, followed by autoxidative polymerization.

Autoxidative polymerization is the process whereby certain monomeric molecules absorb oxygen so that hydroperoxides are formed, which decompose to produce primary valence bonds between the monomers to give a polymeric structure.

Condensation polymerization is the process whereby a small molecule, usually water, is eliminated in the reaction between molecular units so as to produce primary valence bonds between them, resulting in a polymeric structure.

Addition polymerization is the process whereby reactive monomers are added to each other to form homopolymers, copolymers, or heteropolymers. The process is usually activated by a free-radical transfer mechanism.

Homopolymers contain monomeric units which are all identical.

Copolymers contain different monomeric units, but each type must be capable of forming polymers by itself.

Heteropolymers contain different monomeric units, but some of them are not capable of forming polymers alone.

Free radicals are produced by the fission of a covalent bond, i.e. a pair of electrons. The activity of free radicals is due to the unpaired electron on each free radical.

Air-dried films are formed by simple evaporation of the solvent from coatings such as spirit varnishes and nitrocellulose lacquers, and by autoxidative polymerization from oxidizing coatings.

Baked films are formed by thermal polymerization of convertible materials. The process is chiefly condensation polymerization but also includes addition polymerization. Baking temperatures usually range from 200° to 450°F.

Lacquer films are chiefly nitrocellulose coatings which are composed of nitrocellulose, plasticizer, and usually a resin. They dry by simple evaporation of the solvent, which may be at room temperature or "force-dried" at temperatures usually from 140° to 180°F.

Volatile solvents are used to reduce the film-forming components to a satisfactory consistency for application. They should entail the lowest cost possible because they are not usually recovered, but they must have the correct solvent power and rate of evaporation or they will adversely affect the stability of the coating and the properties of the dried film.

Pigments are used primarily to provide color and opacity. They vary considerably in particle size and shape, chemical reactivity, and wettability. They must be dispersed properly in the coating to obtain good stability and film properties. Pigments may contribute strength and hardness to films, and the ratio of pigment to binder affects the gloss and permeability.

BIBLIOGRAPHY

1. H. Mark, *Physical Chemistry of High Polymer Systems,* Interscience Publishers.
2. W. H. Carothers, "Polymerization," *Chem. Revs., 8,* 353 (1931).
3. W. H. Carothers, *Trans. Faraday Soc., 32,* 39 (1936).
4. R. H. Kienle, *J. Soc. Chem. Ind., 55,* 229 (1936).
5. T. F. Bradley, *Ind. Eng. Chem., 29,* 440 (1937).
6. T. F. Bradley *et al., Ind. Eng. Chem., 29,* 579 (1937).
7. G. W. Wheland, *Advanced Organic Chemistry,* John Wiley & Sons, 1949.
8. Wm. A. Waters, *Physical Aspects of Organic Chemistry,* D. Van Nostrand Co., 1950.
9. W. D. Harkins, *J. Am. Chem. Soc., 69,* 1428 (1947).
10. G. F. D'Alelio, *Fundamental Principles of Polymerization,* John Wiley & Sons, 1952.
11. W. P. Hohenstein and H. Mark, *J. Polymer Sci., 1,* 552 (1946).
12. R. S. Stearns and W. D. Harkins, *J. Chem. Phys., 14,* 214 (1946).
13. W. D. Harkins, *J. Am. Chem. Soc., 69,* 1439 (1947).

Man is a composite:
physical to contact the earth,
mental to encompass the cosmos,
spiritual to transcend matter, space, and time.
The ratio of the components determines the type of man.

Chapter 2. VEGETABLE AND MARINE OILS FOR ORGANIC COATINGS

INTRODUCTION

The natural oils which are used in paints, varnishes, and oil-modified resins and as plasticizers in nitrocellulose lacquers are obtained from the seeds and nuts of certain plants and trees and from a few species of fish. They are referred to as vegetable and marine oils. They are further classified as drying, semi-drying, and non-drying oils. This classification is related to their ability to "dry" when spread in relatively thin films and exposed to normal atmospheric conditions. The drying, or the changing from a liquid to a solid film, is due to a combination of absorption of oxygen, colloidal association, and polymerization as outlined in Chapter 1. The semi-drying oils do not form satisfactory films at room temperatures, but these oils may be used in air-drying alkyd resins and in baking finishes. The non-drying oils do not form films under either air-drying or baking conditions; therefore they are used as plasticizers or softeners in nitrocellulose lacquers and in non-oxidizing alkyd resins. The important phenomenon of drying is related to the chemical and physical structure of the oil molecule and will be considered in detail later in this chapter.

Linseed oil is the most widely used drying oil, soybean oil is a typical semi-drying oil, and castor, cottonseed, rapeseed, and cocoanut oils are examples of the non-drying oils. Drying oils are used in paints and varnishes which are required to dry under normal conditions; they may also be used in industrial coatings which are force-dried in shorter time by baking at elevated temperatures. The semi-drying oils find wide application in coatings which are baked, or in combination with certain synthetic resins which impart air-drying properties. The non-drying oils serve as flexibilizers or plasticizers for very hard synthetic resins or for high-polymeric materials such as nitrocellulose. The relative amounts of the oils used in the paint and varnish industry in the year 1951, as given by the Bureau of Census report, are shown in Table 1.

Table 1 Oils Used in Paint and Varnish Industry

Oil	Pounds	Per Cent
Linseed	319,770,000	65.00
Soybean	60,738,000	12.42
Tung	53,991,000	11.02
Dehydrated castor	18,680,000	3.82
Oiticica	7,955,000	1.62
Fish	21,547,000	4.40
Castor	4,843,000	1.00
Cocoanut	3,501,000	0.72
Total	491,025,000	100.00

Although a greater tonnage of drying oils is used in the United States today than ever before the percentage of linseed oil has decreased as shown in Table 2. This is due in part to the technical developments which have improved the drying properties of semi-drying oils, and in part to the extensive use of synthetic resins which are modified wtih semi-drying and non-drying oils in preference to linseed.

Table 2 Per Cent of Drying Oils Used

Oil	1912–14	1929–30	1932–38	1947	1950
Linseed	90	80	60–70	61	55.2
Tung	5–8	15	15–19	12	15.2
Fish	2–5	4–5	4–5	6	4.0
Soybean	...	...	2–3	14	16.5

The drying time of films of drying oils is shortened by addition of metallic driers. The most important metallic driers are the organic salts or soaps of lead, cobalt, and manganese, but those of calcium, zinc, and iron are also used for special purposes. The various types of driers together with the mechanism of their catalytic action are discussed later. In general the drying time is also shortened by an increase in temperature and by exposure to light rather than dark conditions.

Exterior house paint is practically the only widely used coating that has oil as the entire film-forming component. Oleoresinous varnishes are combinations of drying oils and hard resins which are cooked under specified conditions. The ratio of oil to resin may vary from 1 part by weight of oil to 2 parts of resin to as high as 5 to 1, or from 6 to 60 gal of oil to 100 lb of resin. Many oil-modified alkyd resins are combinations of oils and glyceryl phthalate resin. In these combinations the oil may be either the drying, semi-drying, or non-drying type, depending on the end-use requirements of the resin. When non-drying oils serve as plasticizers for high polymers they may be used as the straight refined oils having low viscosity or as blown oils having higher viscosity.

Although the natural oils have many desirable properties, they also have some inherent weaknesses. They are relatively slow drying and produce soft films, hence the use of hard resins in varnishes. Successful processes have been developed for modifying the oils to increase the rate of drying. Such processes may change the structure of the oil molecule, or they may add another constituent such as maleic anhydride or styrene. These oils are then referred to as treated oils, styrenated and other copolymer oils, or synthetic oils. Usually treated oils also have better water and chemical resistance than the original natural oil, as well as faster drying properties.

A thorough understanding of the chemistry of vegetable and marine oils, together with the physical-chemical changes which take place when they are converted into dry films, will be extremely valuable in the study of other film-forming materials, including synthetic resins and high polymers. However, it should be understood clearly that film-formers generally are not pure chemical compounds. Many of them are natural materials and are subject to the variations which that source implies. They cannot be purified by ordinary distillation because their molecular weights are too high. They can be distilled by molecular distillation, but the cost of this process is too high at present for use with drying oils. Nor can they be crystallized from solution economically.

COMPOSITION AND STRUCTURE OF OILS

Vegetable and marine oils are composed of triglycerides of the long-chain fatty acids, together with small percentages of phosphatides, carbohydrates, and other natural impurities. The large majority of the fatty acids have carbon chains containing 18 carbon atoms, although reference to Table 3 will show some with shorter chains and others with longer chains. The chemical structure of the triglyceride is shown on the right-hand side of the equation in Fig. 1. However, it should be kept in mind that the physical structure is more nearly like that shown in Fig. 2. Figure 2 attempts to show the non-planar characteristic of the triglyceride structure. Again this is a simplified version since the carbon atoms in both the glyceryl portion and the fatty acid portion do not lie in a straight line but form zigzag chains. One of the possible arrangements, as illustrated by Farmer (1), is shown in Fig. 3. He also points out the three sections of each chain:

1. The tail, *t*, section, which contains eight carbon atoms and remains constant in most oils.

Fig. 1. Triglyceride: Chemical Formula

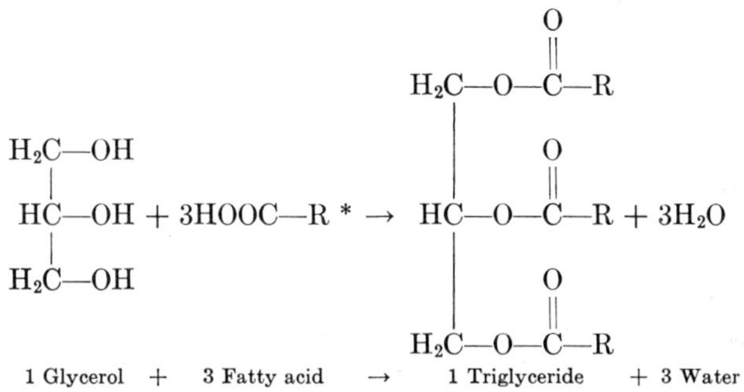

* R is a fatty acid radical. See Table 3 for list of fatty acids.

Fig. 2. Formal Structure of a Triglyceride

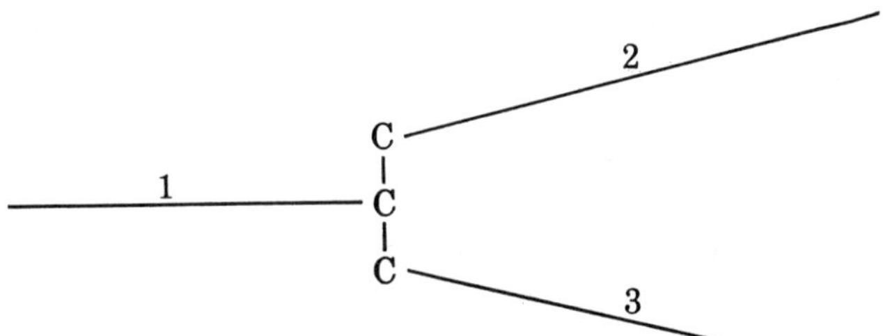

If 1, is in the plane of the paper.
2, is away from the reader.
3, is toward the reader.
The straight lines 1, 2, and 3 represent fatty acid radicals.

2. The functional middle, *f-m*, section, which contains the active double bonds and is different in the various oils as shown in Table 3.

3. The head, *h*, section, which will be longer or shorter depending on the number of double bonds in the middle section.

Such complicated structures are necessary when visualizing some of the reactions which take place in the oxidation, polymerization, and chemical treatment of oils discussed later in this chapter. However, the simplified structure shown in Fig. 2 will be used for all general purposes in this book.

The essential differences in drying, semi-drying, and non-drying properties of the natural oils are due to differences in type and degree of unsaturation in the fatty acid chains of the triglyceride. It is extremely important that the reader obtain a clear understanding of these differences since their significance carries over into other fields of resin chemistry. Table 3 shows the common names and structural formulas for the principal fatty acids in vegetable oils. It is divided into two groups: the saturated acids which are the principal compo-

Fig. 3. Structure of a Triglyceride Oil

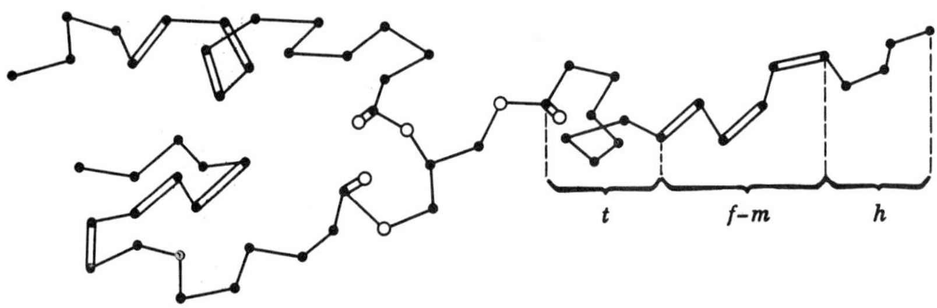

Courtesy of E. H. Farmer, *J. Oil & Colour Chemists' Assoc.*, *31*, 394 (1948).

nents of non-drying oils, and the unsaturated acids found chiefly in drying and semi-drying oils.

TYPES OF FATTY ACIDS

The saturated fatty acids shown in Table 3 range from capric with a 10-carbon chain to behenic with a 22-carbon chain. The three non-conjugated acids, oleic, linoleic, and linolenic, found in the linseed-type oils, have one, two, and three double bonds respectively. Each double bond or ethylenic group (—CH═CH—) is separated from the next ethylenic group by a methylenic group (—CH_2—), which is the characteristic arrangement of non-conjugated unsaturation. Conjugated unsaturation is illustrated by eleostearic acid, found in tung oil, and licanic acid, found in oiticica oil. In both these acids each ethylenic group is adjacent to the next ethylenic group. The relationship between these types and amounts of unsaturation and the drying properties and polymerization rates of the oils will be covered in detail later in this chapter.

Another interesting feature of the unsaturated acids is the arrangement of seven methylenic groups between the carboxyl group and the first reactive unsaturation. The importance of this non-reactive group

Table 3 Fatty Acids

Acid	Structure	Formula
Saturated		
Capric * Decanoic †	$CH_3(CH_2)_8COOH$	$C_{10}H_{20}O_2$
Lauric * Dodecanoic †	$CH_3(CH_2)_{10}COOH$	$C_{12}H_{24}O_2$
Myristic * Tetradecanoic †	$CH_3(CH_2)_{12}COOH$	$C_{14}H_{28}O_2$
Palmitic * Hexadecanoic †	$CH_3(CH_2)_{14}COOH$	$C_{16}H_{32}O_2$
Stearic * Octadecanoic †	$CH_3(CH_2)_{16}COOH$	$C_{18}H_{36}O_2$
Arachidic * Eicosanoic †	$CH_3(CH_2)_{18}COOH$	$C_{20}H_{40}O_2$
Behenic * Docosanoic †	$CH_3(CH_2)_{20}COOH$	$C_{22}H_{44}O_2$
Unsaturated		
Oleic * 9-Octadecenoic †	$CH_3(CH_2)_7CH{=}CH(CH_2)_7COOH$	$C_{18}H_{34}O_2$
Linoleic * 9,12-Octadecadienoic †	$CH_3(CH_2)_4CH{=}CH{-}CH_2{-}CH{=}CH(CH_2)_7COOH$	$C_{18}H_{32}O_2$
Linolenic * 9,12,15-Octadecatrienoic †	$CH_3CH_2{-}CH{=}CH{-}CH_2{-}CH{=}CH{-}CH_2{-}CH{=}CH(CH_2)_7COOH$	$C_{18}H_{30}O_2$
Eleostearic * 9,11,13-Octadecatrienoic †	$CH_3(CH_2)_3CH{=}CH{-}CH{=}CH{-}CH{=}CH(CH_2)_7COOH$	$C_{18}H_{30}O_2$
Licanic * 4-Keto-9,11,13-octadecatrienoic †	$CH_3(CH_2)_3CH{=}CH{-}CH{=}CH{-}CH{=}CH(CH_2)_4C({=}O)(CH_2)_2COOH$	$C_{18}H_{28}O_3$
Ricinoleic * 12-Hydroxy-9-octadecenoic †	$CH_3(CH_2)_3CH(OH){-}CH_2{-}CH{=}CH(CH_2)_7COOH$	$C_{18}H_{34}O_3$
Erucic * 13-Docosenoic †	$CH_3(CH_2)_7CH{=}CH(CH_2)_{11}COOH$	$C_{22}H_{42}O_2$

* Common name.
† Chemical name.

as an internal plasticizer will be discussed later. The arrangement in Table 3 was designed to serve as an aid in visualizing the structure of the fatty acids. For example, if one wishes to visualize an 18-carbon fatty acid containing one double bond, the location of this double bond is found by writing the carboxyl group followed by the seven methylene groups; then the double bond occurs between the next two groups or between the 9- and 10-carbon atoms. The remainder of the fatty acid will consist of CH_2 groups and a terminal CH_3 group to total 18 carbons. Similarly, for an acid containing three non-conju-

Table 4 *   Constants and Composition of Fats and Oils

Acid	Chemical Formula	Vegetable Oils								
		Castor	Cocoanut	Corn	Cottonseed	Linseed	Oiticica	Perilla	Poppyseed	Rapeseed
Saturated acids										
Caproic	$C_6H_{12}O_2$		**0.5** 0.2–0.8							
Caprylic	$C_8H_{16}O_2$		**8** 6–9							
Capric	$C_{10}H_{20}O_2$		**7** 6–10							
Lauric	$C_{12}H_{24}O_2$		**48** 44–51							
Myristic	$C_{14}H_{28}O_2$		**17** 13–18		**0.5** 0.5					**0.5** 0–1.5
Palmitic	$C_{16}H_{32}O_2$	**2** 1–2	**9** 8–10	**8** 7–8	**21** 20–23	**5.5** 4–7	**7** 6–8	**7** 7–8	**5** 3–5	**2** tr–4
Stearic	$C_{18}H_{36}O_2$	**1** 1–2	**2** 1–3	**3.5** 3–4	**2** 1–3	**4** 2–5	**5** 4–6		**3** 3–4	**0.9** 0.4–1.6
Arachidic	$C_{20}H_{40}O_2$			**0.5** tr–0.5	**0.2** 0.2–0.4	**0.3** 0.3–0.9				**0.5** tr–0.5
Behenic	$C_{22}H_{44}O_2$									**1** 0.5–1.5
Lignoceric	$C_{24}H_{48}O_2$			**0.2** 0.2	**0.3** 0.2–1.2	**0.2** tr–0.4				**1** 1–2.4
Unsaturated acids										
Palmitoleic	$C_{16}H_{30}O_2$		**0.2** tr–0.4							**0.1** 0.1
Oleic	$C_{18}H_{34}O_2$	**7** tr–8.5	**6** 5.5–7.5	**46** 43–47	**29** 23–35	**22** 12–34	**6** 4–7	**8** 4–10	**28** 25–30	**18** 14–39

Acid	Chemical Formula	Vegetable Oils—*Continued*								
		Castor	Cocoanut	Corn	Cottonseed	Linseed	Oiticica	Perilla	Poppyseed	Rapeseed
Unsaturated acids—Continued										
Erucic	$C_{22}H_{42}O_2$									**51** 45–55
Ricinoleic	$C_{18}H_{34}O_3$	**87** 86–92								
Linoleic	$C_{18}H_{32}O_2$	**3** 3–6	**2** tr–2.5	**42** 39–42	**45** 42–54	**17** 7–27		**38** 34–44	**64** 58–65	**22** 15–29
Linolenic	$C_{18}H_{30}O_2$				**2** tr–11	**51** 35–65		**47** 44–49		**3** 1–7
Licanic	$C_{18}H_{28}O_3$						**78** 74–81			
							Hydroxy acids **4** tr–4.5			
Analytical constants										
Iodine value (Wijs)		81–91	7.5–10.5	103–128	99–113	155–205	140–160	193–208	130–140	97–108
Saponification value		176–187	250–264	187–193	189–198	188–196	186–193	188–197	189–196	170–180
Melting point, °C		−10 to −12	23–26	−10 to −12	−2 to +2	−20			−15 to −20	−9
Titer, °C			20–24	14–20	30–37	19–21	42–47	12–17	15–19	11–15

NOTE: 1. Boldface type denotes mean values.
2. Lightface type denotes ranges.
3. Iodine value, saponification value, and melting points determined on whole oil.
4. Titer and acid distribution determined on fatty acids only.

* Courtesy of Archer-Daniels-Midland Co.

Table 4 Constants and Composition of Fats and Oils—*Continued*

		Vegetable Oils—*Continued*						Marine Fats and Oils †		
Acid	Chemical Formula	Safflower	Soybean	Sunflower	Tall (commercial)	Tung	Walnut	Herring	Menhaden	Sardine
Saturated acids										
Lauric	$C_{12}H_{24}O_2$								**0.1** tr–0.1	
Myristic	$C_{14}H_{28}O_2$	**0.1** 0.1–1.5	**0.1** tr–0.3				**0.2** tr–0.5	**7** 6–8	**7** 6–7	6
Palmitic	$C_{16}H_{32}O_2$	**4.5** 3–6	**8** 7–10	**4** 3.5–4	**7** 6–8	**4** 1–5	**5.8** 4–7	**14** 12–17	**16** 15–16	10
Stearic	$C_{18}H_{36}O_2$	**2** 1–4	**4** 3–6	**3** 1.5–3		**1** 0.5–2	**1** 1–2.5	**0.5** tr–0.8	**2** 1–3	2
Arachidic	$C_{20}H_{40}O_2$	**0.4** 0.2–0.5	**0.6** 0.3–0.9	**0.6** 0.6–0.8				**Tr** tr–0.1		
Lignoceric	$C_{24}H_{48}O_2$			**0.4** 0.4						
Unsaturated acids										
Myristoleic	$C_{14}H_{26}O_2$		**0.1** tr–0.1					**0.5** 0.2–1.2	**0.1** 0.1	
Palmitoleic	$C_{16}H_{30}O_2$		**0.2** tr–0.4		**1** 1			**6 (−3H)** 5–12 (−2.4 to −3.4H)	**16** 14–17	13
Oleic	$C_{18}H_{34}O_2$	**20** 15–26	**28** 17–57	**33** 22–39	**44** 41–48	**8** 1–19	**22** 12–36	**21 (−3.5H)** 16–22 (−2.9 to −4.8H)	**15** 12–16	24
Linoleic	$C_{18}H_{32}O_2$	**70** 51–79	**54** 28–57	**59** 51–68	**37** 36–39	**4** 1–10	**63** 48–76		**7** 4–15	
Linolenic	$C_{18}H_{30}O_2$	**3** 0.1–8	**5** 2–10			**3** 0.5–10	**8** 3–10		**2** 2–4	

Acid	Chemical Formula	Vegetable Oils—*Continued*						Marine Fats and Oils †—*Continued*		
		Safflower	Soybean	Sunflower	Tall (commercial)	Tung	Walnut	Herring	Menhaden	Sardine
Unsaturated acids—Continued										
Eleostearic	$C_{18}H_{30}O_2$					**80** 74–95				
Licanic	$C_{18}H_{28}O_3$				Conjugated linoleic **11** 9–13					
Arachidonic	$C_{20}H_{32}O_2$							**28 (−4.8H)** 22–30 (−3.9 to −5.5H)	**17 (−10H)** 17–19 (−10H)	26 (−5H)
Clupanodonic	$C_{22}H_{36}O_2$							**23 (−4.9H)** 20–29 (−4.1 to −5.7H)	**11 (−10H)** 11–12 (−10H)	19 (−7H)
	$C_{24}H_{38}O_2$							**tr (−4H)** tr (−4H)	**4 (−10H)** 4 (−10H)	
	C_{26}								**1 (−10H)** 1 (−10H)	
	Over C_{26}				**45** Fatty 36–61 acids				**2 (−10H)** 2 (−10H)	
					42 Rosin 32–46 acids					
					13 Terpenes 7–23					
Analytical constants										
Iodine value (Wijs)		140–150	120–141	125–136	165–170	160–175	145–160	123–142	140–180	170–193
Saponification value		188–194	189–195	188–194	175–185	189–195	189–198	179–194	189–193	189–193
Melting point, °C		−13 to −18	−20 to −23	−16 to −18		4	−12 to −16			
Titer, °C		15–18	20–21	16–20		36–37	14–16	23–27	31–33	27–28

† Unsaturation of fish oil is usually expressed as the hydrogen deficiency. Thus a figure following oleic acid such as 12% (−4.8H) signifies the composition is made up from oleic acid (−2H), linoleic (−4H), linolenic (−6H), and octa-deca-tetrenoic acid (−8H), etc., so as to give the reported composition.

gated double bonds, the first double bond is between the two carbon atoms following the seven CH_2 groups as illustrated above and is therefore between the 9- and 10-carbon atoms. Then, by arranging a CH_2 group between each of the three double bonds, the complete structure can be visualized readily. This acid is, of course, linolenic acid. One exception to the arrangement of seven CH_2 groups is licanic acid, which contains a carbonyl group at the 4-carbon atom.

It should be observed that ricinoleic acid, which is found in castor oil, contains a hydroxyl group on the 12-carbon atom. It is important to remember the location of this hydroxyl group when considering the dehydration of castor oil. The dehydration process converts the non-drying castor oil to a drying oil. The reader should study the structure of each fatty acid until complete familiarity is attained with the variations among the acids.

The average percentages of the various fatty acids in the triglycerides which constitute the many oils of commerce are shown in Table 4. The values are averages and were obtained from the chart issued by the Archer-Daniels-Midland Co. Other sources of such information may vary slightly in the percentages given because of variation in the source of the oil and the method of analysis. Since the oils are natural products they vary somewhat in composition and are affected by the climate and type of soil in which the plants are grown. The oils contain mixtures of fatty acids; therefore the remarks following Table 4 on their distribution in the triglycerides should be kept in mind when relating composition to performance properties of the oils.

DISTRIBUTION OF FATTY ACIDS IN THE TRIGLYCERIDES

It might be inferred from Table 4 that linseed oil contains 51% of the triglycerides of linolenic acid, 17% of linoleic triglycerides, and 22% of oleic triglycerides, with 10% of the triglycerides of saturated fatty acids. However, Hilditch (2) and Bernstein (3) have stated that this cannot be the case because there exists an even distribution of the various fatty acids in the triglycerides. For example, if an oil contained only linoleic acid, or 100% linoleic triglycerides, obviously each triglyceride must have three linoleic acid radicals. However, if an oil contained 33% of oleic acid and 67% of linoleic acid, each triglyceride would have one oleic acid radical and two linoleic acid radicals if the even distribution law were obeyed. This is based on the interpretation of even distribution outlined in Table 5.

Table 5 Even Distribution

When Oil Contains the Given % of Fatty Acid A	All Triglycerides Will Contain the Given Number of A Acids
100	3
66–100	2 or more
66	Not more than 2
33–66	1 but not more than 2
33	Not more than 1
0–33	0 to not more than 1

Some oil chemists believe that the distribution of the fatty acids in the triglycerides does not follow the even distribution but is somewhere intermediate between even and random distribution. However, whether the even, random, or intermediate distribution exists in oils a large majority of the triglycerides are mixed with respect to their fatty acid components. This fact has an important bearing on the drying properties of the oils, and on the qualities of the dried film as discussed on page 110 under mixed glycerides.

SOURCES OF OILS

Linseed oil is obtained from the seeds of the flax plant. Flax is grown in many parts of the world, but the principal producing countries are the United States, Russia, Argentina, Canada, and India. The oil content of the seed varies from 32% to 43%, depending on climatic conditions during growth and on type of seed. The iodine value of the oil may also vary from 170 to 190; this value may be used as a criterion of its drying quality.

Perilla oil is obtained from the seeds of the perilla plant, which is a native of Manchuria, Japan, and India. The seeds contain from 35% to 40% oil which is similar to linseed oil but has better drying properties. The difficulty of harvesting the seed satisfactorily has prevented its extensive cultivation in this country.

Safflower oil first appeared in the American market toward the end of 1949. It is obtained from the seeds of the safflower plant, which is an annual herb growing from 18 to 42 in. in height. The plant is a native of the Himalayan foothills and grows best in semi-arid areas. It is well suited to eastern Colorado, Wyoming, and the Nebraska Panhandle, but good yields have also been obtained in certain parts of California, Oregon, Washington, and Idaho. The seeds contain from 27% to 36% oil, and the seed cake is satisfactory for cattle feed.

The oil has a unique composition (see Table 4) which makes it suitable for alkyd resins and certain varnishes. It is slower drying than linseed but somewhat faster drying than soybean.

Soybean oil is obtained from the seeds of the soybean plant. This plant is indigenous to China but is now cultivated extensively in the United States and other parts of the world. It is an extremely useful legume since it may be used as a source of food as well as many industrial products. The seeds contain from 16% to 20% oil.

Sunflower oil is obtained from the seed of *Helianthusannuus,* which is cultivated in the United States, Russia, India, and Argentina. The seeds contain from 22% to 32% oil which has drying properties similar to those of soybean oil. The seeds also find wide use as a feedstuff.

Tung oil (frequently called China Wood oil) is obtained from the nuts of the tree *Aleurites fordii,* which grows wild in China. Successful plantings have been made in Georgia, Alabama, Louisiana, Texas, Mississippi, and Florida, which yield an average of about 30,000,000 lb of tung oil a year at present. The oil-bearing portion of the nut contains about 50% oil, the quality of which varies somewhat according to the method used for its extraction.

Oiticica oil is obtained from the nuts of the tree *Licania rigida,* which is a native of Brazil. The oil-bearing portion of the nut contains from 60% to 65% oil. The oil is similar to tung oil in that its fatty acid (licanic acid, Table 3) has triple conjugated unsaturation, but is different in that it contains a carbonyl group.

Castor oil is obtained from the seeds of the castor plant, which grows in tropical and semi-tropical countries such as Brazil and India and in parts of the United States. The seeds contain about 45% oil which is non-drying, as indicated by its iodine number of 80–90. About 85% of the total fatty acid of castor oil is ricinoleic acid, which contains a hydroxyl group (see ricinoleic acid, Table 3). This group may be removed together with a hydrogen atom by a dehydration process which is described later. The dehydrated castor oil has good drying properties and is widely used in varnishes, alkyd resins, etc.

Cottonseed oil is practically a non-drying oil obtained from the seeds of the cotton plant. The oil is widely used for edible purposes. In the organic coating industry it is used in the non-drying type of alkyd and other plasticizing resins.

Cocoanut oil is obtained from the nuts of the palm tree, *Cocas nucifera.* The cocoanuts consist of an outer husk of fibrous material used in floor mats and brushes, and an inner part known as the copra.

The copra contains about 65% of a non-drying oil suitable for edible and industrial purposes.

Fish oils are obtained principally from sardines, menhaden, and herring, and are produced on both the Pacific and Atlantic Coasts. The drying quality of the oils varies with the species and condition of the fish and the method of extraction. The oils contain fatty acids longer than 18 carbons, some of which have four double bonds as shown in Table 4.

MANUFACTURE OF OILS

The oil is obtained from the various seeds by either a pressure extraction or a solvent extraction process. Before extraction of the oil, the seed must be cleaned, crushed, and partly cooked. After extraction the oil is filtered to produce the raw oil of commerce. It may then be refined by the acid, alkali, or mechanical process which makes it suitable for varnishes, synthetic resins, and many other uses. The oil may be further treated with maleic anhydride, pentaerythritol, etc., to produce one of the many "treated" oils. Some oils may be copolymerized with such reactive monomers as styrene or cyclopentadiene to obtain the various copolymer oils. The reader should make a thorough study of the oil charts of the various manufacturers since they contain up-to-date information regarding composition, uses, and properties of the constantly increasing variety of oils which are becoming available.

Pressure Extractions. The process of obtaining linseed oil from flax seed is a typical pressure extraction process. The chaff, stalks, husks, and other foreign particles are removed from the seed by a system of blowing and screening. The seed is then ground on vertical roller mills which have three to five rolls. From the mills the crushed seed is conveyed to the cookers, which are cylindrical tanks containing a series of horizontal plates and a rotating shaft with scraper arms attached to it. The crushed seed is fed into the top of the cookers, and the scraper arms move it from each plate to the next lower one. The cookers are steam-jacketed, and in addition live steam is fed into the mass to moisten it to the required extent. The flax "meal" reaches a temperature of 190–200°F at the bottom outlet of the cooker. It is important that the meal be in the correct physical condition and have the proper moisture content for maximum oil extraction in the pressure equipment. The meal is conveyed from cookers to either the Expeller type or the hydraulic type of pressure extractor.

The hydraulic pressure extractors are being replaced by the screw type of expellers, but the hydraulic equipment is still in use in some oil mills. The meal must be formed into cakes and then wrapped in camel's-hair mats before being placed between the plates of the hydraulic presses. Each press contains more than twenty plates. When the press is loaded with meal cakes the pressure is applied, slowly at first, then up to a maximum of about 4000 psi. The time required to

Fig. 4. Expeller Oil Extraction Equipment

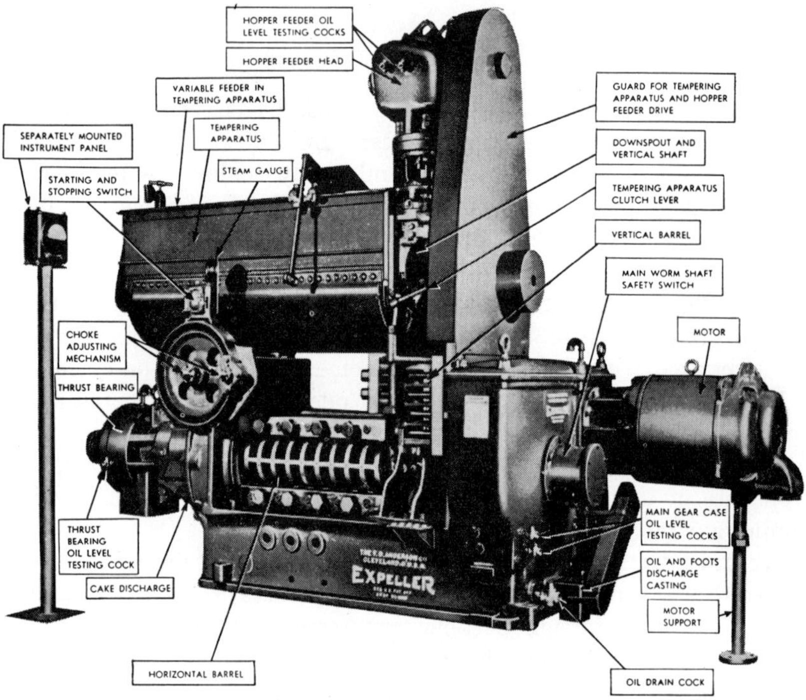

Courtesy of V. D. Anderson Co., Cleveland, Ohio.

press the oil from the cakes is about 1 hr. The oil is run through a coarse filter to settling tanks, and the meal cakes are ground, bagged, and sold for cattle feed.

In the Expeller * equipment the meal from the cookers or conditioners is forced under considerable pressure through a steel barrel by a screw mechanism. The oil flows out through small openings, and is filtered, then run to the settling tank. The cake, containing about 4% oil, is discharged in chip form. It is then ground, bagged, and sold as cattle feed meal. In the interest of increased efficiency the Expeller may have two barrels as shown in Fig. 4. The meal is con-

* Trade name, V. D. Anderson Co.

veyed from the tempering apparatus into the vertical barrel, thence into the horizontal barrel under the continuous pressure flow.

The temperature of the filtered oil from the hydraulic presses or the expellers is about 170–180°F. On cooling in the storage tanks some of the sterols, phosphatides, etc., separate out in a flocculated condition. A filter aid is then added, and the oil is filtered to remove as much as possible of the suspended matter; then it is run to storage tanks. This oil is sold as "raw, double-filtered" linseed oil and is satisfactory for exterior house and barn paints, the manufacture of linoleum, and other uses.

Solvent Extraction. The principal advantage of the solvent extraction process is the higher yield of oil it obtains from the seed. In the pressure extraction processes, a yield of 95% of linseed oil is about maximum, but the yield of soybean oil may be as low as 80%. However, yields of 99–99.5% are obtained with the solvent extraction process. This also means a lower oil content in the meal, which improves it for livestock food. The low labor cost of solvent extraction is another important factor. Practically all plants at present operate some form of continuous extraction with automatic controls. Equipment is available for large-scale production ranging from 300 to 400 tons per day, or for small-scale units designed to handle 10–25 tons per day. The latter type was designed for districts in which both soybeans and cattle are raised. The meal is made available locally as cattle food and saves freight on both beans and meal to and from large mills located in industrial districts.

Experience has shown that soybeans lend themselves readily to solvent extraction. The beans are cleaned, cracked, tempered with steam, flaked, and conveyed to the solvent extraction system. Soybean flakes may be rolled as thin as 0.008–0.010 in. and still be sufficiently durable to withstand handling in the extraction system. This permits a high recovery of oil without formation of "fines" due to breakdown of flakes, which retard the rate filtration of the oil. With seeds which do not form durable flakes, such as flax and cottonseed, it has been found that a prepress operation followed by solvent extraction gives the highest recovery of oil and eliminates the problem of fines. This type of operation is illustrated by the Exsolex Process of the V. D. Anderson Co. and is shown in Fig. 8, page 62. The flakes from soybeans or those prepared from the prepress flax or soybean cakes are conveyed through the extractors countercurrent to the hot solvent. The "miscella," or mixture of oil and solvent, is then stripped of solvent by evaporation and steam stripping. The residual solvent in the flakes is removed by heat and live steam. Several types of extraction equipment are available at present, some of which use a hydrocarbon such as hexane as the solvent, whereas others use trichloroethylene.

The characteristics desirable in a solvent are low cost, narrow boiling range, low latent heat of vaporization, no toxic effect on meal, and non-flammability. Duncan (6) describes the commercial extraction of oils with trichloroethylene. Since this solvent is non-flammable it eliminates the potential fire and explosion hazard of hexane. The Anderson hydrocarbon extraction equipment (Fig. 7, page 61) is practically explosionproof because it is an outdoor unit. Duncan found that the residual trichloroethylene in the meal varied from 0.03% to 0.06%, which is far below the toxicity level of trichloroethylene as reported in the scientific journals.

Sweeney and Arnold (45) at Iowa State College have developed a commercial 25-ton extraction unit based on the use of trichloroethylene. It is manufactured under exclusive patent rights of Iowa State College by the Crown Iron Works of Minneapolis. The initial cost of the solvent is high, but the freedom from fire hazard is a distinct advantage. The efficiency of the process maintains a low solvent loss; therefore the cost of solvent per unit of beans processed is not excessive. The possible toxicity of trichloroethylene to workmen is eliminated by preventing leaks and by maintaining good ventilation. After considerable engineering research the present unit (Fig. 5) was developed; it is very simple to operate and is well suited to small production in farm areas where the farmer can bring the beans to be processed and take home the meal for livestock food.

Kenyon (46) describes the Bollman system of solvent extraction as used by the Central Soya Co. for soybean extraction. The cleaned and cracked beans are flaked at 165–175°F with a moisture content of 9.5–10% to produce flakes 0.008 in. thick. The flakes are sprayed with half miscella and conveyed in baskets in a concurrent direction down the extractor. On their passage up the extractor in a separate section they meet the fresh solvent countercurrently. The solvent is removed from the full miscella by pre-evaporators, followed by falling film evaporators and finally steam stripping under vacuum. The solvent is removed from the meal by a specially designed desolventizer. A mixture of hexanes having a specific gravity of 0.685 at 60°F and a boiling range of 146–156°F is used as the solvent.

Another development of the percolator-type extractor is the Rotocel, Fig. 6, manufactured by the Blaw-Knox Co. It consists of a stationary outer shell containing a rotor divided into sectors which are equivalent to horizontal baskets. The mechanical features are simpler, the total volume is reduced, and extraction efficiency is high. The flakes and solvent are fed continuously and in such a manner that

Fig. 5. Solvent Extraction of Soybean Oil

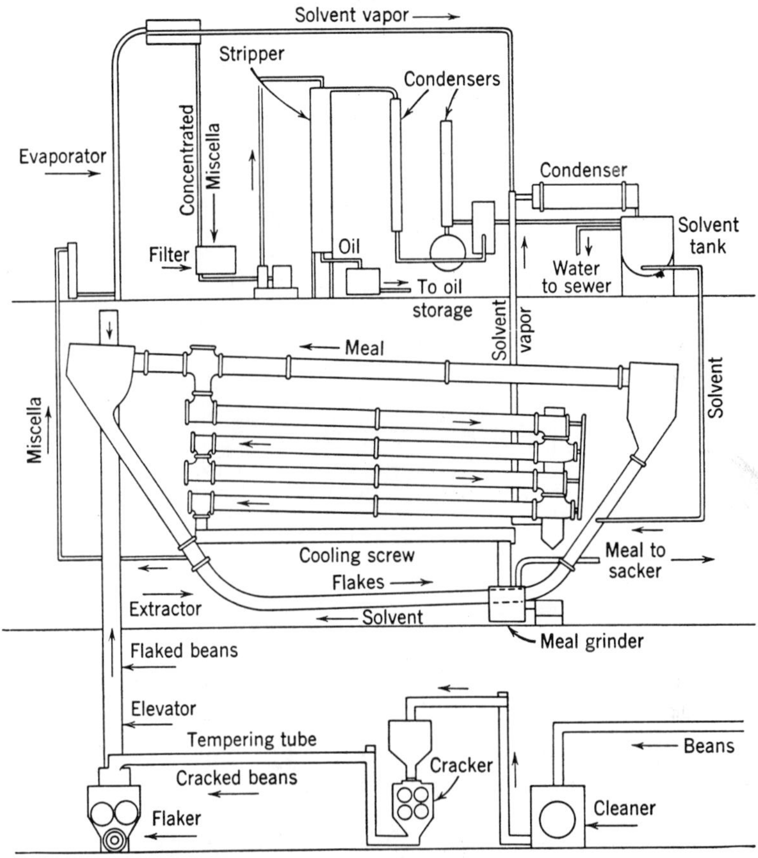

Courtesy of O. R. Sweeney, L. K. Arnold, and E. G. Hollowell, *Bull. 165,* Iowa State College, Ames, Iowa.

completely filtered full miscella is the end product. This further simplifies the apparatus considerably with a reduction in cost. The flakes are desolventized at about 128°F with practically no degradation of the protein. When used for livestock food, the flakes are given a further heat treatment to develop maximum nutritional value.

As indicated previously, the Anderson solvent extraction equipment shown in Fig. 7 is practically explosionproof since it is an outdoor unit.

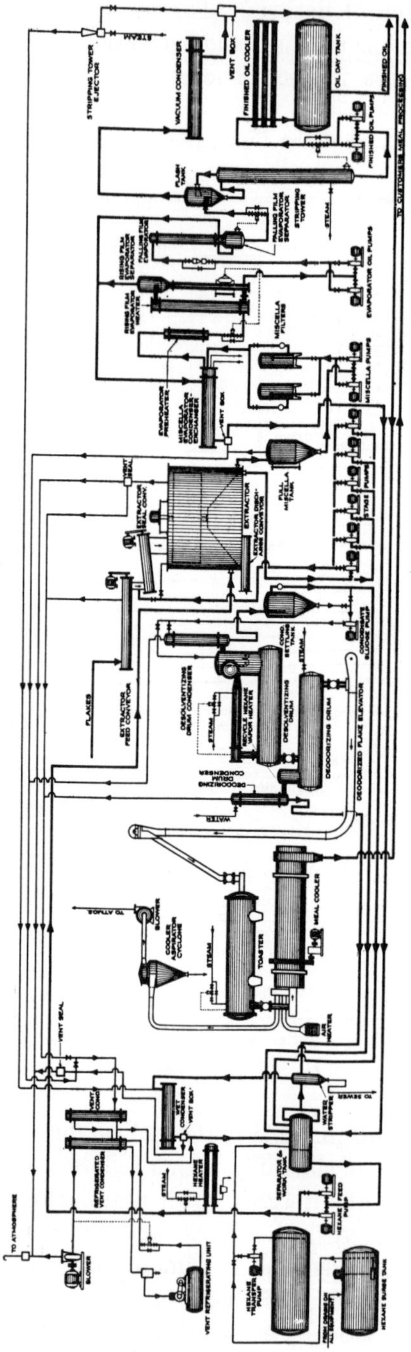

Fig. 6. Solvent Extraction Flow Sheet, Rotocel Extractor

Courtesy of Chemical Plants Division, Blaw-Knox Construction Co.

This prevents the build-up of sufficient hydrocarbon vapor to cause an explosion. A flow sheet of the process is shown in Fig. 8. The flakes pass down the column on a series of plates with adjustable openings, so arranged that the rate of fall may be controlled. The solvent rises countercurrently, and the miscella is removed at the top of the column.

Fig. 7. Outdoor Unit, Solvent Extraction Equipment

Courtesy of V. D. Anderson Co., Cleveland, Ohio.

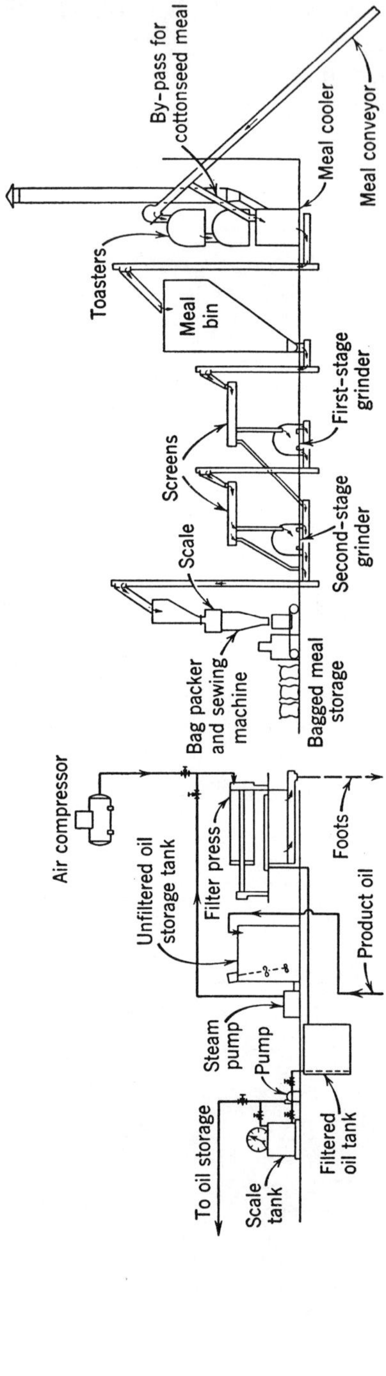

Fig. 8(*a*). Flow Diagram of Complete Plant for Alternate Exsolex Processing of Cottonseed and Solvent Extraction of Soybeans

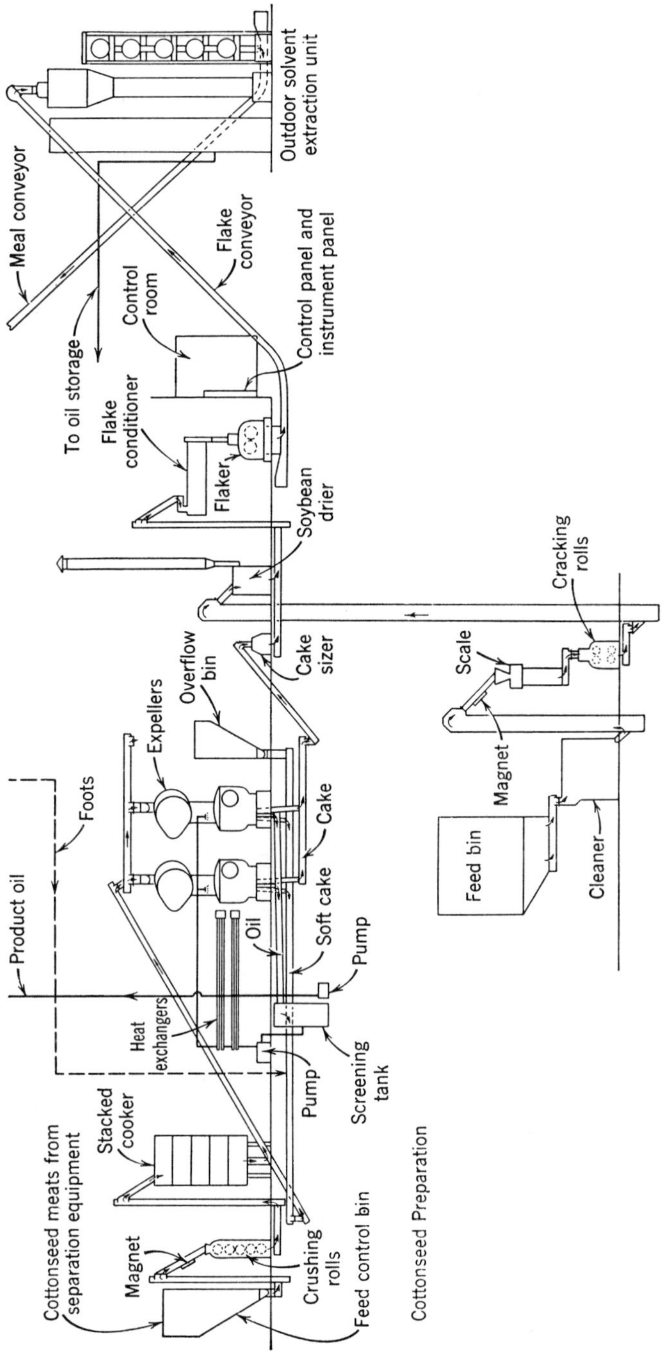

Courtesy of V. D. Anderson Co., Cleveland, Ohio.

Fig. 8(b). Flow Diagram of Solvent Extraction Unit (Vacuum Equipment to Be Added When Necessary)

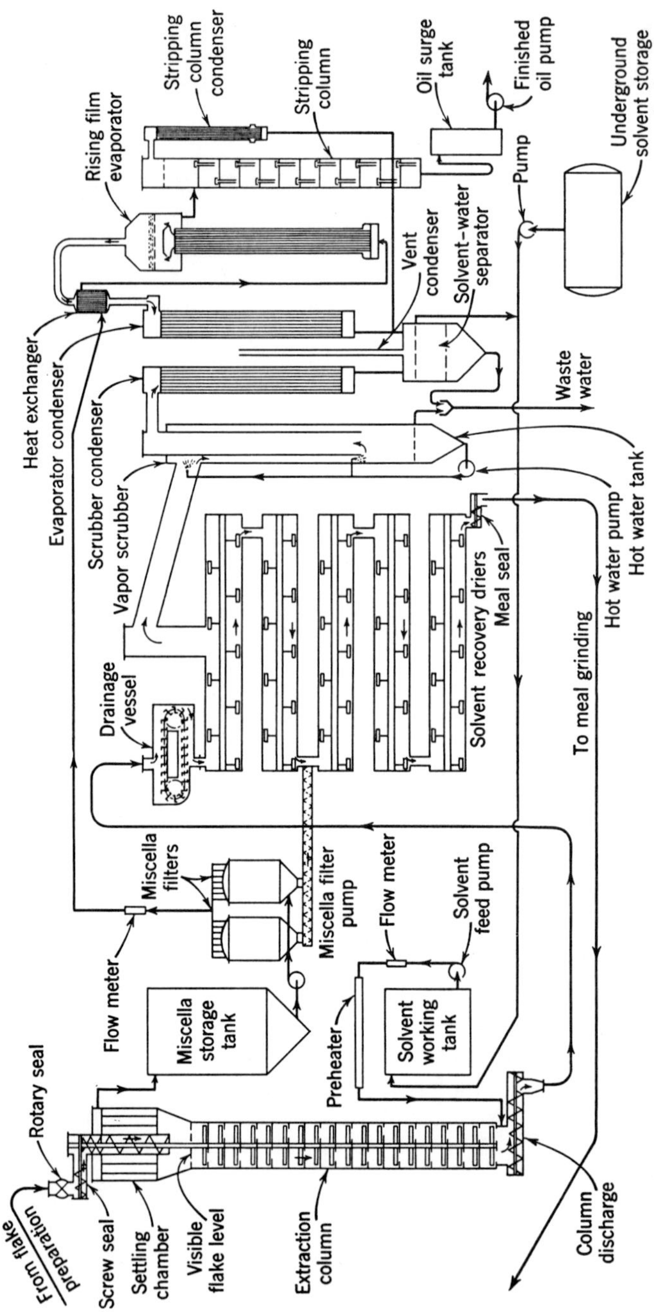

Courtesy of V. D. Anderson Co., Cleveland, Ohio.

The spent flakes are removed from the bottom by a pressure screw device which removes much of the solvent from the discharged meal. The remainder of the solvent is removed as shown in the flow sheet.

The oil from the solvent extraction process is practically the same as that obtained from the pressure extraction process. The sterol and phosphatide content may be slightly higher, depending on the solvent used, moisture content of the flakes, and temperature of extraction, but these materials will be removed in the refining processes.

Refining Processes. Raw linseed and some other oils contain a small percentage, about 0.3%, of sterols, phosphatides, and carbohydrates which "break" or separate out as a mucilaginous sludge when the oil is heated to varnish-making temperatures. See Chapter 15 for description of the break test. The break makes the varnish cloudy and interferes with the drying and the gloss of the varnish film. The break may be removed from the raw oil by one of three refining processes: acid, alkali, or mechanical refining. Mechanically refined oil is satisfactory for certain dark-colored varnishes, baking japans, and grinding vehicles, but it is not used in high-grade varnishes and oil-modified resins. For mechanical refining the raw oil is heated to about 190°F, and an adsorbent such as fuller's earth or activated charcoal is added. After thorough agitation the mixture is filtered to obtain the mechanically refined oil. This oil is "break free" and has paler color and less staining or yellowing than raw oil.

The acid refining process consists of adding strong sulfuric acid to the raw oil in lead-lined tanks with good agitation. Care must be taken to add only enough acid to char the break and cause it to separate. Water is added to stop the reaction, and the acid is removed by washing with hot water or steam. After standing overnight, the oil layer is separated and dried by agitation. Acid-refined oil has a higher acid number than alkali-refined oil because of the formation of some sulfated compounds. The high-acid oil is used to advantage for grinding certain pigments for house paints, but it cannot be used with highly reactive pigments.

The alkali refining process consists of adding a 5–10% solution of sodium hydroxide to the raw oil, sufficient to neutralize its free fatty acid content. The soap formed by this reaction carries down the mucilaginous matter and some of the color-forming material. The treated oil is then washed with water, care being taken not to emulsify the batch. After standing and settling, the oil is removed and a bleaching clay or fuller's earth is added, followed by filtering. The alkali-refined oil can be obtained with an acid number as low as 0.25 max. and is generally considered the best oil for varnish and resin manufacture. The acid number is the number of milligrams of potassium hydroxide required to neutralize 1 gm of sample, and the method of test is given in Chapter 15.

TYPES AND GRADES OF OILS

The oil charts issued by oil-producing companies are excellent sources of information regarding the types and grades of commercial oils. The information is of course more up-to-date than can be found in any book on the subject; therefore the commercial literature should be studied carefully. The raw oil is usually considered the base oil for pricing purposes. A premium is added for all other grades of a particular oil, depending on the type of refining or processing. For some grades of house paint and for linoleum manufacture, raw oil is satisfactory.

Refined Oils. Commercial oil charts show a series of acid, alkali, and mechanically refined oils corresponding to the materials described in an earlier section on the manufacture of oils. The premium on these oils increases with the improved type and degree of refining, which gives the user the opportunity to select the lowest-cost oil consistent with the properties he must have for specific applications.

Boiled Oil. Boiled oil is linseed, and occasionally soybean, oil which contains metallic driers to accelerate the drying rate. Boiled oil is used by painters to mix with white or colored paste paints without the further addition of driers. Boiled oil may be made by adding soluble driers to the cold oil, but it is more common and better practice to heat the oil and drier to produce solution of the drier in the oil. The heating process is known as "boiling," but, of course, the oil is not brought to the boiling point. The usual lead, cobalt, and manganese driers are used in proportions which may range from 0.5% to 1.0% Pb, 0.05% to 0.10% Co, and 0.005% to 0.01% Mn, calculated as per cent metal to oil.

A procedure which could be used to prepare boiled oil is to heat the oil to 450°F and add slowly, and with good stirring, the required amount of litharge as the source of the lead content. Heating is continued to 500°F; then the required amounts of cobalt linoleate and manganese linoleate are stirred in. The oil is allowed to cool and is ready for use. Instead of adding the metal linoleates, the cobalt could be introduced as cobalt acetate and the manganese as manganese borate. In this case the metal salts would be added slowly at about 450°F and the heating continued up to about 525°F. The batch should be held at this temperature until the metal salts have been converted into the soaps or linoleates. The completion of the reaction with the acetate is indicated when no further odor of acetic acid is perceptible in the fumes from the kettle.

There are many variations of the above procedure, some of which include blowing air through the batch while heating. If soluble driers are added to the raw oil to avoid the cooking process as indicated above, the oil will have the necessary drying properties, but will not be a true boiled oil.

Heat-Bodied Oils. Heat-bodied oil is obtained by heating the refined oil to a predetermined temperature and holding at this temperature for the time required to produce a specified increase in viscosity. The usual temperatures at which the oils are bodied are as follows: linseed, 575–600°F; perilla, 575–600°F; soybean, 600°F; fish, 550–560°F. Tung and oiticica oils are rarely used as heat-bodied oils, but when it is necessary to heat-body them, the temperature should be about 450°F, in no case higher than 500°F. Dehydrated castor oil is available commercially at two higher viscosities, V-S and Z3; therefore heat-bodying of the oil is rarely necessary.

The viscosity of oils, varnishes, and resin solutions is usually measured with the Gardner-Holdt bubble-tube method described in Chapter 15. The oil, etc., to be measured is compared with a set of tubes filled with standardized liquids. The standard in which the bubble rises at the same rate as the bubble in the oil being measured gives the viscosity of the oil. The lowest viscosity tube for general use with oils and varnishes is designated as A and corresponds to a viscosity of 0.5 poise, and the highest standard is Z6, corresponding to 148 poises.

The time required to heat-body an oil will depend on the type of oil, the viscosity required, and the temperature. In general, the drying oils body more quickly than the semi-drying oils; there is no significant increase in viscosity when the non-drying oils are held at high temperature. It will be apparent that the rate of bodying is related to the degree and type of unsaturation in these oils, as shown in Tables 3 and 4. The increase in viscosity is due principally to polymerization at the points of unsaturation. The mechanism of the polymerization reaction is treated in detail in a later section. The rate of polymerization increases with temperature; therefore it would appear desirable to body the oils at the highest possible temperature. However, other factors must be considered. Since the flash point of the oils is about 610°F, care must be taken to avoid fires. This danger could be averted to a certain extent by heating in closed kettles with a blanket of inert gas over the oil, but oils are organic compounds and are susceptible to decomposition at high temperatures. The decomposition products are free fatty acids and acrolein from the triglycerides, and some dark-colored compounds. These lower the value of the oil for many paint purposes; therefore the maximum bodying temperature which can be used is governed by the degree of decomposition which can be tolerated.

Another important factor in the heat bodying of oil is the amount of air which is available to the oil during bodying. We have seen that the drying oils form solid films by oxidation at room temperature, and since most reaction rates increase rapidly at high temperature, there will be considerable oxidation of the oil if heating is done in open kettles. Obviously, if the oxidation process converts the liquid oil to a solid film, it will also increase the rate of bodying at high temperature. Unfortunately, oxidized oil is dark colored, and there are the attendant oxidative-decomposition products. Although these are not objectionable for some paints and printing inks, they cannot be tolerated in many architectural and industrial finishes. The oil-bodying equipment described in the next chapter will avoid oxidation by processing in closed kettles, with inert gas supply and under vacuum.

The effect on the various physical and chemical properties of oils bodied at several temperatures with and without vacuum is given in detail by Mattiello and Work (7). The values for alkali-refined linseed oil bodied in air and under vacuum at three temperatures are shown in Tables 6 and 7.

It will be noted that, in all cases, the acid number increases with time of heating and the increase is greater as the temperature is raised. This indicates that the free acids are produced by thermal decomposition, but there are also some free acids formed as the result of oxidative decomposition. These will be reduced when the oil is bodied under vacuum because there is less air in the system; since some of the acids are volatile they will be removed by the vacuum.

The iodine number decreases with increase in viscosity, indicating polymerization or cross-linking at the points of unsaturation. The molecular-weight increase suggests that the polymers are mostly dimers with a small percentage of trimers. There will also be some residual monomers. The color of the oil is darker as the temperature of bodying is increased. Paler color is obtained with the vacuum-bodied oils because there is less oxidation under vacuum.

Table 6 * Alkali-Refined Linseed Oil, Bodied at 535°, 575°, and 600°F in Air

No.	Acid No.	Iodine No.	Vis., poises	Mol. Wt.	Sp. Grav.	Refract. Index	Sap. No.	Color, Gardner	Per. No.	Acetyl No.	% Insol. Acetone
						Temperature: 535°F					
†	0.28	186.2	0.23	738	0.9327	1.4792	191.8	3–4	2.1	3.0	0
1	2.45	159.7	2.2	870	0.9461	1.4831	193.1	3–4	2.6	0.8	0
2	3.00	154.3	3.3	1055	0.9495	1.4841	193.5	3	...	...	6.5
3	3.46	150.0	4.8	1125	0.9528	1.4851	193.2	3–4	1.6	3.5	20.5
4	4.27	143.9	10.1	1200	0.9582	1.4866	193.6	4	...	...	37.5
5	4.70	140.8	15.1	1292	0.9609	1.4873	192.8	4–5	0.9	1.4	47.5
6	5.25	138.0	34.2	1370	0.9644	1.4881	193.5	5	...	...	59.5
7	5.66	136.8	38.4	1535	0.9663	1.4886	193.5	5–6	0.7	3.2	64.5
8	6.01	134.5	56.0	1670	0.9682	1.4892	193.6	5–6	...	...	67.0
9	6.30	134.9	75.0	1775	0.9694	1.4895	193.7	6	0.9	1.3	70.0
10	7.05	133.7	115.0	1890	0.9711	1.4801	193.3	6–7	...	...	71.5
11	7.11	133.6	162.0	1950	0.9715	1.4903	193.0	6–7	0.4	0.2	74.2
						Temperature: 575°F					
1	0.81	180.5	0.58	771	0.9346	1.4805	193.8	3	2.0	2.6	0
2	4.09	151.0	3.5	863	0.9508	1.4845	193.4	3	...	...	10.0
3	5.13	146.0	5.9	908	0.9546	1.4855	193.5	3	1.6	0.0	21.0
4	6.18	141.7	10.5	1165	0.9591	1.4864	193.8	3–4	...	...	35.5
5	6.65	139.8	13.9	1203	0.9609	1.4869	194.3	4	1.4	0.0	42.0
6	8.02	136.6	28.3	1318	0.9650	1.4882	193.6	5	...	...	53.5
7	9.07	136.0	49.9	1590	0.9672	1.4889	193.9	5–6	1.5	0.0	58.0
8	10.07	135.2	68.8	1622	0.9701	1.4896	193.7	6	...	...	
9	11.27	133.3	96.7	1648	0.9713	1.4900	195.4	6–7	0.3	0.0	66.0
10	12.00	132.9	125.2	1700	0.9722	1.4902	194.1	7	...	...	68.7
11	12.55	132.3	235.0	1758	0.9733	1.4908	192.9	7–8	0.0	0.0	70.7
						Temperature: 600°F					
1	0.83	180.3	0.6	727	0.9342	1.4799	191.2	3	1.8	6.2	0
2	15.25	147.3	4.1	936	0.9521	1.4844	191.2	3–4	...	...	15.0
3	6.33	143.8	6.1	1073	0.9554	1.4852	191.7	4	0.9	4.0	25.5
4	7.90	139.3	10.8	1202	0.9596	1.4866	192.3	5	...	...	35.0
5	8.55	138.0	13.6	1261	0.9610	1.4870	192.0	5–6	1.1	0.0	39.5
6	10.7	132.9	28.3	1361	0.9657	1.4882	192.7	6	...	...	51.0
7	7.8	133.3	38.6	1405	0.9677	1.4890	192.0	6–7	0.9	1.4	59.5
8	13.7	131.7	63.8	1445	0.9703	1.4896	194.6	7	...	...	69.0
9	14.7	131.2	83.6	1515	0.9719	1.4901	193.4	7–8	0.4	0.8	66.0
10	15.3	132.1	94.6	1525	0.9717	1.4904	193.9	8	...	...	66.0
11	15.9	132.1	157.2	1595	0.9740	1.4907	193.6	8–9	0.2	0.0	66.0

* J. J. Mattiello and L. T. Work (7).
† Original oil.

Heat-bodied oils are used in varnishes and enamels because they are superior to raw oil for the flowing or leveling property. When raw oil is pigmented to make an exterior house paint, the brush marks may be observed quite readily in the painted surface. These marks may be decreased by partial replacement of raw with bodied oil, and of course they are eliminated entirely in a high-gloss enamel made

Table 7 * Alkali-Refined Linseed Oil, Bodied at 535°, 575°, and 600°F under Vacuum

No.	Acid No.	Iodine No.	Vis., poises	Mol. Wt.	Sp. Grav.	Refract. Index	Sap. No.	Color, Gardner	Per. No.	Acetyl No.	% Insol. Acetone
					Temperature: 535°F						
†	0.28	186.2	0.23	738	0.9327	1.4792	191.8	3–4	1.5	5.3	0
1	1.91	184.6	0.5	790	0.9329	1.4794	192.1	3–4	1.4	2.6	0
2	1.03	152.6	3.9	980	0.9505	1.4845	191.8	1–2	...	...	11.5
3	0.98	147.9	6.5	1100	0.9549	1.4856	191.7	2	1.1	5.8	28.0
4	0.95	142.8	13.5	1235	0.9603	1.4870	191.7	2–3	...	...	47.0
5	0.93	141.2	17.0	1370	0.9616	1.4872	191.7	2–3	1.0	7.3	52.0
6	0.94	138.0	31.0	1460	0.9650	1.4882	191.7	2–3	...	...	60.0
7	0.99	137.3	48.0	1545	0.9672	1.4888	192.4	2–3	1.0	8.7	66.0
8	0.62	135.3	74.2	1800	0.9692	1.4893	191.7	3	...	...	70.5
9	0.76	131.9	114.0	1985	0.9706	1.4898	191.7	3	1.1	6.7	73.0
10	0.84	135.6	138.0	2090	0.9712	1.4899	191.6	3	...	...	72.5
11	1.12	134.5	268.0	2141	0.9728	1.4904	192.4	3–4	0.6	9.0	75.2
					Temperature: 575°F						
1	2.45	150.9	2.9	847	0.9485	1.4839	192.4	1–2	0.3	4.4	0
2	3.03	143.6	5.8	1158	0.9540	1.4853	192.2	2	...	...	7.5
3	2.58	140.5	9.7	1275	0.9585	1.4865	192.4	2	0.8	1.4	28.0
4	2.41	138.9	12.9	1303	0.9604	1.4869	191.9	2–3	...	...	37.5
5	2.12	135.7	24.5	1369	0.9652	1.4881	192.8	3	0.8	0.0	43.5
6	2.09	133.5	40.7	1543	0.9670	1.4890	192.1	3	...	...	56.5
7	2.09	132.9	68.0	1566	0.9691	1.4896	192.5	3–4	0.1	4.7	66.0
8	1.70	130.5	114.0	1703	0.9707	1.4903	191.1	4	...	...	77.0
9	1.66	132.1	115.0	1783	0.9711	1.4901	191.8	4	0.0	4.9	79.0
10	1.39	133.4	150.0	1805	0.9722	1.4904	191.0	4	...	...	78.0
11	1.23	131.4	217.0	1940	0.9726	1.4906	192.5	4	0.0	5.2	80.2
					Temperature: 600°F						
1	0.38	184.0	0.5	705	0.9332	1.4796	193.6	2–3	0.9	5.2	0
2	2.50	149.1	3.7	890	0.9506	1.4844	193.0	2–3	...	...	10.0
3	2.40	143.1	6.6	1065	0.9552	1.4856	193.4	2–3	1.0	5.2	29.0
4	2.28	138.5	11.9	1100	0.9598	1.4867	192.5	3–4	...	...	44.0
5	2.23	138.5	14.3	1382	0.9610	1.4871	192.2	3–4	1.7	2.7	45.5
6	2.20	133.6	34.2	1468	0.9658	1.4886	192.3	3–4	...	...	64.0
7	2.23	132.1	57.3	1530	0.9684	1.4892	192.9	4	0.7	4.2	70.5
8	2.34	130.6	103.0	1910	0.9693	1.4899	192.9	4–5	...	...	74.0
9	2.30	130.0	142.0	2165	0.9710	1.1903	192.6	5	0.7	2.6	75.5
10	2.26	129.9	203.0	2180	0.9714	1.4906	193.3	5–6	...	...	80.5
11	2.22	130.1	278.0	2250	0.9720	1.4910	193.4	5–6	0.8	1.4	80.0

* J. J. Mattiello and L. T. Work (7).
† Original oil.

with a bodied-oil varnish. The highly bodied oils and varnishes require thinning with volatile solvents to reduce them to brushing or spraying consistency. Therefore, films of these materials dry faster than raw oil films because of the rapid evaporation of the solvent, and the fact that the oil is partly converted to the dry stage by the heat polymerization process.

Since linseed oil has a tendency to discolor or "yellow" as the dried film ages, bodied soybean oil is widely used in white enamels. However, the high linolenic acid content of linseed oil which is responsible for the yellowing effect also gives linseed its superior drying property. To overcome the slow drying and bodying of soybean, this oil may be reacted with maleic anhydride and other chemicals as described in a later section.

Limed Oils. Commercial oils are available which have been limed to produce very high viscosity even when reduced to about 35% solids with mineral spirits. These oils are used principally in the "calcicoater" type of flat paint which may be applied directly over calcimine. Small amounts may also be used in lower-grade architectural enamels to give a "false body" which tends to reduce the sagging of the enamel. Usually the oil is linseed or fish, or a combination of these oils. Heat-bodied oils may be limed by adding hydrated lime at about 300°F and continuing the heating to 400° or 450°F. The amount of lime used may vary with specific requirements as indicated in Formulas 1 and 2.

Many variations of these suggestions may be made by changing the type of oil or using calcined magnesium oxide instead of lime. See page 104 for formula and procedure for limed tall oil. The limed oils should not be confused with gloss oils, which are limed-rosin solutions, and are described in Chapter 3 in the sections on rosin.

Formula 1 Limed Linseed Oil

Heat-bodied linseed (Z6)	90 gal
Heat-bodied linseed	10 gal
Hydrated lime	30 lb
Mineral spirits	230 gal

Procedure

Heat the 90 gal of oil to 600°F, and allow to cool to 300°F. Make a paste of the hydrated lime and 10 gal of oil, and add it slowly with good agitation. Raise the temperature to 450°F, then allow to cool to 400°F before thinning with the mineral spirits.

Characteristics

Solids	35%
Oil	96.5%
Lime	3.5%

Formula 2 Limed Fish-Linseed Oil

Refined linseed oil	75 gal
Refined fish oil	25 gal
Refined linseed oil	3 gal
Hydrated lime	25 lb
Mineral spirits	240 gal

Procedure

Heat the linseed and fish oils together to 600°F and hold for a 12-in. string from a cold plate. This will require from 5 to 7 hr. Cool to 350°F, and add the paste made with the 3 gal of linseed and the lime. Stir well, and heat to 350–375°F until a drop on cold glass may be removed without sticking. This usually requires about 30 min, after which the batch should be thinned with mineral spirits.

Characteristics

Solids	35%
Linseed oil	75%
Fish oil	25%
Lime	3%

Blown Oils. Blown oils are made by blowing air through the oil at temperatures from 160° to 250°F. This will induce fairly rapid oxidation of the oil with a consequent increase in viscosity. Therefore high-viscosity oil can be made at lower cost by blowing than by heat bodying, but of course the properties of the two types of bodied oils will be quite different. In general a blown oil yellows more than the same oil which has been heat-bodied, it is not as durable, and it is more reactive with pigments. However, blown linseed and soybean oils have excellent gloss and flow properties and find application in certain enamels and flat paints for architectural purposes. Blown fish oil may be used if it is refined to eliminate the objectionable odor.

Blown linseed oil is also used in the manufacture of linoleum, but for this purpose it is blown to almost a solid consistency often referred to as "sweetmeats." Small percentages of metallic driers or other catalysts may be added to reduce the time required to oxidize the oil to the gel stage. In this condition the oil has excellent binding qualities for the ground cork and other filler used in high-grade linoleum.

Blown castor and rapeseed oils are available in a range of viscosities. Since these are non-drying oils, they are used as plasticizers for high-polymeric film-forming materials such as nitrocellulose.

TREATED OILS

The raw, refined, boiled, heat-bodied, and blown oils have been available commercially for many years. However, over the past decade or so there have become available an increasing number of physically and chemically treated oils. Some of them are the direct result of the shortage of tung oil during World War II, and others are the result of the scientific investigation and the development of improved engineering processes which characterize this particular period of time.

There is every reason to believe that these developments will continue, perhaps to the point of eliminating the natural oils entirely. The cost factor must always be considered as well as the performance factor, as in the parallel development of synthetic rubber versus natural rubber.

The oil treatments and treated oils which have assumed commercial importance are listed below:

1. Solvent segregation.
2. Oil splitting and re-esterification with alcohols other than glycerol.
3. Isomerized oils.
4. Dehydrated castor oil.
5. Maleic-treated oils.
6. Copolymerized oils.
7. Catalyzed oils.

Solvent Segregation. The separation or segregation of a mixture of closely related materials can sometimes be accomplished by selecting a specific solvent which will dissolve one component but not the others. Segregation may also be accomplished with a single solvent provided one component of the mixture becomes insoluble if the temperature is changed (27, 28). The modern methods of segregation are based on the latter phenomenon: a difference in solubility based on a difference in temperature.

It was pointed out in an earlier section that the natural oils are not pure compounds but are mixed glycerides of fatty acids. They also contain some free fatty acid and small amounts of sterols and colored materials. The situation is even more complicated after the oils are heat-bodied or blown. The various components have differences in solubility in solvents such as acetone, methyl ethyl ketone, various alcohols, and furfural. For example, the acetone-solubility test is used to separate the low-bodied portion from the higher-bodied portion in heat-bodied oil or from an oxidized oil film. Hilditch (29) describes the separation of linseed oil glycerides by low-temperature crystallization from ether-acetone solutions. Some early work on segregation of bodied fish oil was reported by Behr (8) in 1940. At first he used a batch process with a mixture of 10% oil and 90% solvent. The later development of a continuous process reduced the cost considerably.

The Pittsburgh Plate Glass Co. (9) reported in 1947 a new selective solvent-distillation process for separating an oil into two fractions one of which is faster drying than the other. The most practical solvent is furfural used in a ratio of 6–14 parts to 1 part of oil. At a temperature of 104°F or above, the entire oil is soluble, but two layers form when the temperature is lowered. One of the layers is mostly solvent containing the fast-drying portion of the original oil; the other is largely the slower-drying fraction of the oil. The solvent is removed from the oil by a continuous distillation process. Several modifications of the process were described. Furfural is widely used as a selective solvent. The separation of colored bodies from FF wood rosin to obtain a pale variety can be made with furfural.

Recently, the use of propane for solvent segregation has been developed by the M. W. Kellogg Co. in the Solexol process (27). Propane is a gas at normal room temperature, but the process operates under sufficient pressure to maintain the propane in the liquid state. Temperatures no higher than 200°F are used in the process, since 206°F is the critical temperature at which all components are precipitated. With liquid propane all components of the oil are soluble at a low temperature, but separation occurs as the temperature is raised. The oil and solvent are fed countercurrently into a column with a temperature differential between the top and the bottom. Therefore the take-off at the top and the bottom will be different.

The Solexol process may be operated under a variety of conditions from simple decolorization to separation of the various drying fractions, vitamins, etc. The process is a continuous one with very low labor cost. Consequently the cost per gallon for segregation is quite low and varies from a fraction of a cent to about 3 cents per gallon. To illustrate the type of segregation which may be obtained typical examples are given in Tables 8, 9, and 10.

Table 8 Fish Oil

	Original Oil	Iodine No. High	Iodine No. Med.	Vitamin	Stearine	Color Bodies	Fatty Acid
Iodine no.	185	240	160	120	50	250	180
% of oil	100	35	47	9	5	3	1

The 35% of the high-iodine-number component has drying properties superior to the 47% of the medium-iodine-number material. The 9% of valuable vitamin component may more than pay for the entire cost of the process. The removal of the non-drying stearine improves the drying and prevents seeding at low temperatures. The removal of the 3% of colored bodies gives the oil a definitely paler color. The removal of the free fatty acid lowers the acid number considerably.

Table 9 Linseed Oil

		First Separation		Second Separation	
				Iodine No.	
	Original	Ref.	Extract	High	Med.
Color, Gardner 1933	11	1	Black	1	1
Iodine no.	185	185	170	210	160
Fatty acid, %	0.8	0.05	1.8	0.03	0.07
% of oil	100	98.5	1.5	48.5	50.0

Table 9 illustrates a two-step process. The first step decolorizes the oil from a color of 11 to a color of 1 in the refined fraction. The acid value is also lowered, but the total material removed from the oil is only 1.5%. The second step separates the refined oil into two fractions: one with a high and the other with a medium iodine number having correspondingly better and poorer drying properties. However, the poorer material may be used to advantage in a semi-drying type of alkyd resin which requires an oil having these characteristics.

Table 10 Soybean Oil

		First Separation		Second Separation		
	Original	Ref.	Extract	Edible	Paint	Sterol
Color, Gardner 1933	12	3	Black	2	6	...
Iodine no.	135	135	100	125	162	115
Fatty acid, %	0.4	0.2	5.0	0.04	0.02	20.0
% of oil	100	98.5	1.5	67.5	30.0	1.0

Table 10 shows a two-step process similar to that for the linseed oil. In this case the second separation gives 67.5% of an oil which is superior to the original soybean for edible purposes because of its lower iodine number. It also gives 30% of an oil which is superior to the original soybean for paint or resin use because of its higher iodine number.

Another modification of the Solexol process is known as solvent crystallization. The oil and solvent are cooled until the high-melting fraction crystallizes, and the slurry is then filtered in a continuous-type filter. It will be apparent that the Solexol process can be adjusted for the segregation of the natural oils and fats into a variety of more useful materials.

Oil Splitting and Re-esterification. The previous section on segregation presents a method for separating an oil into portions having higher and lower iodine values and correspondingly better and poorer

Fig. 9. Polyhydric Alcohols

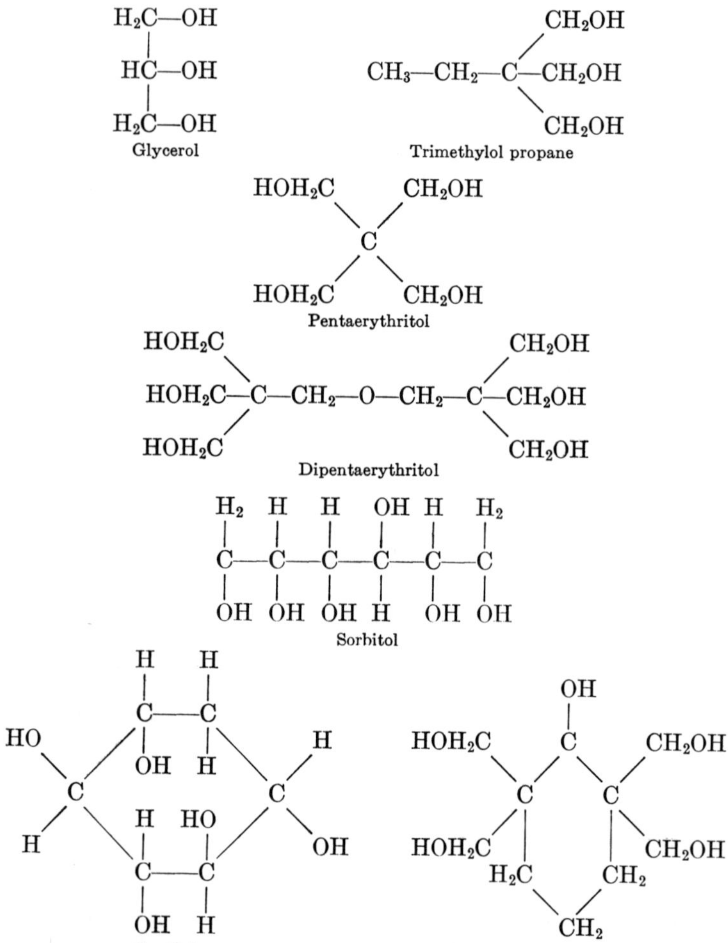

drying properties. However, it will be recalled that triglycerides of most oils are mixed glycerides. This means that in most oil molecules the three fatty acid radicals are different. For example, in linseed oil a particular molecule may contain one or two linolenic radicals, one or two linoleic radicals, and one oleic radical, and some will contain one stearic radical. Of course the total in no case will exceed three, but these will be a mixture of the fatty acids cited. See Table 5 for typical distribution. This also means that, although segregation can produce a better drying fraction of an oil, the result is not the best possible drying oil because some of the molecules contain at least one fatty acid radical which has a low degree of unsaturation.

Obviously, if the oil were split into its components, fatty acids and glycerol, and the fatty acids separated by distillation or crystallization, the separate fatty acids could then be re-esterified with glycerol to produce a series of oils. This series would range from the best possible drying oil to the poorest possible from the original mixed triglycerides. This method of obtaining improved drying oils has been made practical by the development of efficient procedures for splitting fats and oils. It is also possible to improve the drying further by esterifying the fatty acids with alcohols other than glycerol. In general these will be alcohols which have a higher functionality than glycerol. The functionality concept is discussed later in this chapter. The alcohol most widely used for this purpose is pentaerythritol, but others, such as sorbitol, trimethylol propane, inositol (62), and 2,2,6,6-tetramethylolcyclohexanol, are also effective. See Fig. 9 for the structure of these polyhydric alcohols.

Oil-Splitting Process. Oils may be split or hydrolyzed by several processes. One is to saponify the oil with sodium hydroxide and then obtain the fatty acids from the soap by treatment with a mineral acid. Another method is to subject the oil and a small amount of water to high pressure in an autoclave to hydrolyze the oil into fatty acids and glycerol. The Twitchell process, shown in Fig. 11, substitutes a catalyst for the high-pressure autoclave. The Twitchell reagent is made by treating oleic acid and benzene or naphthalene with sulfuric acid. The above three processes are batch processes and therefore not as economical as a continuous process.

The development of the continuous Colgate-Emery process (Fig. 10) for oil splitting has not only improved efficiency with lowered costs but has also made it possible to obtain the products in a higher state of purity. Essentially the process consists of a continuous countercurrent flow of oil and water through a column under heat and pressure and without catalyst. The fatty acids are removed from the top of the column, and the "sweet water" containing the glycerol from the bottom of the column. The central zone of the column is designed so that the water falls slowly, in small drops, through the oil at a temperature of about 500°F. This produces a split or hydrolysis of about 98–99% of the oil.

The direct heat-exchange feature of the process makes it very economical on steam consumption. Since the column is filled with liquid at all times and under sufficient pressure, there is no troublesome boiling. The heat exchange is so efficient that the take-off at top and bottom is at normal temperatures. The details of the process may be seen in Fig. 10.

Fig. 10. Colgate-Emery Fat-Splitting Process

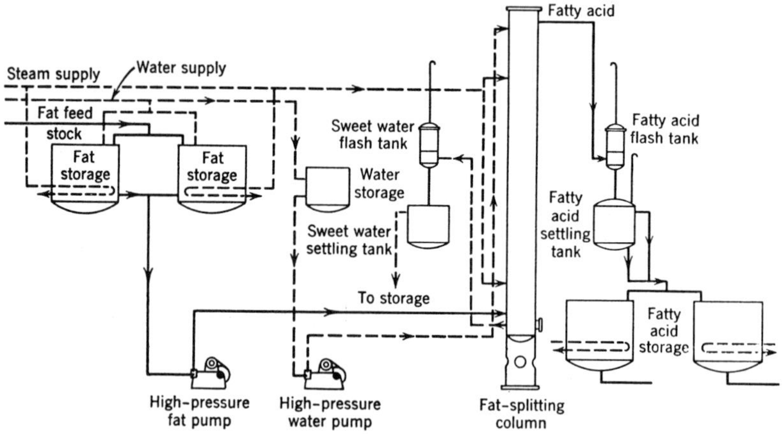

Courtesy of Blaw-Knox Construction Co.

Fatty Acid Separation. The separation of a mixture of fatty acids into its individual components is not a simple operation, because of the relatively close physical properties of the fatty acids. The solvent crystallization Emersol process described by Demmerle (10) uses 90% methanol as the solvent. The equipment has very precise temperature control which enables it to fractionate fatty acids having fairly close melting points. A complete process for obtaining purified fatty acids from fats and oils is illustrated in Fig. 11. It shows the fat split by the Twitchell process, and distillation of the crude fatty acids followed by fractionation by the solvent crystallization process.

Fatty Acids. As a result of improved methods for splitting oil and fractionating the fatty acids, there has become available a wide series of special fatty acids for use in the manufacture of improved drying oils and alkyd resins. The latter application is referred to in detail in Chapter 7 on alkyd resins. Manufacturers such as Emery Industries, Inc., of Cincinnati, General Mills of Minneapolis, and Armour and Co. of Chicago have detailed catalogs describing these products. The fatty acids include those obtained from vegetable and marine oils and also from animal fats and tall oil.

Highly unsaturated materials from fish oil are available, such as Armour's Neo-Fat No. 17, which is a mixture of polymerized glycerides and fatty acids of the C-20 and C-22 unsaturated series, and Neo-Fat No. 19, which is approximately 90% of the C-20 unsaturated acids and 10% oleic and linoleic acids. When these fatty acid materials are esterified with glycerol to form oils or triglycerides, they have very fast drying properties. Similarly, when linseed or soybean oil is split and the fatty acids fractionated and re-esterified with glycerol, a variety of products is obtained. Some of these are better drying than the original oil, and others are slower drying.

Esterification of Fatty Acids. In general the fatty acids may be esterified by heating with the alcohol at a temperature slightly below the boiling point of the alcohol in a closed kettle with a reflux condenser. Various catalysts may be used such as litharge, lime, or calcium naphthenate. A method suggested by the Heyden Chemical Corp. for esterifying linseed fatty acids with pentaerythritol is as follows:

Formula 3 Linseed-Pentaerythritol Ester

Linseed fatty acids	2130 gm
Pentaerythritol	285 gm

Procedure

Heat the linseed acids to 320°F (160°C) in 30 min and add the pentaerythritol. Raise the temperature to 446°F (230°C) in 45 min and maintain temperature until the desired acid number, which is usually from 8 to 10, is reached.

It will be noticed that a 10.5% excess of pentaerythritol is used in Formula 3. It is common practice to use an excess of alcohol in order to obtain more nearly complete esterification in a reasonable time. Since the fatty acids are monobasic, their combining weight will be the same as their molecular weight, which is usually taken as 280. The pentaerythritol is tetrahydric; therefore its combining weight will be the molecular weight of 136 divided by 4, which equals 34. The ratio of 280 to 34 is 8.24, and by dividing the 2130 gm of fatty acids in Formula 3 by 8.24 we obtain 258 gm of pentaerythritol as the amount required by theory.

This oil was heat-bodied by increasing the temperature to 550°F and maintaining this temperature until the desired viscosity was reached. After 4 hr at 550°F, a viscosity of Z3+ or about 55 poises was obtained. The viscosity of a glycerol ester of the same fatty

Fig. 11. Fat Splitting and

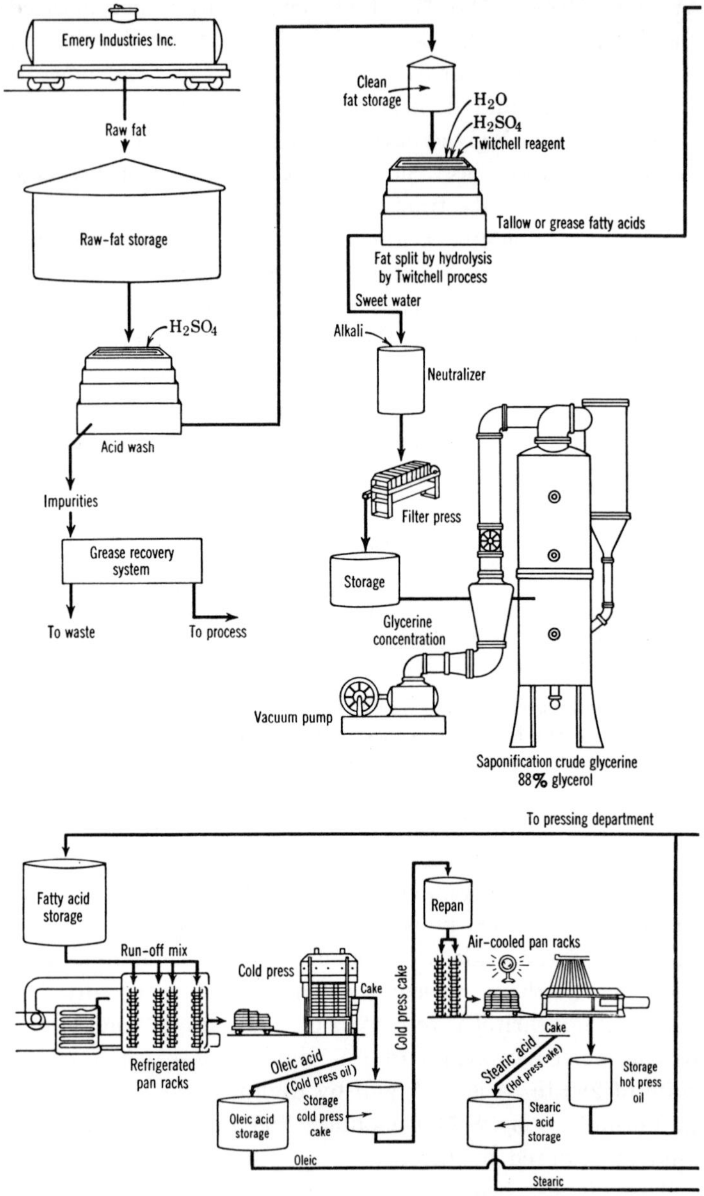

Separation of Fatty Acids

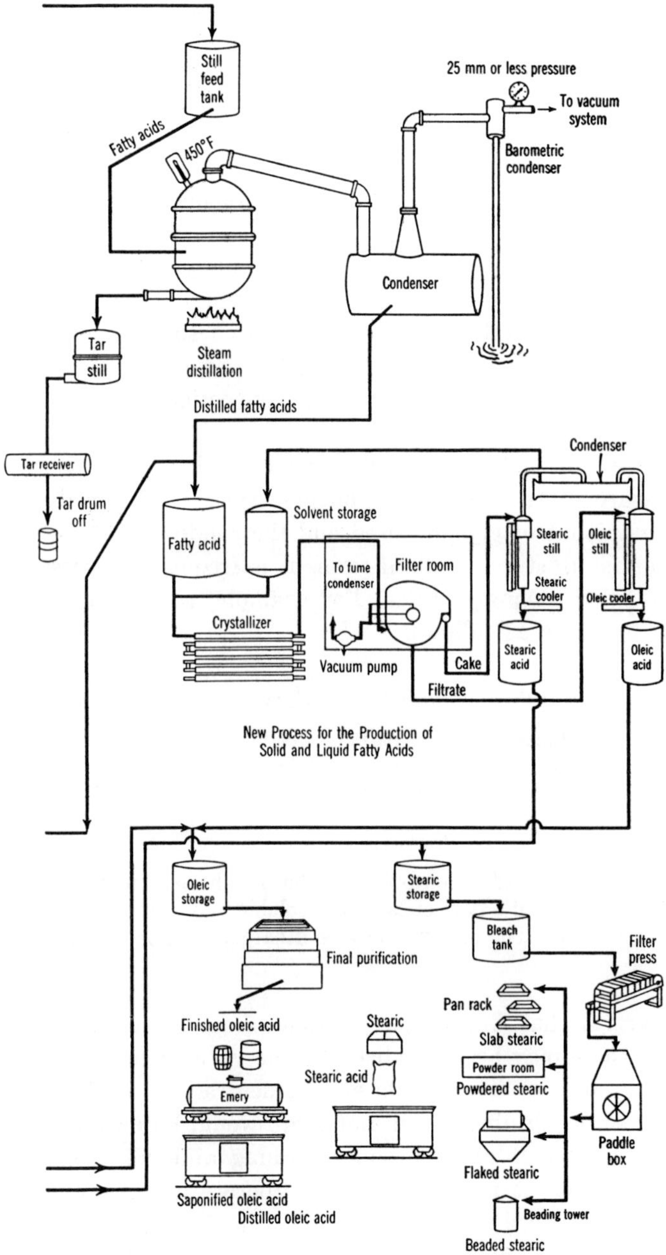

Courtesy of Emery Industries, Inc., Cincinnati, Ohio

acids was only U+ or 7 poises in the same time at 550°F. This illustrates the faster rate of polymerization due to the higher functionality of the tetrahydric alcohol, pentaerythritol. Normally, the glycerol ester would be bodied at a higher temperature; therefore it would have a higher viscosity than U+ in 4 hr, but not as high as Z3+. In addition it would have a darker color and higher acid number from the higher temperature, and a higher heating cost.

Soybean fatty acids may also be esterified with pentaerythritol to obtain oils which have fast drying and bodying characteristics similar to those of linseed oil, but without the objectionable yellowing of linseed. The reduced yellowing is due to the very low percentage of linolenic acid in soybean oil. This type of treated soybean oil may be identified readily in the sales literature of the various oil manufacturers. It will also be discussed further in connection with varnish making and alkyd resin manufacture.

Alcoholysis of Oils. When an oil or triglyceride is heated with glycerol a mixture of tri-, di-, and monoglycerides is obtained, depending on the amount of glycerol used and the conditions of the reaction. In the manufacture of alkyd resins it is frequently necessary to make a so-called monoglyceride. For example, a linseed monoglyceride may be made as follows:

Formula 4 Linseed Monoglyceride

Linseed oil	4 parts by weight
Glycerol	1 part by weight
Catalyst	About 0.1%

Procedure

Heat the oil to 425°F and add the catalyst. Continue heating to 450°F and hold at this temperature until 1 part of sample is soluble in 3 parts of methanol without becoming cloudy. The catalyst may be litharge, lime, or calcium carbonate.

The monoglyceride has free hydroxyl groups from the excess glycerol added, and these may be neutralized by the various acids used in alkyd resin manufacture. They could also be neutralized with linseed fatty acids and so obtain linseed oil, but there would be no object in such a procedure since it could produce nothing different from the original oil. However, it is possible to incorporate a percentage of pentaerythritol into linseed oil by the alcoholysis method and so obtain a faster-bodying oil. Typical examples of such a procedure are as follows:

Formula 5 Mixed Alcoholysis Product

Linseed oil	95 parts by weight
Pentaerythritol	5 parts by weight
Ca naphthenate	0.005% Ca to oil

Procedure

Heat the oil and catalyst to 480°F and add the pentaerythritol. Maintain the temperature until the batch is clear.

Formula 6 Mixed Alcoholysis Product

	Parts by Weight
Oil	200
Pentaerythritol	48
Pentecat* L	4

Procedure

Heat the oil and catalyst to 460°F and add the pentaerythritol all at once, which will lower the temperature to about 400°F. Reheat to 460°F and hold until the batch is clear. The time required will vary from 5 to 20 min, depending on the type of oil.

* Advance Solvents and Chemicals Corp.

The alcoholysis products described above are ready for use in synthetic resin manufacture, or they may be neutralized with fatty acids of the original oil to form mixtures of triglycerides and pentaerythritides. These new oils will have faster drying and bodying characteristics than the original triglycerides.

Isomerized Oils. The two previous sections on segregation and fatty acid separation with re-esterification discussed the making of improved drying oils by separating the better-drying from the poorer-drying constituents of oils. This involved separating the more highly unsaturated molecules from the more saturated or slower-drying molecules. The present section describes a method for improving the drying by changing the position of the unsaturation in the molecule without changing the amount of unsaturation. The term "isomerization" means a molecular transformation whereby the number of atoms in a molecule remains the same but their positions are changed. Such a positional change is illustrated by the shift from a non-conjugated position to a conjugated position shown below:

Isomerization

Non-conjugated: $R—CH{=}CH—CH_2—CH{=}CH—R$

Conjugated: $R—CH{=}CH—CH{=}CH—CH_2—R$

It has been well established that conjugated unsaturation imparts better drying and faster bodying characteristics to an oil than the equivalent amount of non-conjugated unsaturation. This fact provides an incentive to develop a method for changing linseed, soybean, fish, and other non-conjugated oils to their conjugated isomers.

It was observed by Moore (11, 12, 13), Kass and his coworkers (14), Burr (15), and others that isomerization occurred when drying and semi-drying oils were saponified with caustic alkalies in alcoholic solvents. Bradley and Richardson (16) showed that the alkali isomerization occurred in water as well as in alcohols provided the temperature was sufficiently high. Their method for linseed and soybean oils is as follows:

Formula 7 Alkali Isomerization

	Pounds
Linseed oil	25.0
Sodium hydroxide	3.87
Water	25.0

Procedure

Process in an autoclave, with agitation for 4 hr at 225–230°C. Follow by hydrolyzing with sulfuric acid, washing and drying in the usual manner. The yield was 22 lb of fatty acids containing 33.6% of double conjugation and 5.55% of triple conjugation. Similar results were obtained with soybean oil; the isomerized acids contained 35.1% of double conjugation and 1.45% of triple conjugation. The alkali isomerization method is somewhat cumbersome, but it is reported to have been used on a commercial scale.

The so-called "nickel on carbon" catalyst method developed by Cowan (17, 18, 19) and his coworkers appears to be simpler than the alkali method and does not involve splitting the oil. It was reported (17) that, when 6–8% of the catalyst was heated with alkali-refined linseed or soybean oil at 338°F (170°C) for 6 hr, 30–34% of conjugation was obtained. The catalyst was good for 5–10 batches of oil before losing its activity. The isomerized oils were better drying, and their films had improved water and alkali resistance. It was also observed (18) that the rate of bodying of isomerized linseed oil was 3–5 times as fast as that of the untreated linseed, and isomerized soybean was 2–3 times as fast as untreated soybean. The isomerized oils were compared with dehydrated castor oil in varnishes. Although the cooking time was shortened because of the fast bodying rate of the isomerized oils, the varnishes did not have the expected speed of drying or the increased water and alkali resistance. Similar results were obtained with alkyd resins; the isomerized oils reduced the processing time but did not improve drying or water resistance.

Subsequent work (19) showed that the relatively poor drying or "after-tack" was due to another type of isomerization which occurred simultaneously. This was a change from the *cis* form ($\mathrm{R{-}\overset{H}{C}{=}\overset{H}{C}{-}R}$) to the *trans* form ($\mathrm{R{-}\overset{H}{C}{=}\underset{H}{C}{-}R}$); unfortunately, the *trans* form is the poorer-drying form and also the more stable configuration. The mechanism of isomerization will be discussed in more detail under polymerization and film formation. Further work must be done to control the isomerization process so that it will produce the maximum of desirable isomers and the minimum of undesirable ones.

Dehydrated Castor Oil. It has been known for many years that the non-drying castor oil could be dehydrated and thereby changed into a drying oil. However, it was the acute shortage of tung oil during World War II that stimulated research for the commercial production of dehydrated castor oil. This oil is similar to tung oil in that it is a "frosting oil" because it contains conjugated unsaturation, but its air-drying and rapid-bodying properties are not as good as might be expected from this type of unsaturation. Subsequent investigation has shown that the oil contains a mixture of conjugated and non-conjugated unsaturation. However, it is widely used at the present time because it has fairly good drying properties and color retention.

Kraft (20) in 1888 published the fact that ricinoleic acid could be dehydrated. Fokin (21) in 1914 recognized that more than one product was obtained from the dehydration of the ricinoleic acid. He also found that re-esterification of the acid produced an oil which had drying properties. Scheiber (22) in 1928 obtained patents for a process of dehydrating the castor fatty acids and re-esterifying them to produce a drying oil.

Ufer (23) in 1931 obtained patents for the direct dehydration of castor oil, and Colbeth (24) in 1946 patented a continuous process for dehydrating castor oil. Radlove (25) and his coworkers in 1948 described a continuous process based on the "flash principle," using a thin film of the oil, at high temperature for a short time, with catalyst, under vacuum. The temperature was 300–320°C, and the time of the reaction was about 15 min under a vacuum of 30 mm. The catalysts investigated were phosphoric acid, diethyl sulfate, and both concentrated and diluted sulfuric acid. The concentrated sulfuric acid was the most effective catalyst, but it caused considerable darkening of the oil when used in sufficient concentrations. The darkening was overcome by diluting the acid with water to 70–75% concentration.

Elm (26) discusses the chemistry of the dehydration of oils and the mechanism of the catalytic action. After considerable investigation the scheme shown in Fig. 12 was evolved to portray the reaction mechanism.

It is believed that the acid catalyst reacts with the hydroxyl group of the ricinoleic acid to form the ester indicated in Fig. 12. At suffi-

Fig. 12. H_2SO_4 as Dehydration Catalyst

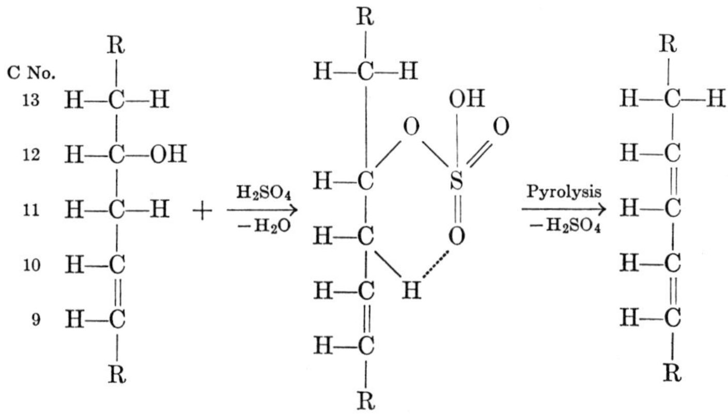

ciently high temperatures pyrolysis occurs and the acid is split off. This reaction would leave a free linkage on each of two adjacent carbons which would then combine to form a new double bond.

The hydroxyl group in ricinoleic acid always occurs on the 12-carbon atom, but there is no directing influence which will ensure that the new double bond will be formed between the 11- and 12-carbon atoms as shown in Fig. 12. In fact it has been quite well established that there is equal chance for the new double bond to be formed between the 12- and 13-carbon atoms. In the first case a conjugated unsaturation is formed which is referred to as the 9,11 unsaturation, and in the second case a non-conjugated unsaturation is produced, known as the 9,12 unsaturation. This partially explains the observed fact that dehydrated castor oil does not dry or heat-body as fast as would be expected if the unsaturation were all of the conjugated type. Another factor which accounts for the slight after-tack which develops on varnish films made with dehydrated castor oil is the change at the double linkage from the *cis* to the *trans* form, as referred to in the previous section on isomerization. This change is illustrated in Fig. 13. Since both isomers are formed with equal ease, it may be expected that the dehydrated castor oil contains about equal parts of *cis* and *trans* isomers. It will be recalled that the *trans* isomer does not oxidize as readily, or dry as fast, as the *cis* isomer.

Kappelmeier (30) observed that the dehydration of castor oil proceeds much more smoothly under vacuum than at ordinary pressure.

Fig. 13. *cis* and *trans* Isomer Formation

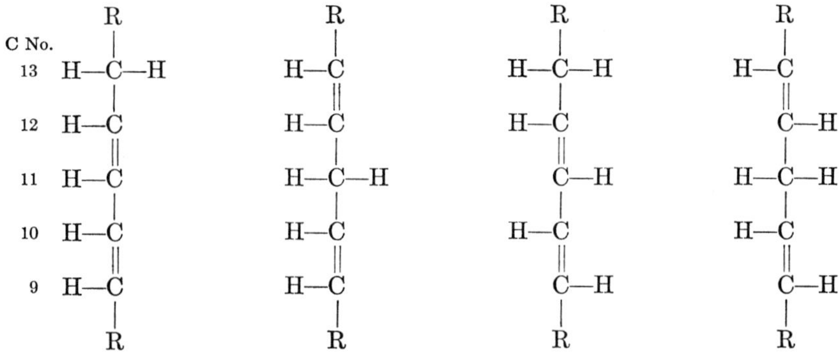

cis 9,11 | *cis* 9,12 | *trans* 9,11 | *trans* 9,12

He attributed this to a partial hydrolysis of the oil by the water produced, followed by an esterification reaction between the liberated fatty acid and the hydroxyl group on the 12-carbon atom. This ester-interchange reaction would block the dehydration reaction because the hydroxyl group is then protected. As the dehydration reaction proceeds the concentration of the ricinoleic hydroxyl group becomes small, which disturbs the equilibrium of the ester-interchange reaction and consequently liberates some of the protected ricinoleic hydroxyls. When the dehydration is conducted under vacuum the water is removed before any significant hydrolysis and ester interchange occur. The evidence for this reaction is the decrease in saponification value instead of an increase as the dehydration proceeds, and a difference in the hydroxyl values for the oil and the methyl esters. Obviously, if the oil contains fatty acid protected hydroxyl groups the hydroxyl value is not a measure of the extent of dehydration.

The research which has been done on dehydration shows that the reaction is not a simple straightforward procedure but is influenced by the conditions under which it is carried out. Many more instances of complicated side reactions will be encountered in the study of oil and resin chemistry. For this reason a simple chemical equation showing the reaction can never be complete because practically every conceivable side reaction takes place to some extent.

Still another factor must be considered in the manufacture of dehydrated castor oil, namely, the possibility of partial polymerization of the dehydrated oil at the high temperature necessary for the reaction. The original castor oil is a non-drying oil and does not

Fig. 14. Oxidation Followed by Dehydration

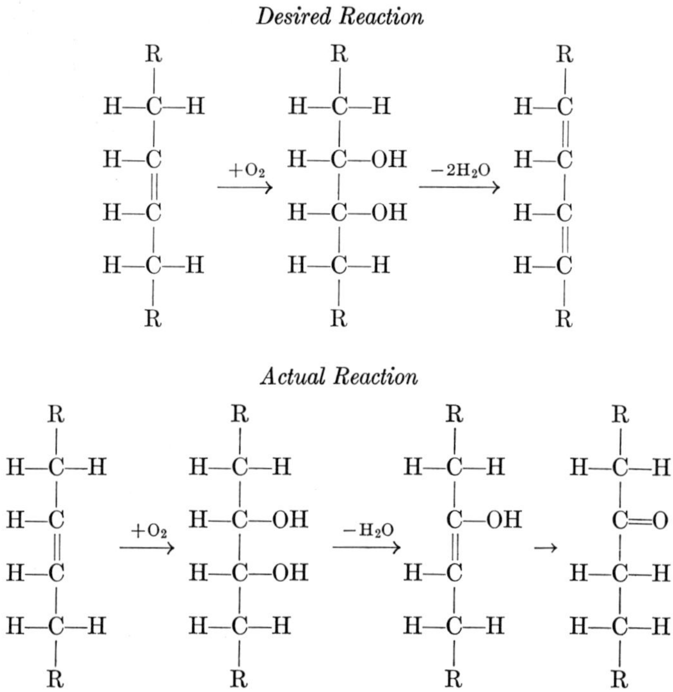

polymerize appreciably, but the 9,11 oil with the conjugated unsaturation polymerizes quite readily. It will also be recognized that it polymerizes more readily than the 9,12 oil with the non-conjugated unsaturation. Therefore, if the oil is permitted to polymerize to any large extent, the conjugated unsaturation will be decreased and the oil will have correspondingly poorer drying properties. It will be recalled that dehydratd castor oil is available at two viscosities, S-V and Z3, the latter being partially polymerized.

Castor oil contains from 80% to 85% of ricinoleic acid, but after dehydration this becomes the two types of linoleic acid, the 9,11 conjugated linoleic and the 9,12 non-conjugated linoleic. Since both types form with equal ease it would be expected that there would be about

40–42% of each. However, commercial dehydrated castor oil contains only about 25% of the 9,11 conjugated linoleic; most probably the remainder has been used in the polymerization reaction.

Attempts have been made to convert oils (or their fatty acids) which have one double bond into oils having two double bonds by the combined processes of oxidation followed by dehydration. For example, if oleic acid is mildly oxidized it can be converted into dihydroxy stearic acid as shown in Fig. 14. If it were possible to dehydrate this material in the manner shown so as to produce a conjugated type of unsaturation, the oil would then have good drying

Fig. 15. Chlorination Followed by Dehydrochlorination

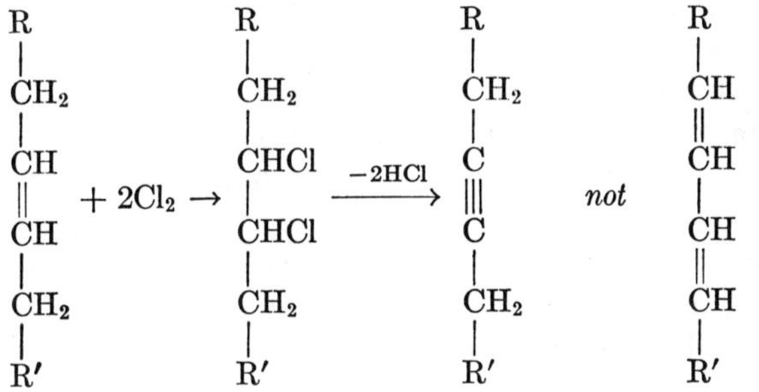

Oleic acid   Stearolic acid *not* linoleic acid

properties. However, extensive studies have shown that the dehydration reaction proceeds in a different manner. One molecule of water is removed easily; then the product rearranges to give the ketone structure indicated in Fig. 14.

Elm (26) also points out that attempts to convert oleic acid esters into linoleic esters by first adding two atoms of chlorine and then removing two molecules of hydrochloric acid are doomed to disappointment. Instead of obtaining the linoleic as might be expected, the actual product is stearolic as shown in Fig. 15. The hydrogen is taken from the same carbon atoms to which the chlorine is attached, with the consequent formation of a triple bond.

Maleic-Treated Oils. The processes for improving the rate of air drying and/or bodying of oils, which were discussed in the preceding sections, involve the separation of the better-drying from the poorer-drying components (segregation), the shifting of non-conjugated double bonds to produce the conjugated type (isomerization), and the removal of water to introduce a new double bond (dehydration). This section describes a process for adding a compound to the unsaturated part of the oil molecule, thus increasing its complexity and heat reactivity. The compound referred to is maleic anhydride, and the oils are known as maleic-treated or maleinized oils. Since the maleic is added at or near the unsaturated section of the fatty acid radical, it retards oxidation slightly so that maleic-treated oils do not show greatly increased air-drying properties. However, they are definitely faster bodying and have better color at equal viscosities and better water resistance in the dried films. The oils usually treated are linseed and soybean; the purpose is to make the treated linseed approach dehydrated castor or tung oil in rate of bodying, and the treated soybean to approach the bodying rate of linseed.

The product obtained from the maleic addition is known as the adduct, and since maleic is a dibasic acid, the adduct can be neutralized with an alcohol such as glycerol or pentaerythritol. If the adduct is used in varnish or resin making the alcohol can be added in the cooking procedure. If the adduct is neutralized with an inorganic alkali, ammonia, or an amine, a water-miscible oil is produced. These "solubilized" oils may be used in water paints of the emulsion type. The dried films have good water resistance because only small percentages of maleic are used in treating the oils.

Usually from 2% to 10% of maleic anhydride is sufficient to produce the desired increase in heat reactivity in the non-conjugated oils. However, when 10% of maleic anhydride is reacted with the conjugated oil, dehydrated castor, the adduct gels rapidly. Useful products are made by blending soybean or linseed oil with dehydrated castor and maleic-treating the blend. The adduct is made by heating the oil and maleic anhydride in a closed kettle with vent and stirrer. The maleic is corrosive to mild steel, iron, or copper; therefore stainless steel or aluminum should be used. The temperature for the reaction will vary from about 200° to 280°C, depending on the type of oil being treated. Bolley (59) suggests the following procedures for making a 10% treated linseed adduct and a 5% maleic-treated linseed-pentaerythritol ester.

Formula 8 Linseed-Maleic Adduct

	Parts by Weight
Refined linseed oil	90
Maleic anhydride	10

Procedure

Heat the oil and maleic anhydride with agitation to 200°C in 1 hr, hold at 200°C for 1 hr, increase the temperature to 230°C slowly (in about 30 min), and hold at 230°C for about 2 hr. The progress of the reaction may be followed by washing the product with water and determining the free acid.

The above adduct may be neutralized with alcohols or alkalies as indicated previously. However, a lower maleic content is desirable, particularly when glycerol or pentaerythritol is used, because of the rapid reactivity of the resulting complex. The above adduct may be diluted with refined linseed oil to give from 2% to 5% maleic, and the required amount of pentaerythritol added slowly at from 190° to 200°C over a period of 4–5 hr. A mild vacuum is then applied for about 2 hr. The slight turbidity which develops may be removed by filtering the product. If a bodied oil is desired it may be obtained by heating at 260°C under an inert atmosphere. A light-colored oil with a viscosity of Z2 and acid value of 10 is obtained in a much shorter time than would be possible with the original linseed oil.

Soybean oil may be maleic-treated in a similar manner but at a temperature of 250–280°C. Mixtures of soybean and dehydrated castor oils also give very good treated oils which have fast bodying rates and improved color and water resistance. See page 106 for a description of maleic-treated tall oil. The use of maleic-treated oils in varnish making is discussed in Chapter 4.

Mechanism of Adduct Formation. Considerable work has been done to explain the addition reaction between maleic anhydride and the unsaturated portion of the fatty acid radicals in drying oils. Clocker (33) in U. S. Patent 2,275,843 proposed that addition takes place at a double bond as follows:

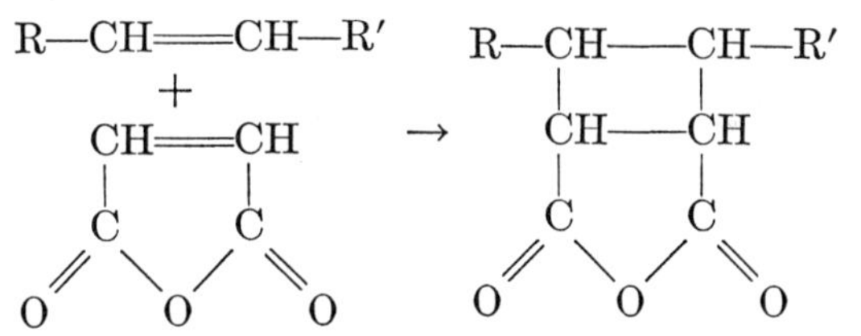

However, this mechanism does not appear to be likely because it would be difficult to form a four-membered ring under such conditions and this type of ring structure is not very stable. A Diels-Alder type of addition is entirely possible for oils with conjugated unsaturation.

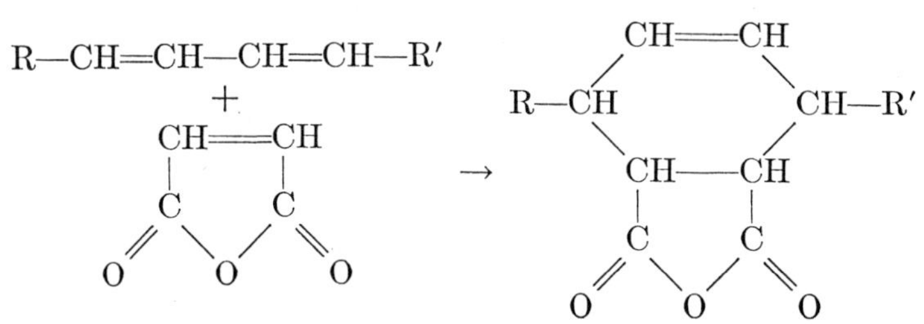

This type of reaction does not take place with oils having non-conjugated unsaturation unless the conditions are such that the oil isomerizes to the conjugated form during the reaction. This type of isomerization and addition was believed to occur with methyl linoleate and maleic anhydride by Wheeler *et al.* (60) in their studies of maleic-treated methyl esters of non-conjugated fatty acids. They also found that maleic anhydride added to the oleic ester without loss of unsaturation in the ester. From this result they concluded that a substituted succinic anhydride was formed by transfer of a hydrogen atom from the fatty acid chain to the maleic anhydride as follows:

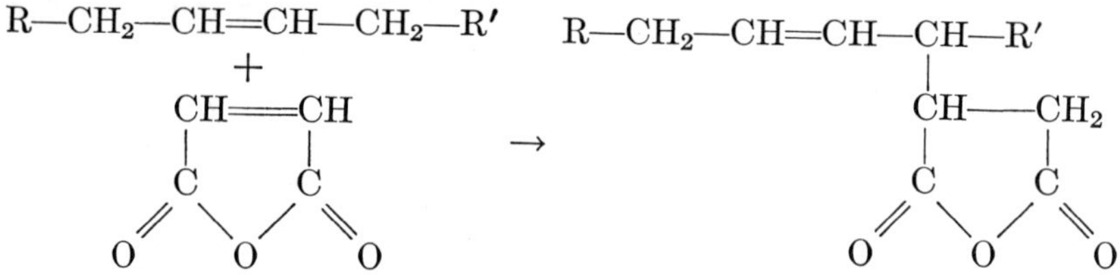

Wheeler *et al.* (60) also found that the first molecule of maleic anhydride reacting with methyl linoleate reacts mostly to saturate one double bond, and the second one adds without affecting the unsaturation. In addition, they also found that the first two molecules of anhydride reacting with methyl linolenate react mostly to saturate one double bond each, and the third molecule adds without affecting the unsaturation. The unsaturation was determined by the Wijs iodine method, and it was recognized that this method is not accurate if conjugated unsaturation has been formed. Therefore no structures for the linoleate and linolenate adducts were given by these workers.

Kappelmeier (61) believes that, with oleic radicals, the addition takes place as indicated above with the formation of a succinic anhydride derivative without affecting the double bond. With the linoleic esters, the first molecule adds to form the succinic anhydride derivative, and a shift in the unsaturation occurs to form the conjugated type. The second molecule of maleic anhydride then adds to the conjugated unsaturation by the Diels-Alder type of reaction. Therefore the structures obtained from the addition of one and two molecules of maleic would be as follows.

One molecule of maleic added:

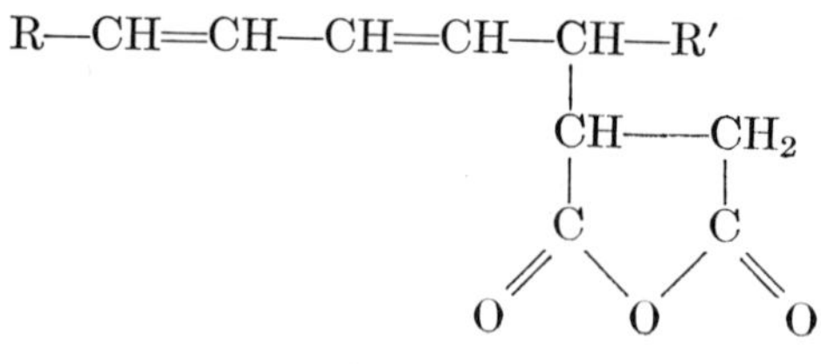

Two molecules of maleic added:

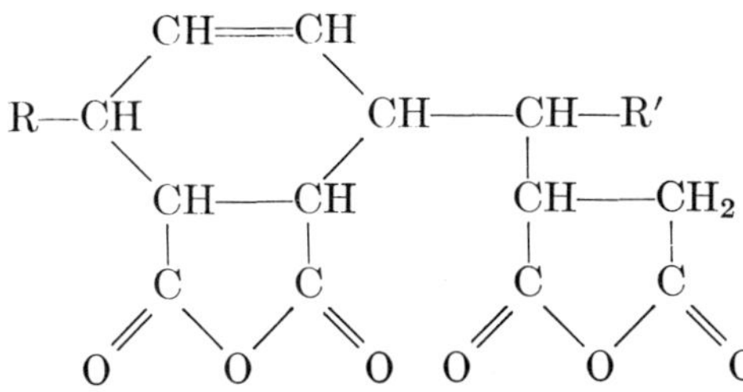

In the case of the linolenic radicals Kappelmeier believes that a molecule of maleic anhydride may be added simultaneously to both ends of the unsaturated portion with the consequent formation of a triply conjugated system. This could be followed by the addition of a third molecule of maleic by a Diels-Alder mechanism to give the structures shown below.

Two molecules of maleic added:

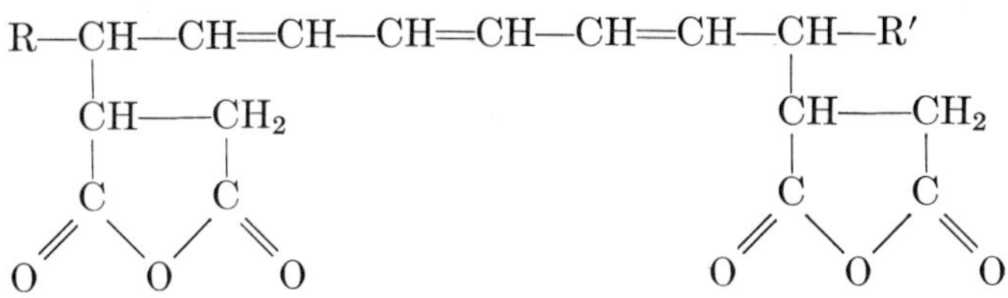

Three molecules of maleic added:

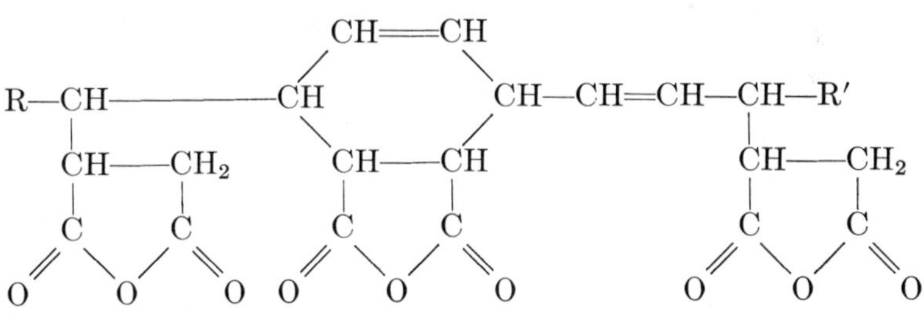

It should be recognized that isomers of the above structures may be formed, but this would affect only the relative position of the groups and not the mechanism of the reaction.

Kappelmeier (61) describes a possible method for determining the extent of the maleic addition from the fact that dimethyl aniline forms a highly colored product with free maleic anhydride. Spot tests against a white background could be made during the reaction until the absence of the characteristic maleic color indicated complete addition of maleic.

Maleinized oils are another example of the chemical treatment of a natural product to improve its properties. Since the treatment is relatively simple and the products have definite advantages over the original natural oil, the commercial success of such a treatment is assured.

Copolymerized Oils. The maleinized oils, described in the previous section, polymerize considerably faster and air-dry somewhat faster than the untreated oils because of their larger molecular weight and the greater number of reactive centers in their molecular structures. In some respects the copolymerized oils are similar to the maleinized oils since they are produced by the addition of another type of molecule to the unsaturated portion of the oil molecule. However, in copolymerized oils, the molecule which is added to the oil is itself capable of polymerizing. Many active monomers capable of forming polymers through their double bonds may be copolymerized with the oil, but styrene, α-methyl styrene, and cyclopentadiene are the ones most widely used at present.

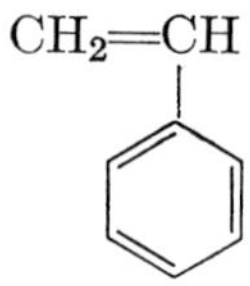

Styrene

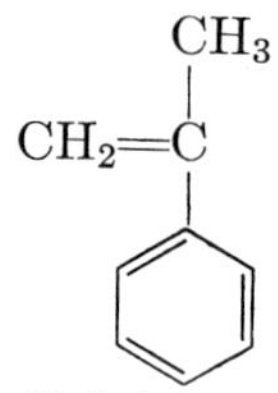

α-Methyl styrene

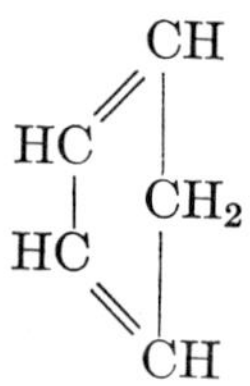

Cyclopentadiene

The tremendous production facilities developed for styrene in the synthetic rubber program during World War II made large quantities available at low cost in the postwar period. With this incentive it was inevitable that styrene would be investigated for use in surface coatings. These appeared first as styrenated oils and later as styrenated alkyd resins. They are described by Hewitt and Armitage (38, 39), Peterson (40), and many others.

Polystyrene plastics are fairly hard and very tough, and it would be highly desirable to add these qualities to oil films. Although polystyrene is a relatively low-melting solid it cannot be incorporated into drying oils by heating, like rosin or some other resins, because it is incompatible on cooling to room temperature. The melt, or solution if dissolved in solvents, separates into two layers, one of oil and the other of polystyrene. However, if monomeric styrene is heated with the oil under reflux and with a peroxide catalyst it will copolymerize with the oil and homogeneous products will result. The copolymerization occurs readily with oils having conjugated unsaturation but very slowly and only to a small extent with oils having non-conjugated unsaturation. For example, the rate of copolymerization of styrene with tung oil is so rapid that the mass gels quite readily. With dehydrated castor oil the reaction is slower because this oil has less conjugation, but with non-conjugated linseed or soybean oil the copolymerization reaction is so slow that the styrene polymerizes and separates because polystyrene is incompatible with the oil. However, mixtures of linseed oil or soybean oil with tung or dehydrated castor oil may be copolymerized satisfactorily. The copolymeric products vary from liquid oils to solids, depending on the ratio of oil to styrene and the degree of polymerization. Bhow and Payne (41) reported that the products obtained from styrenation of dehydrated castor oil were liquids below 70% styrene and solids at 70% styrene and above.

Peterson (40) describes a useful ratio as being 50–55% styrene and 45–50% oil. It will be seen that this corresponds to varnishes having 12.5–15.2 gal of oil to 100 lb of resin since the copolymerized styrene may be considered a hard resin. However, these copolymerized products dry faster, are very much tougher, and are more durable than regular varnishes having the same oil and varnish resin ratio. The regular varnishes are described in a later chapter. Copolymerized oils are available which dry to hard films in 4–8 hr, depending on the type of oil and the degree of polymerization.

The styrene and oil may be copolymerized either by the solvent method or by the mass method. The solvent method gives good control of the reaction, but it is slower and, of course, the solvent must be removed if a solvent-free product is desired. The mass method is considerably faster, but there is a tendency for the styrene to polymerize as well as to copolymerize. This gives cloudy, immiscible products because of the incompatibility of the polystyrene with the oil. The Dow Chemical Co. found that mixtures of 25–35% of α-methyl styrene and 65–75% of styrene instead of straight styrene gave clear, homogeneous products with a variety of oils. The types of oils and processing times are shown in Table 11. A peroxide catalyst such as benzoyl peroxide or *t*-butyl hydroperoxide is used to accelerate the reaction, which is carried out in a closed kettle with reflux condenser. The products are fast drying except the soybean-tung combination. In general they have good water and alkali resistance and fair durability. Improved durability is obtained by copolymerizing styrene with alkyds, which is discussed in Chapter 7 on alkyds. The copolymers retain some of the sensitivity of polystyrene to aro-

Table 11 * Mass Copolymerization Procedure

Type of Oil	DCO, 100%	Linseed, 90% Tung, 10%	Soybean, 80% Tung, 20%	Linseed, 50% DCO, 50%
% oil	45	45	45	50
% mixed styrenes, cat.	55	55	55	50
Heat oil to 160°C				
	Reaction Time, hr			
Add styrenes in . . .	6	6	6	6
Raise temp. to 250°C	5	6	4	5
Hold at 250°C	6	6	6	7
250°–285°C	0.5	....	....	....
250°–300°C	.....	0.5	0.5	1
Hold at 285°C	1.5	....	....	....
Hold at 300°C	.....	2.5	3.5	3
Total time	19	21	20	22
% unreacted styrene	1.2	1.5	2.0	1.2
Color, Gardner 1933	8	9	9	8
Thin with 70% mineral spirits, 30% Solvesso #2 to 60% solids.				
Viscosity	F	C	B	E
Driers: 0.5% Pb, 0.02% Co as metal to oil				
Dry to touch, hr	0.75	1	2.5	0.5
Dry hard, hr	6.0	7	24	5
3% benzoyl peroxide as catalyst				

* Courtesy of Dow Chemical Co.

matic solvents, but they can be used in a variety of architectural and industrial finishes as discussed in Volume II. The copolymers have limited compatibility with other film-forming materials and are frequently used as the entire vehicle in pigmented finishes.

Solvent Copolymerization. The solvent method, as proposed originally by Hewitt and Armitage (38, 39) of Lewis Berger and Sons, Limited, London, involves refluxing the styrene, oil, and solvent until the reaction is complete. The suggested ratios are:

Styrene	20–30%
Oil	30–20%
Solvent	50–50%

With dehydrated castor oil the reflux time was 25 hr and the product had 47.5% solids and a viscosity of 3.5 poises (N-O on the Gardner tubes). The film set to touch in 20 min and was hard overnight. It had good water resistance but poor resistance to aromatic solvents. When linseed oil was used the dried films were opaque, indicating a heterogeneous product, but clear films were obtained with blown linseed. Differences in the rates of reaction and in the products were observed when solvents other than xylene were used. For example, dipentene retarded the reaction from 23 hr with xylene to 49 hr, and the dipentene product had better compatibility but slower drying. A partial replacement of xylene with dipentene is preferred. The solvent is a factor in determining the chain length of the styrene polymers by the free-radical mechanism discussed in the next section. The molecular weight of polystyrene formed in xylene solution is considerably higher than that formed in dipentene as shown in Table 12.

Table 12 * Polymerization of Styrene in Xylene and Dipentene

% Styrene	Temp., °C	Solvent	Reaction Time			
			10 hr		26 hr	
			% Polymer	Average † Mol. Wt.	% Polymer	Average † Mol. Wt.
25	140	Xylene	30	41,600	52	38,000
		Dipentene	Nil	....	Trace	....
50	100–160	Xylene	26	104,000	95	90,000
		Dipentene	24	9,300	36	8,000
80	75–150	Xylene	10	....	86	125,000
		Dipentene	48	10,000	60	17,000
90	80–180	Xylene	20	117,000	94	90,000
		Dipentene	43	19,100	90	14,000

* Armitage, Hewitt, and Sleightholme (39).
† Molecular weight determined by viscosity method.

Since the molecular weight of monomeric styrene is 104.14, the polymers in the table will contain from 100 to 1000 units per polymer chain.

Mechanism of Styrene Copolymerization. In the copolymerization of styrene with drying oils a number of reactions are taking place simultaneously. The reaction with the fastest rate will be the dominant reaction, and frequently it will determine the type of product formed. However, all the possible reactions occur to some extent and so modify the product. It is generally believed that the styrene polymerizes to some degree and the styrene chains then become attached to the unsaturated portions of the fatty acid radicals of the oil. It is also possible that a growing styrene chain may connect two oil molecules by becoming attached to unsaturated sections of each oil molecule. Some of the possible reactions are illustrated by the diagrams which follow.

Polymerization reactions usually involve three steps: initiation, propagation, and termination as discussed in Chapter 1. Initiation is necessary to activate the monomer and may be accomplished by heat, light, catalyst, free radical, or other means. Propagation by a chain reaction assures chain growth and molecular size and usually proceeds more readily than initiation. Termination stops the chain growth and is the result of loss of energy, steric hindrance, reaction of chain end with an inhibitor or a free radical from the catalyst or solvent, or some other mechanism. Benzoyl peroxide, which is widely used as a catalyst, is believed to decompose and form free phenyl radicals as follows:

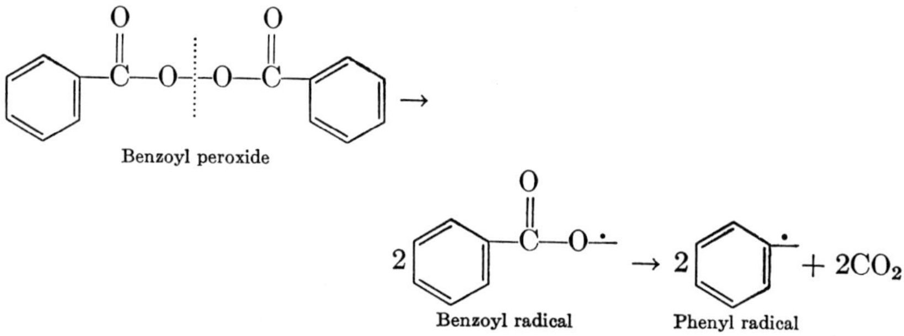

The free phenyl radicals may then activate the styrene monomer and so initiate chain growth. A free radical also may terminate the growing chain as indicated below:

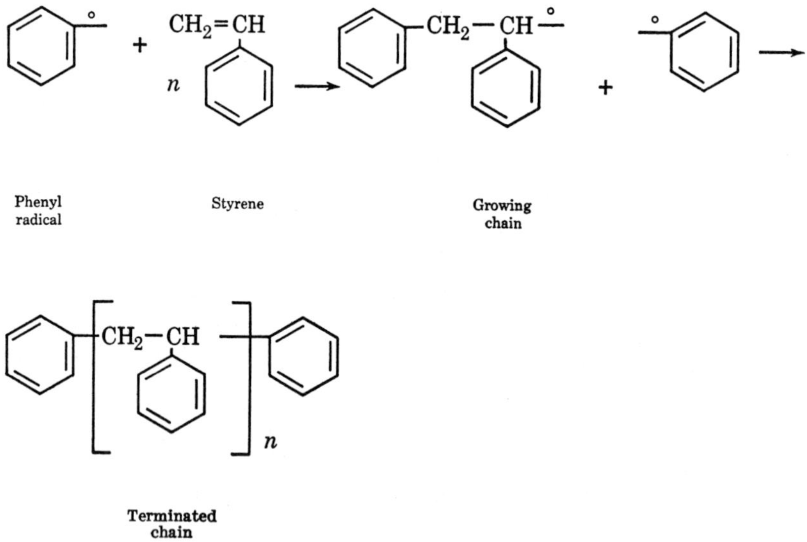

The styrene chains may also grow from both ends of the ethylenic group by addition of monomeric styrene. It is also possible that short growing chains may join to form longer chains. A chain growing from each end may be terminated by addition of a radical as indicated above to form free polystyrene. It may also attach itself to the unsaturated portion of a fatty acid radical in an oil molecule to form the copolymer. A polystyrene chain growing at both ends may copolymerize with two oil molecules to form still another type of product. In any event, the product will be somewhat heterogeneous. The principal factors which control the copolymerization reaction are ratio of reactants, type and amount of catalyst, temperature, and type of solvent if the solvent method is used. Of course, by maintaining all the variables constant, uniform products are made commercially.

The theory of the mechanism of the copolymerization reaction is based on observed reactions with conjugated dienes and styrene. The copolymerization of styrene and butadiene has been studied extensively, and it is believed that both 1,4 and 1,2 addition takes place with the 1,4 addition predominant as indicated.

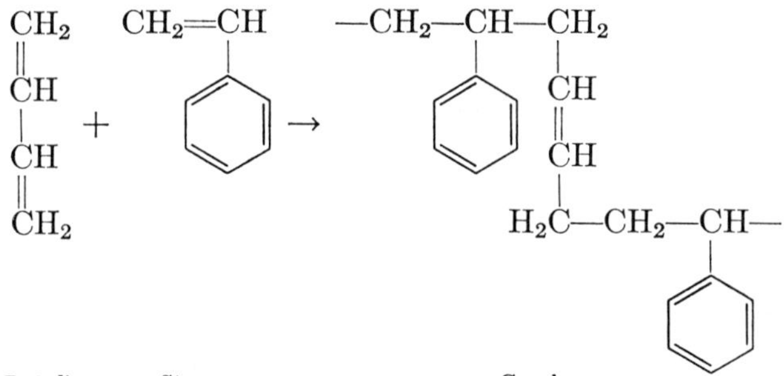

Butadiene Styrene Copolymer

Hewitt and Armitage (38) believed that a similar situation exists when styrene is copolymerized with an oil such as dehydrated castor containing diene unsaturation. Therefore three possibilities exist: (1) addition of chains which have one end terminated with a free radical; (2) addition of growing chains which will be terminated by a free radical; (3) termination of a growing chain by reaction with the unsaturation on another oil molecule. This situation is illustrated diagrammatically as follows:

Styrene chains + Oil molecules → Copolymers

(1) 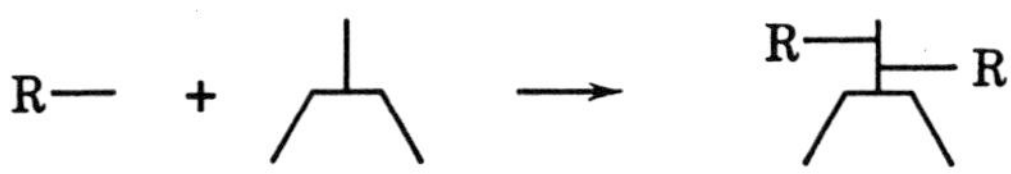

(2) 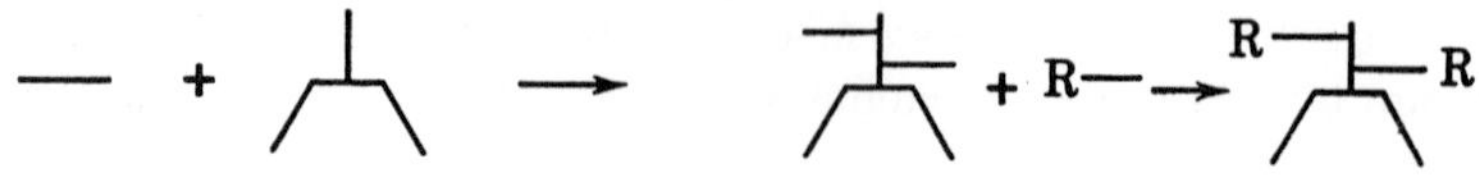

(3) 

With highly conjugated oils, such as tung and oiticica, reaction 3 will be most likely to occur. This will result in a rapid increase in viscosity because of the cross-linking of the oil molecules. It is well known that gelation occurs rapidly when these oils are copolymerized with styrene. The reaction with dehydrated castor oil is controlled more readily because this oil has considerably less conjugated unsaturation. With linseed and soybean oils which have only the nonconjugated unsaturation, the copolymerization reaction is so slow that polystyrene is formed preferentially. Since polystyrene is not compatible with the oils the mixture separates, as mentioned previously, and the product is turbid and unsatisfactory. In the Dow method of mass copolymerization the use of α-methyl styrene retards the rate of styrene polymerization and so permits more copolymers to be formed. However, the methyl branch chain produces a somewhat softer polymer having greater sensitivity to hydrocarbon solvents in the finished film.

Cyclopentadiene is also used in the production of copolymerized oils. This material will polymerize to form cyclic polymers which are stable at room temperature but decompose to give the reactive monomer on heating.

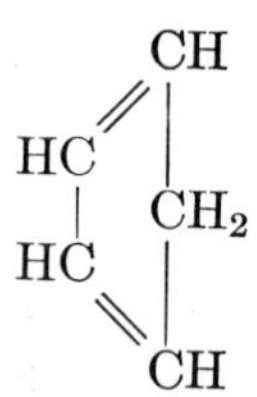

Cyclopentadiene monomer

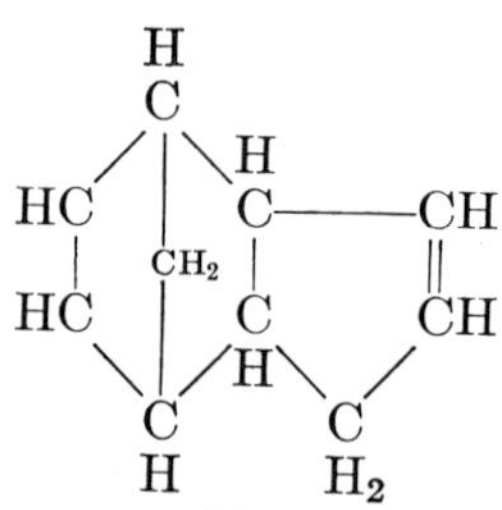

Dimer

It is believed that cyclopentadiene adds to the unsaturated fatty acid radicals of oil molecules in the manner indicated below. It may also form a linkage between two fatty acid radicals in a manner similar to that indicated above for styrene.

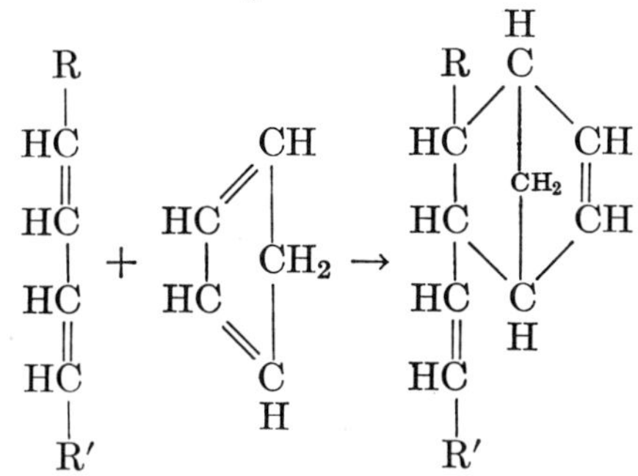

Fatty acid radical Cyclopentadiene Copolymer

A great deal of research is being done on copolymerized oils and resins, and it may be expected that new types will be available and new ideas on the reactions involved will be developed before this book is published. Studies such as those made by Hoogsteen *et al.* (57) show the differences in products obtained by variations in the dehydrated castor oil used. A relationship was shown between the viscosity and degree of dehydration of the castor oil and the homogeneity and film properties of the styrenated product. The most desirable products were obtained from castor oil which had been dehydrated to a minimum hydroxyl content and then bodied to a viscosity approaching 15 poises. Rinse and Korf (58) studied the copolymerization of a variety of oils with styrene and fractionated some of the products. They were interested in producing copolymers of commercial value and also in the mechanism of the reactions.

Catalyzed Oils. Catalyzed oils are drying oils to which has been added a catalyst which will reduce the time required to obtain a given viscosity when the oils are heat-bodied. The catalyst must be such that it will not detract from properties of the oil such as drying rate, pale color, or chemical resistance. Patents (42, 43) have been issued claiming a variety of materials which include β-methyl anthraquinone, diphenol carboxyl anthracene, thio β-naphthol, and "oxidation products of coal-tar hydrocarbons having the general formula $\begin{array}{l} \mathrm{R{-}CH} \\ \mid \quad \Vert \\ \mathrm{R'{-}CH} \end{array}$, where R and R′ are aromatic nuclei."

In general, catalyzed oils heat-body in 50–70% of the time required for the uncatalyzed oil. The shorter time at elevated temperature results in products having better color and lower acid value. The mechanism of the catalytic action has not been fully explored, but it is believed that the catalyst accelerates the isomerization from non-conjugated to conjugated unsaturation. Since polymerization catalysts usually are hydrogen acceptors they may be responsible for production of free radicals on the CH_2 group adjacent to C═C double bonds, thus initiating the polymerization.

An interesting disclosure was made by Berger and Rock (44) in regard to polymerizing a drying oil with a Friedel-Crafts type of catalyst at room temperature or not greater than 300°F, and then neutralizing the catalyst with an inactivator such as an aromatic monocyclic amine. For example, 99.82% of oiticica oil with 0.18% of boron trifluoride at 68–92°F acquired a viscosity of Z3 in 8 hr, Z6 in 37 hr, and gelled in 45.5 hr. When 0.3% pyridine was added to oiticica oil containing 0.1% boron trifluoride and having a viscosity of Z1, the oil remained at this viscosity for 134 days, but gelled in 3 days without the inactivator. An ester gum varnish was made by dissolving 1 part of ester gum in 2 parts of oiticica oil which had been bodied to Z6 viscosity with boron fluoride at room temperature and inactivated with triamylamine. The varnish was thinned with petroleum thinner and produced a clear, hard film after drying overnight.

Tall Oil. The term "tall oil" is an adaptation of the Swedish word *tallolja,* meaning pine oil because the product is obtained indirectly from pine trees. In America we could not use the literal interpretation of *tallolja* because we already had an oil known as pine oil which is different in composition from tallolja. Tall oil is a by-product of the sulfate or "kraft" process for making wood pulp used in the manufacture of paper. The pine wood in the form of chips is digested with sodium hydroxide and sodium sulfide to separate the cellulose fibers from the lignin and other constituents. Filtration removes the fibers as wood pulp, and the filtrate is known as "black liquor." This is concentrated in evaporators, and a curdy mass of "black liquor soap" separates and is skimmed off. The soap is boiled with sulfuric acid to liberate the organic acids, which are chiefly a mixture of rosin acids and fatty acids. The acid mixture is washed with hot water, and the water removed by settling. The product is crude tall oil.

The composition of crude tall oil varies slightly with the condition of the pine wood and modifications in the sulfate process, but it usually falls in the following range:

Rosin acids	42–51%
Fatty acids	42–51%
Unsaponifiable	About 7%

The rosin acids are chiefly abietic acid and its isomers, as described in Chapter 3. The fatty acids are mixtures of unsaturated and saturated acids having the following range of composition:

Oleic acid	40–50%
Linoleic acid	40–45%
Saturated acids	5–10%

Tall oil is produced by about ten manufacturers listed in *Bull. 1* of the Tall Oil Association (64). Information regarding the various grades of tall oil and its use in driers, resins, paints, and other products is available from the Association or its member companies. Crude tall oil is used for varnishes, resins, and paints in which the presence of rosin and dark color are not objectionable. Crude tall oil is refined either by acid refining or by distillation. The refined tall oil may be separated into its components by fractional distillation. In the acid refining process the crude tall oil is dissolved in mineral spirits and concentrated sulfuric acid is added. The resulting acid sludge is separated, the tall oil solution is washed, and the mineral spirits are removed by vacuum distillation. Typical fractional distillation processes are described by Taylor *et al.* (66), in a booklet by the Arizona Chemical Co. (65), and by D. H. Wheeler *et al.* (67). It will be realized that a fractional distillation process may be adjusted to yield a variety of products. These could include distilled tall oil, mixtures of rosin and fatty acids, fatty acids with low rosin content,

Table 13 * Composition of Tall Oil Products

Tall Oil Product	Per Cent by Weight: Fatty Acids	Rosin Acids	Unsaponifiables
Acintol † C crude tall oil			
Georgetown and Panama City plants	42	51	7
Springhill plant	51	42	7
Acintol D distilled tall oil	66	32	2
Acintol DLR distilled tall oil	77	21	2
Acintol FA #1 tall oil fatty acids	92	4	4
Acintol FA #2 tall oil fatty acids	97	1	2
Acintol R-S tall oil rosin	2	94 ‡	4
Acintol R-D tall oil rosin	1.5	92 ‡	6.5
Acintol P tall oil pitch §	29 ‖	31 ‖	33

* Courtesy of Arizona Chemical Co. (65).
† Trade name, Arizona Chemical Co.
‡ Includes rosin anhydrides and esters.
§ Includes esterified acid.
‖ Acids derived from lignin not shown.

rosin with low fatty acid content, and the heavy ends or tall oil pitch. As stated previously the composition of crude tall oil may vary when the black liquor soap is obtained from different paper plants. A typical series of tall oil products produced by fractional distillation at the Arizona Chemical Co.'s plants is shown in Table 13.

The crude tall oil may be used in low-cost coatings where the dark color and presence of rosin are not objectionable. The acidity may be neutralized with lime or alcohols, and the products may be modified with maleic anhydride, phenolic condensates, styrene, or other reactive monomers. A low-cost product is obtained by liming the crude tall oil as indicated in Formula 9. This product should be compared with the limed oils described on page 71 and also with the limed rosin described in Chapter 3, page 151.

Formula 9 * Limed Tall Oil

	Parts by Weight
Acintol C	100
Hydrated lime	9
Calcium acetate	0.1

Procedure

Heat Acintol C to 400°F, sprinkle lime in slowly while increasing the temperature to 520°F, add calcium acetate with the final one-third of the lime, hold at 520–525°F until a drop of the product on cold glass plate is clear or free from turbidity (about 30 min), cool to 350–400°F, and reduce with mineral spirits to desired solid content.

* Courtesy of Arizona Chemical Co.

Procedures similar to those described in the next section for distilled tall oil may be used for reaction of crude tall oil with alcohols, maleic anhydride, etc.

Distilled Tall Oil. The physical and chemical characteristics of the distilled tall oils are shown in Table 14.

It will be realized from previous discussion in this chapter that the distilled tall oil may be esterified with any of the polyhydric alcohols in Fig. 9, with or without modification by maleic or fumaric acid. The product could be thinned with mineral spirits and would be similar to an oleoresinous varnish based on a modified rosin as the resin, as discussed in Chapter 4. The distilled tall oil may also be reacted with polyhydric alcohols and phthalic anhydride to produce rosin-modified alkyd resins, as discussed in Chapter 7. Typical formulations and procedures for these products are given in Formulas 10–12.

Table 14 * Typical Analysis of Acintol † D and Acintol † DLR Distilled Tall Oil

		Acintol D		Acintol DLR
Color, Gardner 1933		9		12
Acid number		190		186
Saponification number		193		192
Iodine number		164		139
Rosin acids, %		32.0		21.3
Unsaponifiables, %		1.4		1.5
Fatty acid, total, %		66.6		77.2
Fatty acid composition				
Linoleic acid, conjugated, %	9		19	
Linoleic acid, non-conjugated, %	37		27	
Oleic acid, %	51		48	
Saturated acids, %	3		6	
	100		100	
Moisture, %		None		None
Ash, %		0.01		0.005
Viscosity, SSU, at 210°F		54		61
Viscosity, Gardner-Holdt, at 25°C		D		E
Specific gravity at 25/25°C		0.9465		0.9336
Weight per gallon at 25°C		7.90		7.75
Flash point, °F, open cup		400		390
Fire point, °F, open cup		435		425

* Courtesy of Arizona Chemical Co. (65).
† Trade name, Arizona Chemical Co.

Formula 10 * Tall Oil-Glycerol Ester

	Parts by Weight
Acintol D tall oil	1000
Glycerol	114 †

Laboratory Procedure

Place ingredients in a three-neck flask equipped with agitator, air condenser, thermometer, and inert gas inlet tube. Heat to 482°F (250°C), pass slow stream of inert gas through the mixture during heating, hold at 482°F (5–7 hr) until the required acid number is obtained, cool, and reduce with mineral spirits if desired. A higher viscosity or shorter reaction time is obtained by increasing the temperature to 525°F.

Characteristics

Viscosity, 100% solids	S
Viscosity, 70% in mineral spirits	A
Color, Gardner 1933	8
Acid number (solid basis)	10

* Courtesy of Arizona Chemical Co. (65).
† 10% excess of theory.

Formula 11 * Tall Oil-Pentaerythritol Ester

	Parts by Weight
Acintol D tall oil	100
Pentaerythritol	120 †

Laboratory Procedure

Place tall oil in equipment as in Formula 9, heat to 400°F, add pentaerythritol, heat to 530°F, hold (about 7 hr) until required acid number is obtained, cool, and reduce with mineral spirits if desired. Maintain good agitation until pentaerythritol is taken up to prevent charring on the bottom of the flask.

Characteristics

Viscosity, 100% solids	Z3
Viscosity, 70% in mineral spirits	B–C
Color, Gardner 1933	9
Acid number (solid basis)	12

* Courtesy of Arizona Chemical Co. (65).
† 3% excess of theory.

The above products may be modified by addition of 2–8% of maleic anhydride. See page 91 for general discussion and mechanism of adduct formation.

Formula 12 * Tall Oil-Maleic, Pentaerythritol Ester

	Parts by Weight
Acintol D tall oil	1000
Maleic anhydride	60
Pentaerythritol	165

Procedure

Heat tall oil and maleic anhydride under inert atmosphere with agitation to 400°F, and hold for about 1 hr to produce the adduct. Add pentaerythritol, raise the temperature to 530°F, hold for acid number of 10–15, cool to 400°F, and reduce with mineral spirits.

Characteristics

Per cent solids	70
Viscosity at 25°C	S–U
Color, Gardner 1933	8–9
Acid number	12–15

* Courtesy of Arizona Chemical Co. (65).

Since the fatty acids in tall oil are not highly unsaturated, pentaerythritol may be preferred to glycerol because of its higher reactivity, which produces higher viscosity and faster drying time. It will be realized that Acintol D may be replaced with Acintol DLR to obtain lower rosin content and a somewhat better-drying oil component. The distilled tall oils may also replace all or part of the fatty acids in the production of rosin-modified alkyd resins, as discussed in Chapter 7.

Fractionated Tall Oil Fatty Acids. The physical and chemical characteristics of the fractionated tall oil fatty acids are given in Table 15. Comparison of their composition with that of the acids in

Table 15 * Typical Analysis of Acintol † FA #1 and Acintol † FA #2 Tall Oil Fatty Acids

	Acintol FA #1	Acintol FA #2
Color, Gardner 1933	9	6+
Titer, °C	3	1
Acid number	189	194
Saponification number	192	196
Iodine number	138	130
Rosin acids, %	3.5	1.0
Unsaponifiables, %	4.0	2.2
Total fatty acid, %	92.5	96.8
Fatty acid composition		
Linoleic acid, %	46.0	48.0
Linolenic acid, %	None	None
Oleic acid, %	51.0	50.0
Saturated acid, %	3.0	2.0
	100.0	100.0
Moisture	None	None
Ash, %	0.005	0.003
Viscosity, SSU, at 100°F	93	93
Viscosity, Gardner-Holdt, at 25°C	A	A
Specific gravity at 25/25°C	0.9005	0.8999
Weight per gallon at 25°C	7.515	7.480
Flash point, open cup, °F	380	380
Fire point, open cup, °F	415	415

* Courtesy of Arizona Chemical Co. (65).
† Trade name, Arizona Chemical Co.

soybean oil in Table 4 shows the similarity. Soybean acids contain a small percentage of linolenic acid, which we shall see on page 127 to be a source of yellowing, and the fractionated tall oil acids contain a small percenatge of rosin acids.

The fractionated tall oil acids may be esterified with the polyhydric alcohols in Fig. 9 to produce a series of oils. They may also be modified with maleic anhydride, styrene, etc., for treated and copolymerized products as discussed previously. These acids are suitable for use in alkyd resins, and typical formulas are given in Chapter 7. The acids may also be used with the Epon resins described in Chapter 7. Acintol FA #1 can be used where a small amount of rosin can be tolerated. Since the rosin acid content of Acintol FA #2 is only about 1% it can be used where rosin is not permissible in more than trace quantities. The composition of this acid indicates its use in products requiring pale color and good color stability.

Tall Oil Rosin. The characteristics of Acintol R-S and R-D tall oil rosins are given in Chapter 3. They comply with the specifications for rosin as defined by the U. S. Dept. of Agriculture pursuant to the Naval Stores Act of 1923. They are suitable for most applications where gum or wood rosin is used.

Tall Oil Pitch. The tall oil pitch Acintol P, listed in Table 13, is a semi-fluid material, soluble in petroleum and aromatic solvents, and compatible with oils, rosin, and other pitches and asphalts. It may be hardened by liming, esterification with polyhydric alcohols, or air blowing. It finds use in dark-colored paints and printing inks and in caulking and roofing compounds. Its characteristics are given in Table 16.

Table 16 Typical Analysis of Acintol * P Tall Oil Pitch

Rosin acids, %	31
Fatty acids, %	29
Unsaponifiables, %	33
Ash, %	0.05
Acid number	63
Saponification number	110
Iodine number	138
Softening point, °F	90
Viscosity, SFU, at 210°F	104
Flash point, open cup, °F	495
Fire point, open cup, °F	545
Pounds per gallon at 25°C	8.41
Color (10% in benzene), Gardner 1933	18

* Trade name, Arizona Chemical Co.

The recovery and utilization of large tonnages of tall oil products constitute another illustration of the efficiency of our industrial civilization. Formerly, tall oil was a waste product; today it is a source of wealth and utility.

FILM FORMATION

The value of drying and semi-drying oils in paints and varnishes lies in their ability to form dry films when exposed to air at normal room temperature and also at baking temperatures. In general, the value of an oil is increased if it can be made to dry faster. In the previous section several methods were described which are designed to reduce the drying time by physical or chemical changes in the oil. A great deal of brilliant research has been done by oil and resin chemists (47) over the past 25 years, but the chemistry and physics of drying oils are so complicated that the mechanisms for all the reactions involved have not been elucidated completely at present. This is due in part to the fact that the oils are mixed glycerides in combination with small percentages of other compounds, and also to the fact that the final dry film is not susceptible to analysis without degradation of its physical structure and chemical composition.

It has been shown that films of drying oils take up from 10% to 12% of oxygen during the air-drying process, that peroxidic structures are formed, and that volatile oxidative decomposition products are given off simultaneously. The films also become insoluble and infusible in the ordinary sense of these terms, which indicates that new primary valence bonds have been formed between oil molecules to give a cross-linked type of polymer.

In this chapter it will be necessary to consider these reactions and mechanisms in greater detail in order to appreciate the various factors which affect the formation of oil films. A more detailed study will also indicate some of the factors which contribute to the deterioration of oil films under the conditions of normal use.

The Ester Linkage. It is shown in Fig. 1, page 45, that the oils are esters of glycerol and various fatty acids, and that the ester linkage may be split or saponified quite readily by steam or alkali. Since the ester linkage is not changed in the film-forming process it will be a part of the final film structure. As such it remains readily susceptible to hydrolysis by steam and saponification by alkalies, and it therefore contributes to the lack of resistance to steam and alkali of the oil film. The poor alkali resistance of oil films is a severe handicap to their use for some applications, but they are improved by dilution or reaction with resinous materials such as the phenolics and the amino resins.

Elm (48) has indicated that the ester linkage is not subject to hydrolysis under conditions of normal interior or exterior exposure to moisture, or to prolonged immersion in fresh or salt water. Therefore, it would not contribute to severe breakdown of the film under these conditions. However, the ester linkage is highly polar, and this would make the film susceptible to moisture absorption and permeation.

The distribution of the three fatty acids in oil molecules as discussed on page 52 shows that most drying oils are mixed, i.e. the three fatty acid radicals are different. In tung and oiticica oils, which contain about 80% of eleostearic and licanic acids respectively, there will be many molecules having all three acid radicals the same. However, even in these oils the triglycerides will be partially mixed because they do not contain 100% of a particular fatty acid. The mixed-glyceride idea assumes considerable importance in the film-forming process. Neither oxidation nor polymerization would be expected to be as rapid in triglycerides which contain one saturated acid radical per molecule as in those molecules having all the acid radicals unsaturated. The extent of mixture may be visualized more readily from Table 17,

Table 17 Distribution of Fatty Acids in Mixed Triglycerides

Oil Molecule No.	Fatty Acids per Molecule			Composition, %	
1	le	lo	sa	Fatty acids	
2	le	lo	sa	Linolenic	40
3	le	lo	sa	Linoleic	30
4	le	lo	ol	Oleic	20
5	le	lo	ol	Saturated	10
6	le	lo	ol		
7	le	lo	ol	Triglycerides	
8	le	lo	ol	30% with 1 saturated acid; relatively poor drying	
9	le	lo	ol	50% with 1 oleic acid; fair drying	
10	le	le	lo	10% with 2 linolenic, 1 oleic; good drying	
				10% with 2 linolenic, 1 linoleic; very good drying	

Abbreviations: le: linolenic; lo: linoleic; ol: oleic; sa: saturated.

which portrays the distribution of fatty acids in an oil similar to linseed which may be assumed to contain 40% linolenic acid (le), 30% linoleic acid (lo), 20% oleic acid, (ol), and 10% saturated acids (sa).

A study of Table 17 indicates at once the importance of the oil-splitting process followed by separation of the fatty acids and re-esterification as described on page 76. The table also shows that in an oil similar to linseed at least 30% of the molecules are relatively poor drying because each one contains a saturated fatty acid. Furthermore, the yellowing of certain drying oils is generally attributed to the presence of linolenic acid, and from the above data it could be predicted that a film of this oil would yellow badly since it has at least one linolenic acid radical in each molecule.

Film Formation. A number of changes may be observed in the study of film formation of drying oils, some of which may be measured quantitatively. Since metallic driers are used in all air-drying oil compositions, they will be assumed to be present in these considerations. It has been shown that driers act simply as catalysts to reduce the drying time, but the actual mechanism of the process has not been demonstrated. One of the important functions of driers is to reduce the induction period, probably by oxidizing the natural oxidation inhibitors present in the oil. When linseed oil with drier is spread in a thin layer the changes given in Table 18 may be observed.

Table 18 Stages in Film Formation

1. Induction period; small change in consistency, natural inhibitors removed by oxidation.
2. Set to touch; marked change in consistency from liquid to firm gel.
3. Hardening; small change in consistency, gel hardening.
4. Aging; change will vary depending on environment.
   (*a*) Diffuse light, interior; slow hardening, relatively little decomposition or deterioration, some yellowing.
   (*b*) Ultraviolet light, interior; rapid hardening and embrittlement with surface deterioration or chalking.
   (*c*) Sunshine, exterior; fairly rapid hardening with embrittlement, severe decomposition and rapid failure, slight yellowing.
   (*d*) Darkness; slow softening, becoming sticky from depolymerization, syneresis of unreacted oil, or solvent effect of oxidation-decomposition products, considerable yellowing.

In addition to the observations in Table 18 the following changes can be measured, as discussed in Chapter 1. At first the film increases in weight, followed by some decrease, but with a final net increase in weight. In general, about 12% oxygen will be taken up, but volatile decomposition products are also given off, including carbon dioxide, water, formic and acetic acids, and hydrogen peroxide. The iodine value decreases, denoting a loss of unsaturation, but this is not proportional to the amount of oxygen taken up. The peroxide value increases considerably in the early stages of film formation, then decreases, but does not become zero. Concurrently, there is some isomerization of the non-conjugated to the conjugated type of unsaturation. There is no significant increase in viscosity until the peroxide value starts to decrease, indicating a connection between peroxide decomposition and cross-linking or polymerization. Non-volatile decomposition products have also been found which include short-chain fatty acids, aldehydes, and their oxidation products—ketones and alcohols. Some of these are believed to have been formed as the result of scission reactions in the film structure.

It will be apparent that many reactions are taking place simultaneously in the film-formation process. Since there is a considerable difference in the rate at which these reactions proceed, the dominant reaction at any time will be the one with the fastest rate. Some of the oxidation-decomposition products are the result of film-structure breakdown; therefore film deterioration is taking place at the same time as film formation. However, the rate is so much slower that film deterioration does not become significant until the rate of film formation has become negligible.

Oxidation. The large increase in peroxide value of an oil during oxidation indicates that oxygen which is added forms peroxidic compounds. One of the earliest postulates was that oxygen added at a double bond to form a cyclic peroxide having the following structure:

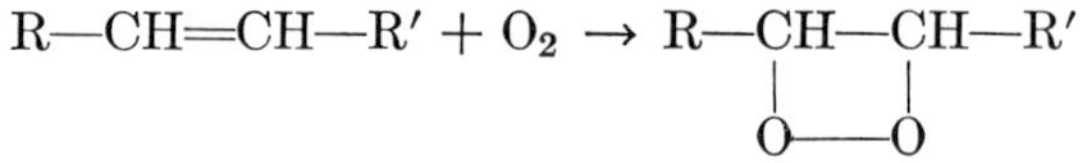

The existence of this four-membered ring structure has not been demonstrated directly, but Franke and Monch (49) believe that the postulate is correct from the relationship they found between the peroxide content and loss of iodine value. The rather remote possibility also exists that this relatively unstable structure may open up and cross-link with another oil molecule if the unsaturated section comes within the correct proximity of the peroxide. However, this six-membered dioxane ring structure has not been proved to exist in oil films.

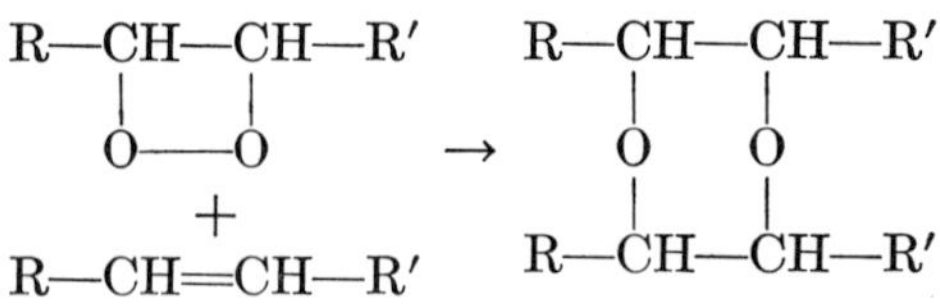

Hydroperoxide Formation. In order to simplify the investigation of the oxidation and polymerization processes many workers replaced the complicated triglycerides with the methyl esters of various fatty acids. The acids can be obtained in fairly pure form by distillation and crystallization, and the methyl esters prepared by standard methods of esterification. Farmer and Sutton (50), working with methyl oleate, isolated a hydroperoxide from partially oxidized methyl oleate. The hydroperoxide group was found to be attached to a carbon atom

Fig. 16. Oxygen Activation and Hydroperoxide Formation

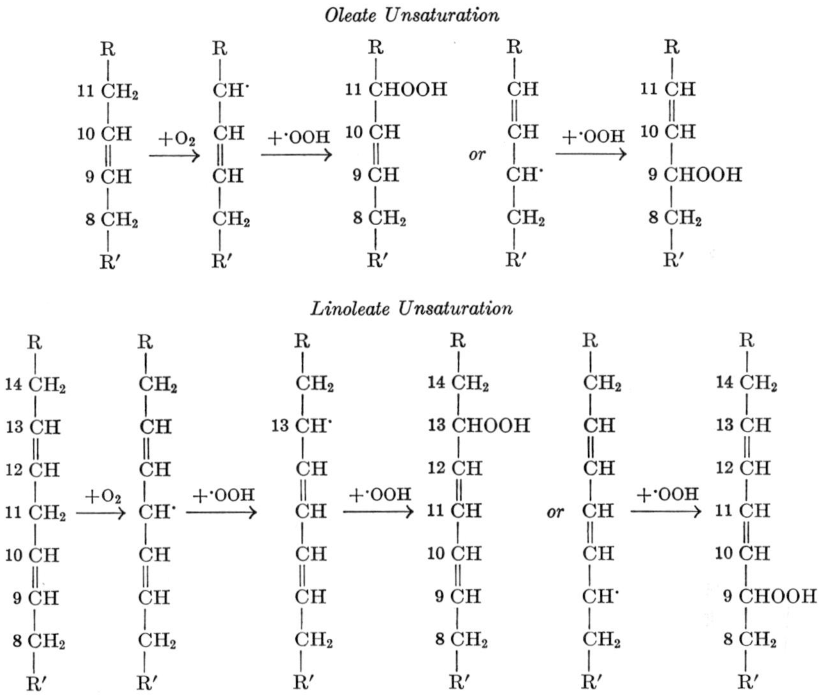

adjacent to a double bond; therefore there was no loss of unsaturation. A possible free-radical mechanism for this reaction is illustrated in Fig. 16.

The final location of the hydroperoxide group in the fatty acid radical of the oil depends on chance activation by the approaching oxygen molecule and the possibility of isomerization along the chain. The reactions for the oleate radical with one double bond and the linoleate radical with two double bonds are indicated in Fig. 16.

Figure 16 shows three possibilities from activation by oxygen:

1. Activation of a CH_2 group adjacent to a double bond with subsequent formation of hydroperoxide at that carbon atom. This may occur at either the 8- or 11-carbon atom in the oleate radical.

2. Activation at the double bond, in which case the hydrogen required for formation of the hydroperoxide radical is taken from the CH_2 group adjacent to the double bond. This is shown in the oleate radical by activation of the 9-carbon atom with a shift of the double bond to the 10:11 position. The hydroperoxide radical then becomes attached to the original carbon atom which was activated to form the hydroperoxide group.

3. Activation of the CH_2 group between two double bonds in the linoleate radical. A free radical is formed at this position, but in this activated condition the unsaturation in the system can shift from the non-conjugated type to the more stable conjugated type. If the shift of the double bond is from the 12:13 position to the 11:12 position, the odd electron moves to the 13 carbon and becomes paired with the odd electron on the hydroperoxide radical to form a normal covalent bond as shown. Also shown is the other possibility of shifting from the 9:10 position to the 10:11 position with the formation of the hydroperoxide at the 9 carbon.

It will be seen that the result of linoleate oxidation is a change from non-conjugated to conjugated unsaturation and the formation of a hydroperoxide on the carbon atom adjacent to this system. It may be possible that, when the hydroperoxide decomposes, it catalyzes the polymerization of the conjugated system with the unsaturation in another molecule to form a direct C—C linkage between the two molecules.

Farmer, Sutton, and Koch (35), working on the oxidation of triple unsaturated ester, methyl linolenate, found considerable isomerization from the non-conjugated to the conjugated type. Both diene and triene conjugation were formed. Bolland and Koch (51), working with methyl linoleate, showed that the primary oxidation product was a monohydroperoxide containing about 70% conjugated diene. This again proved that isomerization from the non-conjugated to the conjugated form was taking place during the oxidation process, and that hydroperoxides were formed. In attempts to explain the isomerization and hydroperoxide formation, the free-radical mechanism was used as illustrated in Fig. 16.

It is believed that the methylene group adjacent to a single double bond, and particularly the one between two double bonds, is more active than the other methylene groups in the fatty acid chain. This suggests the possibility of a free radical being formed on the active methylene group, which would produce a resonating system. Since the conjugated form of unsaturation is the most stable form the tendency of this resonating system would be to isomerize with the odd electron moving to the adjacent carbon atom to form a free radical. The free radical becomes paired with the odd electron on the hydroperoxide group to form a normal covalent bond. However, the actual position of the hydroperoxide will vary; for example, there is about equal chance of it being on the 8 or 11 carbon in the oleate chain if the

Fig. 17. Oxidation, Hydroperoxide Formation, and Isomerization of Linolenate Unsaturation

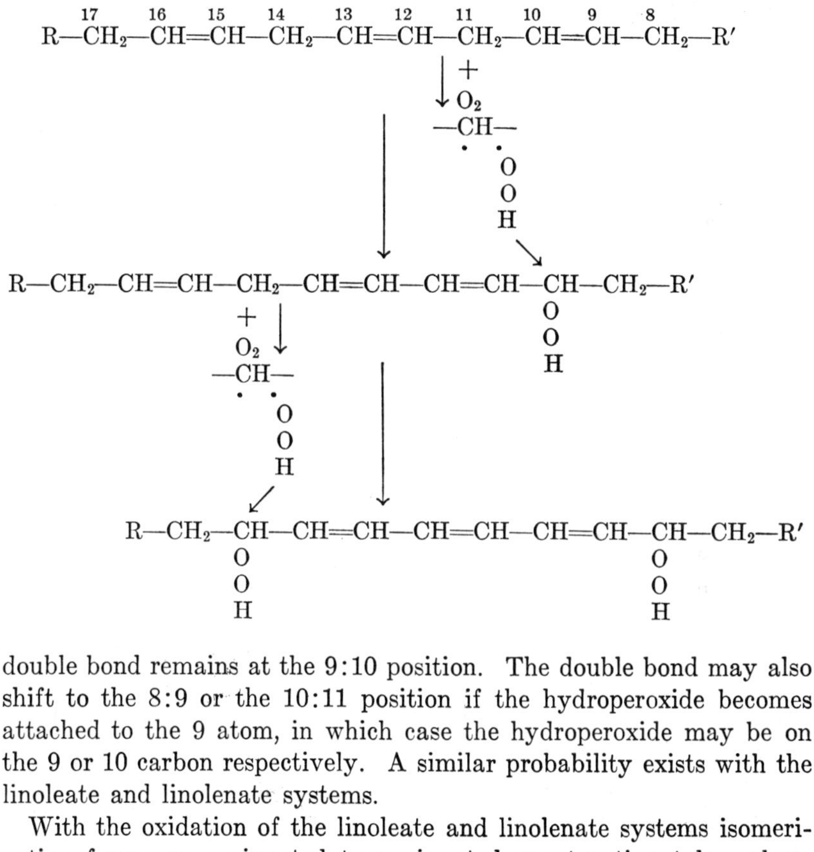

double bond remains at the 9:10 position. The double bond may also shift to the 8:9 or the 10:11 position if the hydroperoxide becomes attached to the 9 atom, in which case the hydroperoxide may be on the 9 or 10 carbon respectively. A similar probability exists with the linoleate and linolenate systems.

With the oxidation of the linoleate and linolenate systems isomerization from non-conjugated to conjugated unsaturation takes place. In the linoleate system the hydroperoxide will be located on the carbon atom adjacent to the unsaturation as illustrated in Fig. 16. In the linolenate system there is the possibility of forming one hydroperoxide group and a conjugated diene, or two hydroperoxide groups and a conjugated triene. A possible mechanism for these reactions is shown in Fig. 17.

In the section on isomerized oils, page 83, it was shown that a change from the normal *cis* form to the more stable *trans* form may also take place when a change in the form of unsaturation occurs. The final configuration would be about equal parts of *cis* and *trans*. The two types are shown in Fig. 18. It will be recalled that the *trans* form is somewhat slower drying than the *cis* form.

Fig. 18. *cis-trans* Isomerization

cis form

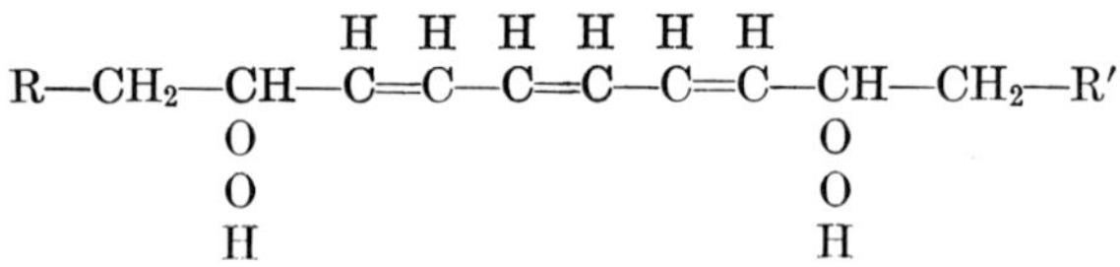

trans form

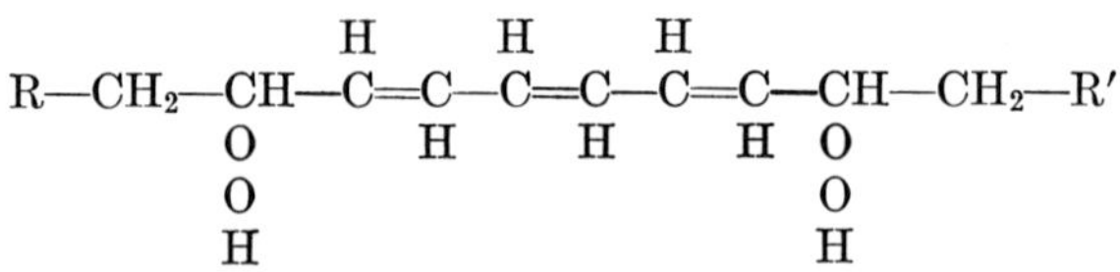

Autoxidative Polymerization. The process whereby drying oils absorb oxygen, form hydroperoxides, isomerize to the conjugated form, and polymerize to the solid state is known as autoxidative polymerization. Neither the mechanism of the polymerization nor the exact structure of the solid film is known definitely. It was stated previously that, in the early stages of oxidation, the peroxide value increases to a maximum, then decreases to almost zero. There is no significant change in the consistency of the oil until the peroxide value starts to decrease. The consistency then increases fairly rapidly until the gel stage is reached. Therefore it would appear that there is a relationship between the decomposition of the hydroperoxides and the polymerization or cross-linking of the system. It is known that autoxidative polymerization is accelerated by the usual metallic driers and also by light and heat, and that both volatile and non-volatile oxidation-decomposition products are formed concurrently. Three possible methods of cross-linking from decomposition of the hydroperoxides are shown in Fig. 19, but it should be remembered that these reactions have not been proved. However, water and hydrogen peroxide have been found in the volatile products of the reaction, and both peroxides and ether linkages have been found in the dry film. After the film has reached the dry stage, oxidation continues, although at a greatly reduced rate. Oxidation is an important factor in film deterioration and is discussed in greater detail on page 126.

Fig. 19. Cross-Linkage from Hydroperoxide Breakdown

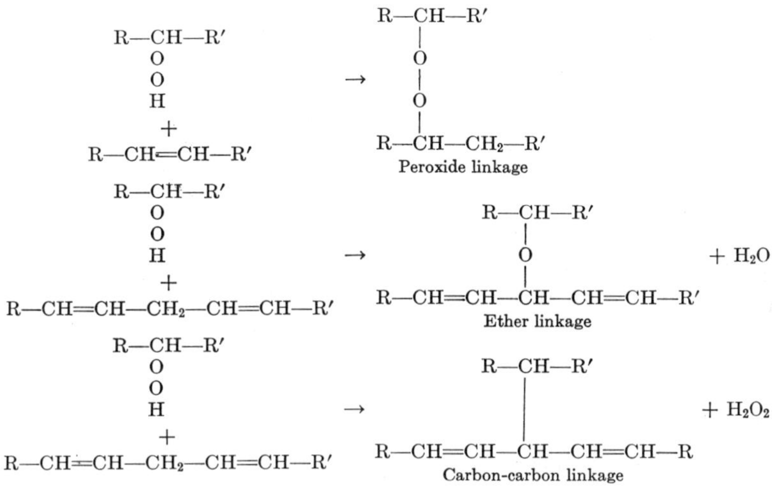

Thermal Polymerization. The previous section described the formation of cross-linked polymeric structures by autoxidative polymerization of drying oils under normal air-drying conditions. Drying oils may also be polymerized by heat, and the methods used for thermal polymerization and the results obtained with different oils are described under Heat-Bodied Oils on page 67. This section will outline the mechanism of thermal polymerization, based on the work of Bradley (52), Wheeler (37), Kappelmeier (30, 31), Scheiber, Diels-Alder, and others.

The three types of polymers, addition, condensation, and copolymers, are described in Chapter 1, together with the significance of functionality. Thermal polymerization of drying oils produces essentially the addition type of polymer by the formation of cross-linkages between the unsaturated section of the fatty acid radicals. If the thermal polymerization is continued long enough, it will result in an insoluble, infusible gel because the oil molecule is a triglyceride, having three fatty acid radicals and a potential functionality of 3. An increased rate of thermal polymerization is obtained when the functionality is increased by replacement of glycerol with the tetra-functional alcohol pentaerythritol or by the use of alcohols with higher functionality.

The type and degree of unsaturation in the fatty acid radical affect the rate of thermal polymerization and the type of gel structure formed. The rate of polymerization is measured conveniently by increase of viscosity with time of heating, as illustrated in Tables 6 and 7. The non-drying oils such as cottonseed and cocoanut show no significant increase in viscosity when maintained at 600°F for prolonged periods of time. These oils contain a high percentage of saturated fatty acids, and the little unsaturation they do contain is of the oleic type, having a single double bond per fatty acid. Soybean oil is classed as semi-drying and shows only a slow increase in viscosity when maintained at 600°F. Linseed oil is a good drying oil and gives a fairly rapid increase in viscosity at temperatures from 575 to 600°F because it contains a considerable percentage of linoleic (two double bonds) and linolenic (three double bonds) types of unsaturation. Fish oil contains some unsaturated fatty acid radicals having four double bonds, and, since these polymerize more rapidly than the above, this oil shows fairly rapid increase in viscosity at temperatures of 550–560°F. However, tung and oiticica oils are not usually prepared as heat-bodied oils because of their very rapid rate of thermal polymerization even at temperatures as low as 450–500°F. Their ability to polymerize rapidly is related to their high content of eleostearic and licanic acid radicals respectively, which contain three double bonds per radical; but of greater importance is the fact that this unsaturation is the conjugated type. Therefore, the rate of thermal polymerization increases and the bodying temperature may be decreased as the non-conjugated unsaturation increases from soybean to fish oil, but there is a rapid increase in rate of polymerization and a sharp drop in bodying temperature when the type of unsaturation is changed from non-conjugated to conjugated.

Although all the drying oils form insoluble gels if held for sufficient time at the above temperatures, the character of the gel differs greatly for the various oils. Linseed oil forms a very sticky, stringy gel and requires many hours at 600°F, but tung oil forms a dry, crumbly gel in not more than 12 min at 540°F (see Chapter 15 on test methods). These gels are three-dimensional, cross-linked structures similar to those obtained in baked films. As pointed out under autoxidative polymerization, these structures are extremely difficult to analyze because of their insolubility; therefore investigations regarding the nature of the structures have been carried out on the simpler monomeric esters. Bradley and Johnson (52) separated the methyl linoleate polymers by molecular distillation and found that they were

Fig. 20. Thermal Polymerization of Conjugated Oils

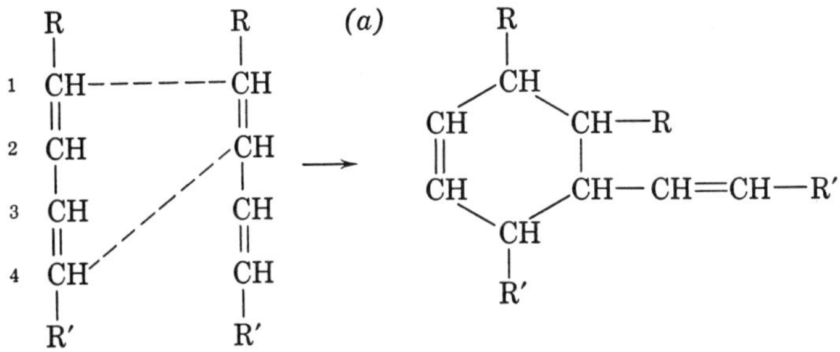

Diene addition

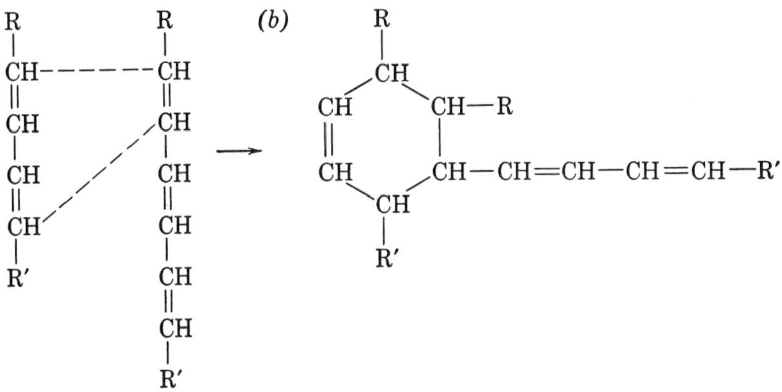

Diene-triene addition

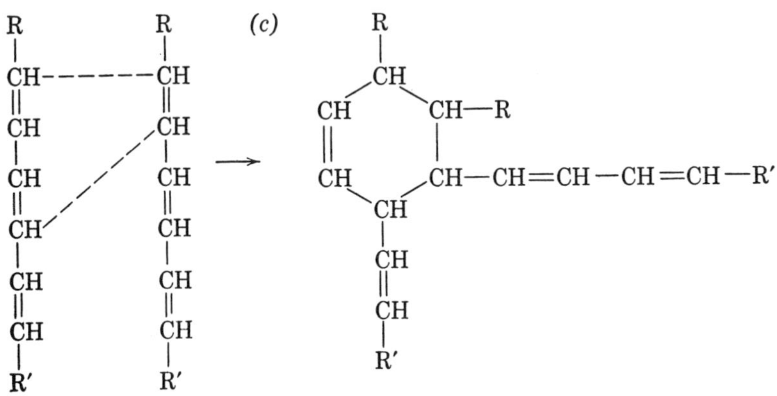

Triene addition

mostly dimers with a small percentage of trimers. Scheiber in 1929 suggested that during the heating of non-conjugated oils there may be some isomerization to the conjugated form, which as we have seen polymerizes more rapidly. Kappelmeier (30, 31) suggested that a Diels-Alder addition may occur with the conjugated unsaturation to give ring structures in the heat-bodied oils or baked films. Bradley and Johnson (52) showed that the physical and chemical constants of the dimers of methyl linoleate were those of a substituted cyclohexene ring structure similar to that obtained from the thermal poly-

Fig. 21. Polymerization of Conjugated with Non-Conjugated Oil

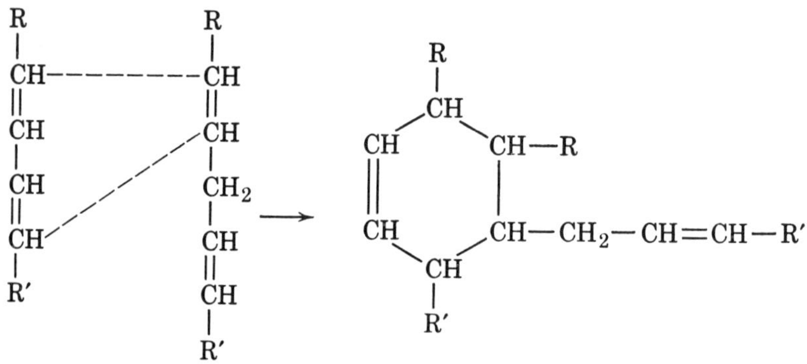

merization of butadiene. From many similar investigations it is generally believed that the thermal polymerization of conjugated diene and triene structures is a Diels-Alder addition at the 1:4 position with a shift of a double bond to the 2:3 position. The mechanism is illustrated in Fig. 20, using R and R′ to indicate the remainder of the fatty acid radical in the drying oil.

The diene structure, Fig. 20(*a*), is present in dehydrated castor oil, which is not as fast bodying as tung oil. A combination of dehydrated castor and tung oils would polymerize more rapidly. The reason will be evident from Fig. 20(*b*) since the side chain on the cyclohexene ring still contains a conjugated diene which can polymerize with another unsaturated molecule. The reason for the extremely fast bodying and gelation of tung and oiticica oils can now be seen from Fig. 20(*c*) since both side chains of the cyclohexene ring contain unsaturation. Further additions may occur at these locations, and an intimately cross-linked structure would be built up rapidly to produce a solid gel.

The structures in Fig. 20 do not explain the thermal polymerization of non-conjugated oils such as linseed and soybean. However, it can be shown by ultraviolet absorption spectroscopy (see Chapter 15) that isomerization to the conjugated form takes place on heating these oils. The thermal polymerization would then follow the procedure given in Fig. 20. Paschke and Wheeler (37) have pointed out that in polymerizing linseed-type oils there will be a much larger amount of the non-conjugated than the conjugated form during the thermal treatment. In view of this fact they suggest that the 1:4 addition takes place between the newly formed conjugation and the original non-conjugation. The resulting structure would be similar to that shown in Fig. 20 but with the double bond in the side chain removed one carbon further from the ring as shown in Fig. 21. This possibility could increase the rate of isomerization since it would remove almost immediately the conjugated unsaturation which was formed.

Direct-Chain Polymerization. In addition to the Diels-Alder type of thermal polymerization, which produces six-membered rings, it is

Fig. 22. Direct-Chain Polymerization

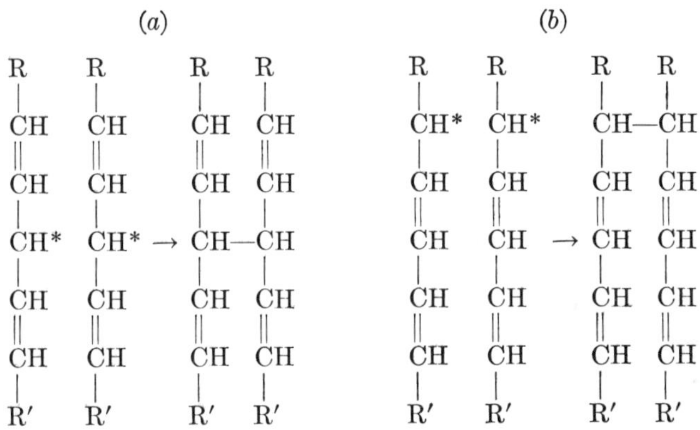

conceivable that direct carbon-carbon linkage may occur between the unsaturated systems in the fatty acid chains. The high temperatures used in thermal polymerization produce violent motion of the molecules, and if two should collide in the proper manner a direct linkage would develop between the carbons of the activated methylene groups. Two such possibilities are illustrated in Fig. 22. However, the dimers formed by this type of polymerization have highly unsaturated systems which may react intramolecularly to produce six-membered rings.

Colloidal Association Effects. The difference between primary and secondary valence forces is described in Chapter 1. When polymeric structures are held together by primary valence forces they show a high degree of insolubility and infusibility. However, strong polymeric structures may also be formed by secondary valence forces, as exhibited by nitrocellulose, polystyrene, and other thermoplastic polymers. These polymers remain soluble in the original solvents used for their application, and they are fusible or heat-softened below the temperature of decomposition. The structures formed by secondary valence forces are sometimes referred to as associated structures or association colloids.

Films of drying oils are relatively insoluble, thereby indicating primary valence forces in their structures. However, the degree of solubility varies with the different oils and with the solvent, temperature, and pressure. For example, linseed oil is completely soluble in acetone, but its solubility decreases as polymerization is increased. However, a sample of polymerized linseed having a given percentage of acetone insoluble under normal conditions may be dissolved in acetone under heat and pressure. It does not seem that acetone would rupture primary valence bonds or depolymerize the oil under these conditions.

In addition, water is a non-solvent for the original oil and the oil film, but oil films will swell considerably when immersed in water. In this respect oil films behave like gelatin and other water-dispersible association colloids, although, of course, not to the same extent. Because of these and other colloidal characteristics some investigators (53, 55, 56) believe that oil films are a special kind of association colloid. Slansky (53) believes that, in the drying of vegetable oils, chemical reactions take place such that more than one phase is produced in the oil. When one of these phases becomes the disperse phase, and when it is produced in sufficient concentration, it coagulates the colloid which then sets to a solid gel. The disperse phase may be produced by oxidation, thermal polymerization, or other means, but the final solid film is the result of colloidal association. This contention would be extremely difficult to prove conclusively, but it will be recalled that actual gelation takes place fairly rapidly in either the air-drying film or in thermal polymerization. Therefore it might appear that the dried oil is an aggregation of polymers held together by both primary and secondary valence forces. If the primary forces predominate the film will be tougher, more cohesive, and less soluble than if the secondary or association forces predominate. In view of the acetone-extractable content of oil films they must be visualized as fairly open structures capable of absorbing considerable amounts of low-molecular-weight materials. Such open structures may shrink on aging and force out some of the low-molecular-weight compounds. Conversely, they may be distended or swollen by other low-molecular-weight materials which have a greater affinity for the inner surfaces of the structures than those materials which were exuded by syneresis of the gel structure. This affinity or inbibition pressure (54) is considered to be produced by absorption forces related to the relative polarity of the internal surface of the structure and the material being absorbed. These forces are therefore forms of van der Waals' or association forces.

Another aspect of the colloid chemistry of drying oils was proposed by Auer (55, 56) which is generally referred to as the gas coagulation theory. It assumes that oxidation is only a secondary phenomenon in film formation and not necessarily connected with it. It also assumes that the oils are lyophilic colloidal systems in which the different triglycerides are in various physical states. A comparatively small proportion of colloid is responsible for the film formation and acts as a gel skeleton in which the remaining part of the oil is dispersed. Heat bodying of oils increases the concentration of this disperse phase and may account for the improved properties of films from bodied oils.

It will be apparent that much work needs to be done before the story of film formation is complete. It will also be seen in the next section that a similar situation exists with respect to film aging and deterioration.

FILM AGING AND DETERIORATION

It was stated previously that the conditions under which a film is aged have a marked effect on the character of the film. Under normal interior conditions with diffused light, a linseed oil film shows slow hardening, some yellowing, but relatively little deterioration; whereas, with ultraviolet light, it shows rapid hardening with considerable surface deterioration. Under normal exterior conditions a linseed oil film shows fairly rapid deterioration which is accelerated by the ultraviolet component of sunshine. Pigmented oil films are much more durable than clear oil films because the pigment protects the oil from the ultra-violet radiation. In general, the greater the light-absorption capacity of the pigment the more durable the film will be. Black paints have outstanding durability. The chemical reactivity of the pigment with the oil is another important factor in durability and is discussed in detail in Volume II.

It was also stated previously that the physical and chemical processes which cause film deterioration occur simultaneously with film-formation processes but at a much slower rate. However, this rate becomes significant when film formation has ceased. Since the useful life of a paint film is dependent principally on the rate of deterioration of the film binder, it is important to know as much as possible about the deterioration processes. The many reactions involved in these processes make exact analysis difficult, but much useful information has been obtained regarding oxidation-decomposition products of drying oils. Elm (48) gives an excellent review of this subject and points to the need for further work in the field.

Many industrial and household paints come in contact with water, organic solvents, grease and lubricating oils, acids, alkalies, and other materials, and the ability of the paint to offer satisfactory resistance to these materials is determined principally by the composition and structure of the film binder. The poor alkali resistance of oil films is due primarily to the fact that the triglycerides are esters and fairly easily saponified, but the alkali resistance may further decrease with age of the film because of the acidic decomposition products which are formed. The water resistance of oil films which have been dried by baking is better than that of films which have been air-dried at normal temperature because the baked films contain less oxidation-decomposition products. Some of these products are water-soluble materials, and all of them are highly polar with an affinity for water. The yellowing of oil films on aging is usually worse with air-dried or oxidized films than with baked films, and it is particularly bad when the film is aged in the absence of light.

Film formation in an oil film may be considered to stop for all practical purposes when the film has dried hard enough for normal use. The film is then a cross-linked, three-dimensional structure held together by a variety of linkages which were described under film formation. When aged under normal conditions the film continues to harden. This may be due to a shrinkage or syneresis of the gel structure, and also to a slow absorption of oxygen at the residual unsaturation, followed with autoxidative type of polymerization. The hardening process makes the film less flexible and less resistant to shock or impact, but it does not account for the severe breakdown or deterioration of the film when exposed to exterior conditions. It would appear that such breakdown could be caused only by a rupture of some of the primary valence linkages which were produced during film formation.

Rupture of Primary Valence Bonds. The ether linkage, described on page 112, is believed to be subject to autoxidation with the formation of hydroperoxides. The decomposition of these hydroperoxides may lead to rupture of the ether linkage and formation of hydroxy compounds, ketones, and hydrogen peroxide according to the following scheme:

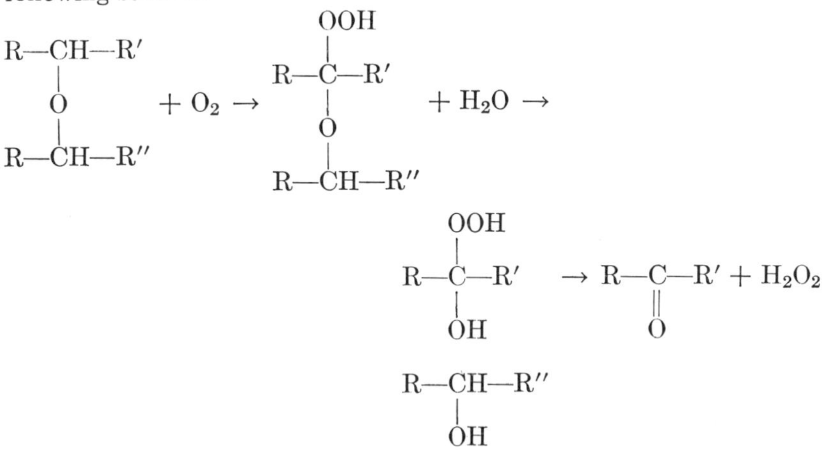

The above reactions lead to film breakdown because of the rupture of one of the cross-links (the ether linkage). The reactions produce hydrogen peroxide, which has been found in drying oil films, and also hydroxy and ketonic groups.

The peroxide linkage, described on page 117, may also be considered to break down, particularly under exposure to sunlight, to produce hydroxy and ketonic groups according to the following scheme:

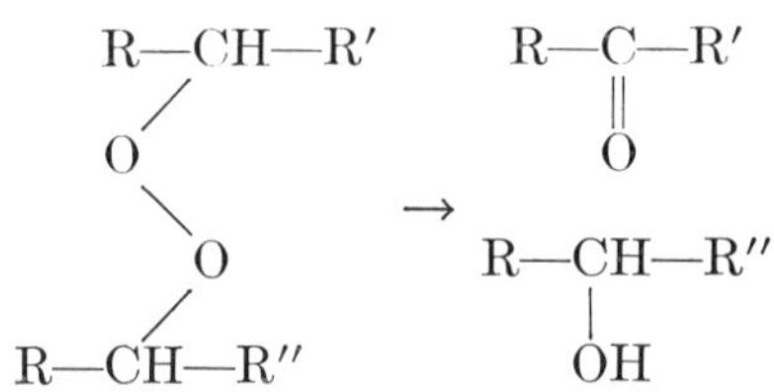

It is unlikely that normal carbon-carbon linkages would be broken during the autoxidative deterioration of oil films since these linkages are very stable, as shown by the durability of various hydrocarbons.

Short-Chain Decomposition Products. The cyclic peroxide, described on page 112, may break down to form two molecules of aldehyde with the rupture of a primary valence bond at the location of the original double bond. This reaction may be visualized as follows:

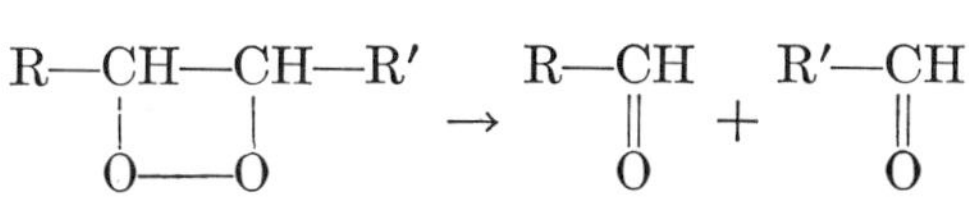

Several investigators have reported the presence of aldehyde in the oxidation-decomposition products of drying oil films. The rupture of the primary valence bond in this case differs from the two previous examples in that the ruptured bond does not connect two different oil molecules. However, the position of the cyclic peroxide in the fatty acid radical determines whether a film rupture or a short-chain radical is produced from the scission reaction. This may be illustrated schematically in Fig. 23, in which (*a*) represents the scission reaction with-

Fig. 23. Oxidative Decomposition of Oils

(a) No Breakdown of Film, Short–Chain Compound Formed

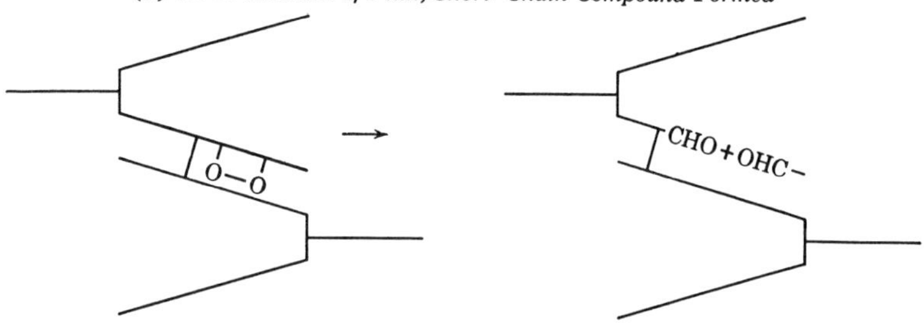

(b) Breakdown of Film

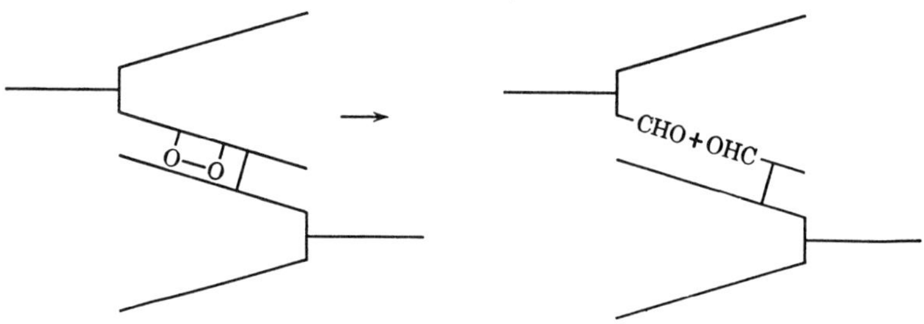

out film breakdown but with the formation of a short-chain aldehyde which may oxidize to form a short-chain fatty acid such as caprylic or caproic. The scission reaction represented by (*b*) involves film breakdown since two oil molecules are separated by rupture of a primary valence bond at the location of the original double bond.

It will be appreciated that rupture of primary valence bonds, as indicated in the last three examples, will result in pronounced deterioration of the film. This may manifest itself as film checking, followed by cracking, or it may produce surface chalking and erosion. These phenomena are discussed in further detail in Volume II in connection with the aging of oil paints for exterior surfaces.

Other Oxidation Products. In addition to the aging reactions which produce rupture of primary valence bonds, there are other reactions which cause chemical changes in the film. These reactions are made possible because the dry film still contains considerable unsaturation and some hydroperoxides, and it is permeable to oxygen, water, and light. Under these conditions a double bond may be oxidized to produce two hydroxy groups, as in the mild oxidation of oleic acid to give dihydroxy stearic acid; also oxygen may add at a double bond or to an adjacent active methylenic group to produce a hydroperoxide. The hydroperoxide group may change to a keto group by loss of a molecule of water, or it may change to a hydroxy group by loss of oxygen. There may also be some isomerization from non-conjugated to conjugated unsaturation.

It will be observed that all the aging reactions referred to result in the production in the film of four types of chemical groupings: hydroxy, carboxy, ketonic, and aldehydic. These groupings would not produce severe breakdown of the film structure, but they would make it more sensitive to moisture permeation because they are highly polar. The carboxyl groups would lower alkali resistance, and the short-chain fatty acids may exert a softening or peptizing action on the gel structure of the film. The keto groups are believed to be connected with the yellowing of the film.

Yellowing of Oil Films on Aging. The yellowing of oil films on aging assumes considerable importance with respect to the production of architectural white paints for interior use. As a result, a great deal of study has been devoted to the fundamental cause of yellowing. The complete answer to this problem has not been obtained to date, but several points seem to be fairly well established. These include the effect of light, the effect of the type and amount of unsaturation, and the effect of keto groups.

It has been observed that little or no yellowing occurs when the film is exposed to direct sunlight or to a source of ultraviolet radiation. Medium yellowing occurs in diffuse light as obtained with ordinary interior room conditions, but severe yellowing occurs in the absence of light. The yellowing is accelerated by temperatures above normal and by high humidity. This suggests that the reactions which are promoted by sunlight (those which lead to the rupture of primary valence bonds) prevent the formation of the chromophoric groups necessary for color. Evidence regarding the production of chromophoric groups may be obtained from the type of unsaturation required to produce maximum yellowing. It is well known that soybean oil films yellow only slightly, whereas linseed oil films yellow badly. Table 4 shows 2.3% linolenic acid in soybean but 45% linolenic acid in linseed oil. It has also been established that tung oil films yellow less than linseed oil films. Both oils have triple unsaturation, but the tung is conjugated and the linseed is non-conjugated. From this it

Fig. 24. Possible Chromophores from Linolenic Radicals

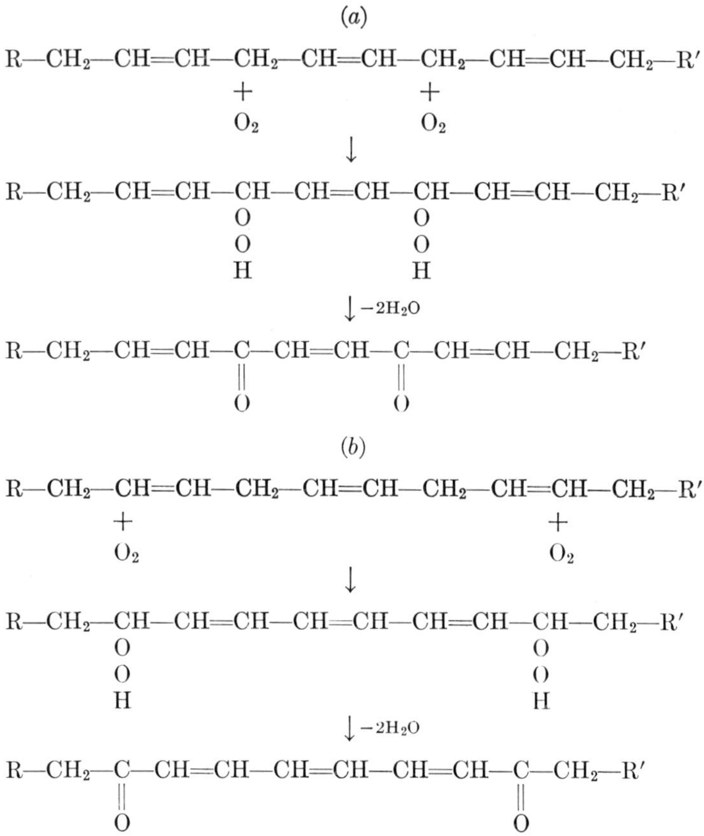

Phorone, a Related Yellow Compound

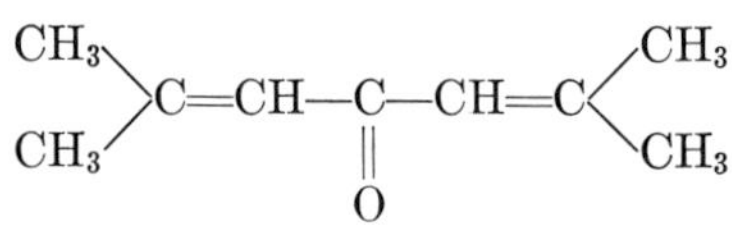

seems most probable that that some modification of the linolenic structure is responsible for the yellowing phenomenon.

According to the Witt theory of color formation a colored molecule must contain one or more chromophoric groups (>C═C<, >C═O, etc.). The color is developed further if a conjugated system is formed with the possibility of resonance. The possibility that such systems could be formed by the oxidation of linolenate unsaturation is indicated in Fig. 17, but the emphasis there is on hydroperoxide formation and subsequent cross-linking. However, because of the large percentage of linolenic radicals in linseed oil it is unlikely that all the hydroperoxides would result in peroxide or ether linkages as shown on page 112. The molecules in the drying film would become immobilized by the gel structure being formed before all of them could cross-link as indicated. Therefore, by continued oxidation and splitting off of water, conjugated chromophoric systems could be formed having the possibility of resonance. A possible mechanism for these reactions is indicated in Fig. 24.

In a part of the extensive investigation of yellowing which has been reported by Elm (48) the compound phorone (Fig. 24) was shown to have a yellow color. From this fact it was deduced that the structures in Fig. 24 would have a deeper yellow or orange color. This work is being continued by Elm at the New Jersey Zinc Co. laboratory, and it may be expected that the results will be of great value in understanding the problem of yellowing.

Another factor which is of significance with respect to color formation is the effect of drier metals, lead, cobalt, and manganese. It may be expected that metallic complexes could be formed, as in the copper phthalocyanines, to give colored compounds. For example, manganese drier causes more yellowing than either lead or cobalt in white finishes, indicating that the metals at least modify the color to some extent.

The use of oils in varnishes and synthetic resins is discussed in later chapters, and their use in paints is included in Volume II.

BIBLIOGRAPHY

1. E. H. Farmer, *J. Oil & Colour Chemists' Assoc., 31,* 394 (1948).
2. T. P. Hilditch, *The Chemical Constitution of Natural Fats,* second ed., p. 15, John Wiley & Sons, 1947.
3. I. M. Bernstein, *J. Polymer Sci., 1–6,* 495 (1946).
4. J. C. Konen, *Official Digest Federation Paint & Varnish Production Clubs, No. 284* (September, 1946).
5. R. D. Bonney in *Protective and Decorative Coatings,* J. J. Mattiello, ed., John Wiley & Sons.
6. I. J. Duncan, *J. Am. Oil Chemists' Soc.,* August, 1948, p. 277.
7. J. J. Mattiello and L. T. Work, *Natl. Paint, Varnish Lacquer Assoc, Circ. 502* (March, 1936).
8. O. M. Behr, *Official Digest Federation Paint & Varnish Production Clubs, No. 180* (April, 1940).

9. Pittsburgh Plate Glass Co., *Paint, Oil, Chem. Rev.*, Oct. 2, 1947.
10. R. L. Demmerle, *Ind. Eng. Chem., 39,* 126 (1947).
11. W. J. Dann and T. Moore, *Biochem. J., 27,* 1166 (1933).
12. W. J. Dann and T. Moore, *Biochem. J., 29,* 138 (1935).
13. T. Moore, *Biochem. J., 31,* 138 (1937).
14. J. P. Kass, E. S. Miller, and G. O. Burr, *J. Am. Chem. Soc., 61,* 482, 3292 (1939).
15. G. O. Burr, U. S. Patent 2,242,230, May 20, 1941.
16. T. F. Bradley and D. Richardson, *Ind. Eng. Chem., 34,* 237 (1942).
17. J. C. Cowan *et al., Ind. Eng. Chem., 38,* 997 (1946).
18. J. C. Cowan *et al., Ind. Eng. Chem., 38,* 1002 (1946).
19. J. C. Cowan *et al., Ind. Eng. Chem., 41,* 294 (1949).
20. F. Kraft, *Ber., 21,* 2731 (1888).
21. S. J. Fokin, *Russ. Phys. Chem. Soc, 46,* 224, 1027 (1914).
22. J. Scheiber, Br. Patent 306,452 (1928) [*Farbe u. Lack, 153* (1929)]; Ger. Patent 513,540 (1930); U. S. Patents 1,979,495; 1,942,778 (1934).
23. H. Ufer, Ger. Patent 529,557 (1931); U. S. Patent 1,892,258 (1932).
24. I. M. Colbeth, U. S. Patent 2,382,119 (1946).
25. S. B. Radlove, W. De Jong, and L. B. Falkenburg, *J. Am. Oil Chemists' Soc.*, August, 1948, p. 268.
26. A. C. Elm, *Paint, Oil, Chem. Rev.*, June 24, 1948, p. 11.
27. H. J. Passino, *Ind. Eng. Chem., 41,* 280 (1949).
28. W. H. Goss, *Ind. Eng. Chem., 40,* 2247 (1948).
29. T. P. Hilditch, *J. Oil & Colour Chemists' Assoc., 32,* 5 (1949).
30. C. P. A. Kappelmeier, *Paint, Oil, Chem. Rev.*, Jan. 6, 1949, p. 22.
31. C. P. A. Kappelmeier, *Paint, Oil, Chem. Rev.*, July 24, 1947, p. 5.
32. O. Diels and K. Alder, *Ann., 460,* 98 (1928).
33. E. T. Clocker, U. S. Patents 2,188,882; 2,188,890 (1940); 2,262,923 (1941); 2,275,843 and 2,285,646 and 2,286,466 (1942).
34. E. H. Farmer, *J. Oil & Colour Chemists' Assoc., 31,* 393 (1948).
35. E. H. Farmer, D. A. Sutton, and H. P. Koch, *J. Chem. Soc., 541* (1943).
36. T. P. Hilditch, *J. Oil & Colour Chemists' Assoc., 30,* 1 (1947).
37. R. F. Paschke and D. H. Wheeler, *J. Am. Oil Chemists' Soc., 26,* 278 (1949).
38. D. H. Hewitt and F. Armitage, *J. Oil & Colour Chemists' Assoc., 29,* 109 (1946).
39. F. Armitage, D. H. Hewitt, and J. J. Sleightholme, *J. Oil & Colour Chemists' Assoc., 31,* 437 (1948).
40. N. R. Peterson, *Am. Paint J.*, June 28, 1948, p. 32.
41. N. R. Bhow and H. F. Payne, *Ind. Eng. Chem., 42,* 700 (1950).
42. Schwarcman, U. S. Patents 2,207,686 and 2,230,470.
43. S. O. Sorenson and J. C. Konen, U. S. Patent 2,213,935.
44. H. G. Berger and G. Rock, U. S. Patent 2,478,451 (1949).
45. O. R. Sweeney and L. K. Arnold, *J. Am. Oil Chemists' Soc., 26,* 697 (1947).
46. R. L. Kenyon, N. F. Kruse, and S. P. Clark, *Ind. Eng. Chem., 40,* 186 (1948).
47. D. H. Wheeler, *Ind. Eng. Chem., 41,* 252 (1949).
48. A. C. Elm, *Ind. Eng. Chem., 41,* 319 (1949).
49. Franke and Monch, *Ann., 556,* 1, 200 (1944).
50. E. H. Farmer and D. A. Sutton, *J. Am. Chem. Soc., 119* (1943).
51. Bolland and Koch, *J. Am. Chem. Soc., 445* (1945).
52. T. F. Bradley and W. B. Johnson, *Ind. Eng. Chem., 33,* 86 (1941).

53. P. Slansky, *Paint Manuf.*, *27*, 5, 144 (1947).
54. H. F. Payne and W. H. Gardner, *Ind. Eng. Chem.*, *29*, 893 (1937).
55. L. Auer, *Ind. Eng. Chem.*, *30*, 466 (1938).
56. L. Auer, *Paint Manuf.*, *18*, 2, 41 (1948).
57. H. M. Hoogsteen, A. E. Young and M. K. Smith, *Ind. Eng. Chem.* *42*, 1587 (1950).
58. J. Rinse and C. Korf, *J. Oil & Colour Chemists' Assoc.*, *32*, 593 (1949).
59. D. S. Bolley, *Ind. Eng. Chem.*, *41*, 287 (1949).
60. W. G. Bickford, P. Krauczunas, and D. H. Wheeler, *Oil & Soap*, *19*, 22 (1942).
61. C. P. A. Kappelmeier, *Paint, Oil, Chem. Rev.*, Aug. 31, 1950.
62. J. P. Gibbons and D. M. Gordon, *Ind. Eng. Chem.*, *42*, 1591 (1950).
63. C. P. A. Kappelmeier, *Paint, Oil, Chem. Rev.*, *16* (Feb. 1, 1951).
64. Tall Oil Assoc., 122 East 42nd St., New York 17, N. Y.
65. Arizona Chemical Co., 30 Rockefeller Plaza, New York 20, N. Y.
66. Taylor *et al.*, U. S. Patent 2,487,000 (1949) to Monsanto Chemical Co.
67. D. H. Wheeler, R. J. Foster, A. P. Berry, *Am. Paint. J.*, *37*, 24 (March, 1953).

The ability to analyze problems, and to synthesize answers, is the acme of intellectual achievement.

Think about problems . . . don't gripe!
Work out the answers . . . don't mope!

Chapter 3. VARNISH RESINS

RESINS AND THE RESINOUS STATE

In the preceding chapter on oils for organic coatings a simple definition for an oil was given because oils are the triglycerides of long-chain fatty acids. No such simple definition is possible for a resin, since resins vary over a very wide range of chemical compositions and physical properties. However, it will be helpful to define the resinous state because it shows the limiting conditions to which a material must conform if it complies with the requirements for a resin.

Conditions for the Resinous State.

(*a*) Resins may range in physical condition from hard, brittle solids to non-volatile viscous liquids.

(*b*) Resins soften on heating without showing a sharp melting point, indicating a non-crystalline condition.

(*c*) Resins may be non-convertible or convertible as defined in previous chapters.

(*d*) Resins form continuous films on cooling from a melt, or after evaporation of the solvent when applied in a relatively thin layer from a solution.

(*e*) Resins are chiefly organic in composition.

(*f*) Resins are clear mixtures of mutually soluble materials. A turbid mass indicates the presence of an insoluble component.

(*g*) Resins are soluble in organic solvents, but there are a few exceptions as noted in the text.

It will be apparent from the above conditions that a great many organic compounds could be made which would be classed as resins. In some cases the cost of producing these compounds would be much higher than that of present resins, and since their properties would not be much superior, they do not appear on the market. Certain resins, such as the alkyds, may be used as the sole binder for clear or pigmented finishes. Some hard resins, like rosin and its modifications, are used in combination with oils or high polymers because alone they produce weak or brittle films. On the other hand some of the high-polymer resins are extremely tough solids and require plasticizers to increase their flexibility, and addition of another type of resin to improve the gloss and adhesion of their films.

Softening Point. The temperature at which a resin will soften is an important characteristic because it gives some indication of the drying speed of the resin and the type of film which will be formed from it. The methods for determining softening point are given in Chapter 15. They are arbitrary methods because resins do not give the sharp melting points obtained with crystalline materials. In general, materials which crystallize from a melt or solution do not form strong films.

Resins may be thermoplastic or the non-convertible type, or they may be thermosetting or the convertible type. In Chapter 1, Fig. 4, it is shown that compounds formed from reactions having 1:1 or 2:1 functionalities are liquid, non-resinous materials. The 2:2 reactions produce resinous materials which are the non-convertible type. It is required that one of the reactants have a functionality greater than 2 in order to produce the convertible-type resins. The degree of polymerization is also a factor in resin properties, but it will be helpful to keep the functionality concept in mind when considering the types of films to be expected from resinous compositions.

Solubility. Resins are almost entirely organic in composition, and even those few resins which are partly inorganic are generally soluble in organic solvents. The silicone resins, for example, are based on a repeating unit of $\begin{array}{c} R \\ | \\ -Si-O- \\ | \\ R \end{array}$, but the R's are organic radicals and impart solubility in organic solvents. Similarly, limed rosin and zinc resinate are partly inorganic, but the large rosin radical determines the solubility characteristics. The relative polarity criterion referred to in Chapter 1 can be applied to the solubility of resins. For example, rosin is soluble in ethyl alcohol because rosin contains the highly polar carboxyl group. When this group is neutralized with lime or glycerol the resulting resins are no longer soluble in alcohol but dissolve in the non-polar hydrocarbons.

It is of interest to compare the solubility behavior of resins and some inorganic materials because the comparison points out a particular characteristic of the resinous state. When a crystalline inorganic salt is dissolved in water the viscosity does not increase appreciably with increase in salt concentration. However, when a specific concentration is reached the salt crystallizes out. It also comes out as crystals when a salt solution is evaporated to dryness. By contrast, a resin solution increases in viscosity as the resin concentration is increased. In very high resin concentrations the "solution" is so viscous that it assumes the physical characteristics of the resin rather than of the liquid, volatile solvent. This condition applies particularly to high-polymer resins, and it may be more nearly correct to say in such cases that the resin is then the "solvent" for the volatile constituent. This may appear to be a purely academic viewpoint, but it has a great deal of practical significance. If the non-volatile resin is the solvent it will tend to hold the volatile constituent very strongly and retard its evaporation. This will slow the drying rate of the film and produce an "after-tack." In some cases it will prolong the time before the finished article can be packed for shipment. Such films are said to have strong "solvent retention" and bad "print resistance" because the paper or other wrapping material will adhere to the finish if the article is packed in the normal schedule. Usually a resin which is less soluble in a solvent will release the last few per cent of solvent, or will have better "solvent release," than a more soluble resin. Advantage is often taken of this fact to improve the drying rate of lacquer films, but, of course, the solubility cannot be reduced to the point where actual separation occurs because of the adverse effects on film strength and gloss. In Chapter 12 under Dispersion Resins it will be found that a non-solvent is used as the volatile component and rapid "solvent release" is obtained. However, in this case the resulting film is weak and must be fused to make it completely coherent and to develop its full strength.

Some resin solutions may be diluted with the solvent indefinitely, similarly to a salt solution. However, this is not true in all cases. A resin may become insoluble and separate out as a gummy mass when diluted beyond a certain minimum concentration. The amount of dilution immediately preceding resin separation is known as the solvent tolerance of the resin. A number of examples are given in the discussion of the various types of resins.

These variations in the behavior of resin solutions seem to indicate that true solution, or complete molecular dispersion, is obtained only in cases of low-molecular-weight resins in strong solvents. In the majority of solutions the resin molecules may be considered as being solvated or dispersed in the solvent as agglomerates. The size of these agglomerates and the relative proportion of resin and solvent molecules in them will depend on the secondary valence forces operating in their surfaces as discussed in Chapter 1. A further discussion of resin solubility is given in Chapter 6 on solvents.

Compatibility. In a great many instances more than one resin is used in the formulation of an organic coating. Such resin combinations must not only be soluble in the volatile solvent but also be mutually soluble or compatible with themselves. Solubility of the resin combination in the volatile solvent is indicated by the clarity or freedom from turbidity of the solution. Mutual solubility or compatibility of the resin combination is determined by pouring some of the solution on a glass panel and allowing the volatile solvent to evaporate. If the remaining film of the resin combination is clear and transparent, the resins are compatible, but if the film is turbid or tends to separate, the resin combination is incompatible.

To say that a resin combination is compatible in the solid state is another way of saying that the materials are soluble in the solid state. Therefore, the same criteria for solubility will be applicable to compatibility as stated previously: similar polarity and molecular weight. For example, it would not be expected that highly polar resins would be compatible with non-polar resins. Also the lower-molecular-weight materials usually have better compatibility than those of higher molecular weight, but no general rule for this can be stated. For example, low-bodied oils are usually compatible with high-bodied oils, but low-viscosity nitrocellulose is not compatible with high-viscosity nitrocellulose. A simple varnish resin like ester gum may be cooked with either low- or high-bodied oils in the varnish-making process. However, more complex resins, such as the pentaerythritol esters of maleic- or phenolic-modified rosin, may not be compatible with bodied oils. Therefore they are solubilized first in an unbodied varnish oil; several examples will be given later.

Two or more materials which are compatible in a specific ratio may not be compatible in all possible ratios. Therefore the compatibility is often determined at the extremes and middle ratios such as 90:10, 50:50, and 10:90. Materials which are compatible in all proportions are, of course, more soluble in each other than those having limited compatibility, and they frequently form better films.

Resins and Gums. An attempt has been made to distinguish between resins and gums by using the following definition:

(*a*) Resins are soluble in organic solvents and melt when heated.

(*b*) Gums are soluble in water and char when heated.

This would be a very convenient differentiation because there are available these two types of resinous materials. (Before the introduction of synthetic resins the natural resins were called gums by the gum importers and varnish makers.) However, certain names have become so well established in the trade terminology that the definition will not be used completely. For example, the glycerol ester of rosin will always be known as ester gum despite its solubility in organic solvents and insolubility in water. Other examples may be found in the literature. There are some water-soluble resinous materials which comply with the above definition, such as gum arabic and gum tragacanth.

Natural and Synthetic. Resins may be classified under the two general categories of natural resins and synthetic resins. Under natural resins should be included only those which are used in the preparation of coatings without being modified by chemical processes. This restriction would not include the physical processes which remove foreign matter, volatile constituents, water, etc. Synthetic resins would then include any chemical modification of natural resins or any resinous material made from non-resinous chemicals. The types, sources, and characteristics of the various resins are given in later chapters under their specific headings, but a chronological table showing the approximate date of first commercial sale is given here.

Approximate Date of First Sale of Resins

	Natural resins *	1933	Vinyl copolymers
1890	Ester gum	1935	Ethyl cellulose
1910	Modified phenolic	1935	Cellulose acetopropionate
1910	Coumarone-indene	1935	Cellulose acetobutyrate
1917	Vinyl acetate	1935	Petroleum resins
1920	Nitrocellulose	1937	Polystyrene plastics
1926	Thiourea-formaldehyde	1939	Melamine-formaldehyde
1927	Air-drying phthalic alkyds	1944	Organosilicon
1928	Oil-soluble phenolic	1946	Polytetrafluoroethylene
1929	Urea-formaldehyde	1946	Polyethylene
1930	Chlorinated rubber	1947	Polyamide resins
1931	Acrylates	1948	Styrenated oils and resins
1931	Chlorinated biphenyl		
1933	Maleic-modified rosin		

* From early recorded history.

VARNISH RESINS

Varnish resins are usually hard, brittle solids which may be cooked with oils to produce oleoresinous varnishes, or dissolved in volatile solvents to produce spirit varnishes. The resin solutions may be used in that form to produce clear films on paper, wood, or other materials. The resin solutions may also be added to oleoresinous varnishes, nitrocellulose lacquers, etc., to increase the resin content of these coating materials. Oleoresinous varnishes are refererd to as "short-," "medium-," or "long-" oil varnishes, depending on the number of gallons of oil used with 100 lb of resin. The usual range is 6–18 gal for short-oil varnishes, 18–30 gal for medium-oil varnishes, and 30–60 gal for long-oil varnishes. The rate of drying, and the physical and chem-

Table 1 General Characteristics of Varnish Resins

Natural resins: This classification does not include rosin and shellac. Natural resins are somewhat uncertain in supply and have variable composition. They find use because of certain specific properties or low cost but have been replaced by synthetics for many applications.

Rosin: Rosin is too soft and slow drying as a varnish resin, but its esters and other derivatives are widely used.

Ester gum: Ester gum is the glycerol ester of rosin. It is used in low-cost varnishes, but it is relatively soft and has poor water and alkali resistance and exterior durability.

Maleic resins: These represent a variety of maleic-modified rosin esters which are harder and paler in color than ester gum. They are intended chiefly for interior finishes and low-cost industrial baking finishes.

Modified phenolic resins: These contain phenol-formaldehyde condensates which are not soluble in drying oils without modification with rosin and subsequent esterification with glycerol. They make fast-drying varnishes with better water and alkali resistance than ester gum or maleic resins. They may be used for exterior finishes if the varnishes are 40 or more gal in length. They have poor color and are not generally used in white finishes.

100% phenolic resins: These are substituted phenol-formaldehyde condensates which are heat soluble in drying oils. They are available in both heat-reactive and heat-non-reactive types. They have excellent water and alkali resistance and good exterior durability. They have fair color but are not satisfactory for high-grade white finishes.

Terpene resins: These resins are polymers of α-pinene and other terpenes which are thermoplastic and have very good water and alkali resistance. They are used chiefly for interior finishes.

Terpene-phenolic resins: These are mixed terpene-phenolic condensates with very good water and alkali resistance and fair color. They are lower in cost than comparable phenolics, and are used chiefly in interior finishes.

Coumarone-indene resins: These are polymers and copolymers of coumarone and indene obtained from the light oils in the by-product coke-oven process. They have excellent alkali resistance but poor color and exterior durability. They have low acid values and are non-reactive with oils. They are used in alkali-resistant finishes, aluminum paints, and ship-bottom paints.

Petroleum resins: These resins are polymers of unsaturated hydrocarbons obtained in the cracking of petroleum oils. They have good solubility and compatibility with drying oils, good resistance to water, alcohol, and alkali, and low acid values. They are sensitive to petroleum hydrocarbons and have only fair color.

ical characteristics of varnishes and their films, depend primarily on the type of resin and oil, and on the ratio of resin and oil or "oil length." Of course, the proper cooking procedure for the various resins and oils must be followed carefully to obtain the best results.

The majority of oleoresinous varnishes are soluble in mineral spirits, but some short-oil varnishes require a proportion of aromatic solvent to ensure stable solutions. The initial drying or "setting" time can be shortened by use of faster-evaporating solvents such as VM & P naphtha or toluene. The term "spirit varnish" was applied originally to solutions of certain resins in alcohol, but the term is now generally used for resin solutions in any solvent. Some of the varnish-resin solutions may also be used in nitrocellulose lacquers, alkyd resin finishes, and other coatings, as described later.

The most common types of varnish resins are described individually in this chapter, and their use in varnishes is discussed in the following chapter. The various types and their general characteristics are given in Table 1.

NATURAL RESINS

The group of resins known in the trade terminology as natural resins are listed in Table 2 with their places of origin and general characteristics. It will be noticed that the list does not contain the two important natural resins, shellac and rosin. This arbitrary separation of shellac and rosin is due to their being found in different geographic locations and consequently being marketed by different organizations from those handling the other so-called natural resins. However, shellac and rosin have sufficient differences in physical properties and applications in industry to justify separate consideration.

Origin of Natural Resins. Natural resins are exudations from trees. The resinous material may exude naturally, or the trees may be "tapped" to increase the rate of exudation and the amount of resin collected. Some of the natural resins may be obtained as fossil, semi-fossil, and recent types. The term "recent type" refers to the resin obtained by tapping the tree and collecting the resin. The term "fossil type" refers to the resin obtained by digging it out of the ground. The fossil types exuded from the trees many years ago and became buried in the natural course of events. The fossil grade has been buried for a longer time than the semi-fossil grade and is harder, more oxidized, and therefore somewhat different in chemical composition, hardness, and solubility.

The resinous material occurs in pockets or passages between the cells of the tree structure; it is not the sap of the tree. It is considered by some authorities to be a foreign matter which the tree exudes to prevent injury to itself. The resinous material which exudes from the tree is a mixture of organic substances some of which are volatile and others subject to oxidation. The loss of volatile constituents and the absorption of oxygen harden the material considerably so that the majority of the natural resins are hard and somewhat brittle solids.

It will be noted in Table 2 that there are also some softer resins often referred to as "balsams." The resinous exudations usually contain a small percentage of water which may be lost by subsequent evaporation or may become occluded in the resin. Occluded water makes the resin turbid or opaque and may cause bad foaming during heat processing in the varnish kettle.

Characteristics of Natural Resins. The collecting of natural resins in tropical countries provides one of the few sources of income for the natives. For this economic reason the European countries which have colonies in the tropics encourage the natural resin industry. They provide headquarters for sorting, cleaning, and grading the resins, and they set up definite standards which are adhered to within the limitations of natural products. The important characteristics of the natural resins, from the standpoint of their application in the coating industry, are their type of solubility and their hardness or melting range. The acid value has some significance when the resins are used without heat processing, but it may be changed considerably by the high temperatures required for varnish making. A simplified classification of the resins is given in Table 2. It is intended to differentiate among those which are cooked into oleoresinous varnishes, those

Table 2 Origin and Characteristics of Natural Resins

Resin	Origin	Melting Range, °F	Acid Value	Solubility Alc *	Solubility Hy *	Use
Congo	Belgian Congo	210–350	100–120	In *	In	Cook in varnish
Congo ester †	Esterified Congo	240–280	5–10	In	S *	Cook in varnish
Kauri	New Zealand	220–280	60–80	In	In	Cook in varnish
Pontianak	Borneo	200–275	120–140	S	In	Cook in varnish
Manila	East Indies	190–250	110–140	S	In	Cook and cold-cut ‡
Boea	East Indies	175–250	120–150	S	In	Cook and cold-cut
Batu	East Indies	250–290	20–40	In	S	Cook and cold-cut
East India	East Indies	200–280	20–40	In	S	Cook and cold-cut
Dammar	Batavia	160–240	25–35	In	S	Cold-cut ‡
Dammar	Singapore	160–240	25–40	In	S	Cold-cut
Accroides	Australia, Africa	170–250	120–130	S	In	Cold-cut
Sandarac	Australia	200–260	120–150	S	In	Cold-cut
Mastic	Island of Chios	130–160	50–70	S	S	Cold-cut
Elemi	Philippines	Balsam	20–35	In	S	Cold-cut

* Alc = alcohol; Hy = hydrocarbons; In = insoluble; S = soluble.
† Included for comparison only.
‡ Cold cut: dissolved in solvent, either cold or by mild heating.

which may be cooked and also "cold-cut" or put into solvent solution for use in spirit varnishes and lacquers, and those which are cold-cut only and used in adhesives and spirit varnishes. The values for melting range and acid value have been chosen arbitrarily to indicate the general range, but, of course, it is possible to obtain grades with closer specifications. It was not considered necessary to include all the details of the types available in this book, since the values may change with market conditions, and the most recent information can be obtained directly in the trade literature.

Chemical Composition. Although a great deal of analytical work has been done, particularly in Europe, there is considerable disagreement regarding the chemical composition of the natural resins. With the development of synthetic resins and the consequent knowledge of resin-forming materials there is less incentive to determine the exact composition of the natural resins. They are certainly of complex rather than simple organic composition. They contain a variety of aromatic acids and various amounts of unsaponifiable material, as well as both saturated and unsaturated compounds related to the resenes and the terpenes.

Uses for Natural Resins. Evidence of the use of natural resins can be found in the earliest recorded history. Usually it was the softer resins which were used because application was made by heat-softening the resin and then applying it to an object by smearing with the fingers or a piece of hot metal. Although turpentine was well known to the Romans there is no indication of its use as a solvent for resins or varnishes until about 1750. The chronological table on page 136 shows that ester gum, the first modification of a natural resin, was used about 1890 and the modified phenolic resins about 1910. Two statements by Dr. H. H. Sabin in his translation of Professor Max Bottler's *German Varnish Making,* published by John Wiley & Sons in 1912, are of interest because they show the attitude of the varnish makers of that period in contrast to those of today. Dr. Sabin states: "Considerable attention has been given in late years to attempts to make synthetic resins; so far as the writer knows these have been intended only for spirit-varnishes; also, in spite of numerous statements to the contrary, they have not proved valuable nor have they come into use except surreptitiously as adulterants for shellac and similar resins." Despite his evident distrust of synthetics Sabin pays tribute to the "inventions" of Dr. L. H. Baekeland as follows: "It should be said that these products are chiefly used in the solid state, as insulating material and for the manufacture of other solid articles, and not as resins; their semi-resinous character leads people to attempt their use in varnishes, and it is by no means unlikely that in time useful preparations of this sort will be made." Today we can state safely that greater advances in resin development have been made since those words were written in 1912 than in all previous history. Despite the developments in synthetic materials the natural resins are still used in considerable quantity because of their low cost and certain good properties.

The column headed "Use" in Table 2 is intended to indicate in a general manner the resins which are cooked with oil to produce oleoresinous varnishes, the resins which are both used in oleoresinous varnishes and cold-cut to make spirit varnishes, and also the resins which are used only as cold-cut solutions. The hard resins are partially broken down by a heat treatment before they can be used with drying oils for varnish making. The heat treatment is known as gum running.

Gum Running. The hard natural resins, congo, kauri, and some grades of pontianak, cannot be solubilized in drying oils by simple heating with the oil as is common practice with synthetic varnish resins. They are first heated to a high temperature which breaks down the resin structure to some extent, and from 10% to 30% is distilled off. The hot oil is then added as required for the various types of oleoresinous varnishes. Congo resin is the most widely used today, since kauri and pontianak are difficult to obtain. When the resin is "run" with the minimum amount of breakdown required to make it soluble in the oil, it is referred to as a "slack-melt" resin. There is a weight loss of 10–20% in a slack melt. A "fine-melt" resin is run to the condition where it is freely soluble in the oil. The weight loss in a fine melt varies from 20% to 30%. Typical procedures for running congo resin are given below.

Procedures for Running Congo Resin. The required amount of congo is placed in a varnish kettle and the lid attached. The lid has a vent for the fumes and holes for thermometer and stirrer. The temperature is raised to 580–600°F as rapidly as the heavy foaming will permit. The batch is allowed to "sweat" for 10–15 min, then reheated to about 650°F. The resin should not be stirred during this heating period, but the condition of the melt may be determined by "feeling" with the stirrer and by removing small samples with the stirrer. At first the original lumps of resin swell considerably and become gelatinous masses. Continued heating converts these gelatinous "floaters" into a liquid state. The slack-melt condition is usually indicated by complete conversion to this liquid state. Continued heating reduces the viscosity of the liquid, and the fine melt is indicated by very free-flowing drops from the end of the stirrer when removed from the kettle for sampling. The time required for obtaining a slack-melt congo will vary from 15 to 45 min at 650°F, depending on the grade and quality of the resin. A fine melt may require as long as 60 or 80 min at 650°F.

It is not practical to give a precise procedure for running natural resins because of the variation in properties among different grades and shipments. The varnish maker soon recognizes these differences and makes the necessary changes in temperature, rate of heating, and time required for the type of melt he desires.

Natural Resin Varnish. After the resin is melted properly the lid of the kettle is removed and the required amount of drying oil is added. The oil should be preheated to about 400°F and added to the melted resin slowly and with good stirring. Cold oil may chill the batch and prevent good solubility of the resin in the oil. When all the oil has been added to the resin, the batch is heated to the required temperature for the type of oil being used and held long enough to obtain complete solubility and the desired viscosity. The types of natural resin varnishes and details of their manufacture are given in the section on oleoresinous varnishes in Chapter 4.

Solubilizing by Mastication. Dr. William Krumbhaar (5) devised a mastication method for solubilizing certain of the hard natural resins, known as copal resins, which is described in U. S. Patents 2,007,333; 2,101,398; and 2,110,803. The resin is forced under pressure into a Banbury mixer at a temperature just high enough to give the resin a plastic consistency. The resin structure is broken down or depolymerized by the strong shearing stress in the mixer. The mass is extruded through a small opening and flattened into thin sheets on calender rolls. On cooling, the thin sheets are broken up readily into small chip or flakes. In this condition the resin can be used for varnish making directly without the necessity of high-temperature running. The mastication process produces resins with better physical properties than resins which have been solubilized by high-temperature degradation.

The masticated copals may be used in combination with ester gum or the phenolic- and maleic-type resins. They may also be thermally combined with these resins by a gelation-degelation process described in U. S. Patent 2,471,629. The masticated copal is processed with rosin, phenolic condensate or maleic anhydride, and an excess of polyhydric alcohol which may be glycerol or pentaerythritol. The excess hydroxyl groups permit the combination to be heated to the gelatinous state without becoming thermoset. Continued heating then depolymerizes the mass and produces resins having different properties from those which would be obtained by simple combinations of masticated copals and the rosin-modified synthetics. Husen (15) describes two commercial resins of this type, Krumbhaar K-333 and K-666, and their uses as cold-cut additions to alkyd resins, and in oleoresinous varnishes and printing inks. The commercial resins are listed on page 146.

Reaction with Abietylamine or Abietyl Isocyanate. Another method for solubilizing congo resin without resorting to thermal degradation is described in U. S. Patent 2,522,584, assigned to the Hercules Powder Co. The congo resin is heated with either an abietylamine or an abietyl isocyanate at a temperature which promotes amide formation between the acid constituents of the congo and the particular amine or isocyanate used. The heating is continued until the resin has an acid number of 25 or less, is compatible with nitrocellulose, and is soluble in regular drying oils for oleoresinous varnishes.

The procedure for one of the examples cited is as follows:

	Parts by Weight
Congo resin	700
Dehydroabietylamine	360

The congo is heated slowly to 275°C, and the amine added over a period of 1 hr while the temperature is increased to 300°C. The resin is then poured and cooled.

The resin had an acid value of 13.4 and a melting point of 132°C by the Hercules drop method. It was soluble in mineral spirits, aromatic hydrocarbons, and a wide variety of organic solvents.

Congo Ester. Congo ester resins are available which may be processed with drying oils for varnish making without the necessity of thermal degradation. The congo ester resins may also be used in nitrocellulose lacquers. The congo ester is made by running the congo resin in the regular manner, and then adding about 6% of glycerol to combine with the free acids. The acid value of run congo is in the range of 70–90, and that of the congo ester is from 5 to 10. The ester may be made from pure congo resin or a combination of congo and rosin or other rosin-modified synthetics.

In general, the congo resin is run to a fine melt before esterifying with glycerol because slack-melt congo tends to gel and become insoluble when esterified. In any case considerable care must be taken to prevent gelation of the pure congo ester. The addition of 10% or more of rosin practically eliminates the tendency to gel, but it detracts somewhat from the quality of the finished ester. Usually the rosin is added after the congo is run because of the severe decomposition of the rosin at the high temperatures used for the running process.

In making the ester the congo is run at 650°F to the desired melt, and rosin added as required. The batch is cooled to 500°F, and the glycerol added slowly with good agitation. The temperature is then raised to 525–530°F and maintained until the acid value drops to the range of 5–10. The process requires from 10 to 18 hr, depending on the type of equipment. When the desired acid value is reached vacuum should be applied to the kettle to remove any free glycerol and traces of moisture. The melting point of the congo ester will vary with the grade of congo but will usually be in the range of 110–130°C. Of course, addition of rosin lowers the melting point in proportion to the amount used. When rosin is added the proper correction must be applied for the amount of glycerol required, since rosin requires almost twice as much glycerol as congo for esterification. The various congo esters are used in oleoresinous varnishes and nitrocellulose lacquers.

Boea, Batu, and East India. The other natural resins which find use in oleoresinous varnishes are Boea, Batu, and East India. Some of the harder grades of Boea require running before they are soluble in the drying oils, but Batu and East India may be used without the running process. These resins are quite dark in color, but the varnish films show some bleaching action when exposed to light. The varnishes are used in traffic paints, printing inks, oilcloth, and adhesives and for other applications where low cost is important.

Some grades of Batu and East India contain a wax-like material which produces turbidity in the varnish and imparts a dull finish to the film if the varnish is not heated too strongly during cooking. These varnishes may be used in furniture finishes to produce a "flat" or rubbed finish. Because of variations in the resins and changes in the solubility of the wax as the varnish ages it is difficult to produce with these materials flat varnishes which will have a uniform degree of dullness.

Natural Resin Spirit Varnishes. Table 2 shows that both the manila and the dammar resins are used in solution for spirit varnishes. The manila resins are the alcohol-soluble type, and the dammar resins are the hydrocarbon-soluble type. Spirit varnishes are widely used as "overprint varnishes" to develop a gloss on printed paper, such as labels and bookbindings. These varnishes also protect the printing from damage by abrasion. Spirit varnishes vary widely in composition, depending on cost and service requirements. The straight resin solutions produce high-gloss films, but they are rather brittle. The brittleness is not a serious defect, however, since the films are very thin and penetrate the coated paper to a considerable extent. Only very small amounts of plasticizers can be added because they retard the rate of drying and produce soft surfaces which may stick the coated papers together. Raw castor oil is generally used as the plasticizer with the alcohol-soluble type of resins. However, much greater strength and toughness may be obtained in paper coatings by blending the resins with ethyl cellulose or the alcohol-soluble type of nitrocellulose. The ratio of resin to cellulosic polymer may vary from 0.5 to 2.0 parts of resin to 1 part of polymer. Details of these lacquer formulations are given in Chapter 11 on cellulosic polymers.

Dammar is used in certain nitrocellulose lacquers to improve adhesion, gloss, and polishing qualities. However, dammar contains a wax-like substance which is not compatible with nitrocellulose and produces turbid solutions and cloudy films. The wax may be removed by dissolving the dammar in toluene and adding denatured alcohol until the wax precipitation appears to be complete. The batch is allowed to stand until the resin solution separates from the wax. The dewaxed resin solution is then obtained by decantation; it is soluble in lacquer solvents and compatible with nitrocellulose.

Elemi. Elemi resin is imported from the Philippines, and, as noted in Table 2, it is a soft material generally referred to as a balsam. It is not soluble in mineral spirits, alcohol, or ketones but dissolves in esters and aromatic hydrocarbons.

Elemi is used as a plasticizer in spirit varnishes and nitrocellulose lacquers. It increases the adhesion of clear lacquers to metal surfaces and is also used in adhesives, printing inks, and a variety of special compositions.

Copal-Type Synthetics. The copal-type synthetic resins are among the recent developments in varnish resins. The general method of producing these resins is described on page 143, and the process is covered by U. S. Patent 2,471,629. They have some interesting properties because of the unique method of combining natural resins and rosin-modified synthetics. The commercial resins which are available and their general uses and characteristics are outlined in Table 3.

Table 3 * Copal-Type Synthetics

Krumbhaar Resin	Type	General Use
K-333	Ester-base copal	Quick-drying enamels, textile coatings, hardening alkyds
K-666	Resin-concentrate copal	Low-cost, general-purpose varnishes and enamels
K-777	Phenol-base copal	Furniture and gymnasium finishes
K-888	Phthalic-base copal	Ink oils and adhesives

Characteristics of Copal-Type Synthetics

Krumbhaar Resin	Melting Point, °C	Acid Number	Viscosity, 60% in Xylol	Color, Rosin Stds.
K-333	142–147	10–12	O–Q	N–M
K-666	140–150	.....	Z8–Z10	N–WG
K-777	167–172	20–25	Z5–Z6	K–I
K-888	100–110	45–55	Y–Z	K–I

These resins can be handled by the usual varnish-making procedures. Their general features are indicated below.

K-333 Best results are obtained with prebodied oils of medium viscosities.

K-666 This resin concentrate should be extended with equal parts of gum rosin. Such a blend can be cooked with either unbodied or bodied straight oils at 575–585°F. With chemically treated oils and tung oil, the rosin and K-666 should be cooked together at 500–525°F before adding the oil and finishing at 575–585°F.

K-777 This resin can be cooked with either soft or hard oils at the usual temperatures.

K-888 This resin is best cooked by solubilizing it first with small quantities of oil (3–4 gal) for a clear pill before adding additional amounts of oil.

* Courtesy of Krumbhaar Chemicals, Inc.

GUM AND WOOD ROSIN

Our natural resources may be divided into two groups: the expendable and the renewable. Coal, oil, and minerals are examples of the expendable type, and products of the farm and forest are examples of the renewable type. A huge crop of rosin may be obtained each year from pine trees provided reasonable care is given to methods of collection and reforestation. The combined gum and wood rosin production in the United States for the 1949–1950 crop year was 2,023,510 drums (1). Gum rosin is obtained by tapping the living tree and collecting the exudation. Wood rosin is obtained from the stumps of pine trees by the mechanical disintegration of the stumps, and extraction of the rosin with hot naphtha.

Rosin is a relatively low-cost material which is widely used in the paint, soap, paper, and chemical industries. Because of its low melting point and high acidity it is usually modified by any one of a number of chemical processes before it is used in paint and varnish. The acidity may be lowered by esterifying with an alcohol. If the alcohol has a functionality greater than 2 the melting point will be raised. Ester gum is the trade name applied to the glycerol ester which has a melting-point range of 90–105°C, compared with about 80°C for rosin. The difference in properties between gum and wood rosin and the methods for modifying these resins are described in later sections.

Gum-Rosin Production. The pine forests of the southern states are the major source of rosin in this country. The gum rosin is collected from April until November by tapping the tree and directing the gum flow by a gutter into a cup. An improved method of tapping is described in *Farmers Bull. 1984,* issued by the U. S. Dept. of Agriculture, and by Nims and Schopmeyer (3). Instead of chipping into the wood, only the bark needs to be removed if the cut is sprayed with a 40–60% solution of sulfuric acid. The acid stimulates the flow and keeps the gum flowing for a longer time. The exudation from the tree has an average composition of:

Rosin	68%
Turpentine	20%
Water	12%

This material is steam-distilled to remove the turpentine and water. The melted rosin remaining in the still is drawn off through a filter and run directly into drums for shipping. The various rosin color grades are given in Table 4. The color will vary from pale yellow to almost black, depending on the method of handling and the type of equipment used for the distillation process.

Wood-Rosin Production. In general, it is not economical to continue the tapping operation longer than 10–15 years. The trees are then cut down and sold for lumber, and the stumps are left in the ground. Since there are many thousands of acres of these stumps, an economical method has been developed for removing them, grinding the wood to uniform size, and extracting the rosin with a solvent. The best results are obtained from stumps which have been in the ground for 10–15 years. This allows oxidation and other chemical changes to take place so that the product of solvent extraction is a mixture of wood rosin, turpentine, dipentene, and pine oil.

The ground wood is extracted with hot naphtha in a series of extractors. The extract is filtered and fractionated to remove the various volatile components, and the rosin is finally stripped of solvent in a stripping column. The crude wood rosin is dark in color, but the relatively small amount of color bodies may be removed by a selective solvent process. If the hot naphtha solution of rosin is treated with furfural, then allowed to cool, the batch will separate in two layers. The color in the fufural layer may then be removed from the rosin-naphtha layer.

Another commercial process for removing the color bodies is to pass the naphtha solution of FF rosin through a series of towers containing absorbent earth. When the earth in the towers has become saturated with color bodies it may be revivified by passing an alcohol-naphtha mixture through the towers countercurrently. The process of color removal and revivification can be repeated almost indefinitely without removing the earth from the towers. Originally, wood rosin was available only in the dark FF grade, but at present it may be obtained in all the grades shown in Table 4.

Rosin Characteristics. In most respects gum and wood rosin are alike; in some of their modifications, however, the gum rosin gives products having slightly higher melting points. Wood rosin has a somewhat greater tendency to crystallize than gum rosin, both in the solid state and from solution. Since the crystalline form has a higher melting point, difficulty may be encountered if it is used in some reactions such as making rosin size. The separation of the crystalline form from solution obviously must be avoided.

Table 4 * Rosin Characteristics

Name	Color †	Softening Point ‡ °C	Softening Point ‡ °F	Acid Value	Sap. Value	Unsaponifiable, %
Penbro	B	76	169	107	135	22.5
Regular	FF	75	167	153	164	12.0
Mary	M	76	169	162	169	9.0
Nancy	N	77	171	164	170	8.5
Window Glass	WG	78	172	166	171	8.0
Water White	WW	80	176	168	172	7.5
Rex	X	81	178	170	172	7.0

* Courtesy of Newport Industries, Inc.
† Rosin cube scale.
‡ Ring and ball method.

Tall Oil Rosin. The products from the fractional distillation of tall oil are listed in Chapter 2, Table 13. These include tall oil rosin, specifications for which were established in 1950 by the U. S. Dept. of Agriculture pursuant to the Naval Stores Act of 1923 and the Regulations thereunder. They require that "such rosin shall have the characteristic form and appearance and other chemical and physical properties normal for other kinds of rosin." The Acintol R rosins (Arizona Chemical Co.) meet these specifications and have the characteristics listed in Table 5.

Table 5 * Typical Analysis of Acintol † R Tall Oil Rosin

Rosin	Type D	Type S
Color grades, cube scale	N, WG, WW	N, WC, WW
Softening point, °C	83	75
Softening point, °F	181	167
Acid number	165	168
Saponification number	175	176
Unsaponifiable, %	6.0	4.0
Fatty acids, %	1.5	2.0
Ash, %	0.002	0.002

* Courtesy of Arizona Chemical Co.
† Trade name, Arizona Chemical Co.

These types of tall oil rosin are suitable for most applications where gum or wood rosin is used. Other types are also available, and further information regarding them may be obtained from the booklet of the Arizona Chemical Co. (16) or from the Tall Oil Assoc. (17).

The color of rosin is measured by comparison with the standard ⅞-in. glass cubes at the U. S. Dept. of Agriculture or the secondary standards of rosin cubes available from the Inspector of Naval Stores, Savannah, Georgia. (See color grades in Chapter 15.)

Rosin is soluble in a wide variety of organic solvents, including mineral spirits and alcohol. The highly polar carboxyl group in rosin is responsible for its solubility in alcohol, since the esterified product, ester gum, is insoluble in alcohol.

Composition of Rosin. Since natural products are rarely found as pure compounds we may expect rosin to be a mixture of mutually soluble materials. These consist of about 90% resin acids and 10% neutral material. The acids are monobasic and contain unsaturation, and these two characteristics are the basis of the various rosin modifications. The principal acid is abietic acid, which may be isomerized to levopimaric acid by heat.

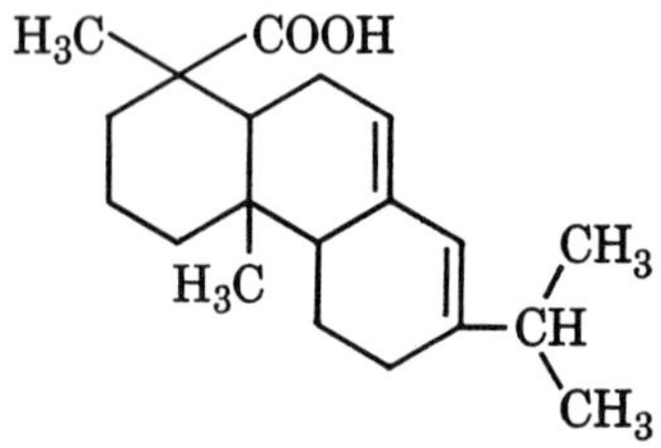

Abietic acid

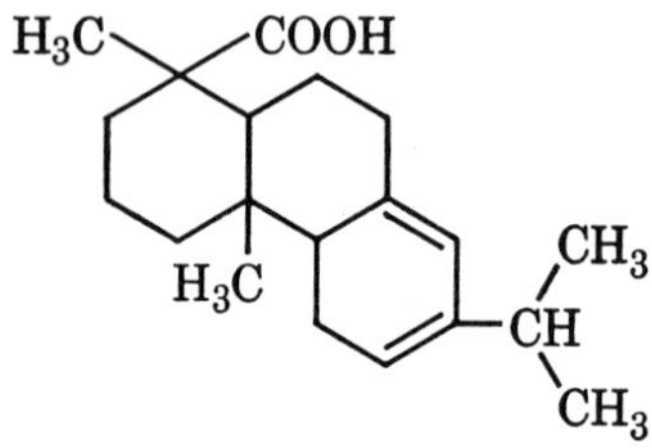

Levopimaric acid

Reactions of Rosin. It will be evident from the previous discussion of functionality and the above formulas that rosin may be used in both condensation and addition reactions. The carboxyl group may be esterified with mono-, di-, or polyhydric alcohols to give non-resinous, resinous non-convertible, and resinous convertible products respectively. Commercial products representing each of the three types are available and are described in a later section. The carboxyl group may also be neutralized with lime or zinc oxide to produce the widely used limed rosin and zinc resinate. The lead, cobalt, and manganese resinates serve as driers for oil and varnish.

Rosin and its derivatives are susceptible to change by oxidation of the unsaturation. However, the unsaturation may be eliminated by hydrogenation. The hydrogenated rosin is more stable since it is not subject to change through oxidation. Polymerized rosin is produced by an addition reaction, but the nature of the unsaturation indicates that the polymers will be chiefly dimers. The polymerizing process as described in U. S. Patent 2,263,915 is to react a solution of rosin or its ester in the presence of an inorganic chloride catalyst, then remove the solvent by vacuum distillation. A specific example was given as follows:

Ester gum	200 gm
Ethyl acetate	400 gm
Anhydrous stannic chloride	15 gm

Procedure

Maintain at 10°C until initial reaction is complete, then raise temperature to 25–30°C, and allow to stand overnight. Wash the batch with water, filter through filter-cel, and remove the solvent by vacuum distillation at 220°C and 20 mm pressure.

A Diels-Alder type of addition is possible with levopimaric acid but not with abietic acid since this reaction does not usually take place when the conjugation is in two rings. However, when the temperature is raised, abietic acid isomerizes to levopimaric, and the addition reaction takes place. The maleic-modified resins described in a later section are produced by this type of addition reaction.

The condensation products of certain phenols and formaldehyde are not soluble in drying oils, but they may be solubilized by combination with rosin. These are known as rosin-modified phenolic resins. Rosin in the form of its ester is used in conjunction with the oil-modified alkyd type of resin. In this case the rosin ester may be used alone, but more frequently the maleic-modified or the phenolic-modified types are used.

A great deal of excellent research on rosin has been done by such companies as the Hercules Powder Co. and Newport Industries, Inc., and a number of their products are used to illustrate the above reactions of rosin and its derivatives.

Limed Rosin. The high acidity of rosin may be lowered and its melting point raised by reacting it with from 4% to 7% of hydrated lime. The rosin is heated to 450°F, and the lime is sifted in slowly with constant stirring. The temperature is then raised to about 525°F, and the batch held at this temperature until a sample drop is clear after cooling. Another method (2) suggests using calcium acetate as a catalyst as follows:

Procedure

Heat rosin to 450°F;
Add 0.1% Ca acetate; sift in lime;
Add 0.1% Ca acetate; heat to 525°F;
Add 0.1% Ca acetate; hold for clear pill.

The "pill" refers to the sample drop removed from the kettle for testing. The holding time varies from 5 to 15 min, depending on the amount of lime being used.

Limed rosin is cheaper than ester gum and is preferred where low cost is important. Although it is easily made by the methods indicated above, it may be more economical to purchase it from a large producer. The purchased product has the added advantage of uniformity in lime content and freedom from unreacted lime.

When limed rosin is used in low-cost pigmented coatings a considerable increase in consistency may occur for one or more of the following reasons:

1. Any unneutralized rosin acids may react with basic pigments, such as zinc oxide, to form soaps.

2. Certain water-sensitive titanium pigments may adsorb sufficient water under humid conditions to react physically with limed rosin to produce flocculation of the pigment and a thixotropic condition in the paint.

3. Soap solutions are sometimes added to low-cost paints to improve brushing characteristics, but these solutions may react with the limed rosin to produce excessive consistency.

These conditions will be discussed in detail in Volume II. Table 6 shows the characteristics of a series of commercially available limed rosins.

Table 6 Characteristics of Limed Rosins

Trade Names *	Color †	Melting Point ‡ °C	Melting Point ‡ °F	Acid No.	% Lime
Rosin §	WW	79	174	168	...
Penbro 70	B	95	203	80	2.5
Double	FF	95	203	111	2.9
Fosfo	FF	110	230	88	4.5
Hifos	FF	126	259	75	5.3
Maroc 6	M	124	255	85	6.0
Helix	WG–WW	133	271	75	5.5

* Trade names, Newport Industries, Inc.
† See Chapter 15 for color ratings.
‡ Ring and ball method.
§ Rosin included for comparison.

In general it is necessary to incorporate at least 4% of lime to eliminate the tendency to crystallize; when amounts greater than 7% are used, there is difficulty with insolubility. The Helix resin in Table 6 is the type most satisfactory for general paint use.

Gloss Oil. Gloss oil is the trade name for a solution of limed rosin in mineral spirits. Gloss oil can be made by thinning a batch of limed rosin after it has cooled to between 450° and 400°F. The gloss oil may be added to varnishes and certain alkyd resins used in making low-cost gloss enamels, or it may be added to the pigment paste when the paste is being reduced to the enamel formulation. Since low cost in combination with uniform performance is essential in the use of gloss oil, it may prove more economical to purchase the gloss oil from the rosin manufacturer, since he can lime the rosin in solution in the extraction process, thereby effecting economies and improved properties. The characteristics of a series of commercially available gloss oils are given in Table 7.

Table 7 Characteristics of Gloss Oils

Solvent: Mineral Spirits

Trade Names *	Pounds per Gallon	% Solids	Vis., Gardner-Holdt	Acid No.	% Metal
Hy-Glo-4	7.97	60–62	X–Z	38–43	4
N-Glo-5	7.94	60–62	J–L	42–46	5
N-Glo-5Y	8.16	63–65	X–Z	44–48	5
T-Glo-8	7.71	65	J–L	45	8
T-Glo-8Y	7.76	65	X–Z	50	9
Zinol †	8.18	64	K	6–10	4.7

* Trade names, Newport Industries, Inc.
† Zinol is a zinc resinate solution included for comparison.

Zinc Resinate. Rosin may be heat-reacted with zinc oxide to form zinc resinate. This reaction is more sensitive than the liming reaction, and difficulty may be experienced with the product. Therefore, commercial zinc resinates usually contain both zinc and calcium. These resins have high melting points and good drying properties when used in coatings. They are used in interior architectural finishes and nitrocellulose lacquers and in rotogravure and heat-set printing inks. The characteristics of a commercial solution of zinc resinate in mineral spirits are shown in Table 7 for the entry Zinol. The characteristics of typical solid zinc resinates are listed in Table 8.

Table 8 Characteristics of Zinc Resinate

Trade Names *	% Zinc	Melting Point † °C	Melting Point † °F	Acid No.‡
Zinar	5.6	182–188	361–370	0
Zitro	4.8	158–164	316–327	15
Zirex	8.75	158–164	316–327	0

* Trade names, Newport Industries, Inc.
† Ring and ball method.
‡ Calculated from known zinc combined.

Rosin Esters. The carboxyl group in rosin may be reacted with a hydroxyl group to form an ester. Since rosin is a monocarboxylic acid, it will have a functionality of 1 with respect to condensation reactions. From the discussion on functionality it will be recalled that 1:1 and 1:2 reactions produce non-resinous materials, and a 1:3 reaction produces a resinous material. These are typified in the commercial materials listed in Table 9, which range from the liquid methyl ester,

Table 9 * Characteristics of Rosin Esters

Hercules Trade Name	Rosin Ester	Softening Point,† °C	Acid No.	Color ‡
	Ester of rosin			
Abalyn	Methyl	Liquid	6	X
	Ethylene glycol	63	12	N
	Diethylene glycol	45	8	M
	Glycerol	91	7	WG
Pentalyn A	Pentaerythritol	112	14	WG
	Ester of hydrogenated resin			
Hercolyn	Methyl	Liquid	5	Z
Staybelite Ester #1	Ethylene glycol	56	11	WG
Staybelite Ester #2	Diethylene glycol	40	7	N
Staybelite Ester #10	Glycerol	83	8	WG
Pentalyn H	Pentaerythritol	105	16	M
	Ester of polymerized rosin			
Polypale Ester #1	Ethylene glycol	80	8	N
Polypale Ester #2	Diethylene glycol	63	8	N
Polypale Ester #10	Glycerol	113	6	WG
Pentalyn C	Pentaerythritol	136	15	M

* Courtesy of Hercules Powder Co.
† Hercules drop method.
‡ U. S. Dept. of Agriculture rosin cube scale.

Fig. 1.

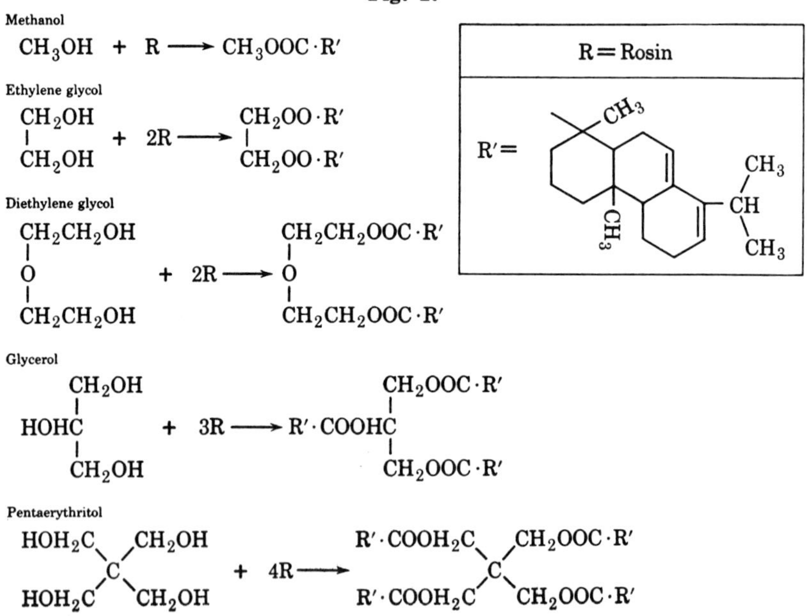

Abalyn, to the solid glyceryl ester, known as ester gum. The esters of hydrogenated and polymerized rosin are included for comparison of physical properties.

Table 9 shows the change in the softening point of the ester due to:

1. Change in the functionality of the alcohol.
2. Change from rosin to hydrogenated or polymerized rosins.

The softening point of the esters increases with increased functionality of the alcohol and with the polymerized rosin in place of the regular rosin. There is a decrease in softening point of the esters in the series made with hydrogenated rosin in comparison with those made from regular rosin. Comparison of the esters of ethylene and diethylene glycol shows a lower softening point for the diethylene ester. This illustrates the effect of structure; the diethylene ester has a more open structure than the ethylene ester. The formulas for the alcohols and rosin esters are given in Fig. 1.

Physical Condition and Molecular Structure. Ester gum and linseed oil are triglycerides; ester gum is the triglyceride of rosin, and

Fig. 2. Structure of Oil and Ester Gum

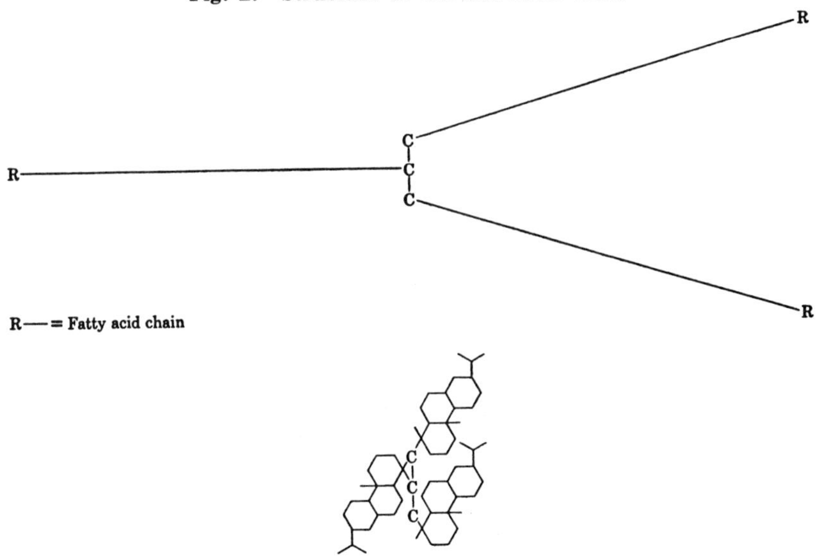

linseed oil is the triglyceride of 18-carbon-chain fatty acids. Ester gum is a hard, brittle solid, and linseed oil is a relatively low-viscosity liquid. This difference in physical condition is due to the difference in structure or shape of the molecules of these compounds. The difference is shown diagrammatically in Fig. 2. The oil molecule is a relatively long, open structure, but the ester gum molecule is a compact, relatively dense structure. There is no significant difference in their molecular weights, the oil being 880 and the ester gum being 944. The relationship between physical properties and molecular structure should be studied at every opportunity since it is invaluable in developing new compounds or in evaluating new resins which appear on the market.

Uses for Rosin Esters. The methyl esters of rosin and hydrogenated rosin are available from the Hercules Powder Co. and are known as Abalyn and Hercolyn respectively. They are used in adhesives and as plasticizers in coatings and printing inks. Because of the unsaturation in Abalyn it tends to oxidize when exposed to air. Where this feature may be objectionable, Hercolyn is recommended.

The glycol esters of rosin are used in adhesives as tackifiers, in textile sizes, printing inks, and water-dispersible paints and inks. The glycerol and pentaerythritol esters are used in varnishes, nitrocellulose lacquers, and a wide variety of paints and enamels.

Manufacture of Ester Gum. Ester gum is made according to the following formula and procedure:

	Parts by Weight
3 moles rosin	906
1 mole glycerol	92
	998
1 mole ester gum	944 yield
3 moles water	54 loss—5.41%

The above theoretical formula shows that about 10% of glycerol is required, but it is customary to use about 12% in general practice.

Procedure

The rosin is placed in an aluminum or stainless steel kettle and heated to 475–500°F, under a blanket of carbon dioxide. The glycerol is then added slowly with agitation; when it is all in, the temperature is raised to 525–550°F. The batch is held until the required acid number is obtained; then a good vacuum is applied to remove traces of water and free glycerol. The finished batch is run through a wire filter into drums ready for shipment.

Types and Properties. Ester gum may be obtained in the regular and high-acid grades; the regular has an acid number of 6–8 and the high acid from 10 to 15. The regular has a slightly higher melting point and better water resistance. The high-acid ester gum is used with tung oil varnishes because it retards gelation better than the low-acid type. It is also preferred in certain grinding vehicles because of its better pigment-wetting property. If the ash content of ester gum is greater than 0.1% it may indicate the use of the lower-cost lime or zinc oxide in the neutralization. Ester gum is soluble in all the usual paint and lacquer solvents except alcohol. The higher-melting-point type is preferred for lacquers because it has better "solvent release." This term indicates the relative ease or difficulty with which the final small percentage of solvent is released from a resin or lacquer film. In general, the higher the melting point of a resin the better will be its solvent release. The degree of solubility is also a contributing factor because, if two resins have the same melting point, the resin with the better solubility will tend to retain the solvent longer than the one with the poorer solubility. This subject will be discussed further in connection with maleic-type resins.

Since ester gum contains unsaturation, as indicated by the formula for abietic acid, it will be oxidized if exposed to the air. Therefore it should not be removed from the metal drums until it will be used, nor should it be stored in finely powdered form. Oxidation darkens the color and reduces the solubility considerably. Ester gum should be stored in a dry place since it will absorb sufficient moisture to produce turbidity in solution or cause bad foaming in varnish making.

Table 9 shows that the glycerol esters of both hydrogenated and polymerized rosin are available. The hydrogenated esters (Staybelite Esters #5 and #10, Hercules Powder Co.) are used where the tendency to oxidize must be avoided. The glycerol ester of polymerized rosin (Polypale Ester #10, Hercules Powder Co.) has a higher softening point and therefore better solvent release, and improved color. It also has less tendency to oxidize than the normal ester because some of the unsaturation has been used in the polymerization. The fact that polymerized rosin behaves as a dibasic acid would indicate that the products of polymerization are chiefly dimers.

Pentaerythritol Ester. An additional increase in softening point is obtained by replacing glycerol with pentaerythritol. This could be predicted from the increase in functionality of the alcohol from 3 to 4, with a corresponding increase in molecular size of the ester as shown in Fig. 1. In addition to better solvent release and harder drying, the pentaerythritol ester is somewhat more stable at the elevated temperatures used in varnish making. This feature is of interest in varnishes made with the slower-bodying oils such as linseed or dehydrated castor, as compared with tung oil. Higher temperatures may be used with less degradation of the resin if it is a pentaerythritol instead of a glycerol ester. Further improvements in this connection are obtained with the maleic- and phenolic-modified esters which are described in the following sections.

Manufacture of Pentaerythritol Ester. Commercial pentaerythritol is usually a mixture of pentaerythritol and dipentaerythritol. A well-known type [Pentek, Heyden Chemical Corp. (4)] consists of approximately 88% pentaerythritol and 12% related alcohols which are chiefly dipentaerythritol. When formulating resins with such material it is necessary to know the combining weight rather than the molecular weight. The value of 36.2 is given by the manufacturer, which means that 36.2 parts of Pentek are required to neutralize a monobasic acid. In commercial practice a slight excess of the alcohol is used; therefore the following weight ratio is suitable:

	Pounds
4 moles rosin	1208
Pentek	145–152

Procedure

The rosin is heated to 390–400°F under a blanket of carbon dioxide. The agitator is started, and the Pentek is added slowly. The temperature is raised to 560–565°F, and the batch held for the required acid number, which is usually from 12 to 15. A partial vacuum applied toward the end of the processing reduces the time required. Aluminum or stainless steel kettles should be used since regular steel tends to darken the color.

In comparison with the pentaerythritol ester of rosin (Pentalyn A), the pentaerythritol ester of hydrogenated rosin (Pentalyn H; see Table 9) has a slightly lower softening point but better stability. Pentalyn H is not of particular interest in varnishes or lacquers but finds wide use in wax compositions and as a tackifier in synthetic rubber adhesives.

MALEIC-MODIFIED ROSIN RESINS

Introduction. Maleic-modified rosin resins are widely used in varnishes, printing inks, and nitrocellulose lacquers. They are available in a range of properties determined by the ratio of maleic to rosin, and the type and amount of alcohol used for esterification. The reactions on which these resins are based are:

1. A Diels-Alder addition of maleic anhydride or fumaric acid to the diene unsaturation in rosin.
2. Degree of esterification of the adduct.

Rosin-Maleic Adduct. The Diels-Alder addition reaction for increasing the molecular complexity of drying oils was described in the previous chapter. A similar 1:4 addition may take place between maleic anhydride and rosin provided the diene unsaturation is in one ring. This type of unsaturation is shown to exist in levopimaric acid but not in abietic acid (see page 150). However, small amounts of levopimaric acid are present in rosin, and it is believed that abietic

Fig. 3. Rosin-Maleic Adduct

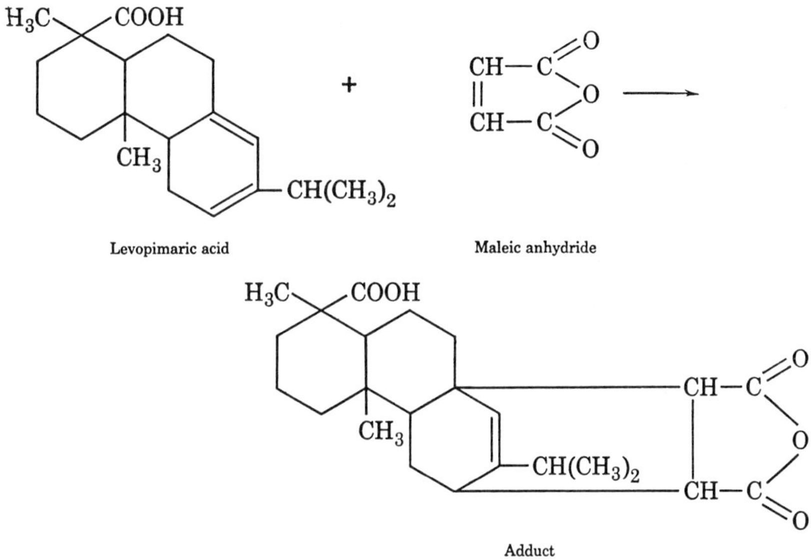

is isomerized to levopimaric under the elevated temperature conditions of adduct formation. The maleic reacts readily with the levopimaric which is present, and the removal of the levopimaric would shift the equilibrium and therefore tend to accelerate the isomerization. The adduct (see Fig. 3) is a high-melting-point, high-acid resin which is soluble in alcohol and the glycols, and finds use in spirit varnishes, steam-set printing inks, and other specialties. However, its high acidity makes it reactive with certain coating materials; therefore it is usually esterified with alcohols for general varnish and lacquer use.

Types of Maleic Resins. Since the adduct is a tribasic acid it would readily form insoluble products with the trihydric alcohol, glycerol. The desired solubility is obtained by using considerably less than the molar ratio of rosin and maleic anhydride. Since the molecular weights of rosin and maleic anhydride are 302 and 98 respectively, the maximum maleic anhydride that would be combined with rosin is about 25%. However, in commercial maleic resins the maleic percentage is considerably less and usually ranges from 5% to 15%. This variation in maleic content is a major factor in the range of hardness and solubility characteristic of maleic resins. A further increase in hardness and reactivity with oils may be obtained by replacing part or all of the glycerol with pentaerythritol. A penta-modified maleic resin would be the type to use with the slower-bodying or "soft" oils in varnish making, but it would not be suitable for lacquers.

A study of the characteristics of the maleic resins available from the various resin manufacturers will indicate the types referred to above. A typical series is given in Table 10.

Table 10 * Characteristics of Maleic-Type Resins

Resin	Softening Point,† °C	Acid No.	Type of Use
Teglac ‡ resin 127	145–155	300–330	Steam-set inks
Teglac resin 128	145–155	260–290	Alcohol solutions
Teglac resin Z152	140–150	25–40	Varnish or lacquer
Teglac resin 161	138–148	25–40	Good solubility
Teglac resin 164	138–150	40–50	Soft-oil varnish

* Courtesy of American Cyanamid Co.
† Ring and ball method.
‡ Reg. U. S. Patent Office.

The degree of esterification of the adduct is indicated by the differences in acid number. The high-acid resins are soluble in alcohol and also in the glycols for use in steam-set inks. Teglac resin 127 was the first resin used in steam-set inks, as cited in U. S. Patent 2,244,103. A 50–60% solution of Teglac resin 127 in diethylene glycol is given as an example of the vehicle for the ink. Immediately after this type of ink is printed on paper it is subjected to an atmosphere containing steam. The resin is precipitated from its solution under these conditions, and the ink is then dry.

Teglac resin 128 is soluble in alcohol and is used in spirit varnishes and shellac substitutes. It is compatible with ethyl cellulose and the alcohol-soluble type of nitrocellulose.

Teglac resin Z152 is a typical general-purpose type suitable for use in nitrocellulose furniture finishes and other interior lacquers, and for varnishes made with "hard" or fast-bodying oils. Teglac resin 161 is designed for maximum solubility in aromatic hydrocarbons or lacquer solvents and for low-temperature incorporation into oils for printing inks. Teglac resin 164 is the type used with soft oils because of its faster rate of bodying. Typical varnish formulations and cooking procedures are given in Chapter 4.

It was indicated above that variations in maleic-type resins can be made by changing the amount of maleic, and the type and amount of the alcohol for esterification. These are the principal variants, but other dibasic acids may be used in partial or complete replacement of the maleic anhydride. For example the alcohol-soluble resins, Krumbhaar 424 and 1515 (Krumbhaar Chemicals, Inc.), are stated to be made with fumaric acid instead of maleic anhydride. Coatings containing maleic are considered to be toxic for some purposes, but it was found that fumaric was relatively non-toxic to rats, whereas maleic was quite toxic. Dr. Krumbhaar's book (5) gives an interesting account of his extensive investigations in resin chemistry and formulation.

Manufacture of Maleic Resins. The reaction between maleic anhydride and rosin is exothermic; therefore care should be taken to avoid overheating and possible fire. The reaction starts about 200°F and is quite violent at 300°F. The adduct may be made by heating the rosin to 200°F and adding the maleic anhydride slowly with a gradual increase in temperature to 300°F until the reaction is complete. The rosin and maleic may be mixed before heating if the temperature is controlled carefully at the critical reaction point.

When making the esterified product, the rosin, maleic anhydride, and glycerol may be placed in the kettle together and the temperature raised carefully to 475–495°F. Under these conditions the maleic reacts with the rosin before esterification takes place. In view of the fact that maleic adds to the diene unsaturation in rosin it will be evident that the hydrogenated rosin would not be satisfactory for this type of resin. Since some of the unsaturation in rosin has become saturated in forming polymerized rosin, this material has only limited use in maleic-type resins.

A Tricarboxylic Acid. An interesting method for making the tricarboxylic acid, instead of the adduct shown in Fig. 3, is described in U. S. Patent 2,522,586. The levopimaric acid-maleic adduct was neutralized and heated in water solution at 150–300°C for sufficient time to isomerize the maleic to the fumaric form. The levopimaric acid-fumaric acid adduct was then recovered by cooling the solution and acidifying it. The product had an acid number of 397–405 and a capillary-tube melting point of 255–257°C. It was combined with fatty acids and alcohols to form a series of alkyd resins having good exterior durability. This material will be referred to again in Chapter 7 on alkyd resins.

PHENOLIC RESINS

Introduction. The most extensive use for phenolic coating resins is in oleoresinous varnishes. Small amounts are also used in the phenolic-modified type of alkyd resins as discussed in Chapter 7. Phenolic resins are used to only a limited extent in nitrocellulose lacquers. Phenolic resins may be classified into four general groups:

1. Modified phenolics; rosin is the usual modifier.
2. 100% phenolics; oil-soluble type.
3. Baking phenolics; alcohol-soluble type.
4. Dispersion resins; phenolic-drying oil polymerizates in xylol.

Phenolic resins is the term which has become general in the trade for the phenol-formaldehyde type of resin. The rosin-modified phenolics were introduced to the varnish industry about 1910 by the late Dr. Baekeland. Varnishes made with the modified phenolics, in comparison with the natural resins and ester gum, were faster drying and had better durability and markedly improved water and alkali resistance. These resins were particularly successful in varnishes made with tung oil and were responsible for the development of the so-called "four-hour enamels." These fast-drying finishes were widely used for kitchen furniture, porch and deck enamels, and similar finishes. The early varnishes were dark in color and tended to yellow or discolor badly on aging. Although great improvements have been made in the color and color retention of phenolic resins they cannot be used in white finishes where color stability is important.

Phenolic resins are the reaction products of a phenol and an aldehyde, with or without a modifier. Hydroxy benzene or phenol,

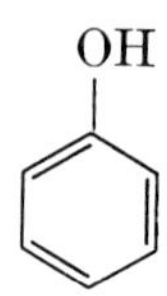

,

sometimes referred to as carbolic acid, was the first phenol to be used in this type of resin, but a number of substituted phenols are also used at the present time. Formaldehyde, HCHO, was the first and continues to be by far the most important aldehyde for this reaction. In the initial reaction formaldehyde adds to the phenol to form a methylolphenol. Condensation polymers are then produced by the elimination of water and the formation of methylene bridges.

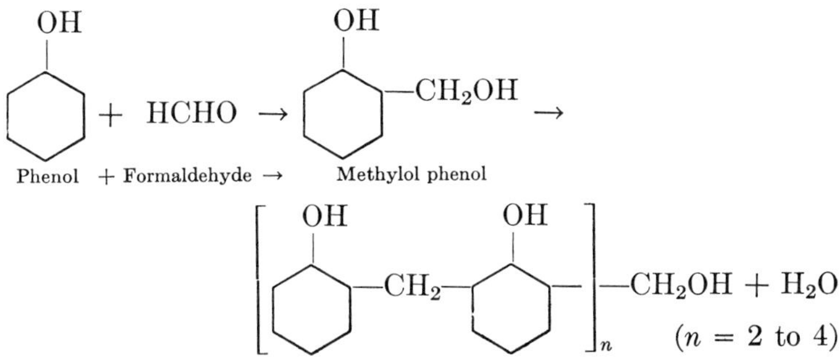

NOTE: The unsaturation in the benzene ring will be omitted except where necessary for clarity.

Types of Phenols. The early phenolic resins were made with phenol and other "tar acids" obtained from the by-product recovered from the coke industry. Although these phenols are still in wide use, the introduction of synthetic substituted phenols has resulted in improved color and has made available the oil-soluble type of 100% phenolic resin. The types of phenols available are shown in Fig. 4.

The condensates of phenol, *m*-cresol, 3,5-xylenol, and *bis*-phenol, are not soluble in drying oils unless they are modified with rosin or its glycerol ester. Run congo or its ester may also serve as the modifier. The last four phenols in Fig. 4 are the type used in the 100% phenolic resins because their condensates are soluble in drying oils without modification.

Within the above classification a number of products may be obtained from each of these phenols and formaldehyde by varying the ratio of formaldehyde to phenol and by using either an acid or an alkaline catalyst. Before describing these various resinous products, we shall consider the properties imparted to the resins by the different phenols.

Position of Substituent. Phenolic resins have a characteristic yellow color, and some tend to yellow badly in the film on aging. The yellow discoloration prohibits their use in high-grade white enamels. In general, the resins made from normal phenol discolor most, whereas those made from *bis*-phenol have much less tendency to yellow. The improved color of *bis*-phenol resins is due mainly to the *para* position being taken up by the substituent. The *para* position is also occupied in the *p*-cresol and the three phenols used for the 100% phenolics. The effect of the position of the substituent in the three cresols on the color of the resin, its solubility in oil, rate of bodying in oil, and drying rate

Fig. 4. Types of Phenols

Phenol
BP: 181°C
MP: 42.5°C

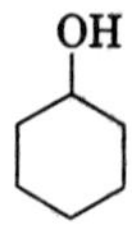

Cresols	*o*	*m*	*p*
BP, °C	191	201	201
MP, °C	30	11	35.5

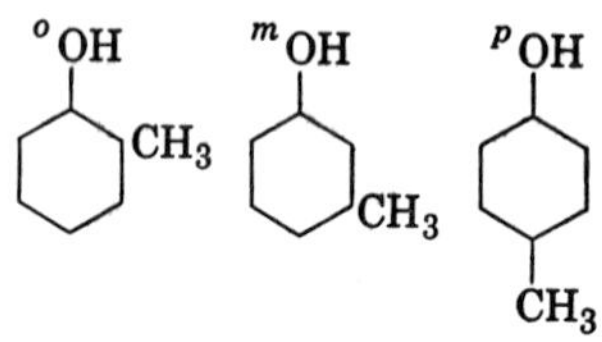

Xylenols

	2,4	3,4	2,5	3,5
BP, °C	211	225	212	219
MP, °C	26	62.5	75	68

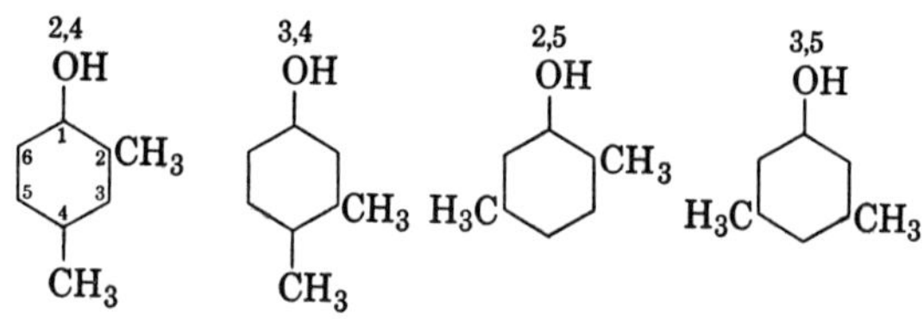

bis-Phenol A

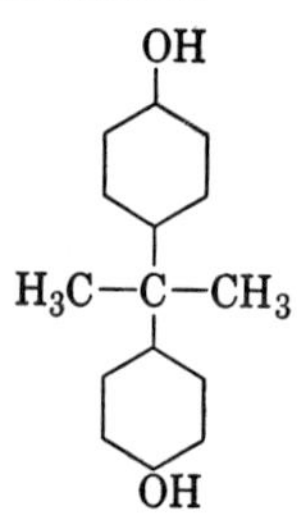

p,p'-dihydroxy diphenol dimethyl methane
or 2,2-diphenol propane
or 4,4'-isopropylidinediphenol

Substituted Phenols for 100% Phenolics

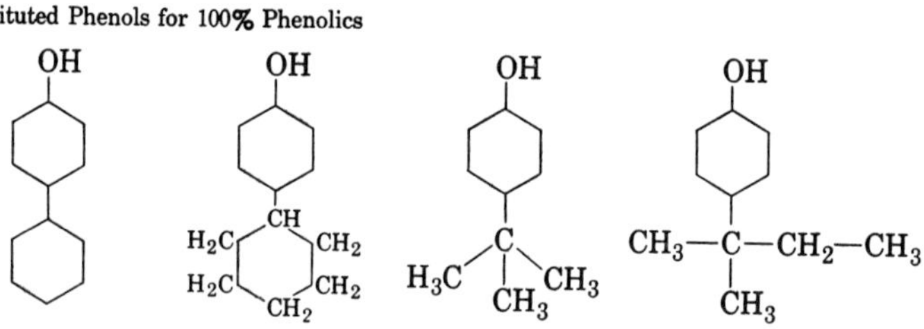

p-Phenylphenol *p*-Cyclohexylphenol *p*-*tert*-Butylphenol *p*-*tert*-Amylphenol

in the film is given in Table 11. It will be apparent from this table that the best results are obtained when the *para* position in the phenol is occupied.

Rate of Reactivity. The rate of reactivity of the phenol is reduced by substitution of an active hydrogen in the benzene ring with another

Table 11 Position of Substituent in Cresol and Its Effect on Resin Properties

Cresol	Color in Resin	Solubility in Oil	Bodying Rate	Drying Rate
ortho	Poor	Good	Poor	Poor
meta	Poor	Very poor	Poor	Poor
para	Fair	Good	Good	Good

group. The reactivity is also affected by the position of the substituent and the type of the substituent group. In the tar acid series, phenol without substituent is the most reactive, cresol is less reactive, and xylenol with two substituent groups is the least reactive. *bis*-Phenol has good reactivity because formaldehyde can add to the two *ortho* positions in each ring. The phenols used for the 100% phenolics are somewhat slow reacting, and their fairly large substituent group is a factor in their oil solubility.

Ratio of Formaldehyde. Resinous materials may be made with ratios of 0.5 to 4 moles of formaldehyde to 1 mole of phenol, but the usual range is from 0.75 to 2 moles of formaldehyde. The reactions in Fig. 5 are given to assist in visualizing the types of resinous materials obtained from the various ratios of formaldehyde. However, it should be kept in mind that the actual materials are more complicated than those shown in Fig. 5.

Catalysts. The reaction between phenols and formaldehyde will proceed at 50–60°C without catalyst, but it is quite slow. Either acid or alkaline catalysts may be used to accelerate the reaction, but by choosing specific catalysts one can obtain some control of the type of product formed. The most effective acid catalysts are oxalic, sulfuric, para-toluenesulfonic, and hydrochloric acids, but acid salts such as stannous chloride may also be used. When the reaction is finished the excess acid may be neutralized with lime and the batch washed with water.

A variety of basic materials may serve as the alkaline catalyst. These include sodium hydroxide, zinc oxide, ammonia, benzylamine, morpholine, and other amines. When the reaction is finished the batch should be neutralized and washed with water. In general, the acid catalysts are used in the ratio of 0.1–0.2% based on the phenol, and the alkaline about 1.0% on the phenol.

The decision to choose the acid or the basic type of catalyst is based on the use for which the resin is intended and the amount of formaldehyde that is to be combined with the phenol. In general, the acid catalysts are used with ratios up to 1 mole of formaldehyde with 1 mole of phenol, and the alkaline catalysts with higher ratios of formaldehyde. The acid catalysts produce insoluble products when high ratios of formaldehyde are used.

Fig. 5. Ratio of Formaldehyde to Phenol

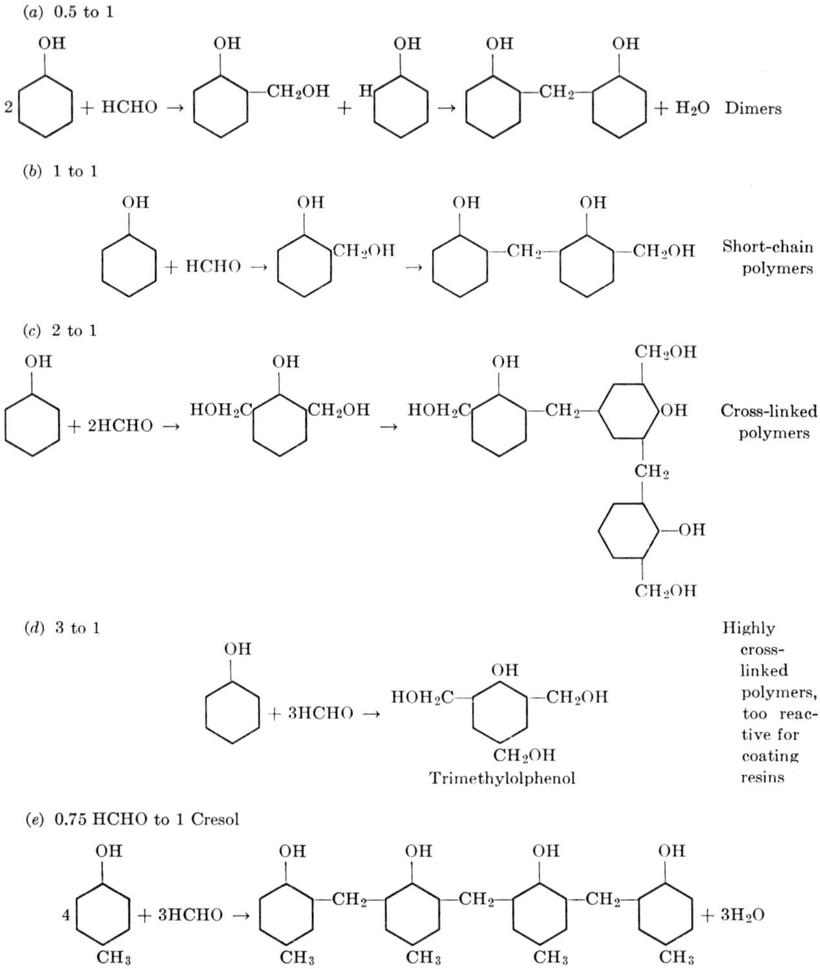

Commercial resins may contain from 4 to 7 cresol units per molecule.

It is generally agreed that with either type of catalyst the first product of the reaction is a methylolphenol. With acid catalysts the reaction proceeds so rapidly to the methylene type of polymer that it has not been possible to isolate the methylol product. However, the methylolphenols have been isolated when alkaline catalysts are used.

The rate of reaction with alkaline catalyst is so much slower that the ratios of formaldehyde may be higher. Some indication of the difference in reaction products is given in Fig. 5. With low ratios of formaldehyde, short-chain polymers are formed with little or no free methylol groups: see Fig. 5, (*a*), (*b*), and (*e*). With higher ratios of formaldehyde the type of phenol used has a bearing on the product. Normal phenol is trifunctional and therefore gives a highly reactive trimethylolphenol. When the *para* position in the phenol is occupied, as in cresol or *p*-*tert*-butylphenol, the functionality is reduced to 2 so that

Fig. 6. Reaction Products of *bis*-Phenol and Formaldehyde

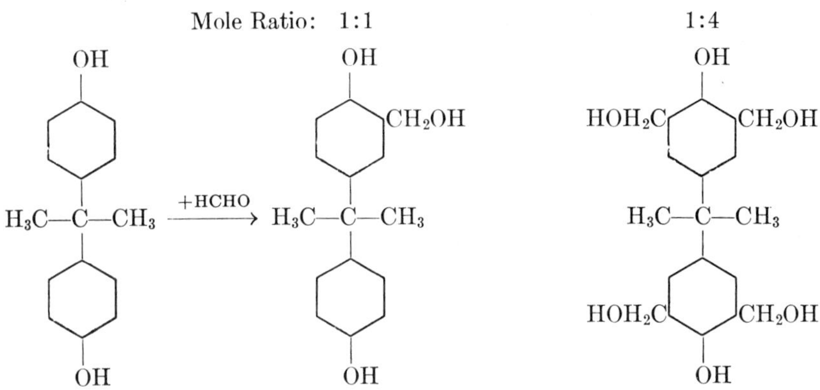

the product is less reactive. However, it will be seen that *bis*-phenol has a potential functionality of 4, as indicated in Fig. 6.

Heat-Reactive Resins. A study of the trade literature on phenolic resins shows that both heat-reactive and heat-non-reactive types are available. The type of heat reactivity referred to is that which takes place in heating the resin with oil for varnish making. The reactive type of phenolic resin is believed to combine with the oil, rosin, or other unsaturated components of the varnish during the cooking of the varnish. There is also some further condensation of the methylol groups in the resin. The heat reactivity of the phenolic resin is therefore due principally to the presence of free methylol groups in its structure. It will be apparent from the discussion of catalysts that the heat-reactive resins are made under alkaline conditions. This subject is discussed in more detail in a later section.

Formaldehyde. Formaldehyde is available at lowest cost as formalin, which is a 37% solution of formaldehyde in water. Formalin is the most widely used source of formaldehyde, but formaldehyde is also available in the water-free state as paraformaldehyde. This material is useful where anhydrous conditions are desired for the reaction between the phenol and formaldehyde, since only the water of reaction will be present. If the reaction is to be carried out under anhydrous alkaline conditions, hexamethylenetetramine, $(CH_2)_6N_4$, may be used as a source of both ammonia and formaldehyde.

Phenolic Condensates. The reaction products of phenols and formaldehyde are referred to as phenolic condensates. They are produced by reacting the components under reflux conditions, in stainless steel or glass-lined equipment. Heat is required to start the reaction, but it is strongly exothermic and cooling facilities are required to control it. The reaction is continued until analysis for free formaldehyde indicates complete formation of methylolphenols. The batch separates into a resinous layer and an upper water layer. After decantation and washing to remove inorganic salts of catalyst the final water is removed under vacuum with a minimum of heating, and the condensate is cooled quickly to prevent further reaction.

Rosin-Modified Phenolics. A series of rosin-modified phenolics may be made by: (1) varying the ratio of condensate to rosin; (2) varying the type of phenol, and the formaldehyde ratio in the condensate; (3) varying the degree of esterification and type of alcohol.

In general, modified phenolics contain from 7% to 12% condensate, although higher percentages are used for special purposes. Usually the alkali resistance and exterior durability increase with increased condensate content.

A range of properties may be obtained by varying the type of phenol in the condensate. From the discussion of the effects produced by the various phenols, it will be recalled that discoloration would be bad with phenol but considerably better with *bis*-phenol or *p*-*tert*-butylphenol. The hardness is also affected by the degree of functionality of the phenol.

The extent of possible reactivity between the condensate and the rosin is determined by the ratio of formaldehyde to phenol, the acid or alkaline catalyst in the condensate, and the degree of polymerization to which the condensate has been taken. The acid-catalyzed, low-formaldehyde-content condensates are probably only dispersed in the rosin modifier, but there is some indication that the alkaline-catalyzed, higher-formaldehyde condensates react chemically with the rosin in the manner shown in a later section.

For most purposes it is desirable to have a low-acid-value resin; therefore the rosin and the condensate are esterified with glycerol, pentaerythritol, or other polyhydric alcohol. For some purposes, such as the manufacture of tung oil varnishes, a high-acid resin is preferred, and the degree of esterification is regulated accordingly. The functionality of alcohol is a factor in the hardness and complexity of the resin structure as discussed previously. The following formulation and procedure for making a pentaerythritol-modified phenolic resin are taken from the booklet *Pentek Resins*, issued by the Heyden Chemical Corp.

Formula 1 Phenolic Condensate

	Parts by Weight
bis-Phenol	1250
Sodium hydroxide	125
Water	960
Formaldehyde (37%)	1110
Conc. HCl	453
Water	960

Procedure

Dissolve sodium hydroxide in water, add the *bis*-phenol with good agitation, and add formaldehyde slowly, maintaining good agitation and temperature below 140°F. When all the formaldehyde has been added, heat batch to 130–140°F, and maintain about 6 hr. Cool to room temperature and neutralize with the HCl and remaining water to a *p*H of 5.5. Wash several times with water and concentrate to 72% solids.

Formula 2 Rosin-Modified Pentek * Ester

	Parts by Weight
Rosin	1000
Phenolic condensate (72% solids)	139
Pentek	110

Procedure

Heat rosin to 375°F under blanket of inert gas. Start the agitator when rosin has melted, and add phenolic condensate slowly to avoid excessive foaming; maintain temperature between 320° and 360°F. After condensate is all in, raise temperature to 390–400°F and add Pentek slowly; then heat to 565°F within 1 hr and hold about 8–9 hr for acid number about 20.

* Trade name, Heyden Chemical Corp.

Commercial Modified Phenolics. The comparison of a series of commercial phenolic resins will illustrate some of the variation in characteristics made possible by the variation in components and processing referred to above. The series of Amberol resins (Rohm and Haas Co.) is selected for this comparison, which is shown in Table 12.

The resins in Table 12 illustrate the manufacturer's attempt to cover all possible uses for modified phenolic resins with the minimum number

Table 12 Characteristics of Modified Phenolic Resins

Amberol * Resin	Soft. Point,† °C	Acid No.	Color, Rosin Stds.	Type of Use
B/S-1	121–129	11–20	M–WG	Interior varnishes and 4-hr enamels
F-7	142–149	11–20	M–WG	General-purpose resin for hard-oil varnishes
F-71	142–149	11–20	I–K	Higher viscosity, more reactive than F-7
F-77	150–160	Under 20	I–K	Faster bodying for soft-oil varnishes
K-12A	150–160	100–132	I–K	Fast-drying, short-oil furniture varnishes
M-21	122–128	11–20	M–WG	High phenol content but pale color
M-82	127–133	11–20	N–WW	More soluble than M-21
M-88	161–169	11–20	K–N	Higher viscosity than M-81 and better with soft oils
M-93	172–180	11–20	M–WG	Highest viscosity, good durability, satisfactory with soft oils
226	66–72	75–85	D–H	Good durability with long-tung-oil varnishes

* Trade name, Rohm and Haas Co.
† Ring and ball method.

of resins. Many other variations could be made, but a wider range would reduce the quantity of each resin used and add to the inventory of both the manufacturer and his customers, which would be uneconomical and poor business. A detailed description of the use of modified phenolic resins in varnish making is given in Chapter 4.

100% Phenolics, Oil-Soluble Type. The substituted phenols used in the manufacture of the oil-soluble type of 100% phenolic resins are *p*-phenylphenol, *p*-cyclohexylphenol, *p-tert*-butylphenol, and *p-tert*-amylphenol. The chemical structures of these phenols are given in Fig. 4. The *p*-cyclohexylphenol and the *p-tert*-amylphenol produce resins which are somewhat slower bodying and slower drying than resins produced from the two other phenols.

Two general types of resins are produced from condensing these phenols with formaldehyde: the heat-reactive type and the heat-non-reactive type. The principal differences in the formulations of these two types are the ratio of formaldehyde to phenol and the use of either acid or alkaline catalysts. The reactivity of the resin increases as the ratio of formaldehyde is increased. In general, a ratio of formaldehyde to phenol of about 1:1 is used for the non-reactive resins, and a ratio of about 2:1 for the reactive resins. The high ratio of formaldehyde permits free methylol groups in the resin structure, which are the source of reactivity since they may be condensed further. In resins with the low ratio of formaldehyde the methylol groups have been condensed to methylene linkages except for the terminal groups.

The various acid and alkaline catalysts which are used were discussed previously. In general, the reactive-type resins are condensed with alkaline catalysts, and the non-reactive type with acid catalysts. Slight differences in the resin characteristics may be produced by varying the amount of catalyst and the particular acid or alkaline compound used.

In the manufacture of these resins the substituted phenol is condensed with aqueous formaldehyde (37% formalin solution), with catalyst, in equipment under reflux conditions. With acid catalysts the products are less stable than when alkaline catalysts are used. The initial condensation product with alkaline catalysts is practically a mixture of monomeric phenolic alcohols. The catalyst is then neutralized, and the product washed with water. The final condensation is usually carried out under acid conditions at a *p*H of about 5–6. The characteristics of the resin will be affected by the operating temperatures and reaction time, as well as by the ratio of constituents as noted above.

Reactive-Type, Oil-Soluble, 100% Phenolic Resin. An interesting variation of the general process has been described by Sharples Chemicals, Inc. It has been called the "dry process" because only part of the aqueous formaldehyde is present in the system at a given time. This process allows good reaction velocity with extremely small amounts of alkaline catalysts, and it eliminates washing and rehandling procedures. A typical laboratory procedure for producing an alkali-condensed, reactive resin, using *p-tert*-amylphenol, is given in Formula 3. Pentaphen 67 is the trade name for *p-tert*-amylphenol manufactured by Sharples Chemicals, Inc.

The resin syrup is rather viscous at room temperature but is quite fluid at 75°C. It may be converted into a hard resin having about the same reactivity by condensing it with more formaldehyde. The cost of the resin may be lowered, and its drying property improved, by replacing 20% of the Pentaphen 67 with phenol on a molar basis. The resin syrup may be converted into a hard, relatively non-reactive resin by adding an acid catalyst, and heating the syrup at 135°C. Of course, the resin syrup may be cooked with drying oils to produce varnishes, or heated with ester gum to produce modified phenolic resins.

In this method some of the toluene is removed continuously from the batch with the condensate. It rises to the surface in the water separator, and arrangement is made to return it to the batch. However, near the end of the processing the toluene is removed instead of being returned to the batch. The "take-off" water from this process will increase in formaldehyde content as the reaction progresses because the concentration of alkylphenol will decrease. Of course, the rate of addition of the formaldehyde could be reduced to compensate for the lower phenol concentration, but that would prolong the processing time beyond the practical limit. To avoid excessive loss of form-

Formula 3 Reactive Type, 100% Phenolic Resin

Materials Used

Pentaphen 67, *p-tert*-amylphenol
Formaldehyde, 37% aqueous solution
Toluene, benzene may also be used
Sodium carbonate, 2.5 *M* solution
Hydrochloric acid, 5 *M* solution
Phosphoric acid, 2 *M* solution
Inert gas

Laboratory Batch

A	Pentaphen 67	820 gm
	Toluene	43 gm
B	Formaldehyde, 37%	25 gm
	Na_2CO_3, 2.5 *M* solution	2 cc
	Formaldehyde, 37%	525 gm
	HCl, 5 *M*	2.2 cc
	H_3PO_4	1.0 cc

Procedure

Prepare a 3-liter, 3-neck flask having ground-glass joints, fitted through the center neck, with a stainless steel agitator having a surface-contact area proportional to those used in production equipment. Fit one side neck with a condenser having proper heat-exchange capacity, and a water separator. Fit the other side with a constant-feed device connected to the formaldehyde charging vessel.

Charge *A* into flask, and heat to 90°C. Add *B* and heat to reflux, about 110°C. Feed in formaldehyde solution at the rate of 105 gm per hr until the 525 gm has been added. Adjust temperature over this period to maintain 103–107°C in order to control the amount of water in the system, the concentration of catalyst, and the reaction velocity. Remove the toluene with the final 5% of the formaldehyde feed. Neutralize the Na_2CO_3 with the 2.2 cc of HCl. Increase the temperature to 110°C, and add the 1.0 cc of H_3PO_4. Provide inert gas stream to remove final traces of formaldehyde, water, and toluene and increase temperature to 130°C in about 30 min. Draw off product, which should be a light-colored, viscous resin syrup, which will remain clear when cold and is substantially free from water.

aldehyde in the take-off water it has been found practical to save the portion of the take-off water which contains more than 10% formaldehyde and use it at the beginning of the next batch. By proper arrangement of a two-reactor system all the formaldehyde can be used, and there is also a considerable saving in heat cost.

The percentage of formaldehyde in the four quarters of take-off water from Formula 3 was:

First quarter:	0–2%
Second quarter:	4–6%
Third quarter:	8–10%
Fourth quarter:	12–14%

A convenient method for the determination of formaldehyde is as follows. To 20 cc of distilled water neutral to bromophenol blue add 1 gm of sample. Add 3 drops of bromophenol blue (1% solution) and neutralize to the end point with one-third normal NaOH solution. Add 10 cc of a 10% solution of hydroxylamine hydrochloride in distilled water and allow to stand for 10 min. Titrate with one-third normal NaOH to the bromophenol blue end point. The number of cubic centimeters required for this last titration is the per cent formaldehyde if the amount of sample is 1 gm and the NaOH solution is exactly one-third normal. Otherwise, the percentage of formaldehyde = $3VN/G$, where V is the volume in cubic centimeters of caustic soda having a normality of N, and G is the weight of the sample.

The resin syrup is pale in color and stable on aging when the formaldehyde is relatively free from iron. However, formaldehyde made by the oxidation of methanol in steel equipment may contain more or less iron formate and formic acid in solution. If present in appreciable amounts, these impurities tend to upset reaction velocity through neutralization of the small amount of alkaline catalyst. The soluble basic iron complexes give a reddish color to the reaction mixture during the process. After neutralization with HCl the iron chloride acts as an acid condensing agent, and the resin syrup, even at room temperature, will split off HCHO and water slowly, and will finally become a rigid resin.

Non-Reactive-Type, Oil-Soluble, 100% Phenolic Resin. An illustration of the formulation and processing log of a non-reactive resin, using the *p-tert*-amylphenol, Pentaphen 67, is given on page 174. The same laboratory equipment and general procedure were used as in the illustration for the reactive resin.

The product of this reaction is a hard resin with a melting point of 125–130°C, having a pale amber color and good color stability when cooked into varnishes. It has only very slight reactivity with drying oils. It is somewhat slower bodying, and its varnishes are about 20% slower drying, than if made with a similar *p-tert*-butylphenol resin. Varnishes made from *p-tert*-amylphenol resin have shown somewhat better water and alkali resistance than comparable varnishes made from *p-tert*-butylphenol resin.

Formula 4 Non-Reactive Type, 100% Phenolic Resin

A	Pentaphen * 67	820 gm
	Benzene	43 gm
B	Formaldehyde, 37%	25 gm
	HCl, 5 *M*	1 cc
	Formaldehyde, 37%	432 gm
	H_3PO_4, 2 *M*	1 cc

Procedure

Charge *A* into flask and heat to 90°C. Add *B* and heat to 120°C. Feed in formaldehyde solution at the rate of 100 gm per hr and maintain times, temperatures, and conditions indicated in the following log.

* Trade name, Sharples Chemicals, Inc.

Typical Log * for Pentaphen-Formaldehyde Non-Reactive Resin

Time, hr	Temp., °C	Grams H_2O HCHO Off	Grams Dry HCHO Off	Grams 37% HCHO In	Grams Dry HCHO Reacted	Additions
0	90	0	0	25		
1	120	76	0.8	125		0.5 cc 5 *M* HCl
2	115	156	3.0	225		
3	125	250	10.0	325		0.5 cc 5 *M* HCl
4	135	340	26.0	425		
4.3	140	383	35.0	457	135.0	1.0 cc 2 *M* H_3PO_4 plus CO_2 atm.
5.3	145					
6.1	150					Resin is poured

* Courtesy of Sharples Chemicals, Inc.

When making varnishes with the heat-reactive resins, usually there is considerable foaming in the kettle. This is due to elimination of water from the continued condensation of the resin and possible reaction with the oil. The remarkable fortifying effects of these resins on oils are shown by comparison with ester gum in a linseed oil varnish. With a resin-oil ratio of 1:2 the varnish made with ester gum is relatively slow drying; its films have poor durability, they turn white in boiling water in 5 min, and they are destroyed by 5% NaOH solution in 15–30 min. By comparison a 100% phenolic resin varnish is fast drying; its films have good durability, and they withstand boiling water and 5% NaOH solution for 7 hr without being seriously affected.

Similarly, the heat-reactive phenolic resins, in combinations with ester gum in amounts varying from 5% to 25% phenolic, raise the softening point of the ester gum 30–50°C. Varnishes made with these combinations have improved drying time, durability, and water and chemical resistance. Of course, the modified phenolic resin varnishes are not as resistant as the 100% phenolic varnishes. The use of these resins in varnishes is described in more detail in Chapter 4.

A number of variations of the 100% phenolic resins are available commercially. Some of these are illustrated by the series of Bakelite resins described in Table 13.

Table 13 Characteristics of 100% Phenolic Resins

Bakelite * Resin	
	Heat-Non-Reactive Type
BR-254	Softening point, 195–225°F. The first 100% phenolic resin made available commercially; it is based on *p*-phenylphenol. Accelerates the bodying rate of both hard and soft oils and imparts durability and water and alkali resistance.
BR-4036	Softening point, 185–210°F. Intended for use with hard oils (tung and oiticica). The varnishes are more flexible than those made with other resins, have good color, are soluble in mineral spirits, and are stable with basic pigments such as red lead.
BR-9400	Designed for use with pentaerythritol-modified rosin resins as a fortifier to improve durability and alkali and water resistance. Somewhat reactive with basic pigments.
BR-9432	Softening point, 250–280°F. Similar to BR-254 but better color and solubility in mineral spirits. Does not accelerate the bodying rate of oils as much, nor does it dry quite as fast.
BR-103	Softening point, 215–245°F. Better solubility than BR-254 but somewhat slower drying. Varnish films darken on drying. Used in primers, especially under-water type, and in dark-colored finish coats.
	Heat-Reactive Type
BR-18036	Liquid resin. Used to fortify rosin in low-cost varnishes. Also used with tall oil and as a phenolic modifier for alkyd resins. Improves water and alkali resistance.
BR-3360	Softening point, 175–205°F. Less reactive than the following two resins. Is used alone with oils, or as 5–20% additions to ester gum for varnishes.
BR-14634	Softening point, 190–215°F. More reactive than BR-3360; used for similar purposes but better with soft oils.
BR-10282	Softening point, 180–210°F. Most reactive of the three resins; used to fortify ester gum but especially good with blown or bodied oils for coil-insulating varnishes.

* Trade name, Bakelite Co., Div. Union Carbide and Carbon Corp.

Table 12 indicates in a general manner the variation in resin characteristics obtained from varying the type of phenol, ratio of formaldehyde, type of catalyst, degree of polymerization, and processing conditions.

Heat Reactivity. A great deal of research has been done on the mechanism of the reaction between phenolics and drying oils and rosin, but all the details have not yet been worked out. Turkington *et al.* (6) showed that the heat-non-reactive Bakelite resin BR-254, when heated with linseed oil, accelerated the changes in viscosity, density, and refractive index to a greater extent than could be expected from the simple additive effect of the individual components. Later it was reported that the accelerating effect decreased as the unsaturation in the oil decreased from tung to linseed to soybean oil.

Hultzsch (8) studied the reaction between methylolphenols and unsaturated compounds such as styrene and maleic esters and proposed the following mechanism for the reaction between methylolphenol and the unsaturation in drying oils:

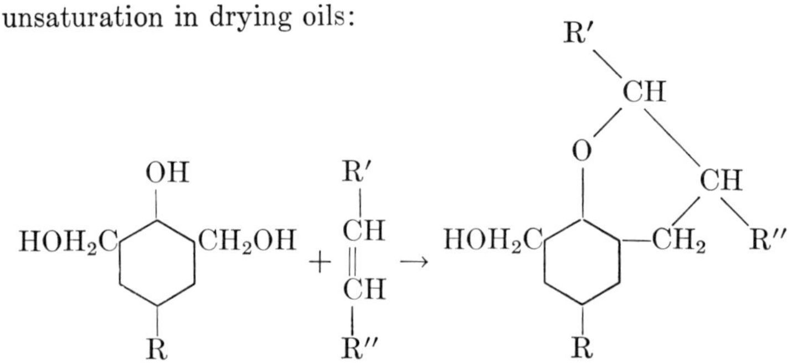

Chroman ring structure

Singer (9) proposed that the chroman ring structure was formed through an intermediate quinone methide. Sprengling (10) proved

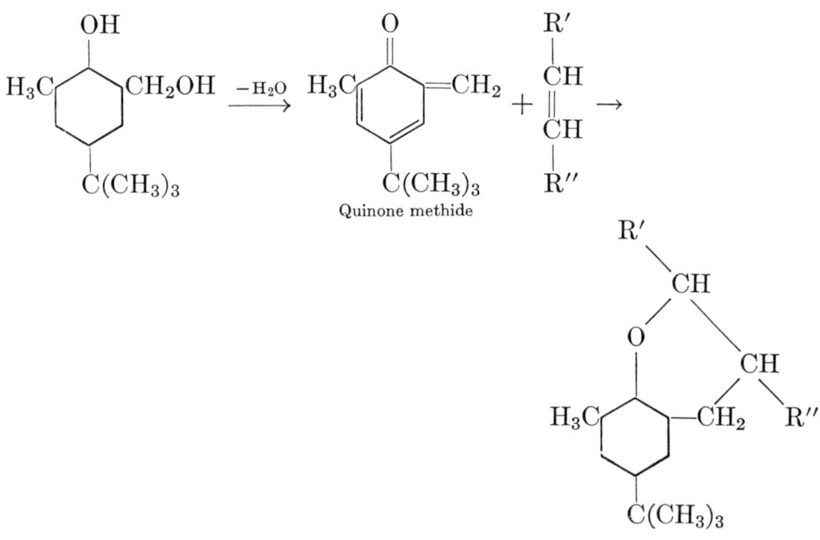

Quinone methide

this hypothesis by working with the *o*-methylol derivative of *p-tert*-butyl-*o*-cresol and oleic acid or its methyl ester.

Krumbhaar (5) reported the reaction between 75 parts of an alkali-condensed *p-tert*-butylphenol and 25 parts of ester gum to give a product which had a melting point higher than either compound and 50°C higher than that of the ester gum. He also illustrated the difference in heat reactivity of an acid-condensed and an alkali-condensed *p-tert*-butylphenol condensate by determining the molecular weight increase and final condition of the resin with continued heating as shown in Table 14.

Table 14 * Heat Reactivity, *p-tert*-Butylphenol-Formaldehyde Condensates

Acid-Condensed			Alkali-Condensed		
	Mol. Wt.	Units		Mol. Wt.	Units
Reflux to clear	500–600	3	Reflux to clear	400	2
Heat to 170°C	850–900	5	Heat to 130°C	750–800	3–4
Hold at 170°C	1100	6	Heat to 230°C	1150	5–6
				Gels at	6–7

Further heating at 175°C: no change. Decomposes above 200°C to give phenol and formaldehyde.

* Courtesy of Dr. Wm. Krumbhaar (5).

The data in Table 14 indicate the possible molecular growth of both acid and alkali condensates during varnish making to 6 or 7 units of the phenol and formaldehyde. In general, with larger molecular size the resin becomes harder and more resistant to water and alkali but less soluble in oils and solvents. However, an important consideration in this respect is the size distribution of the molecular aggregates. If the resin is a mixture of large and small aggregates, it will be less soluble and the resulting varnish will be less stable than if made with a resin having a more uniform molecular size. The two resins may have the same softening point and average molecular weight, but their useful properties will be quite different. Krumbhaar has developed a maturing process which increases the molecular weight uniformly. It consists of holding the resin under mild temperature conditions and good vacuum for 18–24 hr. This removes traces of volatile materials and allows the resin aggregates to grow slowly and uniformly.

The data in Table 14 also illustrate the difference in heat reactivity of the acid- and alkali-condensed resins. The greater reactivity and final gelation of the alkali-condensed resin are related to the presence of free methylol groups in its structure due to the higher formaldehyde ratio used in its formulation. This arrangement increases its functionality and permits cross-linking to take place. However, the exact mechanism for this cross-linking reaction is not known at present. The possibility of this type of resin reacting with drying oils to form chroman structures was discussed previously.

In the acid-condensed resins there are few or no free methylol groups, therefore no cross-linking and gelation. This would also eliminate the possibility of oil reactivity of the type proposed for the alkali-condensed resins. However, the vast improvement in film properties of varnishes made with this type of resin and linseed oil over the oil film itself, or the oil and ester gum, seem to be much greater than can be accounted for by mere dispersion of the phenolic resin in the linseed oil. The improvement in durability and water and alkali resistance was referred to earlier in this section. The limitations of our present chemical and physical methods for determining the structure of these complex resinous structures have retarded such investigations.

Runk (11) made combinations of *p-tert*-butylphenol and phenols having higher functionality to impart some reactivity in the acid-condensed type of resin. He was interested in developing maximum "through-drying" qualities in heavy films of insulating varnishes. He found that 12–18% of *bis*-phenol with 82–88% *p-tert*-butylphenol condensed under *p*H of 5.5 gave the best results with tung, Y-bodied linseed, or maleinized linseed oil. This illustrates the possibility of using mixtures of phenols to produce specific results. Turkington and Allen (7) reported on the properties of a large number of phenols when used in resins for varnishes.

Phenolic and Alcoholic Hydroxyl Groups. There are two types of hydroxyl groups in phenolic resins: the phenolic hydroxyl and the alcoholic hydroxyl. The phenolic hydroxyl, attached to a carbon in the benzene nucleus, is quite stable under heat, does not react with fatty acids, but reacts with the unsaturation in drying oils in the manner indicated previously. The phenolic hydroxyl group is not affected in the regular condensation reaction of either acid- or alkali-condensed resins.

In the condensation reaction the alcoholic hydroxyl group on the methylol substituent of the benzene nucleus is removed together with an atom of hydrogen to form the molecule of water. This reaction forms the methylene bridges illustrated in Fig. 5. There is also the possibility of ether formation in the presence of sufficient methylol groups. The ether linkage may be decomposed at elevated temperature or by application of a high vacuum with the elimination of formaldehyde and the formation of a methylene bridge as follows:

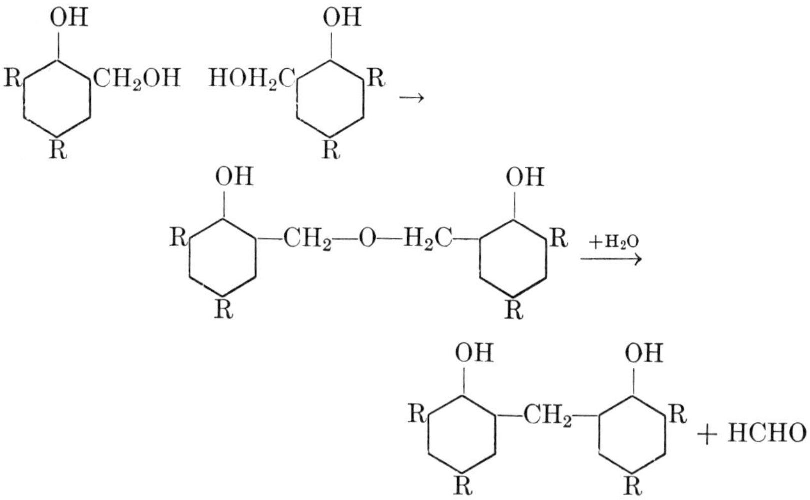

The alcoholic hydroxyl groups can react with any free fatty acids in the oil during the varnish-making operation; therefore these varnishes may have low acid numbers and good stability. The acid value of the 100% phenolic resins has no practical significance and is not given in the trade literature. When the acid value is determined in the normal manner by titrating with KOH, it indicates only the free phenolic hydroxyl groups since these behave as acidic groups. The total hydroxyl content of the resin may be obtained from the acetyl value. Therefore, the free alcoholic hydroxyls may be determined from the difference between total hydroxyl value and phenolic hydroxyl value. It is sometimes difficult to obtain an accurate value for total hydroxyl content from the acetyl value because of the possible side reactions in the determination of the acetyl value.

Acetyl-Value Determination. The acetyl value may be determined by refluxing 5 gm of resin with 15 gm of acetic anhydride for 2 hr. The excess acetic anhydride is hydrolyzed by boiling with water, and the sample is washed until neutral. The sample is then saponified with 100 cc of I *N* alcoholic KOH by refluxing for 1 hr. The excess alkali is determined by titration with I *N* hydrochloric acid. Since both types of hydroxyl groups are acetylated in this reaction, the KOH used is the saponification value for both, which may be converted into the total hydroxyl value.

Baking Phenolics. Solutions of special, alkaline-catalyzed, heat-reactive phenolic resins are used to a considerable extent as baking coatings. In order to obtain these resins in soluble form the resinification reaction is stopped when the resin is still soluble in solvents. The resin solutions are applied to the objects to be coated and then baked at a specified time and temperature. This completes the resinification or condensation reaction so that the resin coating becomes infusible and insoluble in solvents.

From the previous discussion of the many variables entering into the phenolic condensates it will be apparent that a series of coating materials may be made having a range of physical and chemical properties. A typical series is the BV baking solutions of the Bakelite Co., Div. Union Carbide and Carbon Corp. They range in solubility from ethyl alcohol, through alcohol + hydrocarbon, to the Cellosolve type of solvent. The baked films have excellent solvent and corrosion resistance. They also have very good chemical resistance, but they should be selected for specific applications. For example, some have better alkali resistance than others, and some are more flexible than others. Although these materials cannot be classed as varnish resins, they are mentioned here because of their relation to the phenolic condensation reaction. Their application and use will be described in detail in Volume II under the general heading of chemical-resistant baking finishes.

Dispersion Resins. The phenolic dispersion resins are prepared by heat-processing phenolic resins and drying oils in the presence of a dispersing agent. In the process the oils are polymerized far beyond the gelled state, but because of the presence of the dispersing agent the composition may be dispersed in solvents, usually of the aromatic hydrocarbon type. These dispersions, when applied as films, are substantially non-oxidizing and dry rapidly by the evaporation of solvent. The films have excellent durability and resistance to water and alkalies. Although they may be used as the sole film-former, they are usually blended with alkyd resins or varnishes to shorten the drying time and improve water and chemical resistance.

The BK series of dispersion resins of the Bakelite Co., Div. Union Carbide and Carbon Corp., is a typical group. They will not be discussed further in this section because they cannot be classed as varnish resins, but they will be referred to again in Chapter 7 on alkyd resins and also in Volume II in connection with their use in a variety of organic coatings.

TERPENE RESINS

Introduction. Terpene resins are of interest in varnish making because of their relatively low cost, pale color and color retention with heat, and very good alkali, water, and alcohol resistance. They are thermoplastic resins and are non-reactive with drying oils. They

Fig. 7. Terpene Hydrocarbons

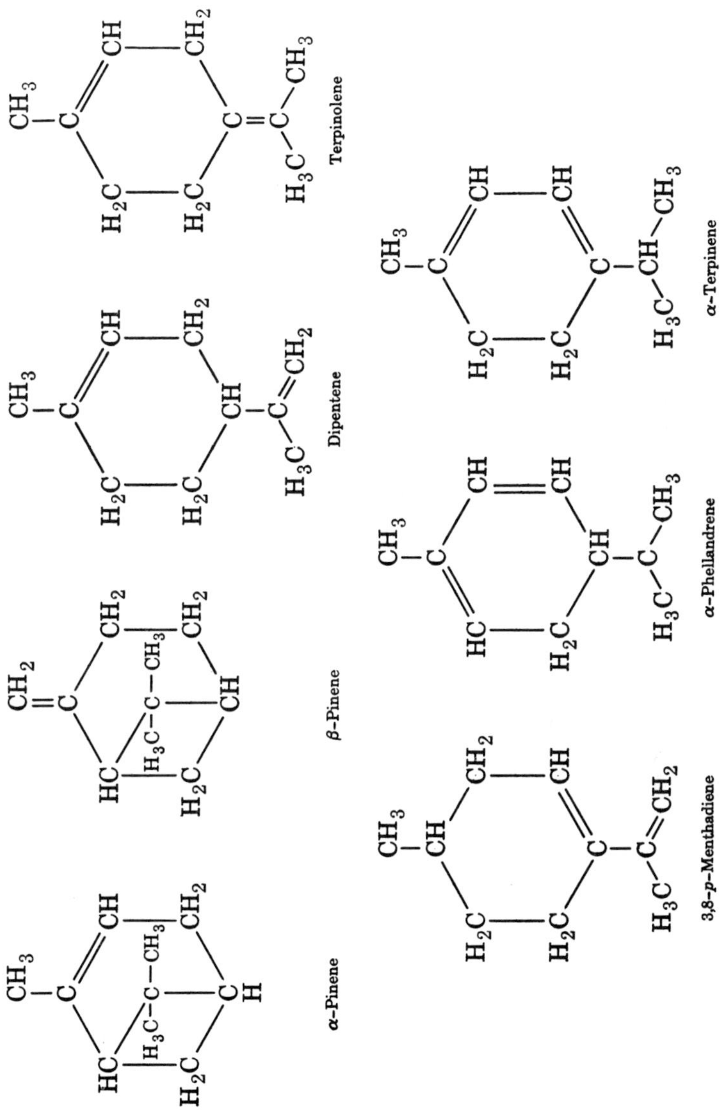

have low acid values and therefore do not have any retarding effect on the gelation of hard oils. Although they are satisfactory for interior finishes they usually do not have good durability in exterior finishes. Since they are the polymerization products of the terpene hydrocarbons they may vary considerably in properties, depending on the type of raw material, polymerization conditions, and degree of polymerization.

Raw Materials. The terpenes are a series of unsaturated cyclic hydrocarbons which includes the compounds shown in Fig. 7. They are liquids with fairly close boiling points and are difficult to separate precisely by distillation. The technical materials usually show a distillation range of 10–20°C between the first-drop and the 95% distilled, indicating a mixture rather than a pure compound. Since the terpenes are unsaturated they tend to oxidize slowly, which is another source of raw-material variation. The unsaturation also permits modification with such materials as maleic anhydride and the phenols. The maleic adducts are discussed in Chapter 7 on alkyd resins.

Turpentine is one of the principal raw materials and is obtained from both gum and wood rosin production as discussed previously. Turpentine is also obtained from the kraft paper industry as "sulfate" turpentine. The gum turpentine contains about 60% of α-pinene and 30–35% of β-pinene; the wood turpentine contains about 80% of α-pinene but no β-pinene. β-Pinene is by far the most useful material for the preparation of terpene resins.

Commercial Types. The range of terpene resins available commercially is illustrated in Table 15 by the series of Piccolyte resins made by the Pennsylvania Industrial Chemical Corp. The wide range of

Table 15 Characteristics of Piccolyte * Resins

Resin	S-10	S-25	S-40	S-55	S-70	S-85	S-100	S-115	S-125
Softening point,† °C	10	25	40	55	70	85	100	115	125
Color	Very pale amber								
Acid number	Less than 4								
Sap. number	Less than 4								
Solubility	Completely soluble in aliphatic hydrocarbons								
Ash, %	Less than 0.1								

* Trade name, Pennsylvania Industrial Chemical Corp.
† Ring and ball method, tolerance 3°C.

softening points and good solubility of the terpene resins permit their use in wax compositions, rubber compounding, printing inks, and surface coatings. They find only limited application in cellulosic compositions because of restricted compatibility. However, one of the large uses for terpene resins is in the manufacture of adhesives.

TERPENE-PHENOLIC RESINS

A variety of resinous materials for varnishes may be made by reacting terpenes or terpene alcohols with phenol. These resins may be complicated further by condensing with formaldehyde and modifying with rosin and maleic anhydride. Phenol may be condensed with α-terpineol to give a substituted phenol:

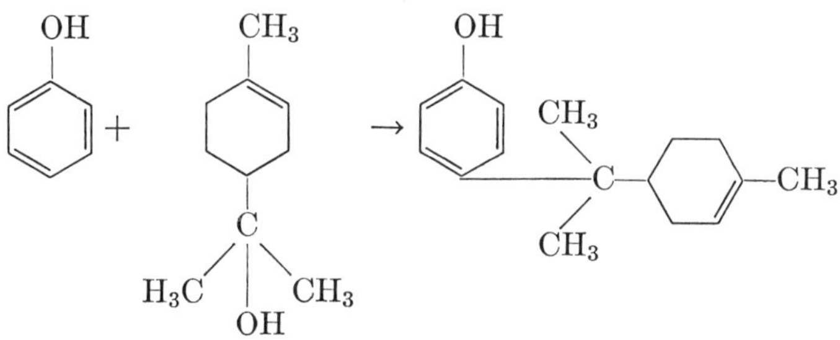

Phenol + α-Terpineol

Addition products may be made with phenol and α-terpinene which may be varied by changing the ratio of α-terpinene to phenol as indicated:

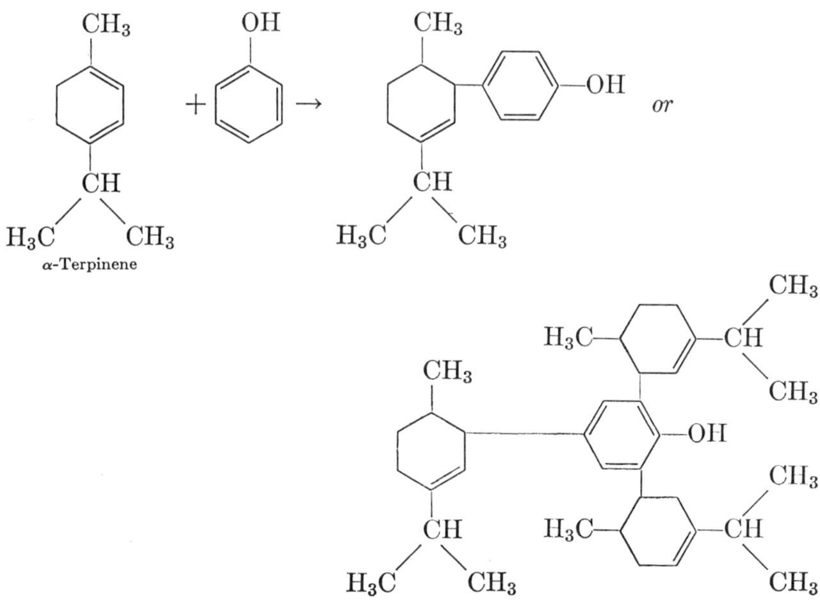

α-Terpinene

It will be apparent from previous discussion that a great variety of products may be obtained from the types of reactions illustrated on page 183. The terpene may be partially polymerized before reacting with the phenol, or the polymerization may be carried out after the reaction. Rosin may be added to lower the cost. This would result in a high acid number unless the product is esterified with a polyhydric alcohol. The inclusion of maleic anhydride to produce a maleic adduct would also enlarge the resinous molecule and increase the hardness.

As may be anticipated, the terpene-phenolic resins have excellent water and alkali resistance and somewhat better color than resins made with straight phenol. Both reactive and non-reactive types are possible. In general, these resins do not retard the gelation of tung oil; therefore mixed oils are used in long-oil varnishes. The resins are excellent in medium- and short-oil varnishes as illustrated in Chapter 4.

The series of terpene-phenolic varnish resins in Table 16 shows a range of hardness and solubility obtained from variation in molecular weight and molecular complexity.

Table 16 * Terpene-Phenolic-Type Resins

Durez Resin	209	219	220	220-V	225	240
Melting point,† °C	116–122	148–156	158–164	157–167	169–179	178–194
Acid number	103–117	60–70	15–21	5–13	50–60	52–68
Color (rosin scale)	WG–1	N–H	N–H	N–H	M–H	M–H
Pounds per gallon	9.1	9.1	9.1	9.1	9.1	9.1
Viscosity, Gardner-Holdt						
60% in toluol	B–C	D–F	H–J	W	N	Z3–Z4
50% in min. spirits	B–C	C–E	H–J	W–X	P–R	Z2–Z5

* Courtesy of Durez Plastics and Chemicals, Inc.
† Ring and ball method.

COUMARONE-INDENE RESINS

Introduction. The coumarone-indene resins are polymers and/or copolymers of coumarone and indene. They are often referred to as Cumar resins, but this term is the trade name of the Barrett Div. of Allied Chemical & Dye Corp. for this class of resins. They are available in a hardness range from brittle solids to viscous liquids. In varnishes they are neutral, thermoplastic, non-reactive resins, and they impart a high degree of alkali resistance and good dielectric properties. Coumarone-indene varnishes are used in aluminum paints because they promote good leafing of the aluminum, thus producing a more brilliant finish. Special grades of coumarone-indene resins are available for varnish making and also for use in adhesives, printing inks, rubber compounding, linoleum, and floor tiles.

Raw Materials and Manufacture. The coal-tar naphthas obtained in the by-product coke-oven process contain a variety of related unsaturated cyclic compounds which include the following:

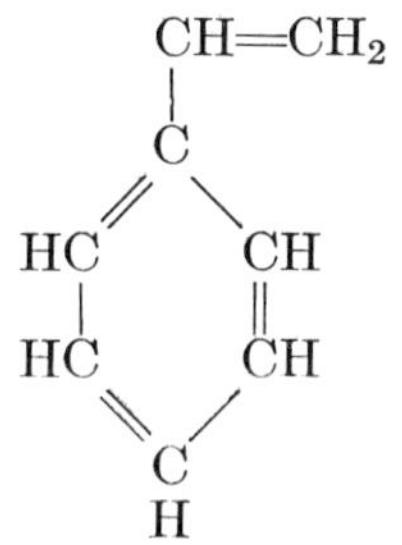

Styrene

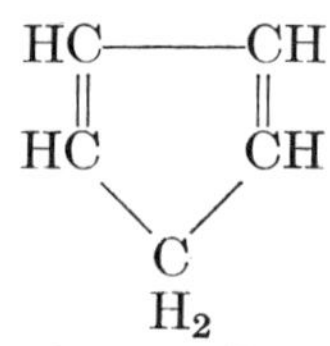

Cyclopentadiene

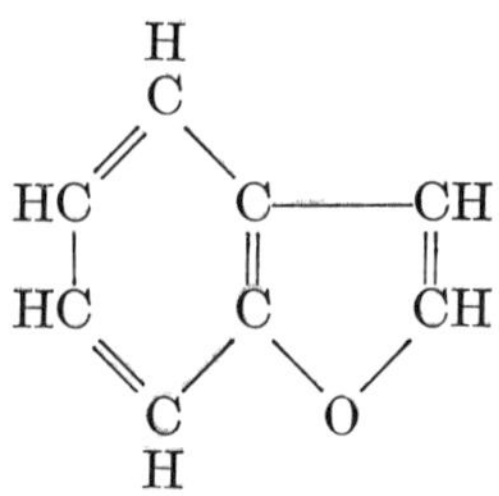

Coumarone

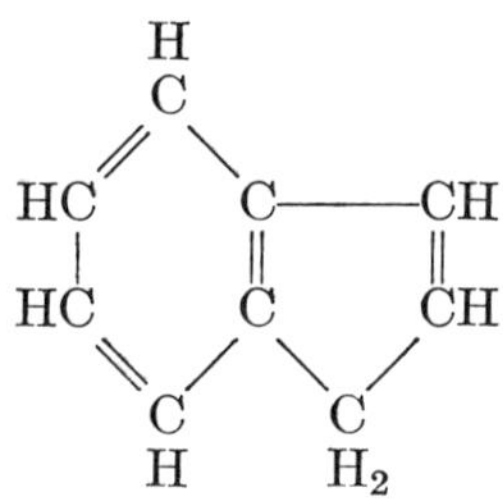

Indene

By careful distillation of the naphthas, fractions are obtained which contain the various unsaturated compounds. The unsaturates are then polymerized in solution in the naphtha by slow addition of sulfuric acid as the polymerization catalyst. Other catalysts, such as stannic chloride, aluminum chloride, or antimony chloride, may be used. Temperatures in the range of $-20°$ to $+30°C$ are maintained to control the type of polymer and to improve the color of the resin. When the desired degree of polymerization has been reached, the acid is neutralized and the batch washed with water. The batch then separates into two layers, and the naphtha layer containing the polymerized resin is steam- or vacuum-distilled to remove the solvent. The molten resin is run from the still onto a cooled roll from which it is removed as flakes.

The characteristics of the resinous product are determined by:

1. Degree of purification of the crude naphtha.
2. Type of unsaturated compound in the naphtha fraction.
3. Type and amount of catalyst and rate of addition.

4. Temperature control during polymerization.
5. Extent of neutralization and washing of product.

A very careful separation of the unsaturates in the naphtha fraction would be expensive and not entirely justified. It was stated previously that one of the characteristics of resins is that, quite often, they are mixtures of mutually soluble compounds. Frequently better results are obtained from mixtures or copolymers, but, of course, it is essential that uniformity be maintained in resin characteristics from batch to batch. The degree of polymerization is a factor in the hardness of the resin, but the distribution of molecular size in the polymer is also important.

Molecular Weight. The coumarone-indene polymer does not attain a high molecular weight such as may be obtained with polystyrene. Dostal and Raff (13) studied the polymerization of indene and found that the heats of activation of chain propagation and chain termination were smaller than that of chain initiation. Also, the velocity constant of chain termination was large compared to that of chain propagation. Therefore the average polymer chain is stopped before it can become very long, and no high-molecular-weight polymers will be formed. Zettlemoyer and Pieski (14) polymerized coumarone-indene under typical conditions and removed the monomers and dimers by distillation. They found the remainder of the polymers to be mostly tetramers with a considerable amount of trimer. Commercial resins vary in polymer size from 3 to 12 units. The polymers are chains of the various original monomers as shown for coumarone.

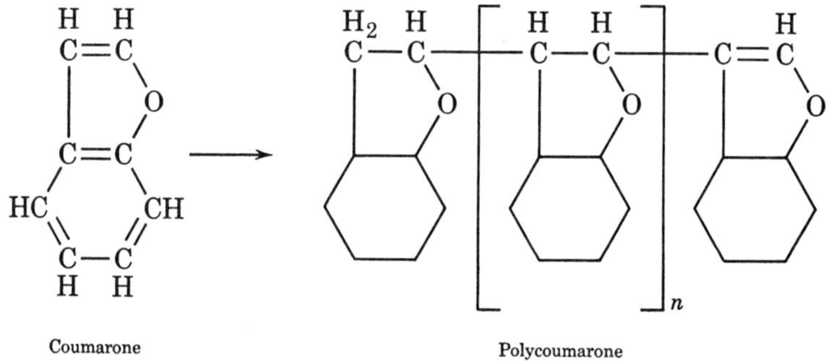

Coumarone Polycoumarone

The resin characteristics are affected by the molecular size as indicated above, but they may also be varied by introducing various amounts of different monomers. For example, the methyl and dimethyl coumarones are present in the crude naphtha, and additions of these materials may be expected to increase the solubility of the resin because of the methyl side chains, but they would tend to lower the hardness. However, lower hardness is not always a criterion of increased solubility.

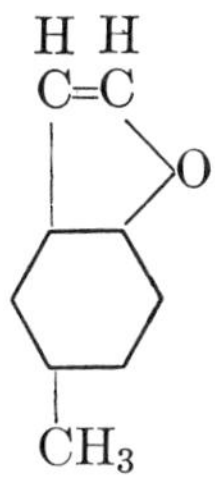

Methyl coumarone

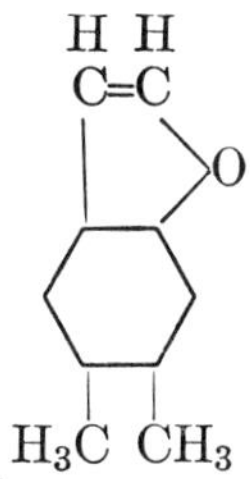

Dimethyl coumarone

Hardness and Solubility. The varnish grades of coumarone-indene resins are available in three hardness grades: hardest, 128–136°C; hard, 99–117°C; and medium hard, 99–107°C. The hardness is determined by the shouldered ring and ball method. The hardest grade is used for best chemical resistance and through-drying of a film, and also for rotogravure inks. The hard grade is the most widely used for varnish work, and the medium grade is preferred where maximum solubility is required.

The coumarone-indene resins are soluble in a wide range of organic solvents, including aromatic hydrocarbons. They are also soluble in mineral spirits and VM & P naphtha but insoluble in the alcohols.

Color and Yellowing. Great improvements have been made in coumarone-indene resins with respect to their original color and their tendency to discolor or yellow on aging. However, their discoloration, especially when exposed to sunlight or ultraviolet, is still a limiting factor to their choice for applications where pale color and color retention are important. The yellowing appears to be a photochemical reaction since it does not occur as badly in the absence of strong light.

The original color of coumarone-indene resins is measured in solution in pure benzene. Two grams of the resin are dissolved in 25 ml of benzene, and the solution is placed in a standard bottle. It is then compared with standard solutions of mixtures of cobalt chloride, ferric chloride, hydrochloric acid, and distilled water. The approximate relationship between the color grades and the rosin standards is as follows:

Coumarone Color Rating	Rosin Standards
½	WW
1	WG–N
1½	K
2	I–H
3	F
4	D

The use of coumarone-indene resins in varnishes is described in the following chapter. They impart excellent chemical resistance and good water resistance. They are thermoplastic, non-reactive resins which discolor when exposed to sunlight and have only fair exterior durability unless protected with an opaque pigment such as flake aluminum powder. Varnishes having satisfactory exterior durability may be obtained by combining coumarone-indene resins with phenolic resins. The extremely low acid value of coumarone-indene resins is an advantage for use in aluminum paints because it does not deleaf the aluminum. However, it is somewhat of a disadvantage if litharge or manganese oxide is to be cooked into the varnish. The low acidity does not promote good solubility of these driers, but this difficulty may be overcome by using soluble driers such as the naphthenates.

PETROLEUM RESINS

Petroleum oils are subjected to a thermal cracking process to increase the yield of gasoline. During the cracking process a variety of unsaturated straight-chain, branched-chain, and cyclic hydrocarbons is formed. These include the following:

$CH_2{=}CH_2$

Ethylene

$CH_2{=}CH{-}CH_2{-}CH_3$

Butylene

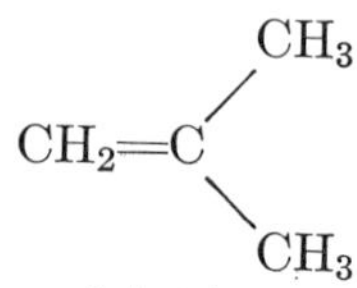

Isobutylene

$CH_2{=}CH{-}CH{=}CH_2$

Butadiene

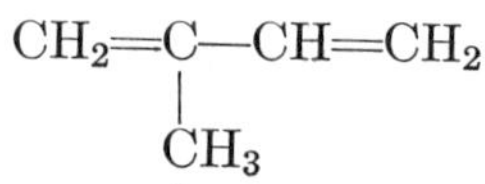

Isoprene

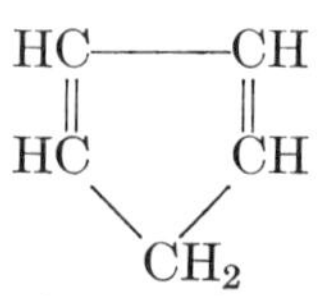

Cyclopentadiene

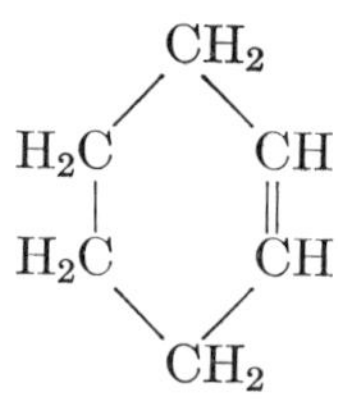

Cyclohexene

These materials form gummy polymers on storage; therefore they must be removed from the gasoline. They may be removed by an absorption process, or they may be polymerized in the gasoline and the polymeric materials obtained by distillation. By careful control of materials and processes polymers are obtained ranging in molecular weight from 250 to about 900 and in softening point from 50° to 105°C. Higher-softening-point resins may be produced for special requirements. The products vary from rubbery polymers to hard, friable solids. Some of these polymers contain considerable unsaturation, as indicated by iodine numbers from 190 to 300, which imparts a

Table 17 * Typical Properties of Hydrocarbon Drying Oils

Grade	Panapol † 1B	Panapol † 1C	Panapol † 2C	Panapol † 1E
Form	Liquid	Liquid	Liquid	Liquid
Color	Black	Black	.....	Black
Color, Gardner 1933	.....	.....	12	.....
Specific gravity	0.966	0.9752	0.9725	1.000
Weight, lb/gal	8.044	8.122	8.10	8.328
Iodine number, Wijs	200	230	180	240
Molecular weight	240	300	330	375
Viscosity, Gardner-Holdt	F–G	N	U–V	Z6+
Flash, PMCC, °F	186	198	168	225
Non-volatile matter, ASTM D-154	70.0	80.0	96.8	95.0
Saponification number	Negligible	Negligible	Negligible	Negligible
Acid number	Negligible	Negligible	Negligible	Negligible

Solubility and compatibility

These hydrocarbon drying oils of the Panapol series are soluble in aliphatic and aromatic naphthas. They are compatible with most vegetable drying and non-drying oils and many commonly used synthetic drying oils.

* Courtesy of Pan American Chemicals Div., Pan American Refining Corp.
† Trade name, Pan American Chemicals Div., Pan American Refining Corp.

Table 18 * Typical Hydrocarbon-Resin Specifications

Panarez † Resins	#1	#3	#6	#11	#12
Grade	1–190	3–210	6–210	11–210	12–210
Form	Solid	Solid	Solid	Solid	Solid
Specific gravity at 60/60°F	1.044	1.056	1.106	1.050	1.040
Weight, lb/gal	8.704	8.796	9.220	8.747	8.663
Color					
Barrett	23	1	2	7	4
50% sol., Gardner 1933	....	8	11	....	17
Softening point, ASTM, °F	180–200	200–220	200–220	200–220	200–220
Iodine number, Wijs	195	195	175	60	40
Molecular weight	540	650	540	575	780
Saponification number	Negligible	Negligible	Negligible	Negligible	Negligible
Acid number	Negligible	Negligible	Negligible	Negligible	Negligible
Non-volatile matter, wt. %	100	100	100	100	100
Ash	Trace	Trace	Trace	Trace	Trace

Solubility and compatibility

These hydrocarbon resins of the Panarez series are soluble in aliphatic and aromatic naphthas. They are compatible with most vegetable drying and non-drying oils and many commonly used synthetic resins.

* Courtesy of Pan American Chemicals Div., Pan American Refining Corp.
† Trade name, Pan American Chemicals Div., Pan American Refining Corp.

degree of reactivity on heating with drying oils, and also allows some oxidation to take place when the polymers are spread out as thin films for coatings. Others have iodine numbers in the range of 40–60 and are classified as non-reactive when heated with drying oils. They vary in color from almost black to pale amber; consequently some are suitable for light colors but not for pure whites. They are soluble in aliphatic hydrocarbons and have good compatibility with vegetable oils and many synthetic resins. They have very low acid and saponification numbers and good resistance to water, alcohol, acid, and alkali.

When used in long-oil varnishes, particularly with phenolic resins, the petroleum resins show satisfactory durability on exterior exposure. They do not retard the gelation of tung oil; therefore mixtures of tung oil with linseed or soybean oil are often used. They may be dissolved in solvent and blended with bodied oils for aluminum-paint vehicles. Blends of petroleum resins and drying oils may be emulsified and combined with casein in emulsion paints. Some grades are also used in alkyd resins as described in Chapter 7. The general characteristics of petroleum drying oils and resins are illustrated in Tables 17 and 18 by the series of Panapol oils and Panarez resins manufactured by the Pan American Chemicals Div., Pan American Refining Corp.

BIBLIOGRAPHY

1. E. T. Holdman, *Am. Paint J.*, Dec. 25, 1950.
2. *Newport Products,* Newport Industries, Inc., New York, N. Y.
3. L. W. Nims and M. C. Schopmeyer, *Ind. Eng. Chem., 39,* 1504 (1947).
4. *Pentek Resins,* Heyden Chemical Corp., New York, N. Y.
5. Wm. Krumbhaar, *Coating and Ink Resins,* Reinhold Publishing Co.
6. V. H. Turkington, R. C. Shuey, and L. Schlecter, *Ind. Eng. Chem., 30,* 984 (1938).
7. V. H. Turkington and I. Allen, *Ind. Eng. Chem., 33,* 966 (1951).
8. K. Hultzsch, *J. prakt. Chem., 158,* 275 (1941).
9. R. J. R. Singer, *Kemisk, 23,* 49 (1942).
10. G. R. Sprengling, Preprint Booklet, Div. Paint, Varnish, and Plastics Chemistry, American Chemical Society, September, 1950.
11. R. H. Runk, *Ind. Eng. Chem., 42,* 2110 (1950).
12. H. S. Lilley, *J. Soc. Chem. Ind., 67,* 196 (1948).
13. H. Dostal and R. Raff, *Z. physik. Chem., B-32,* 417 (1936).
14. A. C. Zettlemoyer and E. T. Pieski, Preprint Booklet, Div. Paint, Varnish, and Plastics Chemistry, American Chemical Society, September, 1950.
15. W. Husen, *Paint, Oil, Chem. Rev.,* June 8, 1950.
16. Arizona Chemical Co., 30 Rockefeller Plaza, New York 20, N. Y.
17. Tall Oil Assoc., 122 East 42nd St., New York 17, N. Y.

Express important ideas in the minimum of words if maximum effectiveness is desired.

Chapter 4. VARNISHES: COMPOSITION, MANUFACTURE, AND USE

INTRODUCTION

There are two major classifications for varnishes: oleoresinous varnishes and spirit varnishes. The oleoresinous varnishes are mixtures of drying oils and varnish resins, usually combined by a heating or "cooking" process and then thinned or "reduced" to normal viscosity with a volatile solvent. Metallic driers may be cooked in during the thermal processing, or they may be added as solutions to the finished varnish. For some purposes, oleoresinous varnishes are made by simple blending of bodied oils and resin solutions. Spirit varnishes are resin solutions, with or without small amounts of oil or other type of plasticizer.

Varnishes may be used as clear, transparent coatings or as vehicles for a wide variety of pigmented, opaque coatings for architectural and industrial purposes. Most architectural coatings are air-dried, but many industrial coatings are dried by a baking process. After a general discussion of the fundamentals of varnish making, a series of varnishes will be described according to their use classification. Today, varnishes are in close competition with many types of synthetic resins and cellulosic lacquers.

COMPOSITION

In general, the physical and chemical properties of the oils and resins used in varnish making carry over into the finished varnish. The exceptions to this statement are the heat-reactive resins, which change in the cooking process and combine with the oil chemically, as described in the previous chapter. The reasons for adding hard resins to oils are to:

Shorten the drying time.
Increase the hardness of the film.
Improve the gloss.

Improve water and chemical resistance.
Increase durability in some cases.

The degree of change produced by the resin will be related to the type and properties of the resin and to the proportion of resin to oil used in the varnish. The trade refers to varnishes as being short-oil, medium-oil, or long-oil varnishes. This terminology indicates the number of gallons of oil used with 100 lb of resin. The general designation corresponds to the following proportions:

Short-oil: 6–18 gal oil to 100 lb resin.
Medium-oil: 18–30 gal oil to 100 lb resin.
Long-oil: 30–60 gal oil to 100 lb resin.

Obviously short-oil varnishes will be harder and more brittle than long-oil varnishes. The flexibility of varnish films may be determined by the kauri reduction test (see Chapter 15 on test methods). Since resin solutions dry simply by evaporation of the solvent, whereas drying oils require considerable time to dry by oxidation, it will be apparent that the shorter the oil length the faster will be the drying of the varnish. It will also be seen that the hardness of the resin, as indicated by its softening point, is a factor in fast-drying varnishes. However, the type of oil used is a more important factor, as should be apparent from the discussion of the various drying oils in Chapter 2. Clear films of drying oils do not have a sharp, full gloss but tend to appear dull or "shrunken." Addition of resin to the oil increases the "fullness" of the film and develops a sharp gloss which is desirable for many applications. Addition of resin will increase the water and alkali resistance of a varnish film in proportion to the amount used, but to a greater extent in accordance with the type of resin. For example, ester gum offers little improvement because it is not much better than the oil, but the 100% phenolic resins effect a definite improvement as indicated in the previous chapter. Films of drying oils are not very durable on exterior surfaces unless they are protected from ultraviolet absorption by an opaque pigment. Most of the varnish resins do not improve the durability of a clear oil film to any appreciable extent, but the phenolic resins are an exception. The modified phenolics in oil lengths of 45 gal or greater have good durability, and the 100% phenolics in oil lengths of 25–35 gal constitute the most durable varnishes. It should be kept in mind that the exterior durability of any clear film will be improved by the addition of the correct amount of opaque pigment.

OILS USED IN VARNISHES

All the drying oils are suitable for varnish making, but practically no semi-drying or non-drying oils are used. Occasionally, small amounts of bodied soybean oil are included in pale-color baking varnishes and in varnishes for architectural white enamels. This is contrary to the formulation of alkyd resins, where large proportions of semi-drying oils may be used in air-drying alkyds and the entire oil content of some baking alkyds may be the non-drying type.

It was pointed out in Chapter 2 that oils containing conjugated unsaturation increase in body, or polymerize, faster than oils having non-conjugated unsaturation. Therefore, tung oil is the best oil for varnish making because of its very fast bodying rate, resulting in a short cooking period. Oiticica oil is almost as fast as tung oil, but dehydrated castor oil is slower because of its decreased conjugation. Linseed oil requires a long cooking period by comparison with the conjugated oils and also a slightly higher temperature. Three means have been used to reduce the time for varnish making with linseed oil: (*a*) prebody the oil in large batches, then simply heat the bodied oil and resin for varnish; (*b*) modify the oil as indicated in Chapter 2 by maleic addition, penta treatment, isomerization, etc.; (*c*) modify the resin so that it accelerates the bodying of the oil. Modified oils and resins are available from a number of oil and resin manufacturers, and some of them were described in previous chapters; their use in specific varnishes is given later.

Soybean oil may also be modified by such treatments so that it becomes a drying oil rather than a semi-drying oil and is therefore satisfactory for varnish making. It was stated previously that bodied soybean oil is sometimes used in varnishes where pale color and color retention are important.

Safflower oil is faster drying and bodying than soybean but not as fast as linseed. It has shown good performance in alkyd resins but would be slow for varnish without some modification of the oil or the use of resins which accelerate the bodying of the oil.

CHILL-BACK OILS

Conjugated oils not only body rapidly in varnish making, but also change from the viscous-polymer to the insoluble-gel state almost instantaneously. Since this rapid polymerization is exothermic the heat released aggravates the situation. Therefore practically it is necessary to have some material which can be added to the varnish when the desired body has been reached to lower the temperature, or chill it back to the point where it will no longer polymerize spontaneously to the gel state. Frequently this material is a portion of the oil intended for the varnish formulation and is called the "chill-back oil." If unbodied oil is used as the chill-back it will not receive the same cooking as the remainder of the oil and will lower the viscosity of the varnish unduly; therefore chill-back oils are usually bodied oils. Since it is not practical to make a bodied tung oil, the chill-back oil may be bodied linseed, combinations of linseed and tung, bodied dehydrated castor, and occasionally bodied soybean. Certain resins may also be used as the chill-back, as discussed under coumarone-indene resins, and this will be illustrated further in the section on varnish making.

GASPROOFING

The conjugated oils are sometimes referred to as "frosting oils." This is due to the fact that these oils will dry with a partially crystallized or frosted effect if they are not cooked sufficiently. The frosting in the varnish film is accelerated if the varnish is dried in an oven which has bad fumes from a gas burner. The frosting may show as slight "gas checks" or "crow's footing" throughout the film, or the film may be completely dull and crystallized. Therefore frosting is often referred to as "gas checking," and a varnish which is processed so as to be free from this difficulty is referred to as a "gasproof" varnish. In very short-oil varnishes, 10 gal or less, there is usually no difficulty with gasproofing because of the high ratio of resin. In longer-oil varnishes the type of resin is an important factor in controlling the gasproofing. For example, ester gum, coumarone-indene resins, or the maleic resins have very little effect in preventing gas checking, but some of the phenolic resins are very beneficial. Gas checking is usually prevented if the varnish can be heated to top temperatures of 575–585°F during the cooking. This will be discussed in further detail in the section on varnish making.

DRYING TIME

As stated previously, short-oil varnishes dry faster than long-oil varnishes but the drying time is also influenced by the type of oil used. The fastest-drying varnishes are made with tung oil, but oiticica oil varnishes are almost as fast. Dehydrated castor oil varnishes are considerably slower because of reduced conjugation, and sometimes they show an "after-tack," probably due to the *cis-trans* isomerization discussed in Chapter 2. Usually, linseed oil is used in combination with other oils and quite often as a bodied oil for chill-back. It will be recalled from Chapter 2 that raw linseed should be avoided in varnishes because of the "break" when the oil is heated. The break-free oils are satisfactory, and the many modified linseed and soybean oils give faster bodying rates, although they do not improve the drying time to the same extent. However, if the varnish is well bodied it will have a higher viscosity at equal solids content and will set and dry faster after application. The type and amount of driers are also factors in the drying time and are discussed in more detail later.

YELLOWING

For many purposes the original color of a varnish and the color retention of its film after drying are of little consequence. However, both these characteristics are important if the varnish is to be used as a paper finish, a finish on blonde maple furniture, a vehicle for white enamels, etc. The original color and after-yellowing are influenced by both resin and oil and to a lesser extent by the type of drier. It will be recalled that the yellowing of oil films is considered to be due to the linolenic acid content, and, since linseed oil has the highest percentage of this acid, it is the worst oil for after-yellowing. The conjugated oils produce pale varnishes with a minimum of after-yellowing when made with the maleic-type resins. There is a wide variation in yellowing with the phenolic resins, but even the palest phenolics are not suitable for varnishes used in high-grade, non-yellowing, white finishes.

Normal amounts of lead and cobalt driers do not affect the color or after-yellowing to any appreciable extent, but manganese drier has a pronounced after-yellowing effect. In fact, an excess of manganese drier is used in certain tin decorating varnishes to produce a brass effect when a clear coat is baked on tin plate for cans and bottle caps.

WATER AND ALKALI RESISTANCE

Oil films have very poor resistance to water and alkali, but the conjugated oils are somewhat better than the non-conjugated oils. Ester gum and the maleic resins are also poor in water and alkali resistance, but tung oil varnishes made with these resins show quite good resistance. However, the best resistance is obtained with the 100% phenolic resins and tung oil in short-oil varnishes. The poor resistance of linseed oil can be improved tremendously with the heat-reactive 100% phenolics. Coumarone-indene resins have excellent alkali resistance, and in some cases they may be blended with the phenolics to lower cost. Highly water-resistant varnishes are often called "spar" varnishes and are described in more detail in another section.

MANUFACTURE OF OLEORESINOUS VARNISHES

Only the general principles of varnish manufacture and types of equipment used will be described. The reader should supplement this information with the latest catalogs from equipment manufacturers. Varnish plants vary in size from a few open kettles to the completely automatic plant of the Pacific Paint & Varnish Co., Berkeley, Calif., described by A. T. Challman in the First Quarter, 1949, issue of *Instrumentation.*

Equipment. The oil and resin are cooked in either movable or stationary kettles, thinned or "reduced" with volatile solvent, filtered, and pumped to storage tanks. The movable kettles range in size from 250 to 500 gal, and the stationary or "set" kettles from 1000 to 3000 gal. Iron kettles produce dark-colored varnishes but are satisfactory where color is not important. Many movable kettles are made of copper, which has excellent heat conductivity; although the color of the varnish cooked in copper kettles is better than when made in iron, it is not as good as that of varnishes made in aluminum or stainless steel. Aluminum gives pale color but has only fair heat conductivity. It is damaged easily by careless hand stirring, may crack if chilled rapidly, and cannot be cleaned with alkali. Stainless steel is very good but has relatively poor heat conductivity.

The movable kettles are placed on three- or four-wheel trucks for moving on or off the fire pit and to the pump for emptying. The kettles are usually straight sided but may be tapered or hemispherical as shown in Fig. 1. If desired, covers are available which have small openings for thermometer and stirrer and a vent for fumes. The vent may be connected to the fume-control system, which eliminates the objectionable odors before ejection into the atmosphere. When varnishes are cooked in open kettles without a cover, a heavy lid should be kept handy to smother a fire if the varnish flashes or catches fire. If the varnish takes fire, the thermometer and stirrer should be removed and the lid placed in position on top of the kettle as rapidly as possible. Prompt but calm and careful action when a fire starts will prevent serious damage.

Fig. 1. Varnish Kettles

Straight-side kettle with removable lid. Six lugs welded to the kettle support it on the four-wheel truck.

Tapered-side kettle. Tapered sides facilitate pouring; lid is removable.

Hemispherical kettle, designed for easy stirring. Covers are available if desired. Note the three-wheel truck.

Courtesy of Brighton Copper Works, Cincinnati, Ohio.

Stationary or set kettles may be built into a firebrick structure over a fire box or may be individual units as shown in Fig. 2. Unit *A* is a 1000-gal stainless steel kettle with a self-contained gas furnace. It is equipped with mechanical stirrer and explosionproof motor, a manhole with cover for loading, inlets for thermometer and inert gas supply, and a vent for fumes. Kettle *B* is a 1500-gal stainless steel, gas-fired

Fig. 2. Stationary or "Set" Kettles

A. Stainless steel kettle, 1000-gal capacity, with built-in gas-fired furnace.

B. Stainless steel kettle, 1500-gal capacity, furnace adjustable for gas, oil, or electric heating.

Courtesy of Brighton Copper Works, Cincinnati, Ohio.

unit, fully equipped. Suitable adjustments may be made for oil firing or electric heating. The type of fuel used will depend on the economical advantage of one over the other in the particular geographic location.

Set kettles may be operated without mechanical stirrer, relying on convection currents and inert gas for movement of the material in the kettle. Set kettles are always equipped with covers, and the contents should also be covered or "blanketed" with carbon dioxide, nitrogen, or other inert gas. In most cases the inert gas is also taken to the bottom of the kettle and released through a series of fine holes in the pipe; it then passes through the varnish as small bubbles.

Inert-Gas Generator. The inert gas may be purchased in cylinders or may be made more economically by a gas generator. Compact and fully automatic inert-gas generators are available such as the one

Fig. 3. Inert-Gas Generator

Side view of 5000-cfh inert-gas producer, showing indirect heat exchanger with water separator and trap. On the left can be seen the inert gas storage tank operating at 100 psig.

Courtesy of C. M. Kemp Manufacturing Co., Baltimore, Md.

shown in Fig. 3, manufactured by C. M. Kemp Manufacturing Co., Baltimore, Md. It uses either natural or manufactured gas as fuel which is burned under carefully controlled conditions to deliver a mixture of 87–88% nitrogen and 12–13% carbon dioxide as the inert gas. The inert gas is stored under pressure of about 100 psig; in addition to varnish cooking, it can serve to blanket the varnish storage tanks and

Fig. 4. Varnish Stack

Six-kettle stack with Selas gas burners.

Gas burner in operation.

Courtesy of American Lithographic Varnish Co. and Selas Corp. of America.

to force the varnish from the tanks to the point of use in the plant. The use of inert gas in the storage tanks, pipe lines, etc., reduces the problem of varnish skinning in the tanks and subsequent breaking up of the skins into small gel particles. These particles are very difficult to filter from the finished paint because they are soft and will deform and pass through fine wire or cloth screens.

Fire Pits. The source of heat for movable kettles may be coke, oil, or gas. In any case the fire is contained in a pit over which the kettle may be moved on the truck. Coke is a good fuel, but maintaining a uniform fire requires considerable attention. Oil is a widely used fuel; the circular pit may contain from one to three burners. The pits are of firebrick construction, and the oil flame enters the pit tangentially so that the kettle is heated by radiant heat rather than by direct contact with the flame. Gas is an excellent fuel where it is available at low cost. Care should be taken not to allow the gas flames to come into direct contact with the kettle because of local overheating. A radiant-type heater such as that made by the Selas Corp. and shown in Fig. 4 gives excellent results. A specially designed refractory tip, mounted in the center of a ceramic cup, breaks the gas-air mixture into fine streams which burn rapidly and completely. The many small cups in a burner make the entire refractory surface into an excellent source of radiant heat.

Fume Control. In many cities the health authorities will not permit the acrid fumes from varnish cooking to be vented freely into the atmosphere. At present there is no simple, inexpensive method for destroying the fumes. The two most widely used methods are burning and scrubbing. The fumes may be drawn by suction from the kettles through a pipe system to a gas burner. With a well-designed burner the fumes may be burned almost completely to carbon dioxide and then vented to the atmosphere.

The fumes may also be scrubbed with water and the residue separated from the water layer. This requires somewhat more complicated apparatus; as stated above, neither method is cheap and simple to operate. Considerable work is being done on this problem at the present time by the Chicago Paint and Varnish Production Club, and it may be expected that a satisfactory answer to fume disposal will be developed from this work. It has not been found economical to recover the material in varnish fumes as a by-product.

Thinning or Reducing. Many varnishes are finished cooking at a temperature of 560–590°F, and most of them are chilled back to 500–525°F with an oil or resin. In some large installations the varnish is pumped through a heat-exchanger into the thinning tank containing the solvent. The tank should be equipped with a condenser and a good agitator. However, the heat-exchanger is not a general feature. The solvent loss may be kept as low as 1% with a condenser in the thinning tank, and the fire hazard is reduced considerably over the open-kettle method of thinning.

In the open-kettle method the varnish is cooled to 450–400°F and the solvent run into the kettle directly with careful stirring. This operation should be done in a room with good ventilation and separated from the varnish-cooking room by a fire wall and a fire door. The fumes from the solvent are heavier than air and will settle to the floor. The draft from the varnish fires has been known to draw the fumes from a considerable distance to the fires, and a flash-back to the thinning kettle results in a disastrous fire.

Filtering. The suspended solids in the thinned varnish are removed in a filter press or by a centrifuge. A plate and frame press may be used, but it is difficult to clean and there is some loss of volatile from the varnish. The Sparkler type of horizontal-plate filter shown in Fig. 5 gives good results and is relatively simple to clean and operate. It is completely enclosed and therefore avoids the loss of volatile material. A bed of filter aid is built up on a paper filter which is readily removed at the end of the run. The scavenger-plate feature practically eliminates hold-up, thereby delivering the entire batch with no unfiltered hold-over.

The centrifuge type of filter gives good results, and the equipment is fairly easily cleaned after the run. However, there is a tendency to force the varnish through the centrifuge too rapidly, which, of course, does not give good results. There is also a tendency for particles of soft skin to slide through the machine when the rate of flow of varnish through the machine is high because the skin particles have nearly the same specific gravity as the varnish.

Storage. The storage tanks for varnish should be the closed type with manhole covers for occasional cleaning. The closed type of tank permits a positive pressure of inert gas to be maintained over the varnish, thereby reducing the tendency of the varnish to form a heavy skin on its surface. This skin breaks when the varnish is withdrawn and contaminates the finished product.

Opinion differs with respect to the position of the draw-off pipe. In most cases it is at the bottom of the tank so that the contents may be emptied completely. However, an additional draw-off pipe may be located about 6 in. from the bottom for regular use. This arrangement

Fig. 5. Sparkler Horizontal Plate Filter

CROSS SECTIONAL VIEW

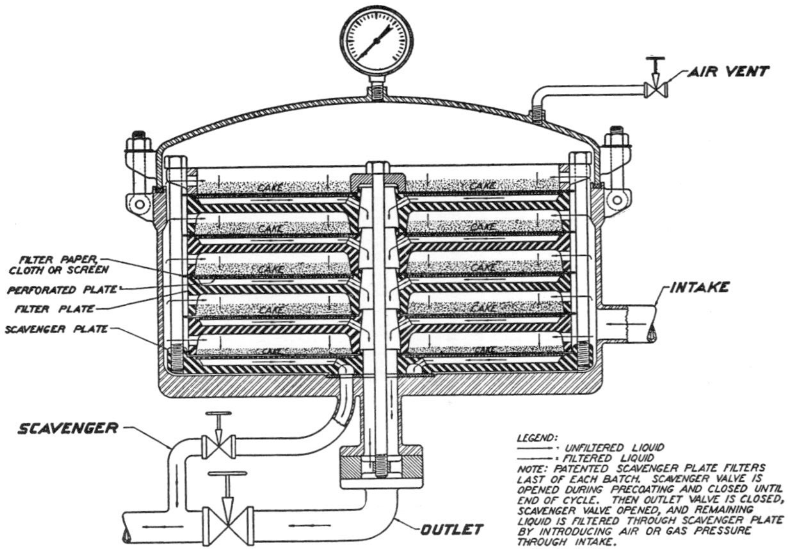

Courtesy of Sparkler Manufacturing Co., Mundelein, Ill.

leaves a layer of varnish in the bottom of the tank, containing any sludge which may have settled from the varnish during storage. The bottom layer may be removed occasionally and refiltered, or it may be used in second-grade products.

Sampling and Testing. In most cases the skill of the varnish maker is depended on to determine when the varnish is properly cooked. However, after the batch is thinned a sample should be taken to the laboratory for the regular control tests. These include the determination of viscosity, specific gravity, color, acid number, and total solids content. Some of these values should be checked again after filtering or from the storage tank. The air-drying time should be checked, or, for a baking varnish, the hardness should be checked at the required baking schedule. Many varnishes must be tested for gasproofness, exterior durability, and compatibility with other film-forming materials or with certain basic pigments. These tests are described in Chapter 15 on test methods.

Fire Protection. Adequate fire-protection apparatus should be maintained near the source of possible fires. This should be supplemented with proper training of employees regarding the correct procedure in case of an emergency. The actions taken in the first few minutes frequently determine whether the fire will be brought under control rapidly or whether it will get out of hand and do extensive damage. The carbon dioxide and foam types of extinguishers both give good control of varnish fires, but they must be used properly to obtain satisfactory results. A stream of water will serve to cool the outside of a kettle on fire, but the water should not be directed into the kettle. The water produces heavy foaming so that the contents would overflow the kettle, thereby spreading the fire considerably.

FORMULATION AND USE

From a knowledge of the performance characteristics of the various drying oils and varnish resins, the formulator may write the formulas for a series of varnishes designed to meet specific requirements. However, it is also important to have the optimum cooking procedure for any given oil and resin combination. Another complicating factor is that, although most varnishes can be cooked on fairly precise time-temperature schedules, the end point of any particular batch depends on the skill and judgment of the varnish maker. He determines when the cooking is finished by noting the relative hardness of a sample drop or "pill" removed from the kettle and placed on a steel or glass panel, or by noting the length of the "string" he obtains when he puts his finger on a soft pill and draws it away rapidly. Obviously such empirical test methods are susceptible to human error, but experience will develop a remarkable degree of accuracy.

Frequently, several small kettle batches are blended to make a large batch. In such cases, each kettle batch should be tested separately for viscosity at the required solids content. It is bad practice to blend an undercooked batch with an overcooked batch to compensate for the viscosity differences. Such blending always results in poor stability of the varnish, particularly when it is used in pigmented finishes.

VARNISHES VS. ALKYDS

The alkyd resins, which are described in Chapter 7, have replaced varnishes for some purposes. In general, alkyds are made from somewhat higher-priced raw materials, and they require a longer processing time in more complicated equipment. This economic disadvantage is offset by improvement in certain properties, notably, paler color and better exterior durability. However, varnishes can be made which have water and alkali resistance greatly superior to that of alkyds and better "through-dry." The latter term indicates the ability of varnish films to air-dry more uniformly through their entire thickness. All air-drying films tend to form a dry surface more quickly than they become

Table 1 Composition and Performance of Varnishes vs. Air-Drying Alkyds

Ingredients and Characteristics	Varnish	Alkyd
Type of oil	Good drying oils, little or no semi-drying	Considerable use of semi-drying oils
Type of resin	Considerable variation	Chiefly glyceryl phthalate
Solvents	Chiefly aliphatic	Aliphatic and aromatic
Driers	Pb, Co, Mn, Ca, cooked in batch or added afterwards	Pb, Co, Mn, Ca, added to finished alkyd
Resistance to skinning	Bad to good	Fair to very good
Air-drying	Very good through-drying	Fair through-drying
Water resistance	Poor to excellent	Fair
Alkali resistance	Very poor to excellent	Very poor to fair
Color	Very poor to good	Good to excellent
Gloss	Good to excellent	Good
Adhesion	Excellent	Excellent
Embrittlement with aging	Poor to good	Fair to very good
Exterior durability	Poor to good	Good to excellent

dry or oxidized in the interior of the film. Varnishes have better through-dry chiefly because of the difference in type of oil used. Varnishes are made with the good drying oils, tung, linseed dehydrated castor, etc., but many air-drying alkyds are made with large percentages of the semi-drying soybean oil. A very general comparison of the composition and performance of varnishes and alkyds is shown in Table 1.

TYPES OF VARNISHES

A very wide variety of varnishes may be made to meet the various requirements of cost and performance. Since it would not be practical to describe all possible variations, it was thought desirable to outline the types used for specific applications. These will fall into the two general categories of architectural and industrial finishes and will include those listed in Table 2.

Table 2 Applications for Varnishes

Furniture finishes
Metal-primer varnishes
Water-resistant durable varnishes
Alkali-resistant varnishes
Spar varnish
Aluminum-paint varnishes
Tin decorating finishes
Architectural enamel varnishes
Shelf goods
Floor varnishes

In many of the varnish formulations given in the following section the specific trade name of the resin will be given for convenience of identification. Most of these resins are described in Chapter 3. Of course, it should be understood that similar resins made by other manufacturers may be expected to give comparable results. Furthermore the cooking schedule should be used as a guide rather than a definite requirement because the top temperature and particularly the time of processing will vary with the size of the batch and type of equipment. Since there is considerable variation in price and availability of drying oils the varnish formulator must be prepared to make substitutions accordingly. He cannot obtain identical results from different oils, and frequently a change in type of oil will necessitate a change in resin to obtain the best compromise. An attempt will be made to indicate some of these possibilities in the following section.

Furniture Varnishes. The furniture business is highly competitive, and a sacrifice is often made in the quality of finish in the interest of lower cost. Although low price of finishing materials is a factor in reducing the cost of furniture finishing, it is not as large an item as the savings obtained from faster drying time and easier sanding, rubbing, and polishing. Therefore, less durable finishing materials are often used, not so much because they are inexpensive, as because they reduce the time and cost of these finishing operations. Complete details of furniture finishing are given in Volume II, where comparison is made of varnish, nitrocellulose lacquer, and low-bake finishing systems.

In general the wood is stained and filled, a sealer coat applied and sanded, and then one or more coats of finishing varnish applied. If more than one coat of varnish is applied the first coat is sanded lightly. The finish may be left in its original gloss, but more often it is rubbed to a dull or satin finish with pumice and oil. In high-grade work the finish may be polished to a somewhat higher gloss with rotten stone. The hand operations of sanding, rubbing, and polishing, even with mechanical aids, are some of the most expensive in furniture finishing; consequently, every effort is made to reduce them to a minimum. In some cases the finish coat of varnish is dulled or flattened by inclusion of zinc or aluminum stearates or colloidal silica, but this does not give as desirable a finish as hand rubbing.

It should be apparent that speed and ease of rubbing will depend on hardness of the film. Since short-oil varnishes become dry and hard more rapidly than long-oil varnishes and also rub more easily to a better-looking finish, they are used despite their lower flexibility and less resistance to scratching and marring. In addition, furniture is bulky and requires considerable floor space in finishing rooms and storage. Therefore, the furniture manufacturer wants the fastest possible finishing schedule so that he can ship his stock quickly. Furniture must be packed in paper, burlap, etc., to protect it from damage during shipment, and unless the finish is thoroughly dry and hard these materials will adhere to it and leave an ugly print mark when removed. This places an additional problem on the varnish formulator to develop rapid-drying, print-resistant finishes and still retain some degree of toughness and durability. The following examples indicate the types of varnishes in general use. The differences in cooking times and top temperatures should be noted together with differences in color and viscosity at 50% or 60% solids content. The viscosity is measured with the Gardner-Holdt bubble tubes unless noted otherwise, and the color determined by the Gardner 1933 standards.

Formula 1 High-Grade Rubbing Varnish

Modified phenolic resin
Oil: tung-linseed combination 12 gal

Amberol * K-12A	100 lb
Tung oil	8 gal
Linseed, body Q	4 gal
Mineral spirits	30 gal

Cooking Procedure

Heat resin and 4 gal of tung oil to 550°F. Add 4 gal of tung oil and reheat to 550°F. Check with 4 gal of linseed and hold at 500°F for required body as indicated by a firm pill. Cool to about 450°F and reduce. Add naphthenate driers to give the following percentages by weight of metal to oil in the varnish: 0.3% Pb, 0.015% Co, 0.015% Mn.

Characteristics

Total solids	50%
Viscosity	G–H
Color	11–12

* Trade name, Rohm and Haas Co.

Formula 2 Furniture Rubbing Varnish

Maleic-modified rosin resin
Oil: tung-linseed combination 12 gal

Teglac * resin Z152	100 lb
Tung oil	8 gal
Linseed, body Q	4 gal
Mineral spirits	30 gal

Cooking Procedure

Heat resin and tung oil to 450°F to complete solution, then continue heating to 575°F. Hold about 5 min, then check with linseed oil to 500°F and hold for required body. Cool to about 450°F and reduce. Add soluble driers to give the following percentages of metal to oil in the varnish: 0.5% Pb. 0.05% Co.

Characteristics

Total solids	50%
Viscosity	R–T
Color	5–6

* Reg. U. S. Patent Office.

This varnish is paler in color than that made by Formula 1, but it is not as water or alkali resistant. The following formula illustrates an all-linseed-oil type for use when tung oil is not available or is too high in price.

Formula 3 Soft-Oil Rubbing Varnish

Phenolic-modified, pentaerythritol-rosin ester
Oil: bodied linseed 10 gal

Pentalyn * M-802A	100	lb
Linseed, body Q	10	gal
Mineral spirits	27.5	gal

Cooking Procedure

Heat resin and oil to 585°F in 1 hr; hold at 585°F for body (about 40–50 min); cool to 450°F and reduce. Add soluble driers to give 0.5% Pb, 0.08% Co, 0.05% Mn as metal to oil in varnish.

Characteristics

Total solids	50%
Viscosity	C–D
Color	12

* Trade name, Hercules Powder Co.

Despite the use of the fast-bodying penta resin, the cooking time for varnish 3 is longer than that for varnish 1 or 2. Furthermore, varnish 3 could not be expected to dry through as rapidly or as completely as varnishes made with some tung oil. Oiticica and dehydrated castor oil may be used with appropriate modifications of cooking time and temperature and type of hard resin.

Formula 4 Mixed Resin-Mixed Oil Rubbing Varnish

Phenolic-modified pentaerythritol rosin ester—100% phenolic resin
Oil: tung-linseed combination

Pentalyn * 802A	66.6	lb
Bakelite 9400	33.4	lb
Tung oil	8	gal
Linseed, alkali-refined	2	gal
Mineral spirits	33	gal
Toluol	7	gal

Cooking Procedure

Heat resins and oils to 565°F; shut heat off and allow to coast to 575°F. Cool to 450°F and reduce with mineral spirits followed by the toluol. Add soluble driers to give 0.5% Pb, 0.05% Co as metal to oil in varnish.

Characteristics

Total solids	40%
Viscosity	D–F
Color	12

* Trade name, Hercules Powder Co.

This varnish is fast drying and gasproof, and it illustrates the addition of a 100% phenolic resin to the modified phenolic resin to improve water and solvent resistance. The small amount of linseed oil assists in solubilizing the resins in the cooking process.

Furniture such as serving tables and bar tops requires a finish which will withstand water-alcohol mixtures without softening or showing a white ring from contact with the bottom of wet glasses. Varnishes 2 and 3 have poor resistance to such conditions, and varnishes 1 and 4 have only fair resistance. A varnish which will meet these requirements may be made with a 100% phenolic resin and 12½ gal of tung oil as in varnish 18. This type of varnish is frequently referred to as a bar-top finish.

Metal-Primer Varnishes. Medium-oil-length varnishes are used as vehicles for air-drying and baking primers for metal finishing. Usually, they are pigmented with corrosion-inhibiting pigments, but they may also be pigmented with white pigments for use under white enamels. Excellent results are obtained with a varnish of 25 gal tung-linseed combination and a modified phenolic resin as in Formula 5. This varnish is fast air-drying or baking and has excellent toughness and adhesion and good water resistance. Formula 6 shows the different cooking procedure required when the type of oil is changed. Formulas 7 and 8 illustrate the use of a resin which will gasproof the oil at low temperatures, 465–475°F. They also show the difference in cooking times for tung and oiticica oils with this resin. Formulas 9 and 10 illustrate the use of resins designed especially for soft oils. It will be noted that the soft oils are used as bodied oils to reduce the cooking time of the varnish. It should also be noted that the cooking procedure is modified to assure proper dispersion or solubility of the resin in the oil.

Formula 5 Medium-Oil, Modified Phenolic Varnish

Oil: tung-linseed combination

Amberol * F-7	100 lb
Tung oil	18 gal
Linseed, body Q	7 gal
Mineral spirits	46 gal

Cooking Procedure

Heat resin, tung oil, and 3 gal of linseed oil to 560°F in 30–35 min, check immediately with 4 gal of linseed oil, and cool to 500°F. Hold for required body. Cool to 450°F and reduce.

* Trade name, Rohm and Haas Co.

Formulas 5–10 represent a very useful group of varnishes which may serve many purposes other than as vehicles for metal primers. Reference will be made to them under the other applications.

Formula 6 Medium-Oil, Modified Phenolic Varnish

Oil: linseed-dehydrated castor combination

Amberol * F-7	100 lb
Linseed, body S–V	10 gal
Dehydrated castor, body Z3	15 gal
Mineral spirits	46 gal

Cooking Procedure

Heat resin and 10 gal of bodied linseed oil to 580°F and hold for a clear, cold pill on a glass plate. Add the 15 gal of bodied dehydrated castor oil and reheat to 580°F. Hold for a clear, cold pill on glass. Cool to 450°F and reduce.

* Trade name, Rohm and Haas Co.

Formula 7 Medium-Oil, Modified Phenolic Varnish

(Gasproof at low temperatures)

Pentalyn * 25B	100 lb
Tung oil	25 gal
Mineral spirits	41 gal
Turpentine	4 gal

Cooking Procedure

Heat resin and oil to 475°F, hold for 45 min, cool to 450°F, and reduce with mineral spirits and turpentine. Add soluble driers to give 0.5% Pb, 0.05% Co, and 0.05% Mn as metal to oil. Care should be taken not to exceed 475°F in the processing. The holding time will vary somewhat with the grade of oil, and it may be necessary to lower the holding temperature to 450°F for large batches. The addition of the National Aniline Co.'s anti-skinning agent ASA will reduce the tendency to skin in the container. Usually 0.5% of ASA on varnish solids is sufficient.

Characteristics

Total solids	50%
Viscosity	G–H
Color	10–11

* Trade name, Hercules Powder Co.

Formula 8 Medium-Oil, Modified Phenolic Varnish

(Gasproof at low temperatures)

Pentalyn * 25B	100 lb
Oiticica oil	25 gal
Mineral spirits	41 gal
Turpentine	4 gal

Cooking Procedure

Heat resin and oil to 465°F, hold for 115 min, cool to 450°F, and reduce with mineral spirits and turpentine. Add soluble driers to give 0.5% Pb, 0.05% Co, and 0.05% Mn as metal to oil. Add 0.5% ASA based on varnish solids.

Characteristics

Total solids	50%
Viscosity	G–H
Color	10

* Trade name, Hercules Powder Co.

Varnishes 7 and 8 illustrate the ability of certain resins to gasproof tung or oiticica oil varnishes at low temperatures in reasonable cooking time.

Formula 9 * Medium-Oil, Modified Phenolic Varnish

Oil: bodied linseed

Amberol † F-77	100 lb
Linseed oil, body Z4	25 gal
Mineral spirits	46 gal

Cooking Procedure

Heat resin and 10 gal of oil to 580°F and hold for clear, cold pill on glass. Add remainder of oil, reheat to 580°F, and hold for clear, cold pill. Cool to 450°F and reduce.

* Courtesy of Rohm and Haas Co.

† Trade name, Rohm and Haas Co.

Formula 10 Medium-Oil, Modified Phenolic Varnish

Oil: bodied dehydrated castor

Amberol * M-88	100 lb
Dehydrated castor oil, body Z3	25 gal
Mineral spirits	46 gal

Cooking Procedure

Heat resin and 10 gal of oil to 540–550°F and hold for clear, cold pill on glass. Add the remainder of the oil, reheat to 550°F, and hold for clear, cold pill on glass. Cool to 450°F and reduce.

* Trade name, Rohm and Haas Co.

Driers. The driers for varnishes 5–10 vary with the type of pigmentation and air-drying or baking conditions. The general range for air-drying conditions is as follows: 0.5–0.8% Pb, 0.05–0.10% Co, 0.01–0.02% Mn, with the soft-oil varnishes requiring the upper end of the range. For baking conditions of 200–250°F about 0.02% Mn is desirable, but at 275°F or higher the drier is not necessary unless a very short baking time is required.

If lower-cost primer vehicles are required, the modified phenolic resin may be replaced entirely or in part by ester gum with the hard oil, or by penta-modified ester gum with the soft oils. The cooking time will be increased, and the varnishes will be slower drying and have less resistant films. Another possibility is the use of a petroleum resin as in Formula 11.

Formula 11 Medium-Oil, Petroleum Resin Varnish

Oil: dehydrated castor-fish combination

Panarez * 6-210	100 lb
Dehydrated castor oil, body Z3	16 gal
Alkali-refined sardine oil	9 gal
Mineral spirits	40 gal

Cooking Procedure

Heat 50 lb of the resin and the dehydrated castor oil to 585°F in 35–40 min and check with the sardine oil. Reheat to 550–560°F, hold for body, and check with the remaining 50 lb of resin. Cool to 450°F and reduce. Add soluble driers to give 0.5% Pb and 0.05% Co as metal to oil in the varnish.

Characteristics

Total solids	53.4%
Viscosity	E–F
Color	13

* Trade name, Pan American Chemicals Div., Pan American Refining Corp.

Varnishes of Maximum Water Resistance and Durability. For the combination of maximum water resistance and exterior durability, it is necessary to use medium- to long-oil varnishes made with 100% phenolic resins and tung oil, or a mixture of tung and linseed oils. In general, resins are more water resistant than oils; therefore short-oil varnishes are more water resistant than longer-oil varnishes. However, short-oil varnishes are not flexible enough for exterior durability. The following six formulas illustrate the use of heat-reactive and heat-nonreactive pure phenolic resins with tung oil and tung-linseed combinations.

Formula 12 25 Gallon Tung, Heat-Non-Reactive Phenolic Varnish

Bakelite * BR-9432	100 lb
Tung oil	25 gal
Mineral spirits	30.5 gal

Cooking Procedure

Heat resin and 12½ gal tung oil to 560°F in 30–35 min, hold 15 min, add remainder of the oil. Reheat to 450°F, hold about 60 min for body, reduce, add soluble driers to give 0.4% Pb, 0.05% Co, 0.02% Mn, metal to oil.

Characteristics

Total solids	60%
Viscosity	M
Color	11

* Trade name, Bakelite Co., Div. Union Carbide and Carbon Corp.

This varnish is gasproof, non-reactive with zinc oxide, has no skins in the skinning test in 2 weeks, passes 110% kauri reduction, fails at 120%. The dried film shows no failure after immersion in 5% NaOH solution for 24 hr.

Formula 13 25 Gallon Tung, Heat-Reactive Phenolic Varnish

Bakelite * BR-14634	100 lb
Tung oil	25 gal
Mineral spirits	52.5 gal

Cooking Procedure

Heat oil to 450°F in 20 min, remove from fire, and allow to cool to 445°F. Add the resin and stir continuously until melted, which takes about 15 min. Reheat to 425°F in 20 min, hold about 5 min for body, then reduce. Add soluble driers to give 0.3% Pb, 0.03% Co, 0.015% Mn as metal to oil.

Characteristics

Total solids	47%
Viscosity	E
Color	10

(Note the difference in cooking procedure between varnishes 12 and 13.)

* Trade name, Bakelite Co.

This varnish is gasproof and is slightly reactive with zinc oxide. It skins within 24 hr; therefore it requires about ½ oz of antiskinning agent per gallon. It passes 80% kauri reduction but fails at 90%. The dried film shows no failure after immersion in 5% NaOH for 48 hr.

Formula 14 25 Gallon General-Utility Varnish

Bakelite * BR-14634	25 lb
WW rosin	10 lb
Ester gum	65 lb
Tung oil	20 gal
Linseed oil, body Z1	5 gal
Mineral spirits	44 gal

Cooking Procedure

Heat rosin, ester gum, and tung oil to 400°F. Remove from fire, add BR-14634, and stir continuously until complete dispersion is obtained. Heat to 560°F in about 20 min; check at once with the linseed oil. Cool to 450°F in about 35 min and reduce. Add soluble driers to give 0.5% Pb, 0.05% Co, and 0.02% Mn as metal to oil. Add 0.2% antiskinning agent by volume of varnish.

Characteristics

Total solids	51%
Viscosity	E
Color	13

* Trade name, Bakelite Co.

The varnish is gasproof, is moderately reactive with zinc oxide, does not skin in 2 weeks, passes 70% kauri reduction, but fails at 80%. The dried film shows no failure after immersion in 5% NaOH for 4 hr. Note the lower alkali resistance due to the rosin modification. This formula illustrates the use of heat-reactive 100% phenolic let down with rosin.

Formula 15 25 Gallon Linseed, Heat-Non-Reactive Phenolic Varnish

Bakelite * BR-254	100 lb
Alkali-refined linseed oil	25 gal
Mineral spirits	23.5 gal
Xylol	6 gal

Cooking Procedure

Heat oil and resin to 585°F in 40–50 min, hold about 60 min for body, cool to 450°F, and reduce. Add soluble driers to give 0.5% Pb, 0.1% Co as metal to oil.

Characteristics

Total solids	60%
Viscosity	M
Color	12

* Trade name, Bakelite Co.

The varnish shows no skinning in 2 weeks, is not reactive with zinc oxide, passes 55% kauri reduction, fails at 65%. The film is resistant to 5% NaOH for 8 hr. Compare the toughness and flexibility with those of tung oil varnish 12. Also compare the alkali resistance of varnishes 12, 13, 14, and 18. Note the use of the aromatic solvent.

Formula 16 33 Gallon Tung-Linseed, Heat-Non-Reactive Phenolic Varnish

Bakelite * BR-254	100 lb
Alkali-refined linseed oil	13 gal
Tung oil	20 gal
Mineral spirits	26 gal
Cellosolve solvent	3 gal
Hi-Flash naphtha	1 gal
Dipentene	4 gal
Toluol	2 gal

Cooking Procedure

Heat resin and linseed oil to 550°F in 35 min and hold for 15 min. Add tung oil, reheat to 465°F in 10 min, hold about 25 min for an 18-in. string from a cold pill. Reduce and add soluble driers to give 0.2% Pb, 0.02% Co, 0.01% Mn as metal to oil.

Characteristics

Total solids	60%
Viscosity	D
Color	11

* Trade name, Bakelite Co.

The varnish is gasproof, has no skinning in 2 weeks, shows no reactivity with zinc oxide, passes 130% kauri reduction, fails at 150%. The film is resistant to 5% NaOH for 5 hr. Note the solubilizing of the resin in the linseed oil and the use of a mixed solvent.

Formula 17 50 Gallon Tung, Heat-Non-Reactive Phenolic Varnish

Bakelite * Br-254	50 lb
Bakelite * BR-4036	50 lb
Tung oil	50 gal
Mineral spirits	54.5 gal

Cooking Procedure

Heat oil and resins to 450°F in 45 min, hold about 25 min for body, reduce, and add soluble driers to give 0.1% Pb, 0.01% Co, 0.007% Mn as metal to oil.

Characteristics

Total solids	58%
Viscosity	D
Color	10

* Trade name, Bakelite Co.

The varnish is gasproof, has no skinning in 2 weeks, shows no reactivity with zinc oxide, passes 170% kauri reduction, fails at 180%. The film is resistant to 5% NaOH for 10 hr. This illustrates the use of 100% phenolic resin and tung oil to produce a varnish having excellent flexibility but retaining good resistance to alkali and water.

Alkali-Resistant Varnishes. A comparison of the alkali resistance of varnishes 12–17 shows the effect of type of resin and oil. The heat-reactive resin in varnish 13 is better than the heat-non-reactive resin in varnish 12, and both are much superior to the rosin-modified resin in varnish 14. The comparison between tung and linseed oils with heat-non-reactive resins in varnishes 12 and 15 shows the superiority of tung oil for alkali resistance. The increase in oil length from 25 gal to 50 gal, varnishes 12 and 17 respectively, shows the decrease in alkali resistance with more oil. For maximum alkali resistance a short-oil varnish must be used, but, of course, this advantage cannot be obtained without a sacrifice of flexibility. However, flexibility is not important on many rigid surfaces. The following varnish gives excellent results in alkali-resistant finishes.

Formula 18 12½ Gallon Tung, Heat-Non-Reactive Phenolic Varnish

Bakelite * BR-254	100 lb
Tung oil	12.5 gal
Xylol	27.5 gal

Cooking Procedure

Heat oil and resin to 450°F in 30 min, hold about 25 min for body, reduce, and add soluble driers to give 0.05% Pb, 0.006% Co as metal to oil.

Characteristics

Total solids	50%
Viscosity	D
Color	10

* Trade name, Bakelite Co.

The varnish is gasproof, does not skin in 2 weeks, is not reactive with zinc oxide, and fails at 0% in the kauri reduction test. The film is resistant to 5% NaOH for 96 hr.

Coumarone-indene resins also have excellent alkali resistance, but they do not retard the gelation of tung oil. This defect can be overcome by using part of the resin as a chill-back, and in addition such materials as litharge and cobalt acetate in the longer-oil varnishes. The procedure for varnish 19 illustrates the general method.

Formula 19 16 Gallon Tung, Coumarone-Indene Varnish

Cumar * Resin V	100 lb
Tung oil	16 gal
Cobalt naphthenate (6%)	12 oz
Mineral spirits	28 gal

Cooking Procedure

Heat oil very quickly to 400°F, add 75 lb of resin, and continue heating rapidly to 550°F. Take kettle off the fire, allow temperature to rise to 570–575°F, check with 25 lb resin, and water-cool the outside of the kettle to a batch temperature of 500°F. Hold for body between 500–450°F, reduce, and add drier.

* Trade name, Barrett Div., Allied Chemical and Dye Corp.

Combinations of heat-reactive phenolic resins and coumarone-indene resins may be made to lower the cost over straight phenolic resins and still maintain very good alkali resistance. However, this type of varnish does not have the durability of a straight phenolic varnish.

Formula 20 27 Gallon Tung-Linseed, Cumar-Heat-Reactive Phenolic Varnish

Cumar * Resin W	75 lb
Heat-reactive phenolic	25 lb
Tung oil	24 gal
Linseed oil, body Q	3 gal
Mineral spirits	33 gal

Cooking Procedure

Heat tung oil and phenolic resin to 450°F in 25–30 min and hold until reaction begins. Add 35 lb Cumar, heat to 500°F, add at once 40 lb Cumar and the linseed oil, obtain body at 480–450°F, reduce, and add driers to give 0.5% Pb, 0.03% Co as metal to oil.

* Trade name, Barrett Div., Allied Chemical and Dye Corp.

The cooking procedure will vary slightly with different types of heat-reactive phenolic resins, and considerably if the heat-non-reactive type is used. A great many combinations of Cumar resins and the various drying oils may be made to produce varnishes having better alkali resistance than if they were made with ester gum or the maleic-type resins.

In the varnishes which have been discussed the resin-oil relationship has varied from simple dispersion of the resin in the oil to actual chemical reaction of the resin with the oil. In general, improved film properties are obtained from varnishes in which the resin has reacted with the oil. A similar situation exists in the oil-modified alkyd resins; the oil content is chemically combined with the other constituents, thereby giving better film properties.

The description of alkyd resins in Chapter 7 shows that they are usually processed in more complicated equipment than is required for varnish making. In order to obtain some of the advantages of alkyd resins while using varnish-making equipment the Hercules Powder Co. developed a chemical intermediate known as Pentalyn B-56. When this intermediate is cooked in a varnish kettle with various unbodied oils an ester-interchange reaction is believed to occur which produces a semi-alkyd type of resin. Fast-drying, durable finishes may be produced from this type of resin because of the composition of the intermediate and its ability to react with the oil.

In general, the refined unbodied oils are used with Pentalyn B-56. Oils such as soybean oil may be used with Pentalyn B-56 even though they are not satisfactory with the usual varnish resins because of their slow-drying feature. An indication of the processing times and temperatures for a series of oils with Pentalyn B-56 is given in the following outline. The data are based on a combination of 100 lb of Pentalyn B-56 and 20 gal of the various oils except the soybean oil, for which the amount was 15 gal.

Outline * of Processing Schedule

Pentalyn B-56 with Various Oils

Oils	Fish	Soybean	Safflower	Linseed	Dehyd. Castor, G–H
Time to top temp., min	45	50	50	60	45
Top temp., °F	600	585	600	585	575
Time at top temp., min	90	100	90	85	45
Viscosity at 50% solids in mineral spirits	G	G †	D–E	G–H	E–F
Color at 50% solids in mineral spirits	12	10–11	11	9	9

* Courtesy of Hercules Powder Co.
† Viscosity at 55% solids.

Spar Varnishes. The term "spar varnish" indicates a varnish which has good water resistance. Spar varnishes are used as clear finishes for marine work and as vehicles for marine enamels and floor and porch enamels. The best water resistance is obtained with 100% phenolic resin and tung oil in oil lengths from 25 to 50 gal. Typical examples of this class are varnishes 12, 13, and 17. Where lower cost requirements must be met the modified phenolic resins may be used, but the oil length should not be less than 40 gal for exterior applications.

The modified phenolic resin Amberol 226 is designed especially for this type of varnish, as shown in Formulas 21 and 22.

Formula 21 44 Gallon Tung, Modified Phenolic Spar Varnish

Amberol * 226	100 lb
Tung oil	44 gal
Mineral spirits	64 gal

Cooking Procedure

Heat resin and oil to 500°F, hold about 40–50 min for a 12-in. string from a cold pill, cool to 450° F, and reduce.

* Trade name, Rohm and Haas Co.

Formula 22 44 Gallon DCO, Modified Phenolic Spar Varnish

Amberol * 226	100 lb
Dehydrated castor oil, body S–V	44 gal
Mineral spirits	64 gal

Cooking Procedure

Heat resin and oil to 560–575°F, hold about 2 hr for a 10- to 12-in. string, cool to 450°F, and reduce.

* Trade name, Rohm and Haas Co.

Formula 23 is for another modified phenolic resin spar varnish. This varnish may be made gasproof at a relatively low-temperature cook because of the type of resin used. (See also Formulas 7 and 8.)

Formula 23 40 Gallon Tung, Modified Phenolic Spar Varnish

Pentalyn * 25B	100 lb
Tung oil	40 gal
Mineral spirits	46 gal
Turpentine	5 gal

Cooking Procedure

Heat resin and oil to 500°F, hold about 15 min, allow to cool to 450°F, reduce with mineral spirits and turpentine. Add soluble driers to give 0.5% Pb, 0.05% Co, and 0.05% Mn as metal to oil. Add 0.5% antiskinning agent ASA based on varnish solids. The batch should be watched carefully on the down heat; if a slight string develops the batch should be water-cooled and reduced.

Characteristics

Total solids	55%
Viscosity	D–E
Color	10–11

* Trade name, Hercules Powder Co.

Aluminum-Paint Varnishes. The brilliant finish of aluminum paint is produced by the flakes aligning themselves in almost a continuous layer in the surface. The effect is known as "leafing"; if anything interferes with proper leafing, the finish is much duller. Varnishes promote good leafing provided they have low acid number, do not contain lead drier, and are not overcooked. A small percentage of aromatic solvent also promotes good leafing, but a solvent such as butanol will "deleaf" the aluminum and produce a duller finish.

Aluminum paints are used on a very wide variety of both interior and exterior surfaces and also on surfaces such as metal smokestacks and automobile exhaust pipes which are subjected to elevated temperatures. Finishes based on oil varnishes tend to blister and peel off the exhaust pipe, but a gloss oil will volatilize without blistering and leave the layer of aluminum flakes tightly bonded to the pipe. However, for exterior use on smokestacks a more durable long-oil finish must be used. Considerable success has been obtained with silicone resins for this work, but their cost is quite high.

Varnishes 16 and 17 give excellent results where good durability and water resistance are required. Varnishes 5 and 6 are entirely satisfactory for interior surfaces and for exterior applications where exposure conditions are not severe. Aluminum paints are sometimes used as primers for exterior wooden surfaces. These are usually long-oil varnishes ranging from 60 to 80 gal of oil to insure sufficient flexibility to withstand the dimensional changes in the wood due to moisture gain and loss. These varnishes may be made of simple mixtures of ester gum and bodied linseed oil thinned with mineral spirits. The coumarone-indene resins likewise give good results because of their low acid value and good solubility. It is also important that the oil be not highly bodied and that it have a low acid value. The medium-bodied oils which have been heat-bodied under vacuum give the best result. A typical example is shown in varnish 24.

Formula 24 60 Gallon Aluminum-Paint Vehicle

Cumar * Resin W	100	lb
Bodied linseed OKO, S-70	60	gal
Xylol	10	gal
Mineral spirits	79	gal
Cobalt naphthenate (6%)	3.2	lb

Cooking Procedure

The ingredients may be dissolved in the solvents by agitation at room temperature. Instead, the oil may be heated to 470°F, and the resin added; when it is melted, the batch is cooled to 400°F, thinned with the solvents, and the driers are added.

* Trade name, Barrett Div., Allied Chemical and Dye Corp.

Variations in the proportions of the above materials may be made to give a series of aluminum-paint vehicles ranging from very short-oil varnishes to the long-oil type.

Tin Decorating Varnishes. Tin decorating varnishes are used as clear coatings over lithographed tin plates and also as vehicles for the pigmented first coat. These coatings are applied to flat sheets of tin plate which are cut and formed into decorative metal containers, bottle caps, etc.

Another type of tin decorating finish is applied on the inside of food containers. These coatings must withstand the processing conditions used to cook or preserve the foodstuff. They are highly specialized coatings because of the various types of food and the fact that they must be odorless, tasteless, and non-toxic. Some are based on both pure and modified phenolic resins with tung oil in lengths of 20–30 gal. The processing coatings are baked at schedules ranging from 275°F for 30 min, through 300°F for 20 min and 350°F for 10–15 min, to as high as 400–425°F for 5–10 min. The range of varnishes including 12, 13, 14, 17, 21, and 22 is the type used. The lead drier should be omitted, and only cobalt or manganese used in amounts ranging from 0.001% to 0.005% as metal to oil in the varnish. Manganese yellows considerably, and sufficient can be included to impart a yellow brass effect to the tin metal if desired. A brighter brass color can be obtained from small amounts of tartrazine yellow or orange pigment.

Although the varnishes indicated above give good results for many foodstuffs they are not sufficiently resistant to beer. For beer-can linings a vinyl resin gives excellent resistance, but it is usually applied over a baked varnish coating to assure good adhesion and to prevent contact of the vinyl coating with the tin. The tin acts as a degradation catalyst for vinyl coatings, particularly at elevated temperatures.

Since many of the decorative coatings are white or pale tints the varnishes for this class must be made of non-yellowing oils and resins. The baking schedules range from 200°F for 30 min, through 250°F for 20–25 min, to 275°F for 15–20 min. Because of the pale colors obtained with alkyds and alkyd-amino combinations they are sometimes used for this work, but satisfactory results can be obtained in many cases with combinations of maleic-type resins with the conjugated oils, tung, oiticica, and dehydrated castor, with or without some bodied soybean oil. Manganese drier cannot be used for pale baking finishes, but about 0.002–0.04% cobalt is generally satisfactory.

A mar-resistant surface or "slickness" can be imparted to the finish coat by incorporating about 1 oz to the gallon of paraffin wax, petrolatum, or similar material. The wax is not compatible with the film; therefore it exudes to the surface and gives the "slick" feel. It cannot be used in air-drying finishes because it retards oxidation tremendously. Also it should not be used in baking undercoats because it detracts from the adhesion of subsequent coats. The following varnishes are typical of those generally used for metal decorating purposes.

Formula 25 Metal Decorating Varnish

Amberol * 801P	100 lb
Dehydrated castor oil, body Z3	10 gal
Mineral spirits	28 gal

Cooking Procedure

Heat resin and oil to 580° F, hold for clear pill on glass, cool to 450°F, and reduce. Add soluble driers to give 0.03% Co, as metal to oil.

Characteristics

Total solids	50%
Viscosity	G
Color	9

* Trade name, Rohm and Haas Co.

This varnish gives excellent results as a clear gloss finish coat over white and colored lithography. The following varnish is more flexible and suitable as both a clear and a pigmented finish. It is made with the penta-modified type of maleic resin for fast cooking time with the soft oils.

Formula 26 Metal Decorating Varnish

Teglac * resin 164	100 lb
Dehydrated castor oil, body S–V	12 gal
Soybean oil, body X	3 gal
Mineral spirits	33 gal
Zinc naphthenate (8%)	12 oz
Cobalt naphthenate (6%)	16 oz

Cooking Procedure

Heat resin and dehydrated castor oil to 575°F in 45–60 min, check with soybean oil, cool rapidly to 450°F, and reduce. Add soluble driers as indicated to give 0.05% Zn, 0.05% Co as metal to oil.

Characteristics

Total solids	50%
Viscosity	D–E
Color	9–10

* Reg. U. S. Patent Office.

The bodied soybean oil assists the good flow-out which is essential to tin decorating enamels. These enamels are applied by a roller coating machine, and the enamel surface is badly ribbed or striated from contact with the soft delivery roll. These striations must flow out rapidly because the tin sheet enters the tunnel baking oven almost immediately after coating. The zinc drier serves the double purpose of assisting the wetting of the pigment and keeping the film "open," or preventing rapid surface drying in order to give better through-drying in the short baking schedule.

Architectural Enamel Varnishes. Architectural enamels are chiefly white or light tints; therefore their vehicles should be nonyellowing. Alkyd resins modified with soybean and dehydrated castor oils are used for high-grade white enamels because they are satisfactory for both interior and exterior purposes. For interior use very good vehicles can be made with maleic-type resins and dehydrated castor oil with or without some bodied soybean oil. The penta-modified type of resin gives faster bodying time in the kettle, but care must be taken to solubilize the resin completely in the oil. The following varnish illustrates the use of a "split cook" to assure solubility of the resin in the low-viscosity oil.

Formula 27 Architectural Enamel Vehicle

Teglac * resin 164	100 lb
Dehydrated castor oil, body S–V	8 gal
Dehydrated castor oil, body Z3	12 gal
Soybean oil, body X	10 gal
Mineral spirits	44 gal

Cooking Procedure

Heat resin and the S–V-body dehydrated castor oil to 575°F in 40–45 min, add the Z3-body oil and reheat to 575°F in 15–20 min, check with the X-body soybean oil to 450°F, and reduce.

Characteristics

Total solids	55%
Viscosity	E–F
Color	9

* Reg. U. S. Patent Office.

Driers. Soluble driers may be added to the varnish or to the finished enamel. For normal conditions add 0.5% Pb, 0.04% Co as metal to oil in the varnish. However, if a fumeproof enamel is required the lead should be replaced with 0.04–0.08% Ca. The type of pigment used has some effect on the amount of drier required. This type of varnish gives easy brushing, good leveling, and a higher gloss than can usually be obtained with an alkyd resin.

There is a large demand for low-cost interior enamels in metropolitan areas for redecorating apartments. The situation is highly competitive with the result that little attention is paid to after-yellowing or other aging characteristics of the film. Vehicles for this purpose may be made by mixing resin solutions and bodied oils in the proportions to give 25–50 gal varnishes. The resins include ester gum, zinc resinate, and limed rosin, and the oils range from high-grade, heat-bodied, vacuum-processed linseed oil to blown or oxidized linseed or fish oil. The catalogs of oil manufacturers list a variety of special oils designed for this general use.

Shelf Goods. "Shelf goods" is the term applied to paints and enamels sold by the local hardware store. They are designed for use by the homeowner rather than the master painter. The latter usually buys in 5-gal pails and from the paint manufacturer or his distributor. Shelf goods include a range of white and bright-colored enamels and porch and floor finishes. These materials must have good package stability since they may be on the dealer's shelf for 6 months or more before they are used. They should not skin or change in consistency because of reactions in the vehicle or with the pigment. The reactions in the vehicle are usually due to overbodying with the formation of incipient gel particles. These particles tend to grow on aging and become less soluble, finally separating as visible specks. The reactions with pigments are of two types. The pigment may flocculate around the gel particles to form "seeds" which settle to a gummy mass in the can and form specks in the film. Instead, the vehicle may have a high acid value and react with basic pigments through a chemical process to form soaps, some of which increase the consistency a great deal. The worst condition is obtained from a combination of overbodying and high acid value. The high acid peptizes the gel condition in the clear vehicle to some extent, but if the acid reacts with a pigment the entire paint may develop a colloidal condition having the consistency of liver. This process, which is usually referred to as "livering," may exist in a mild form which gives bad thickening, or in a severe form which makes the paint useless.

The most satisfactory type of vehicle for shelf goods is the modified phenolic varnishes from 20 to 30 gal in length. They are fast drying; in fact they made possible the so-called "four-hour enamels." They have good can stability when processed properly, but their color may not be suitable for certain grades of white enamels. The phenolic resin may be replaced by the maleic-type resin for white finishes, but, in general, the maleic resin varnishes are not as stable with colored pigments as the phenolic varnishes. Varnishes 5, 6, 9, and 14 are typical of the varnishes used as shelf-goods vehicles. The differences in drying time and film characteristics among these varnishes should now be apparent to the reader. The situation in regard to the availability and prices of the various oils is such that the formulator must be familiar with the modifications necessary to change from one oil to another with a minimum difference in the performance of the product.

Floor Varnishes. Considerable difference of opinion exists with respect to the choice of varnish or shellac for floors. The matter can be resolved by considering the type of use that the floor must withstand. For normal use in homes shellac has the distinct advantage of ease of repair. The scratches which are made rather easily in its surface by particles of grit, etc., may be removed readily by "touching up" with a dilute solution of shellac. The dried film remains soluble and is merged completely in the new application. Varnish films become insoluble in subsequent coats; therefore the surface scratches and worn spots tend to show through the new coat. However, varnishes are necessary on floors which must withstand considerable traffic, such as in gymnasiums, offices, and industrial buildings, because varnishes have greater toughness and resistance to wear.

Consider the requirements for a floor varnish: it must dry as quickly as possible and be hard enough to walk on overnight, but be sufficiently tough or elastic to resist abrasion and scratching. It should also be resistant to water and alkaline cleaning solutions. Obviously, a phenolic resin-tung oil varnish would be preferred if conditions permit. Compare varnishes 1, 5, 12, and 13 for use on floors. Varnish 1 would meet the requirements except toughness on aging; it would become too brittle to withstand abrasion without bad scratching. Varnish 5 would be slower drying than varnish 1 but would be satisfactory in most respects. It would be somewhat better if made with all tung oil. Varnishes 12 and 13 would give excellent results, but the price may be too high for general use.

If the soft oils must be used instead of tung oil, one cannot expect equal results but good performance would be obtained from varnish 6, 9, or 10. For this purpose the oil length could be reduced to 20 gal in the interest of faster and harder drying.

The reader should review the properties which may be expected from the series of varnishes mentioned. It would be a good exercise to list the varnishes in tabular form and assign numerical values for the various characteristics, including cost.

Set a goal in life,
strive constantly toward it.
Life will be truly successful
when the goal remains on the horizon.

Chapter 5. DRIERS

Driers are used in some types of organic coatings to accelerate the conversion of the liquid coating to the dry film. They are of special interest in the oxidizing, air-drying coatings, but they are also included in many baking coatings. The use of driers with drying oils and varnishes is described in previous chapters, and their use with alkyds and other synthetic resins will be described in subsequent chapters. The properties and manufacture of "boiled oil," which is linseed oil with driers cooked into it, are described on page 66. The historical development of driers is outlined by Elm (8) and Greenawald (9).

COMPOSITION

Driers are heavy metal soaps of organic acids. The early driers were principally the lead, cobalt, and manganese soaps of linseed fatty acids or rosin acids; these are known as linoleate driers and resinate driers respectively. At present a variety of organic acids is used, and their metal soaps are available as driers, as listed in Table 1. They are marketed as solutions in mineral spirits with a specified metal content, and also without solvent for use in printing inks, and in coatings where solvent must be avoided.

Table 1 gives the general characteristics of commercial driers; however, they may vary somewhat with different manufacturers. Special concentrations are available in some cases. Commercially, the linoleate and resinate driers are referred to by these names, but the other types are sold under the trade names given in Table 2.

It is extremely important that driers maintain good solubility and stability in solution and in combination with the various coatings. Since the linoleate, resinate, and tallate driers are made from oxidizable acids they show some tendency to change on aging. The naphthenate and octoate driers are made from saturated organic acids which are non-oxidizing; therefore they are more stable. As stated previously the linoleates are made from the mixed acids of linseed oil (see Chapter 2 for composition), and the resinates from the natural

Table 1 Characteristics of Commercial Driers

Type	% Solids	% Metal	Specific Gravity	Pounds per Gallon	Viscosity, poises
Linoleates					
Pb	100	13.0	1.05	8.75	Heavy liquid
Pb	100	26.5	1.25	10.45	Solid
Co	100	4.26	0.97	8.05	Heavy liquid
Co	100	8.50	1.03	8.58	Solid
Mn	100	4.35	0.98	8.15	Heavy liquid
Mn	100	8.20	1.03	8.58	Solid
Resinates					
Pb	100	16.0	1.27	10.63	Solid
Pb	100	24.0	1.40	11.70	Solid
Co	100	3.0	1.11	9.23	Solid
Mn	100	3.5	1.13	9.40	Solid
Zn	100	1.0	1.08	9.02	Solid
Naphthenates					
Pb	42	16.0	1.00	8.33	0.5
Pb	62	24.0	1.15	9.6	0.5
Pb	100	34.0	1.44	12.0	Paste
Co	56	6.0	0.96	8.0	0.5
Co	100	10.0	1.15	9.6	Paste
Mn	62	6.0	0.97	8.1	0.5
Mn	100	10.0	1.15	9.6	Paste
Zn	60	8.0	0.98	8.2	0.5
Ca	62	5.0	0.95	7.9	0.5
Ca	58	4.0	0.93	7.76	0.5
Fe	72	6.0	1.01	8.4	1.4
Octoates					
Pb	54	24.0	1.12	9.33	0.5
Co	40	6.0	0.90	7.50	0.5
Mn	45	6.0	0.90	7.50	0.5
Zn	41	8.0	0.90	7.50	1.6
Ca	43	4.0	0.90	7.50	0.65
Tallates					
Pb	57.6	16.0	1.04	8.70	0.5
Pb	71.3	24.0	1.16	9.70	1.25
Co	45.4	4.0	0.90	7.50	0.5
Co	74.6	6.0	0.98	8.15	3.7
Mn	48.2	4.0	0.90	7.50	0.5
Mn	72.4	6.0	0.97	8.10	2.5

rosin acids. The acids in the tallates are obtained from tall oil, which is a mixture of fatty acids and rosin acids having the following range of composition, depending on the type of wood and method of production: 42–51% fatty acids, 42–51% rosin acids, and about 7% unsaponifiable. The fatty acids are chiefly oleic and linoleic. (See Chapter 2 for further details on tall oil.)

Table 2 Trade Names of Commercial Driers

Manufacturer	Type of Drier: Naphthenate	Octoate	Tallate
Advance Solvents and Chem. Corp.	Soligen	Hexogen	Advasol
Ferro Chemical Corp.	Naphthenate	Catalox	Tallates
Fred. A. Stresen-Reuter, Inc.	Naphthenate	Octoate	....
Nuodex Products Co.	Nuodex	Octoate	Nuolates
Harshaw Chemical Co.	Uversol	Octasol	Linoresinate

The naphthenates are made from the naphthenic acids obtained as a by-product in petroleum refining. They correspond to these general formulas: $C_nH_{2n-2}O_2$, $C_nH_{2n-4}O_2$, and $C_nH_{2n-6}O_2$, and contain both cyclopentane and cyclohexane derivatives.

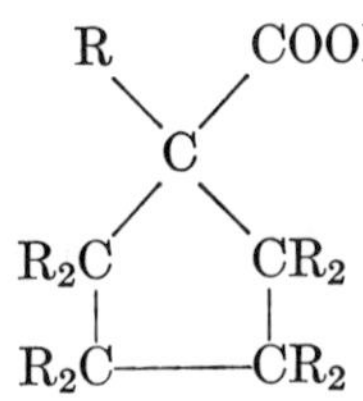

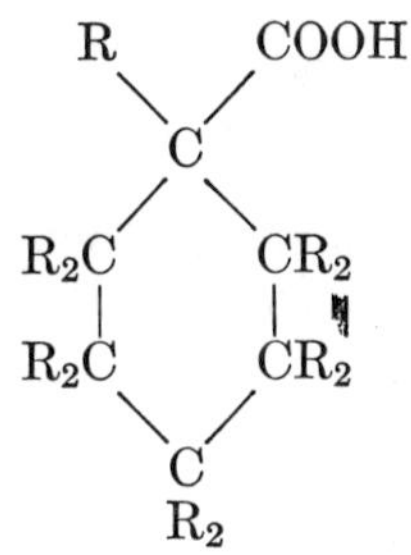

R = H or alkyl radical.

The naphthenic acids are saturated compounds not readily susceptible to oxidation, and therefore they produce more stable soaps. However, they contain a very small amount of color bodies, and they have a somewhat objectionable odor. (See page 235 for the recent development of odorless driers.) Both these features are eliminated in the octoates which are made from octoic acid, also known as 2-ethyl hexoic acid:

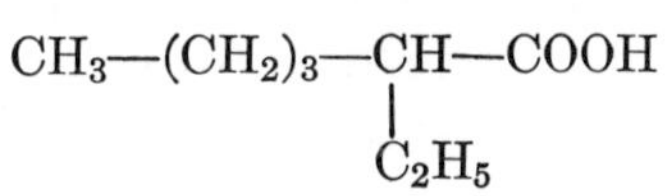

This acid is also non-oxidizing, and excellent driers are made from it, but it is more expensive than naphthenic acid.

MANUFACTURE OF DRIERS

The driers described above are the standard commercial products having definite metal contents. However, some varnish manufacturers make a line of driers by cooking the metallic salts or oxides into drying oils and resins and thinning the product with mineral spirits. These are known as "liquid driers" or "japan driers," and they vary considerably in composition and in metal content. They also tend to "sludge out" on aging because of oxidation with resulting decrease in solubility.

The oil is usually linseed or fish oil, and rosin is the principal resin used. In some driers a small amount of lime is added to improve dispersion and stability. Litharge is the chief source of lead, although basic lead carbonate and lead acetate may be used. Cobalt acetate is the chief source of cobalt because the oxides are too insoluble. Manganese is used as the oxide, MnO_2, or the borate. Since the oxide is very slow in reacting with the oil, a high temperature is required, resulting in a very dark color. A finely powdered grade of MnO_2 is available which reacts more readily, but it is more expensive. When pale color is required, manganese borate is used.

In general, the oil and resin are heated to from 350° to 500°F, depending on the metallic salt to be added, and the salt is sifted in slowly with good agitation. The heating is continued until the reaction is complete; but this end point is difficult to determine, and the final metal content may not be uniform from batch to batch. The batch is cooled to about 400°F, thinned with mineral spirits, and pumped to storage. The draw-off pipe in the storage tank is usually located about a foot from the bottom, which allows the inevitable sludge to settle. The sludge is removed periodically through a large draw-off pipe in the bottom, or by removal of a manhole cover.

The standard commercial driers are made by either one of two processes: fusion or precipitation. In the fusion process the calculated amount of metallic salt is heated with the organic acid. Since the reaction takes place more readily with the acid than with the oil the reaction temperature is lower, and the color of the product is considerably better. The metallic salt is added to the acid slowly with good agitation, and the temperature rises from liberation of the heat of reaction. After the initial reaction is complete the temperature is increased to 225–300°F to eliminate the water of reaction. The heat must be applied slowly to prevent excessive foaming. If the drier is to be marketed at 100% solids the batch is cooled and packaged after adjusting its metal content to specification. When drier solution is required the batch is pumped into a thinning tank containing the solvent, and agitated to complete solution. A small amount of filter aid is added, and the batch filtered into a blending tank. A sample is taken and the metal content determined. The metal content as made is usually higher than the specification and is adjusted by addition of solvent.

The precipitated driers usually have better color and uniformity in metal content than the fused driers. The precipitated driers are made by adding a water solution of a heavy metal salt to a water solution of the sodium soap of the fatty acids, and washing the precipitated driers, followed by filtering and drying. The sodium soap is made by reacting sodium hydroxide with the fatty acid to produce a neutral soap. This soap should be as free from carbonates as possible because the drier metals form carbonates which are insoluble in oils. The soap solution is filtered, and usually a slight excess of acid is added to improve the stability of the finished drier.

A water-soluble metal salt, such as lead acetate, lead nitrate, cobalt sulfate, or manganese sulfate, is added slowly to the hot soap solution, with agitation, until precipitation is complete. The water layer is drawn off, and the precipitation is washed several times with hot water. Final drying is accomplished by heating at temperatures from 210° to 300°F, depending on the type of drier being made. The finished drier may be marketed at 100% solids or dissolved in mineral spirits and sold as a solution. The characteristics of typical commercial driers are given in Table 1.

A detailed account of drier-soap manufacture is given by Kastens and Hansen (5), and a comparison of tallates and naphthenates is presented by Unkefer (6).

An interesting development in the production of naphthenate driers is disclosed in two patents (15, 16) assigned to the Nuodex Products Co. The first describes the invention of the two-phase liquid system for the double-decomposition process, and the second discloses the process for the direct reaction of metals with organic acids to form organometallic salts. The advantage of the later process is the elimination of the by-product, sodium sulfate, which is difficult to remove entirely from the finished drier in the double-decomposition process. The reactions in the two processes may be illustrated by the two equations for the production of cobalt naphthenate given below:

Double-decomposition process:

$$CoSO_4 + 2(R\text{—}COONa) \rightarrow Co(R\text{—}COO)_2 + Na_2SO_4$$

Direct metal-reaction process:

$$2Co + 4(R\text{—}COOH) + O_2 \rightarrow 2Co(R\text{—}COO)_2 + 2H_2O$$

R = naphthenic acid radical.

Before the development of the two-phase liquid system the water solutions of cobalt sulfate and sodium naphthenate were reacted at elevated temperature under strong agitation. The resulting cobalt naphthenate formed a pasty mass from which it was practically impossible to wash out the sodium sulfate completely because of strong occlusion. In this process the cobalt naphthenate was dried and dissolved in mineral spirits to produce the commercial drier solutions.

In the two-phase liquid system (15) the cobalt sulfate and sodium naphthenate are reacted in the presence of both water and mineral spirits by means of an ingenious mechanical device. The resulting cobalt naphthenate migrates into the mineral spirits, and the sodium sulfate into the water. After separation the mineral spirits solution is washed, then freed from moisture by vacuum at low temperature. This system is a decided improvement, but it does not eliminate the occluded sodium sulfate entirely.

The above equation for the direct-metal process (16) shows that there can be no contamination with sodium sulfate because it is not present in the reaction at any time. The direct process involves heating under reflux the organic acid, hydrocarbon solvent, water, and finely powdered metal until the reaction is essentially complete. The water is then distilled off, and the product is filtered and adjusted for metal content. Several modifications of the process are described (16). An improvement in the drying properties is claimed from the elimination of traces of sodium sulfate.

USE OF DRIERS

It is becoming standard practice to specify the amount of drier metal to use with oils, resins, etc., rather than the amount of drier. The amount of the various driers, naphthenate, linoleate, etc., may then be calculated from their known metal content. In drier specifications for oils the requirements are usually stated as a given percentage of metal to oil. For varnishes the drier requirements are based on a percentage of metal to the oil content of the varnish. However, for various synthetic resins, alkyds, silicones, etc., the drier is usually specified as a percentage of metal to resin solids. Typical drier specifications appear as follows:

1. 0.6% lead, 0.04% cobalt, 0.004% manganese as metal to oil.
2. 0.3% lead, 0.03% cobalt as metal to resin solids.

Although driers are used most extensively in air-drying, oxidizing-type coatings, they are also of advantage in many baking coatings. In any case the amount of drier used should be the minimum consistent with meeting the required drying schedule. Excess drier adds to the cost, increases the tendency of the coating to discolor on aging or baking, and detracts from retention of flexibility on aging. An excess of cobalt and manganese driers may produce a wrinkled finish in both air-drying and baking coatings, particularly if the film is slightly thicker than normal.

Gregg and Anderson (11) studied the effect of various amounts and combinations of driers on the water and alkali resistance of varnish films. Some improvement in water resistance was obtained by using the optimum drier. However, it was pointed out that such films do not develop good water resistance until after drying for about 14 days, and the resistance improved on further aging. This is in contrast to alkali resistance, which increases as the film dries to a certain point, then decreases as the film ages, because of the development of free acids from oxidation decomposition.

Drier recommendations for specific oils and resins are given in other chapters of this book and may also be obtained in considerable detail from manufacturers of driers, oils, and resins.

DRIER METALS

Considerable work has been done on the drying effect of various metals. Klebsattel (1) reports the drying behavior of twenty metals and their various soaps. He found a wide range of activity of the metals; cobalt, vanadium, and manganese are generally the most active. Although the drying action of metals such as nickel or copper is practically zero, no metal has been reported to have a retarding action on the drying rate. In general, the combination of two or more metals is preferred because each has a specific action in the drying process. For example, lead is usually considered to be a "through" drier, and cobalt a "surface" drier; manganese is more surface than through. Zinc retards surface drying but promotes through drying by keeping the surface open for continued oxygen penetration and absorption.

Iron driers are very dark in color and are not particularly effective at room temperature. They are excellent at elevated temperatures and were widely used when black baking japans were in vogue. It has been reported (10) that iron naphthenate can be used in aluminum paints to advantage since lead driers must be avoided in these paints because of their deleafing action.

Calcium does not appear to have much drying power alone, but it is effective in the presence of cobalt. It has also been suggested as an addition agent to reduce drier adsorption as discussed later. There is a tendency for lead drier to become unstable in certain vehicles, and the addition of calcium naphthenate appears to exert a stabilizing action on the lead. Lead drier darkens considerably in the presence of hydrogen sulfide fumes; therefore it should not be used in fumeproof enamels. A combination of calcium and cobalt driers has proved satisfactory for these coatings. Also, the calcium does not have the toxic property of lead and may be used in non-toxic, lead-free coatings. However, it has been observed that the water resistance of a coating containing calcium drier is not as good as if lead drier were used; therefore the calcium should be held to a minimum at all times.

Recently, the Nuodex Products Co. has produced a 6% solution of cerium naphthenate. In many cases cerium has no advantage over lead or calcium, and its drying action varies with different vehicles such as oils, varnishes, or alkyds. However, in fumeproof paints, with cobalt as the active drier, cerium is about twice as effective as calcium, and it produces better initial color and color retention. In white baking alkyd resin finishes, cerium alone is not quite as good as either cobalt or manganese in developing film hardness, but it is better in color. The combination of cerium with either cobalt or manganese appears to give equal hardness with better color.

Another development, announced after this chapter was written, is the production of rare earth naphthenate and rare earth octoate solutions, and the solution of a "zirconium chemical complex," by the Advance Solvents and Chemicals Corp. These new driers are known as Soligen rare earth naphthenate 4%, Hexogen rare earth octoate 4%, and Zirco drier catalyst respectively. In the rare earth driers the metals are mainly lanthanum and cerium in equal proportions. These driers are recommended chiefly for baking finishes and should not be considered replacements for cobalt in air-drying finishes. They are reported to produce better color retention than cobalt in baking finishes and improved film properties. It is believed that they catalyze polymerization reactions, thereby producing more cross-linkages in the film structure and correspondingly increased hardness and toughness. Their use with alkyd-amino resin baking finishes may make it possible to obtain equal film hardness with less of the higher-cost amino resin. These driers are supplied as solutions in mineral spirits containing 4% by weight of the rare earth metals.

The Zirco drier catalyst is recommended as a partial replacement for cobalt or manganese driers in air-drying finishes. However, it is not effective in vehicles containing phenolic resins. In times of national emergencies there is a shortage of both cobalt and manganese.

The following typical replacements have been suggested on a metal basis to obtain equal drying time, but greater overnight hardness and improved adhesion of the coatings:

1. Replace 0.05% cobalt with {0.03% cobalt; 0.02% Zirco drier.
2. Replace 0.05% manganese with {0.025% manganese; 0.025% Zirco drier.
3. Replace 0.50% lead with 0.05–0.2% Zirco drier in fumeproof paints.

The Advance Zirco drier catalyst is supplied as a solution in mineral spirits containing 6% zirconium as metal.

It will be seen from this discussion of the drier action of the various metals that it is important to have the right combination for best film properties. If too much surface drier is used the film tends to skin over and retard the penetration of oxygen and the drying of the subsurface of the film. Under such conditions strains are set up in the dried film which weaken it, and these may cause premature failure of the film, particularly under exterior exposure to sun, rain, etc. Rapid surface drying may also cause a wrinkled appearance in air-dried coatings, especially if the application is slightly thicker than normal. Of course, this condition is used to advantage in the baked wrinkle finishes, where thick films are essential to the development of a good wrinkle pattern.

ODORLESS PAINT DRIERS

Recently, there has developed considerable interest in the production of odorless paints, particularly for interior architectural work. The odorless solvents are discussed in Chapter 6, and odorless alkyds in Chapter 7. Odorless driers were marketed in 1952. The Advance Solvents and Chemical Corp. produces Odorless Hexogen driers and a series of special Od-R-Free driers. The Nuodex Products Co. has produced a series of lead, cobalt, and manganese Odorex driers derived from specially treated fatty acids and using the odorless-type solvents. These driers are not naphthenates or tallates, but they are equivalent to the naphthenates in all properties except odor.

DRIER ADSORPTION

It is well known that coatings based on carbon black, and those containing large proportions of titanium dioxide, tend to dry more slowly after they have aged. This loss of drying power on aging is believed to be caused by adsorption of drier on the surface of the pigment. In this condition the drier is not available to the oil to accelerate the oxidation process. The adsorption appears to be somewhat selective since the top driers (cobalt and manganese) seem to be removed more than lead.

This phenomenon has received considerable attention from both pigment and drier manufacturers. The surface characteristics of the pigments may be changed so that less adsorption occurs, but this is expensive and may interfere with other desirable properties of the pigments. Of course, the difficulty may be avoided by adding a large excess of drier initially so that enough will be available after the adsorption capacity of the pigment has been satisfied. This is not good practice, however, because drier adsorption is slow; therefore an excess of drier may be present if the coating is used before it has aged very long. The best method for reducing this difficulty is to add a material which will be selectively adsorbed by the pigment when it is being ground or dispersed in the vehicle.

It has been found that from 2% to 4% of litharge on the weight of the pigment, added to the batch before grinding, is of considerable benefit. The Nuodex Products Co. recommends the addition of 1–2% of "Nuact Paste," based on the weight of vehicle solids, to the batch before grinding. This material is a lead-vanadium compound which is believed to be selectively adsorbed by the pigment during grinding, thereby reducing the subsequent adsorption of cobalt drier. The Ferro Chemical Corp. investigated the use of calcium naphthenate for this purpose. It was found that addition of calcium naphthenate seemed to facilitate wetting or dispersion of the pigment with a consequent reduction of drier adsorption.

A recent development is the drier stabilizer and activator known as Activ-8, produced by R. T. Vanderbilt Co. This material is covered by U. S. Patents 2,526,718 and 2,565,897 and has the following composition:

	Per Cent by Weight
o-Phenanthroline	38
2-Ethyl hexoic acid	22
n-Butanol	40

Only small amounts of Activ-8 are required to stabilize the drier or to prevent loss of drying on aging. The actual amounts required will vary with different paint formulations, but the general range is from ¼ lb to 2 lb per 100 gal of paint. In some formulations the use of Activ-8 permits drastic changes in the type and amount of drier without loss of drying efficiency. Activ-8 may also be used in baking finishes to reduce the baking time. It is reported that the addition of 1 lb of Activ-8 to 750 gal of an alkyd automotive finishing enamel reduced the baking time from 35 min to 22 min. This interesting material appears to be almost unique in its action since only one other compound has been found to act similarly among 400 compounds which are related to it constitutionally.

The problem of drier adsorption does not lend itself to a simple solution because of the large variation in type of vehicle used with these pigments. Adsorption is known to be readily affected by trace amounts of materials which may vary from batch to batch. For example, a variation in the acid value, or type of acid in the vehicle, would be expected to make a considerable difference. In some coatings the variation in moisture content which may be expected from handling the materials in the low humidity of winter or the high humidity of summer would be sufficient to change the adsorption characteristics. A further discussion of this problem will be given in Volume II.

"SEEDING"

"Seeding" is the term applied to a rough condition in the surface of clear and pigmented coatings. At first glance the condition might appear to be caused by small specks of foreign matter deposited on the coating during drying, but such specks may be distinguished readily from seeding. True seeding may be caused by at least three conditions: (1) by a separation or insolubilization of the drier; (2) by gel particles in the oil or resin; (3) by accidental addition of small amounts of water to the coating. If the drier has poor solubility characteristics it will precipitate in the paint on aging, particularly if it is cooled below normal temperature during shipment or storage. The precipitate does not redissolve readily on returning to normal temperature and will appear as "seeds" in the coating after it is applied.

Gel particles in the oil or resin of the vehicle may be produced by overpolymerization during processing. These are soft, slimy particles which pass through many filter media and consequently appear in the coating. They also seem to grow through agglomeration on aging, and are large enough to project above the surface of the coating. A small percentage of moisture in a paint may cause flocculation of the pigment with the formation of flocculates large enough to project above the surface of the coating. Gel particles are discussed further in Chapter 7 on alkyd resins, and the flocculation effect in Volume II.

MECHANISM OF DRIER ACTION

Although driers have been used successfully for many years the precise mechanism by which a drier accelerates the drying of a coating has not been demonstrated conclusively. It was stated in Chapter 2 that when a natural drying oil is applied as a film there is an initial induction period during which no apparent change takes place. However, if a highly purified glycerol ester of linolenic acid is used instead of the natural oil the induction period is practically eliminated. It is believed that the natural drying oils contain antioxidants which must be oxidized by the drier before oxidation of the oil can take place to form the film.

It has been observed that the rate of oxygen absorption by an oil film is increased when driers are present. This indicates that the driers accelerate the oxygen activation of the α-methylene groups referred to in Chapter 2 and later in this chapter.

It has also been observed that the total oxygen absorbed by a film during drying is less when driers are present. This suggests that driers accelerate the decomposition of the hydroperoxides which are formed by oxygen absorption, thereby increasing the rate of cross-linking and consequent film formation.

CONDITION OF DRIER METAL

In order that a metal function as a drier it must be in a form which is soluble in the oil, resin, and solvents with which it is to be used. In 1936 Krumbhaar (2) prepared colloidal dispersions of lead, cobalt, and manganese in butanol and tested them for their drying action on linseed oil. The finely dispersed lead formed a true solution rapidly because of reaction with the fatty acids in the oil; therefore no drying tests could be made with it. However, the cobalt and manganese dispersions showed no drying effect on the oil, indicating the need for a true solution. Solubility is not the only criterion since compounds such as lead tetraethyl have no drying action despite their good solubility in oils and resins. Apparently the metal must be present as the cation for it to be effective.

The metals which are the most effective driers may exist in more than one state of oxidation. This has led to the belief that the catalytic effect on the drying rate is due to the ability of the drier compound to become oxidized readily to an active form. In this condition it transfers the oxygen to the oil continuously without necessarily returning to the lower state of oxidation. In fact, if the metallic compound is more stable in the lower state of oxidation it usually does not function as a drier. Skellon (3) investigated the catalytic action of sixteen metals as oleates in the oxidation of monoethenoid fatty acids. He found a relationship between catalytic action and atomic structure of the metal. When an inner sheath of electrons is incomplete the metal is more active.

For example, nickel is similar to cobalt in many respects, and its naphthenate soap is readily soluble, but it does not exert a drying action. It is more stable to oxygen in its lower valency state because its inner sheath of electrons becomes completed, and the metal is nonreactive as a drier.

Williamson (4) has given consideration to the possibility that the drier may add directly to a double bond to form a relatively unstable complex with the oil. In the presence of oxygen the complex is presumed to decompose to regenerate the drier compound, and the oxygen then adds to the oil.

In the discussion of the oxidation of oils in Chapter 2 it was pointed out that hydroperoxides are formed at the α-methylene groups. The α-methylene group is activated by the oxygen with the abstraction of hydrogen to form a free radical on the methylene group and a hydroperoxide radical. These radicals combine to place the hydroperoxide on the α-methylene group. From the previous discussion it appears probable that driers accelerate these reactions.

Skellon (3) has shown that metallic salts accelerate the decomposition of oleic acid hydroperoxide. It is known that when the hydroperoxides in a drying film decompose chemical linkages are formed between the oil molecules and the film dries more rapidly. It would appear therefore that another function of driers is to accelerate the decomposition of the hydroperoxides and so hasten the drying of the film.

ANALYSIS OF METAL CONTENT

Two contributions from the laboratories of the Nuodex Products Co. to the determination of metal content of driers have appeared in recent publications. A method for the determination of cobalt is described (13) and also a potentiometric method for the determination of manganese (12). Technical information and literature are available from drier manufacturers on the general subject of drier-metal determination. There are also standard methods of test for driers, such as ASTM Designation D564-43, "Standard Methods of Testing Liquid Driers" (1943), and the Federal Specifications for Driers: Naphthenate, Liquid, Concentrated, TT-D-643 (March 30, 1949).

This chapter has been restricted intentionally to a discussion of the metallic soaps which are used as driers. The naphthenates of other metals are available for specific purposes. For example, chromium naphthenate is used as an antichalking compound, copper naphthenate as a fungicide for wood and fabrics, and mercury naphthenate as a mildewcide.

The aluminum, zinc, calcium, and magnesium stearates represent another group of metallic soaps. These are used not as driers but as flatting agents, lubricants, and water repellents, and in cosmetics and other compounds.

BIBLIOGRAPHY

1. C. A. Klebsattel in *Protective and Decorative Coatings,* Volume I, J. J. Mattiello, ed., John Wiley & Sons.
2. Wm. Krumbhaar, *By Gum!,* January and February, 1936.
3. J. H. Skellon, *J. Soc. Chem. Ind., 69,* 116 (1950).
4. L. Williamson, *J. Oil & Colour Chemists' Assoc., 34,* 119 (1951).
5. M. L. Kastens and F. R. Hansen, *Ind. Eng. Chem., 41,* 2080 (1949).
6. G. G. Unkefer, *Official Digest Federation Paint & Varnish Production Clubs, No. 316,* 301 (May, 1951).
7. G. Fearnley, *Official Digest Federation Paint & Varnish Production Clubs, No. 244,* 132 (March, 1945).
8. A. C. Elm, *Ind. Eng. Chem., 26,* 385 (1934).
9. F. S. Greenawald, *Official Digest Federation Paint & Varnish Production Clubs, No. 284,* 467 (1948).
10. C. A. Klebsattel, *Official Digest Federation Paint & Varnish Production Clubs, No. 284,* 686 (1948).
11. G. W. Gregg and L. E. Anderson, *Official Digest Federation Paint & Varnish Production Clubs, No. 270,* 421 (1947).
12. Nuodex Products Co., *Paint, Oil, Chem. Rev., 14* (Sept. 14, 1950).
13. H. B. Gordon and W. D. Lang, *Paint, Oil, Chem. Rev., 14* (Jan. 19, 1950).
14. E. F. Bennett, *A Review of Driers and Drying,* Chemical Publishing Co., 1941.
15. Roon, Leon, and Gotham, U. S. Patent 2,113,496 (April, 1938) to Nuodex Products Co.
16. Nowak and Fischer, U. S. Patent 2,584,041 (January, 1952) to Nuodex Products Co.

Build a mansion for your life on earth;
provide a spacious room for Thought,
and one with many doors for Philosophy;
include a workshop for Science,
an elegant hall for Recreation,
a sanctuary for the Spirit,
And you will enrich the earth, and fulfill your purpose.

Chapter 6. VOLATILE SOLVENTS

SOLUTIONS

A solution of one material in another may be considered an intimate mixture of the molecules of one material with the other. Such an intimate mixture is most readily attained when the chemical character, size, and shape of the molecules of the two or more materials are practically identical. As the molecules of the materials vary more widely in these chemical and physical characteristics so their ability to mix intimately, or to form a solution, will become less and less.

There are, of course, gaseous, liquid, and solid solutions. Gaseous solutions are most readily attained because the molecules of all ordinary gases are small and the distance between molecules is so large compared with their size that the molecules of various gases mix intimately without interference. In addition, the force of attraction between gaseous molecules is quite small, compared with their kinetic energy. In general, as the force of attraction between the molecules of a material increases in comparison with the kinetic energy of the molecules, so the physical condition of the material changes from gaseous to liquid or to the solid state.

Under normal bulk conditions the mutual attraction between the molecules of a liquid must be sufficiently large to prevent complete volatilization, yet not large enough to prevent the movement of the molecules past one another. A liquid flows more or less readily. The ability to flow is a measure of this force of attraction and other related factors such as molecular size and shape.

However, all liquids are not soluble in each other; kerosene and water separate into two distinct layers immediately after intimate mixing. The mutual attraction among the polar water molecules for each other, and among the non-polar kerosene molecules for their own kind, is large enough to prevent one liquid from mixing with the other under normal conditions. As the degree of polarity, or chemical and physical similarity, is changed from water, through the lower alcohols, to the higher alcohols, so the ability of these compounds to mix freely with kerosene is increased. The same general principles apply when organic solids are dissolved in organic liquids, which will be the main subject of this chapter.

Solubility in the solid state is also attained when a high degree of homogeneity exists in a mixture of various solid components. Glass is an example of a mixture which forms a liquid solution at high temperature and maintains sufficient homogeneity at normal temperature to form a transparent, solid solution. Many metals also form solid solutions because of the ability of the component metals to mix intimately when cooled from the liquid to the solid state. We are apt to refer to the ability of organic coating materials to mix intimately with one another in the solid state as their compatibility more frequently than as their solubility. This is due to the fact that the organic molecules may be so large that we do not visualize complete solubility or a very intimate mixture, but rather, a tolerance of one material for the other because of chemical and physical similarities. This tolerance or compatibility is often limited to specific proportions of one material in the other material. It will also be seen that the introduction of a third material, such as a volatile solvent, may increase the compatibility of the mixture to the point where liquid solution can exist, but the solution may not be maintained in the solid state after the solvent has evaporated.

A true solution may be considered to be a molecular dispersion. We also recognize the range of dispersions listed below with typical examples of each.

Dispersions

Size Range	Type
Molecular	Solutions
0.005–0.2 μ	Colloids
0.2–1.0 μ	Pigments in gloss enamels
0.5–3.0 μ	Pigments in flat finishes
Coarse	Concrete

SOLVENT AND SOLUTE

If a solution is a molecular dispersion, how do we determine which component is the solvent and which is the solute? In solid solutions the component having the greater percentage is usually considered to be the solvent. Iron dissolves a small percentage of carbon to form steel, but larger amounts of carbon separate as in cast iron. Nitrocellulose dissolves a small percentage of castor oil, but larger amounts of castor oil separate or "sweat out" to give a greasy film. We may also say that nitrocellulose dissolves a small percentage of ethyl acetate, but larger amounts of ethyl acetate dissolve the nitrocellulose. Of course, when we refer to a "volatile solvent," obviously the volatile component is the solvent. However, it was stated in Chapter 1 that a very good solvent has poor "solvent release." Under this condition the final few per cent of volatile solvent may become the solute and be held very firmly by the solid component. Conversely, a poor solvent under this condition may not dissolve in the solid; therefore it will evaporate from the film much more readily. Further discussion of this condition is given under the resinous state in Chapter 3, as well as under nitrocellulose lacquers in Chapter 11, and in connection with organosols in Chapter 12.

RELATIVE POLARITY AND MOLECULAR SIZE

The effect of relative polarity and molecular size on solubility was discussed in Chapter 1. The following polar systems may be used to illustrate this effect.

1. Water → Glucose → Carboxymethyl cellulose (CMC) → Cellulose

The small glucose molecular is soluble in water, but the large cellulose molecular, made up of glucose anhydride units, is insoluble. However, if the cellulose is made more polar as in CMC, then it has considerable solubility in water even though its size is relatively large.

2. Solubility of alcohols in water, gm/100 ml

Methyl	Ethyl	Propyl	*n*-Butyl	*sec*-Butyl	*tert*-Butyl	*n*-Amyl
100 *	100 *	100	8.3	29	100	2

* These have infinite solubility in water.

The solubility of the alcohols in water decreases as the non-polar carbon chain increases, but the effect of shape on polarity may be seen with the increasing solubility of normal, secondary, and tertiary butyl alcohols.

Butyl Alcohols

Normal: $CH_3CH_2CH_2CHOH$

Secondary: $CH_3CH_2CH(OH)—CH_3$

Tertiary: $(CH_3)_2$ $CH_3—C—OH$ (i.e., $(CH_3)_3C—OH$)

A typical non-polar system may be illustrated as follows:

Mineral spirits → Mineral oil → Paraffin wax → Polyethylene

The relatively small molecules of mineral oil are soluble in mineral spirits, but the paraffin wax is poorly soluble, and polyethylene is insoluble. However, if hydrogens on alternate carbons in polyethylene are replaced with hydroxyl groups to give the highly polar polyvinyl alcohol, this alcohol then has a degree of solubility in water.

EFFECT OF VOLATILE SOLVENT ON FILM PROPERTIES

When an organic coating is applied as a solution in a volatile solvent the physical properties of the dry film may be affected considerably by the properties of the solvent. This statement applies more particularly to coatings such as lacquers which dry by simple evaporation. However, it also applies in a lesser degree to coatings which change chemically or physically on drying, such as air-drying oxidizing coatings or heat-polymerizing materials.

The rate of change in consistency of the coating as it dries is a function of the rate of evaporation of the solvent and its solvent power. If the solvent evaporates very rapidly a large difference in solvent concentration will develop between the top and bottom of the film. This may be sufficient to cause blistering as the final solvent evaporates. It may also produce stresses in the film which become a source of weakness in the completely dried film. The large molecules of a lacquer do not have an opportunity to align themselves in the most favorable position for maximum film strength if the solvent evaporates too rapidly.

The film-forming solids in lacquers are mixtures of high polymers, resins, and plasticizers, and the volatile portion is a mixture of true solvents, latent solvents, and diluents. The ratio of the volatile mixture changes as the film dries because of the more rapid evaporation of its lower-boiling components. If the final volatile material is not a solvent for the high polymer it may precipitate the polymer, and if it is not a solvent for the resin it may cause turbidity or "resin blush."

FLOW, LEVELING, AND GLOSS

The ability of a coating to flow or level out to a smooth surface depends on the inherent flow property of the solid binder, the character of the solvent, and the pigment concentration. Paints made with unbodied oil do not flow well; the brush marks are quite evident in the dried coating. Enamels and lacquers made with resinous vehicles and low pigment concentration flow out to smooth surfaces. However, if these coatings are applied by the spray method, and a highly volatile solvent is used, they may dry with an irregular or "orange peel" surface. The rapid loss of solvent during application of the coating increases its consistency to the point where it can no longer flow properly. An "orange peel" surface results, which detracts from the gloss considerably. The condition may be avoided by using a balanced solvent with the correct rate of evaporation and solvent power.

SOLVENT CHARACTERISTICS

The general properties of solvents which determine their usefulness in organic coatings are listed below:

1. Solvent power.
2. Boiling point.
3. Rate of evaporation.
4. Flash point.
5. Flammability.
6. Chemical stability.
7. Color.
8. Odor.
9. Toxicity.
10. Corrosion resistance.
11. Pounds per gallon.
12. Cost per pound.

Since the business world must be based on sound economics, the cost factor is always an important one. Solvents vary considerably in specific gravity or pounds per gallon. This factor must be considered carefully, because many solvents are purchased by the pound and sold by the coating manufacturer by the gallon. Two solvents having equal costs per pound may be quite different in costs per gallon.

Solvent Power. The solvent power of a solvent for a solute is its ability to disperse the solute. A solvent has maximum solvent power when it disperses a solute as completely as possible, over a wide range of concentration.

A number of tests have been established to evaluate the solvent power of various solvents. They include the kauri-butanol, aniline point, and dimethyl sulfate tests for hydrocarbons, and the dilution ratio and constant viscosity tests for lacquer solvents. Although these tests are very useful it should be remembered that they provide arbitrary criteria only. They should not be expected to give absolute solvent-power values for specific solvents when used with the wide range of coating materials available today.

The solvent index (1) is an excellent criterion of solvent power, applicable to any solvent or coating combination, but it does not lend itself to a general comparison which may be used by solvent manufacturers. It may be expressed as the ratio of the viscosity of the coating in the standard solvent to the viscosity in the new solvent:

$$\text{Solvent index} = \frac{\text{Viscosity in standard solvent}}{\text{Viscosity in new solvent}}$$

Since a better solvent usually produces a lower viscosity in the coating, the solvent index will be a value greater than 1 for better solvents, and less than 1 for poorer solvents. This method has the distinct advantage of rating the solvent power of a new solvent in the actual resin combination used in the coating. The best results are obtained when the index is determined for a range of solid contents (20%, 30%, 40%, 50%, 60%, and 70% solids), and a curve of the data is plotted.

The kauri-butanol value, usually abbreviated KB, is determined by titrating 20 gm of kauri-butanol 33% solution with a hydrocarbon solvent until the mixture becomes so turbid that a printed sheet cannot be read clearly through it. Normal mineral spirits has a KB value of about 35–38, and the KB for toluol is 105. The values for a series of commercial solvents are given in Tables 4 and 5 (pages 258 and 259). The KB value is a useful criterion of the solvent power of hydrocarbons. However, the relative solubility of a kauri resin solution in these solvents may be expected to be different from the solubility of the wide variety of oils, varnishes, alkyds, and other resins in use today in the hydrocarbons.

The aniline point is the temperature at which turbidity occurs in a mixture of equal volumes of aniline and solvent. In ASTM Designation 1012-49T the end point is determined by cooling the clear mixture slowly, and noting the temperature at which turbidity occurs. The aniline point for mineral spirits is in the range of 110–150°F, but it would be at an inconveniently low temperature for toluol or xylol. Therefore the mixed aniline point test was devised (2, 3) for such solvents. It consists of diluting the good solvent with an equal volume of *n*-heptane of specified purity and using this combination with an equal volume of aniline to determine the temperature at which separation occurs. The mixed aniline point value of toluol is about 50°F, and of xylol about 60°F. Again the aniline point values are useful criteria of solvent power, but it will be apparent that they cannot be absolute values.

The dimethylsulfate test indicates the percentage of aromatic and unsaturated aliphatic hydrocarbons in a hydrocarbon solvent. Since these compounds are known to have higher solvent power than the saturated aliphatic hydrocarbons the test is of value. However, it is not widely used at present because the dimethylsulfate is very toxic, and it irritates the eyes, nose, and throat unless extreme care is taken.

The dilution ratio is used chiefly to evaluate the solvent power of the hydrocarbon diluents employed in nitrocellulose lacquer thinners. The test may also serve to indicate the relative solvent power of true solvents. For example, a solvent which permits greater dilution with toluol will be a better solvent than one which permits less dilution with toluol. As stated previously, lacquer thinners are mixtures of true solvents, latent solvents, and diluents. The diluents are non-solvents for nitrocellulose and are used chiefly to lower the cost. Aromatic hydrocarbons can be used in much larger ratios than aliphatic hydrocarbons. The dilution ratio is the ratio of diluent to true solvent which will just produce turbidity in a nitrocellulose solution having a final concentration of 10% nitrocellulose. It is important that the ratio be recorded at a standard final concentration of nitrocellulose. Two or more values are obtained at different concentrations, and the results extrapolated to 10% because the ratio is a linear function of concentration. Some specifications require a final concentration of 8%.

The dilution ratio is a useful criterion, and the values for a wide variety of hydrocarbons are given in Tables 4 and 5. Of course, the test is made with nitrocellulose as the only solid component. Practically all commercial lacquers contain plasticizers and a wide range of types and amounts of resins in addition to nitrocellulose. Since many of the diluents are solvents for these materials it might be expected that the actual ratio of diluent in a commercial lacquer would be different from that shown by the standard test. However, experience has shown that dilution ratio values may be used in the general formulation of lacquers. The dilution ratio of a specific hydrocarbon varies with the type of true solvent used in the test, as indicated above. A range of values is given under hydrocarbon solvents.

Frequently the end point is difficult to determine in tests which depend on the development of turbidity from the incipient precipitation of solid components. When a series of such tests is run, additional data may be obtained by allowing the samples to stand after the end point has been determined. The condition of the turbidity should be examined at different time periods, or the series may be allowed to stand overnight. In some cases the turbidity may redissolve, and in others the precipitation may become more severe, indicating differences in solvent power not given by the original end-point values. Of course, the values in the dilution ratio test must be those obtained at the end of the titration, and not after the sample has aged.

The constant viscosity procedure (4) evaluates mixtures of true solvents and diluents in terms of the viscosity they produce with several concentrations of nitrocellulose. When the data are plotted showing viscosities obtained at several concentrations with a variety of solvent-diluent combinations, the lowest cost combination which gives a solution of constant viscosity may be determined. However, the data obtained from this procedure do not lend themselves to the expression of solvent power in a simple form.

In the above tests for solvent power, the results are based on precipitation or viscosity values at a known solvent combination. When the coatings are applied in actual use, the known solvent combination changes in the ratio of its components because some evaporate faster than others. The constant viscosity procedure when made over a range of solvent combinations and solids content approaches most closely to the actual film condition, particularly at high solid concentrations. However, considerable time is required for such a series of tests, and the simpler dilution ratio values are more generally used.

Boiling Point. The boiling point of a solvent is the temperature at which its vapor pressure equals that of the atmosphere. Atmospheric pressure is 760 mm Hg at standard conditions. The boiling point is not a true criterion of the rate of evaporation as shown in the next section. When the boiling point of a mixture of two solvents is higher or lower than that of either of the components the mixture is known as an azeotropic mixture. Many examples are listed in the International Critical Tables, the more familiar of which are ethyl alcohol and benzene, ethyl alcohol and ethyl acetate, and many mixtures of non-polar hydrocarbons with polar liquids. When the boiling point of a mixture is lower than that of either of the components it means that the vapor pressure is greater; therefore the rate of evaporation will be faster.

A pure liquid having a single chemical composition, such as ethyl acetate or CP benzene, has a definite boiling point at a specified pressure. However a liquid composed of a mixture of similar chemical compounds, such as mineral spirits or commercial xylene, boils over a range of temperatures. The distillation range is often stated as the temperatures of the initial boiling point, the points at which 50% and 90% are distilled, the dry point, and the final end point. Typical distillation ranges are shown in Tables 4 and 5.

Rate of Evaporation. All liquids and solids have vapor pressures the values for which vary with the temperature of the surrounding atmosphere. Volatile liquids have high vapor pressures under normal conditions, and therefore they evaporate readily. The vapor pressure and rate of evaporation are indications of the ease with which the molecules of a liquid leave its surface. In general, small molecules evaporate faster than large molecules. In non-polar liquids, such as the hydrocarbons, the molecules exist as individuals, but in many polar liquids the molecules are highly associated because of the secondary valence forces between them. This association effectively increases their size, so that the rate of evaporation will be retarded. The degree of association decreases as the temperature increases, and at the boiling point the association is practically zero. Therefore, two liquids may have equal boiling points; but, if one is highly associated and the other is not, their rates of evaporation at normal room temperature will be very different. Table 1 contains some examples taken from Bogin's work (5) which illustrate this feature of volatile liquids. The rates of evaporation are comparative only, based on *n*-butyl acetate being equal to 100.

Table 1 Boiling Point vs. Evaporation Rate

	Molecular Weight	Boiling Point, °C	Evaporation Rate *
n-Butyl acetate	116.15	126.5	100
Cellosolve	90.12	135.0	40
Methyl Cellosolve	76.09	124.5	55
Ethyl acetate	88.10	77.1	525
Ethyl alcohol	46.07	78.5	203
Benzene	78.11	79.6	500
Toluene	92.13	111.0	195
Xylene	106.16	135–45	68

* Relative to *n*-butyl acetate = 100. Larger values mean faster evaporation rates.

Cellosolve evaporates much more slowly than butyl acetate despite its lower molecular weight and higher boiling point because it is a highly associated polar liquid having an ether-alcohol structure, $C_2H_5OC_2H_4OH$. Similarly, ethyl alcohol is much slower evaporating than either ethyl acetate or benzene despite the nearly equal boiling points of these compounds. By contrast, the non-polar hydrocarbons, benzene, toluene, and xylene, decrease in rate of evaporation with increase in boiling point in a normal manner.

It will be seen that if a mixture of two solvents is made, one of which is polar and the other non-polar, the degree of association in the mixture may be quite different from that in the original components. The boiling point will be reached at a lower or a higher temperature, depending on the direction of change in the degree of association. The rate of evaporation at normal room temperature will also be different for these mixtures.

Since the rate of evaporation is an indication of the ease with which molecules leave the surface of a liquid, it will be apparent that the rate from a solvent will be different from the rate from a solution. There is a greater concentration of solvent molecules in the same area of surface of the solvent than there is in the surface of a solution. Also, the solute will retard the evaporation of the solvent, depending on its affinity for the solvent molecules. This phenomenon was referred to previously under "solvent release," and it will be seen that the evaporation of a good solvent may be retarded to a greater extent than that of a poor solvent. The vinyl resin dispersions in practically non-solvents may be used as an extreme example; the volatile component evaporates very rapidly because there is little affinity for it by the resin. This is discussed further under organosols in Chapter 12 on vinyl resins.

It is difficult to relate precisely the results from the various methods for measuring rate of evaporation to actual experience in organic coatings. The rate will be affected not only by the type and concentration of solid components, but also by the thickness of the film and the circulation of air across the surface of the film. Various methods for measuring rate of evaporation are described by Gardner (6).

Flash Point. The flash point is the temperature at which enough vapor is being evaporated from a liquid to propagate a flame above its surface. It is usually measured by the Tagliabue closed-cup tester (ASTM D-56). This instrument gives more easily reproducible results than the open-cup tester, although the latter is used by the Interstate Commerce Commission in its rating for shipping requirements. Low-flash-point materials are required to be shipped in heavier-gage steel drums than higher-flash-point materials. Red warning labels must also be placed on products with flash points below 80°F, open-cup method. The results from the open-cup method are from 5–10° higher than those from the closed-cup. The flash point of a solvent or a liquid coating is a good indication of its flammability or its susceptibility to becoming ignited.

Flammability. The flash point, as described above, is the temperature at which a minimum concentration of vapor is available to ignite the mass by propagation of a flame. There is also a maximum concentration of vapor for ignition above which the mixture of vapor and air does not ignite. If the vapor is heated to the auto-ignition temperature it will ignite spontaneously without application of a flame. Table 6 shows the data for the flash points, explosive limits with air, and auto-ignition temperatures for a series of solvents.

The flash point of a mixture of liquids may be higher or lower than that of its lowest-flash-point component. Usually, the flash point may be raised by including a small proportion of a low-boiling non-flammable liquid such as methylene chloride or carbon tetrachloride. This procedure is used more generally in the cleaning industry than in the coatings industry. Whether or not it is practical depends on cost, toxicity, and actual need for reduction of fire hazard.

Although there is always a potential danger when handling flammable liquids it can be eliminated for practical purposes by taking adequate safety measures. Two of the important measures are: (1) provide good ventilation so as to avoid vapor-air mixtures which are flammable, and (2) eliminate all possible sources of static electricity, sparks, or flames. Two bulletins on safety procedures available from insurance companies are listed in the bibliography of this chapter.

Toxicity. Considerable work has been reported on the toxicity of solvents. Most of this work was done with small animals and is therefore not directly related to human beings, but some results are based on industrial experience. Solvents may be taken into the human body by inhalation of the vapor, absorption through the skin, or ingestion through the mouth. The effects on the body vary with different solvents and individuals.

Certain solvents produce dermatitis, or inflammation of the skin; some produce narcosis, or a state of unconsciousness; and others seriously affect the vital organs such as the blood or kidneys. There is a wide variance among individuals in susceptibility to these toxic effects. There must also be a definite amount of solvent vapor in the air before its effects are felt; therefore good ventilation will usually prevent toxicity from inhalation.

Actual contact with liquid solvents should be avoided as much as possible. Preparations are available for application to the hands to retard solvent attack on the skin. Creams based on materials such as lanolin may be applied before and after contact with solvents to restore the fats removed from the skin by the solvents. These preparations may be of no value to persons with specific allergies; in fact, they may even accentuate the difficulty.

It should be kept in mind that a very high concentration of any solvent vapor in the atmosphere will have a serious physiological effect simply by depriving the person of sufficient oxygen for normal requirements. However, some solvents have highly toxicological effects even in small amounts. It is a general opinion in the industry that volatile liquids which have a maximum allowable concentration well below 100 ppm are quite toxic and should be handled with extreme care. Those which have a maximum allowable concentration in the range of 100–200 ppm are moderately toxic. These still require care in handling, and good local ventilation should be provided where they are used. The volatile liquids listed below are among those considered to be toxic, but the last three are only moderately toxic:

Benzene	Butyl formate
Carbon disulfide	Methyl alcohol
Cyclohexane	Methylene chloride
Dioxane	Trichloroethylene
Tetrachloroethane	

The references given in the bibliography should be consulted for further details and specific effects.

Corrosion and Acidity. Solvents should be free from acidic materials and sulfur compounds which are corrosive to metals or which may react with certain pigments and resins used in paints. Good grades of solvents are purified to the extent that they do not present corrosion or reaction problems. Any free acidity may be determined by titration with potassium hydroxide solution, using phenolphthalein as the indicator and expressing the acid number in the normal manner.

It is difficult to remove the last trace of sulfur from some hydrocarbon solvents, but the sulfur may be detected by the copper-corrosion test (ASTM D235-39). In this test a strip of bright copper is immersed in solvent and refluxed on a water bath for 30 min. The copper should remain bright; any darkening indicates the presence of corrosive sulfur compounds.

Odor. A strong or annoying odor in a solvent may have a serious psychological effect on workmen making and using organic coatings. The odor is sometimes caused by very small amounts of impurities which are not easily removed from the solvent. The odor may also be an inherent characteristic of the solvent. Individuals vary considerably in their reaction to odors. For example, the odor of butyl alcohol may not be offensive to some persons, but it is quite nauseating to others. Many other such examples may be found in the industrial use of solvents.

The importance of solvent odor in interior house paints was recognized in 1951, as indicated by the fact that two "odorless mineral spirits" were placed on the market. One of these was the Shell 410 E.P. Thinner, Shell Oil Co., and the other was Amsco Odorless Mineral Spirits, American Mineral Spirits Co. The odor may be removed from such thinners by solvent extraction or other purification processes which add somewhat to the cost. Both thinners have low kauri-butanol values, indicating slightly reduced solvent power. The Amsco thinner has about the normal distillation range for mineral spirits, but the Shell thinner has a somewhat higher end point and slower evaporation rate. It will be noted in Chapter 5 that odorless drier solutions have been made available, and odorless alkyd resin solutions are described in Chapter 7. These products use the odorless-type mineral spirits as solvents, but other changes are also made in their solid components.

TYPES OF VOLATILE SOLVENTS

The solvents in general use by the coatings industry may be classified as follows:

- Terpene solvents
- Hydrocarbon solvents
    - Aliphatic
    - Naphthenic
    - Aromatic
- Oxygenated solvents
    - Alcohols
    - Esters
    - Ketones
    - Ether-alcohols
- Chlorinated compounds
- Nitroparaffins
- Furans

Terpene Solvents. The terpene solvents are the oldest in use by the paint industry and are obtained from pine trees. They have been replaced by lower-cost aliphatic hydrocarbon solvents in many coatings. Their chemical properties make them quite valuable as raw materials for synthetic resins and other compounds. The chemical structures of important constituents of terpene solvents are shown in Fig. 1. Turpentine is the most widely used terpene solvent, its principal use being in house paints and in some varnishes. The production of gum turpentine from the exudation of the pine tree is described on page 147. The production of wood turpentine, dipentene, and pine oil from solvent extraction of pine stumps followed by steam distillation is described on page 147.

Gum turpentine contains 60–65% α-pinene and 30–35% β-pinene. Wood turpentine contains about 80% α-pinene, the remainder being dipentene, terpinene, and terpene alcohols. Dipentene has a higher boiling point than turpentine and excellent solvent properties. It is also used to retard the skinning of varnishes, synthetic resins, and enamels. The heavy fractions obtained from the production of wood turpentine are known as pine oil. These fractions consist of terpene

Fig. 1. Some Constituents of Terpene Solvents

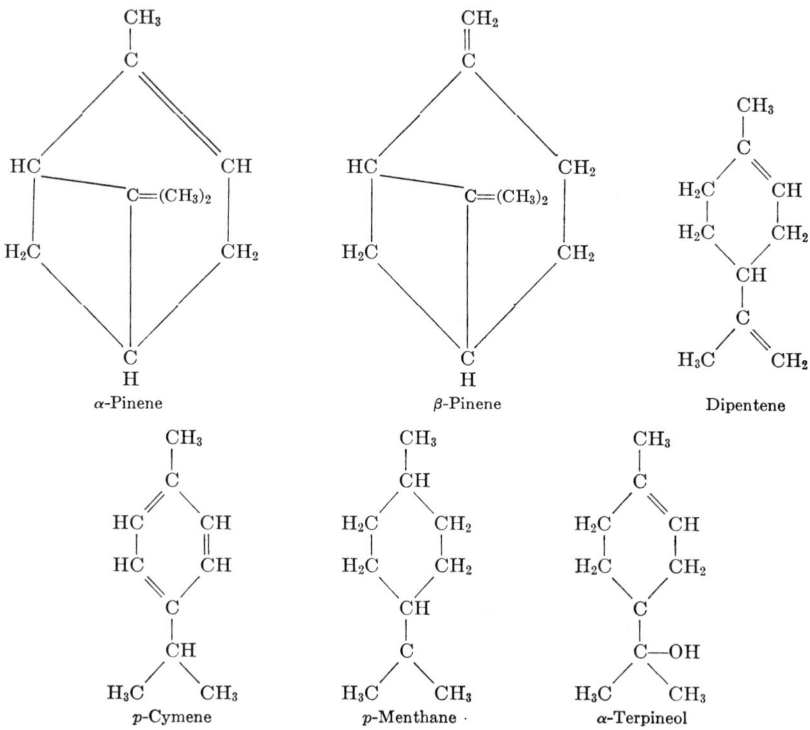

Other terpenes are shown in the section on terpene resins in Chapter 3.

tertiary and secondary alcohols, plus varying percentages of high-boiling terpene hydrocarbons. Small percentages of phenol ethers and ketones are also present. The polar-non-polar structure of pine oil makes it suitable for a wide range of uses. It has excellent solvent properties, improves the flow of enamels, retards skinning, is an anti-foaming agent, and has some bactericidal action. A typical group of terpene alcohol solvents is given in Table 2.

p-Cymene and *p*-menthane are obtained from the catalytic disproportionation of dipentene. As a result they have a higher degree of purity and are used as chemical raw materials as well as solvents.

Table 2 * Typical Terpene Alcohol Solvents

Solvent	Engler Distillation, °C 5%	50%	95%	Flash Point, open-cup	KB Value	Sp. Gr. at 15.5°C	Pounds per Gallon
GNS † No. 5	194.6	211.4	221.5	167	Inf.	0.931	7.75
Water white	195.0	212.0	220.0	163	Inf.	0.932	7.76
White	202.0	212.0	218.0	173	Inf.	0.934	7.78
Beta pine oil	207.0	216.0	225.0	174	Inf.	0.9302	7.75

* Courtesy of Newport Industries, Inc.
† Trade name, Newport Industries, Inc.

p-Menthane is a saturated terpene and therefore not susceptible to oxidation like the unsaturated terpenes.

The specific uses for terpene solvents in varnishes, resins, and coatings are given in other sections of this book and in Volume II. The physical characteristics of a typical group of commercial terpene solvents are given in Table 3.

Table 3 * Typical Terpene Solvents

Solvent	Engler Distillation, °C 5%	50%	95%	Flash Point,† °F	KB Value	Sp. Gr. at 15.5°C	Pounds per Gallon
Turpentine S.D.	156.0	157.0	162.0	101	54	0.8615	7.17
Dipentene	174.0	177.0	183.0	130	63	0.8510	7.09
p-Cymene	177.0	177.0	177.5	138	73	0.8602	7.17
p-Menthane	169.0	171.3	172.2	120	42	0.8200	6.80
Gensol ‡ #2	153.0	161.0	172.0	105	53	0.8520	7.09
Solvent § #30	174.0	180.0	200.0	128	Inf.	0.8800	7.33

* Courtesy of Newport Industries, Inc.
† Tagliabue open-cup method.
‡ Gensol (trade name, Newport Industries, Inc.) #2 is a special cut of terpene hydrocarbon from which the extraction solvent has not been removed entirely. It is an excellent replacement for turpentine and dipentene in most formulations.
§ Solvent #30 is a synthetic material obtained from the manufacture of other products. It contains chiefly high-boiling terpene hydrocarbons plus 1,4-cineole.

Hydrocarbon Solvents. The petroleum and coal-tar hydrocarbon solvents are used extensively because of their low cost, good solvent power for oils and resins, and effectiveness as diluents for nitrocellulose lacquers. The petroleum solvents are the lighter fractions obtained by the distillation and fractionation of the crude oil. They include the original turpentine substitutes, "Varnish Makers' and Painters' Naphtha" (VM & P Naphtha) and mineral spirits. Vast improve-

Fig. 2. Aliphatic, Naphthenic, and Aromatic Hydrocarbons

Aliphatic

C_nH_{2n+2}	BP, °C
n-Hexane, $CH_3CH_2CH_2CH_2CH_2CH_3$	68.7
2-Methyl pentane, 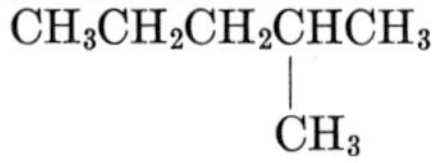	60.3
2,2-Dimethyl butane, 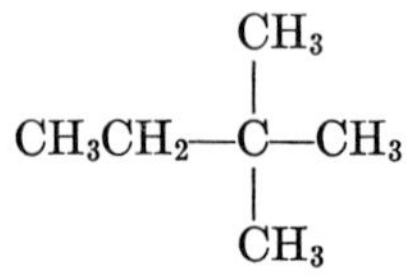	49.7
n-Hexadecane, $CH_3(CH_2)_{14}CH_3$	280

Naphthenic

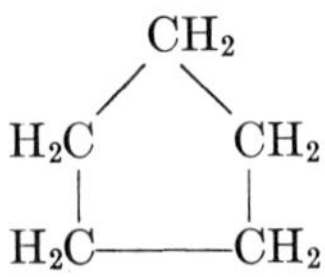

Cyclopentane
BP, °C: 49.5

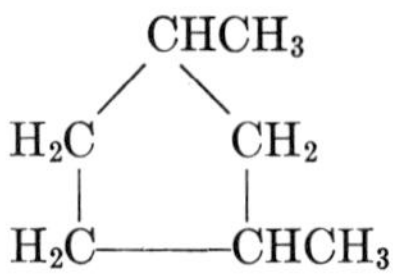

1,3-Dimethyl cyclopentane
90.9

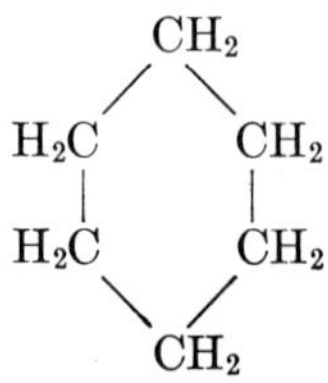

Cyclohexane
80.8

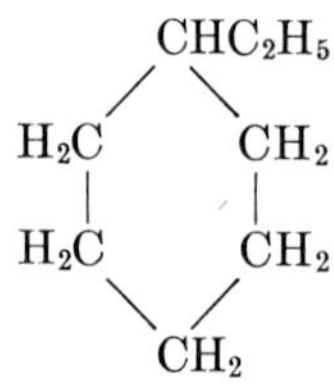

Ethylcyclohexane
131.8

Aromatic

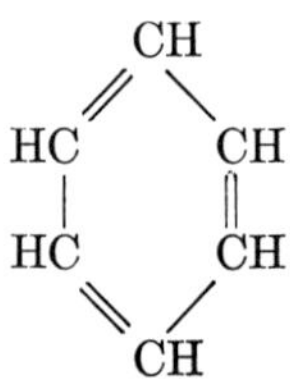

Benzene
BP, °C: 80

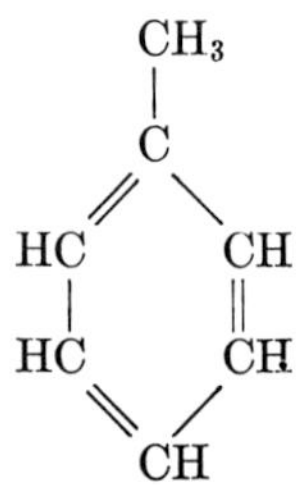

Toulene
111

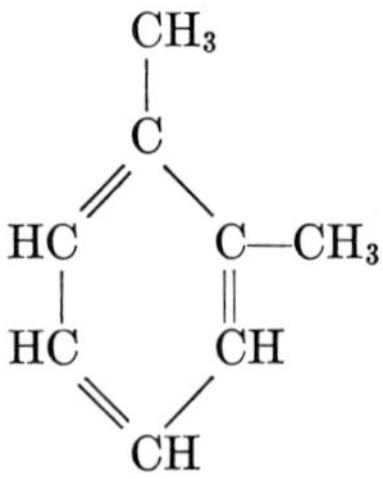

o-Xylene
BP, °C: 144

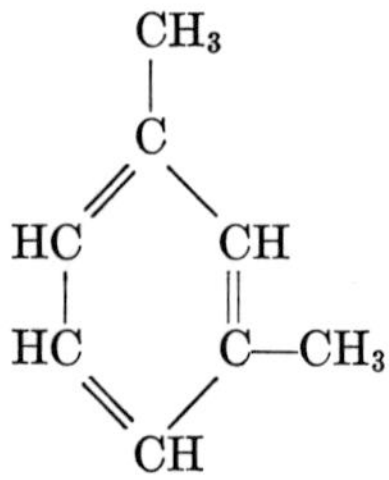

m-Xylene
139

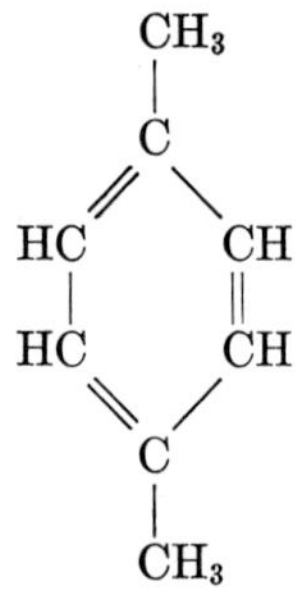

p-Xylene
138

ments have been made in large-scale fractionation apparatus with the result that today many grades of hydrocarbon solvents are available, as shown in Tables 4 and 5.

The coal-tar hydrocarbons are obtained by distillation of the material from the coke oven by product-recovery process. They include benzene (benzol), toluene (toluol), xylene (xylol), and other aromatic hydrocarbons. The tremendous wartime demand for toluene to make explosives stimulated research to produce it from other sources. Today more aromatics are produced from petroleum than from coal tar.

The hydrocarbon solvents may be classified chemically in three groups:

1. Aliphatics: straight- or open-chain, saturated hydrocarbons.
2. Naphthenics: cyclic, saturated hydrocarbons with or without alkyl side chains.
3. Aromatics: cyclic hydrocarbons containing the benzene ring structure.

The commercial hydrocarbon solvents are usually mixtures of closely related compounds and isomers, hence the range in distillation temperatures for single solvents shown in Tables 4 and 5. The structures of typical hydrocarbons are shown in Fig. 2, with the boiling points of the pure chemicals.

Fig. 3. Increase in Boiling Point with Chain Length for Normal Alkanes

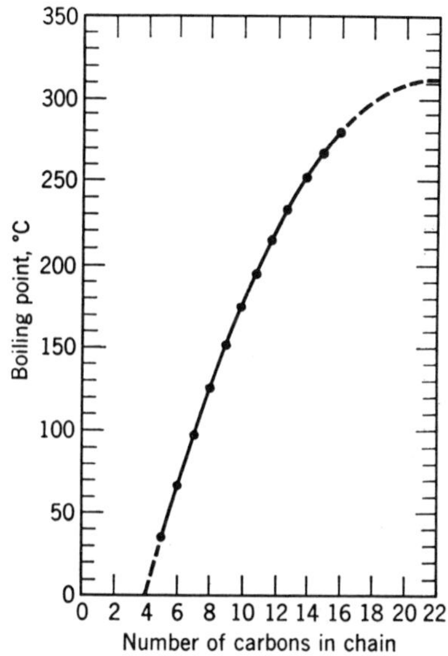

The boiling points increase with increase in molecular weight in a given series, but the effect of molecular shape is shown by the first three compounds in Fig. 2. They are isomers of hexane and therefore have the same composition and molecular weight, but the boiling point decreases with the shortening of the main carbon chain. The increase in boiling point of the normal straight-chain saturated hydrocarbons or alkanes is shown in Fig. 3. The normal alkanes containing from 5 to 16 carbon atoms in the chain are liquids at room temperature. The solid paraffin wax contains from 18 to 25 carbons per chain, and polyethylene contains several hundred carbons. This topic was discussed in Chapter 1 with respect to the secondary valence forces and chain length, and their effect on the physical properties.

The wide range of properties in hydrocarbon solvents which are available commercially is illustrated in Table 4 with the Amsco solvents of the American Mineral Spirits Co. ASTM Designation D86-46 gives the procedure for distillation of petroleum hydrocarbons. The report usually contains the temperatures of the initial boiling point (when the first drop falls from the end of the condenser), the points at which 50% and 90% by volume have been distilled, the

Table 4 * Amsco Hydrocarbon Solvents

	ASTM Distillation, °F						Solvent Power			
Solvent	Init. BP	50%	90%	Dry	End	Flash Point,† °F	KB Value	Aniline Point	NC ‡ Dil. Ratio	Pounds per Gallon
1. Textile spirits	150	159	163	172			33.5	133	1.25	5.82
2. Rubber solvent	105	180–90	220–30	240–60	250–70		32	137	....	5.75
3. S. textile spirits	170	195	210	222	225		36	130	1.32	6.02
4. S. extraction solv.	205	210	220	228	235		38	125	1.40	6.05
5. Lactol spirits	200	206–20	216–40	225–55			33–40	120–35	1.40	6.05
6. S. naphtholite VM & P	215–35	250	270	295	305		35–39	125–35	1.30	6.15
7. Naphtholite super	250	264	277	290	293		33	139.5	1.25	6.2
8. Naphthol mineral spirits	295	310	317	324	328	95	34	139	....	6.4
9. Mineral spirits	310	335	355–70	380–90	385–95	105	37	130	....	6.5
10. #46 spirits	310	340	370	390	395	105	41–44	115–25	....	6.65
11. Hi-Flash mineral spirits	330	350–60	365–75	380–90	400–05	115–20	33–37	140	....	6.5
12. #140 solvent	370	380	489	398	402	145	32	147	....	6.6
13. #460 solvent	360	395	425	450	455	140	37	135	....	6.8
14. Retardsol	350	430	475	505	510	140	32	150	....	6.75
15. Stoddard solvent	...	340	370	380	385	105	....	...	....	....
16. Pentane	93	95	97	99			25.0	...	....	5.25
17. Isohexanes	135	140	143	144			27.5	150	1.25	5.56
18. Hexane	152	154	154	157			30	140	1.25	5.70
19. Isoheptanes	185	190	193	194			35	132	1.29	6.03
20. Heptane	202	207	208	209			35.5	132	1.30	6.07
21. Iso-octanes	210	220	225	230			38	122	1.35	6.22
22. Octane	235	245	250	255			32.0	145	1.26	6.10
23. Petroleum ether	90	98	116	138			25.0	...	....	5.30
								Mixed Aniline Point		
24. Amsco Solvent A	205	240	265	275	280	..	63	100	2.00	6.60
25. Amsco Solvent A-80	212	235	255	270	275	..	80	73	2.25	6.90
26. Amsco toluol	228	230	231	232	233	..	105	50	2.60	7.21
27. Amsco Solvent B	280	300	340	370	375	85	75	80	2.00	7.01
28. Amsco Solvent B-90	277	287	310	350	355	83	98	66	2.30	7.10
29. Amsco xylol	275	282	286	290	293	82	97	55	2.40	7.23
30. Amsco Solvent C	250	275	285	295	300	..	72	88	2.15	7.00
31. Amsco Solvent D	310	340–50	365–75	380–90	395	105–25	70	85	....	7.05
32. Amsco Solvent E	350	400	410	485	490	140	72	90	....	7.38
33. Amsco Solvent F	345	365	390	405	410	130	70	88	....	7.15
34. Amsco Solvent F-80	335	350	375	400	405	125	80	80	....	7.16
35. Hi-Flash naphtha	320	333	345	363	365	112	89	64	....	7.23
36. Super Hi-Flash	315	325	335	348	350	110	94	58	....	7.27

* Courtesy of American Mineral Spirits Co.

† Tagliabue closed-cup method, ASTM D56–36.

‡ At 10% nitrocellulose final concentration with butyl acetate as the true solvent.

dry point (at which the bottom of the flask becomes dry), and the end point or temperature at which the last drop is obtained. Additional heat must be applied after the bottom of the flask is dry to obtain the last drop. It is advantageous in many cases to have the spread in distillation temperatures kept as small as possible, but this requires closer fractionation with an increase in cost.

In Table 4 the solvent power is indicated by the range of KB values from 34 to 37 for the regular mineral-spirits-type of solvent to 105 for toluol. The straight aniline point is used for the aliphatic or mineral-spirits-type solvents, and the mixed aniline point for the aromatic type, as explained previously. The results from the nitrocellulose dilution ratio test are given for the solvents having fast enough evaporation rates and sufficient solvent power to be used as diluents. The test was run with butyl acetate as the true solvent portion. The values would be different if another solvent were used, as explained previously. It will be noted that the aromatics are higher in weight per gallon than the aliphatics.

A typical set of characteristics of aromatic solvents as produced by the coal-tar industry is given in Table 5.

Table 5 * Aromatic Solvents

	Distillation						
Solvent	First Drop, min. °C	% Recovered at T °C	Dry Point, max. °C	Flash Point,† °F	KB Value	Mixed Aniline Point	NC Dil. Ratio
Thiophene-free benzol	...	100 ‡	...	..	...	..	....
Nitration benzol	...	100 ‡	...	..	...	..	....
Industrial pure benzol	...	100 §	...	<10	107	14	....
Industrial 90 benzol	78	Min. 90–100°	120	<10	106	14	2.88
Nitration toluol	...	100 ‖	...	..	...	..	....
Industrial pure toluol	...	100 ¶	...	40	100	10	2.93
Nitration xylol	137	100 within 3°	140.5	..	...	..	....
Five-degree xylol	137	100 **	143	..	...	..	....
Ten-degree xylol	135	100 within 10°	145	..	...	..	....
Industrial xylol	...	Min. 90–145°	155	80	92	12	2.73
Hi-Flash	150	....	193	105	88	20	....
Crude light solvent	...	Min. 90–160°	180	80	100	16	....
Wire enamel solvent	...	Max. 10–160° Min. 70–190° Min. 95–235°	250	145	125	17	....

* Courtesy of the Barrett Div., Allied Chemical and Dye Corp.
† Tagliabue open-cup method.
‡ Within 1°, including 80.1°.
§ Within 2°, including 80.1°.
‖ Within 1°, including 110.6°.
¶ Within 2°, including 110.6°.
** Within 5°, including 139.1°.

Fire and Explosion Hazard. A comparison of the flash point, autoignition temperature, and explosive limits in air for a variety of hydrocarbon solvents may be obtained from Table 6. The fire hazard is related to the volume of vapor in the air; the volumes of solvent vapor per gallon of solvent evaporated at 80°F and 212°F are given. Also given is the solvent-vapor density in comparison with air.

Table 6 * Physical Characteristics of Various Solvents

Solvent	Flash Point, °F	Auto-ignition, °F	Explosive Limits, % by vol. in air		Cu Ft Vapor per Gallon		Vapor Density Air = 1	Liquid Pounds per Gallon at 60°F
			Lower	Upper	80°F	212°F		
Gasoline	0	495	1.3	6.0	..	..	3–4	6.15
Benzol	13	1076	1.4	8.0	38	47	2.77	7.36
Methyl alcohol	52	1065	6.0	36.5	83	107	1.11	6.63
Ethyl alcohol (99%)	57	1034	4.0	14.0	58	72	1.59	6.61
Isopropyl alcohol (99%)	59	1148	...	...	44	55	2.07	6.57
Ethyl acetate	26	1130	...	...	34	43	3.04	7.40
sec-Butyl acetate	60	...	...	...	25	31	4.00	7.22
Acetone	6	1340	2.15	13.0	27	35	2.00	6.64
Methyl ethyl ketone	24	...	...	...	37	47	2.41	6.75
Hexane	0	450	1.2	6.9	27	34	2.97	5.77
Heptane	22	450	1.0	6.0	24	30	3.51	6.07
Octane	45	450	0.9	6.0	23	28	3.65	6.05
Process naphtha	21	450	1.0	6.0	24	30	3.51	6.10
VM & P naphtha	55	450	1.2	6.0	21	26	4.04	6.11
Varsol #1	105	450	1.1	6.0	19	24	4.76	6.53
Varsol #2	110	450	1.1	6.0	20	24	4.73	6.67
Varsol B	105	450	1.1	6.0	18	23	4.90	6.55
Varsol C	105	450	1.1	6.0	18	23	4.90	6.47
Varsol Hi-Flash	112	450	1.1	6.0	17	22	5.10	6.46
H.S.D. naphtha	122	500	1.1	6.0	17	21	5.41	6.55
Solvesso toluol	45	1026	1.3	6.7	31	39	3.17	7.23
Solvesso xylol	81	900	1.0	5.3	27	34	3.65	7.25
Solvesso 100	110	900	1.0	6.0	23	29	4.24	7.25
Solvesso 150	149	900	1.0	6.0	21	27	4.80	7.42
High aromatic naphtha	140	...	...	...	20	25	5.14	7.53

* Courtesy of Esso Standard Oil Co.

The volume of solvent vapor at a given temperature may be calculated from the relationship between molecular weight and volume. For example, when the molecular weight is expressed in pounds, 1 lb mole of vapor occupies 400 cu ft at 80°F and 500 cu ft at 212°F. These factors are expressed in the following formula:

$$\left\{\begin{array}{l}\text{Cubic feet of vapor per}\\ \text{gallon of solvent evap-}\\ \text{orated at 80°F}\end{array}\right\} = 400 \times \frac{\text{Pounds per gallon}}{\text{Molecular weight}}$$

Using this formula to calculate the concentration of toluol at 80°F, one obtains about 31 cu ft of vapor per gallon evaporated. When using a factor of safety of 4 this means that $4 \times 100 \times 31 = 12{,}400$ cu ft of air must be supplied for each gallon of toluol evaporated to keep the concentration safely below the explosive point.

The toxicity is also related to the volume concentration of solvent vapor as explained previously. Benzol is generally considered more toxic than toluol, partly because of the fact that its volume concentration is much greater than that of toluol under normal working temperatures, since its vapor pressure is higher.

Tables 2–6 should be supplemented with the most recent information from solvent manufacturers regarding new or improved products. However, the data show that hydrocarbon solvents are available having a wide range of solvent properties and rates of evaporation. The lower-solvent-power types are quite satisfactory for many oleoresinous varnishes, long-oil alkyd resins, and similar coating materials. The higher-solvent-power types, based on all or part aromatic hydrocarbon, are used in the less easily soluble short-oil alkyds and as diluents for nitrocellulose lacquers, amino resin solutions, etc. The use of specific hydrocarbon solvents with various coating materials is given in detail in other chapters.

Oxygenated Solvents. In contrast with the non-polar hydrocarbons the oxygenated solvents are polar compounds. Many of them have considerable solubility in water and some are completely soluble, such as the lower alcohols. By the same criterion of polarity, they are better solvents for the more polar film-forming materials such as shellac, cellulose esters, urea- and melamine-formaldehyde resins, and vinyl resins. The oxygenated solvents are alcohols, esters, ketones, ether-alcohols, and related compounds. A few aldehydes are used, but in general they are somewhat reactive and have strong odors. A representative list of oxygenated solvents and their characteristics are given in Table 7. Many others are available, and their properties may be obtained from handbooks and manufacturers' literature.

In general, the oxygenated solvents are more expensive than the hydrocarbons; therefore mixtures are made whenever possible to lower the cost. Of course, in some cases this is not feasible. Shellac is soluble in ethyl alcohol, but only the waxy portion is soluble in hydrocarbons. However, ethyl cellulose is soluble in either alcohol or toluol, but the solvent which gives the lowest viscosity is a mixture of alcohol and toluol. Another example of the use of solvent mixtures is the manufacture of nitrocellulose lacquers. The regular soluble type of nitrocellulose may be dissolved in esters and ketones, but it is insoluble in alcohols. However, a mixture of ester and alcohol frequently is a better solvent than the ester alone. The esters are known as true solvents, and the alcohols as latent solvents. In addition, considerable amounts of hydrocarbons may be added to lacquer solvents, thereby

Table 7 Oxygenated Solvents

Solvent	Formula	Mol. Wt.	BP, °C	VP at 30°C	Evap. Rate	Sp. Gr. at 20°C/4	Pounds per Gallon
Alcohols							
Methyl	CH_3OH	32.04	64.5	165	370	0.793	6.6
Ethyl	C_2H_5OH	46.07	78.5	79	203	0.791	6.6
Isopropyl	$(CH_3)_2CHOH$	60.09	82.5	60	205	0.789	6.58
n-Butyl	C_4H_9OH	74.12	117.1	10.5	45	0.810	6.75
sec-Butyl	$CH_3CHOHC_2H_5$	74.12	99.5	26	115	0.808	6.72
Esters							
Methylacetate	CH_3COOCH_3	74.08	60	265	1040	0.935	7.8
Ethyl acetate	$CH_3COOC_2H_5$	88.10	77.1	120	525	0.886	7.37
n-Butyl acetate	$CH_3COOC_4H_9$	116.15	126.5	16	100	0.875	7.29
sec-Butyl acetate	$CH_3COOCH(CH_3)\cdot C_2H_5$	116.15	111.5	29	180	0.860	7.16
sec-Amyl acetate	$CH_3COOC_5H_{11}$	130.18	135	10	87	0.862	7.18
Ketones							
Acetone	CH_3COCH_3	58.08	56.1	280	720	0.792	6.6
Methyl ethyl ketone	$CH_3COC_2H_5$	72.10	79.6	119	465	0.805	6.72
Diethyl ketone	$C_2H_5COC_2H_5$	86.13	105	46	...	0.815	6.78
Methyl isobutyl ketone	$CH_3COC_4H_9$	100.15	118.0	25	145	0.802	6.67
Cyclohexanone	$CH_2(CH_2)_4CO$	98.14	156	7	25	0.946	7.87
Miscellaneous							
Tetrahydrofuran	$(CH_2)_4O$	72.10	65.4	150	630	0.888	7.40
Cellosolve	$C_2H_5OC_2H_4OH$	90.12	135	7	40	0.931	7.75
Methyl Cellosolve	$CH_3OC_2H_4OH$	76.09	124.5	...	55	0.964	8.04
Cellosolve acetate	$CH_3COOC_2H_4C_2H_5$	132.16	156.2	3.5	24	0.974	8.10
Carbitol	$C_2H_5OC_2H_4OC_2H_4OH$	134.17	201.9	...	..	0.990	8.25
Ethyl lactate	$CH_3CHOHCOOC_2H_5$	118.13	155.0	4	22	1.030	8.60
Butyl lactate	$CH_3CHOHCOOC_4H_9$	146.18	188.0	...	6	0.980	8.16

lowering the cost. The nitrocellulose dilution ratios of the hydrocarbons are given in Tables 4 and 5. Typical lacquer-solvent combinations are given in a later chapter.

Chlorinated Hydrocarbon Solvents. The chlorinated hydrocarbons listed in Table 8 are excellent solvents for a wide range of organic materials. Reference was made on page 58, Chapter 2, to the use of trichloroethylene in the solvent extraction process for vegetable oils. It is also utilized in the vapor-degreasing process for cleaning metal objects before finishing, which is described in Volume II.

Methylene chloride is used in a non-flammable type of paint remover. It is also combined with other solvents to raise the flash point. The mixture by volume of 25% methylene chloride, 5% perchloroethylene, and 70% Stoddard solvent, known as Cleaning Mixture #49, does not flash below its boiling point of 107.6°F. The relative nonflammability of the chlorinated solvents is one of their valuable characteristics. However, care should be taken to provide good ventilation where they are used because of their toxicity. The toxic action varies with different members of the group, but all produce narcosis. Because of their relatively high cost they have found only limited use in coatings.

Table 8 * Solvent Characteristics

Solvent	Formula	Mol. Wt.	BP, °F	VP, mm Hg, at 68°F	FP,† °F	Flam ‡ Rating	Max. Allow. Conc.,§ ppm	Vapor Density Air = 1	Sp. Gr. of Liquid at 68°/39°F
Methylene chloride	CH_2Cl_2	84.94	103.6	355	..	Non	200–500	2.93	1.326
Chloroform	$CHCl_3$	119.39	142.2	162	..	Non	75–100	4.13	1.489
Trichloroethylene	$CHCl{=}CCl_2$	131.40	188.4	57.8	..	3	100–200	4.54	1.464
Perchloroethylene	$CCl_2{=}CCl_2$	165.85	250.2	14.4	..	0	200	5.83	1.623
Methyl chloride	CH_3Cl	50.49	−10.2	3672	<32	Mod.	100–500	2.52	0.920
Carbon tetrachloride	CCl_4	153.84	169.7	90.0	..	0	50–100	5.32	1.594
Ethylene dichloride	$C_2H_4Cl_2$	98.97	182.3	62.5	56	60–70	75–100	3.42	1.253
Propylene dichloride	$C_3H_6Cl_2$	112.99	205.5	41.7	62.6	Mod.	....	3.90	1.155
Tetrachloroethane	$C_2H_2Cl_4$	167.86	295.3	5.1	..	Non	0.5–10 ‖	5.79	1.596
Pentachloroethane	C_2HCl_5	202.31	320.9	3.1	..	Non	....	6.99	1.681
Acetone	$(CH_3)_2CO$	58.08	133.0	186	0	90	200–1000	2.00	0.792
Methyl ethyl ketone	$(CH_3)COC_2H_5$	72.10	175.3	76.5	30	Flam.	....	2.41	0.805
n-Butyl acetate	$CH_3COOC_4H_9$	116.16	259.7	10.0	72	Mod.	150–400	4.00	0.88
Methyl alcohol	CH_3OH	32.04	148.0	98.3	52	Mod.	100–300	1.11	0.792
Ethyl alcohol	C_2H_5OH	46.07	173.0	44.0	55	70	100–1000	1.59	0.79
Isopropyl alcohol	$(CH_3)_2CHOH$	60–09	180.4	34.0	53	70	100–400	2.07	0.789
Tetrahydrofuran	$(CH_2)_4O$	72.10	149.7	146	−4	Flam.	<100	....	0.888
n-Hexane	C_6H_{14}	86.17	155.7	122	−15	90–95	1000	2.98	0.659
n-Heptane	C_7H_{16}	100.20	209.2	36.2	25	Flam.	400–1000	3.45	0.683
Benzol	C_6H_6	78.11	176.2	75.5	12	95–100	50–100	2.77	0.88
Toluol	$C_6H_5CH_3$	92.13	231.1	22.0	40	75–80	100–200	3.14	0.886

* Courtesy of Electrochemicals Department, E. I. du Pont de Nemours & Co., Inc.

† Tagliabue closed-cup method, unless otherwise indicated. Wherever possible the data were taken from *National Fire Codes*, Volume I (Flammable Liquids, Gases, Chemicals, and Explosives), National Fire Protection Association, Boston, 1948.

‡ Those substances rated as moderately flammable by the Underwriters' Laboratories are stated to be much less hazardous than gasoline, but more hazardous than ammonia. Trichloroethylene, methylene chloride, chloroform, and tetrachloroethane are practically non-flammable and non-explosive at ordinary temperatures, but at higher temperatures and under favorable conditions they may form weakly combustible mixtures with air. The rating numbers accord with the Underwriters' Laboratories standard of classification in which:

Ether rates	100
Gasoline rates	90–100
Alcohol (ethyl) rates	60–70
Kerosene (100°F flash) rates	30–40
Paraffin oil rates	10–20

The Underwriters' Laboratories numerical ratings were not available for all the solvents; therefore relative ratings are included for purposes of general comparison.

§ With the exception of tetrahydrofuran, the data given represent the range of maximum allowable concentrations of solvent vapors in air (in parts per million) for an eight-hour day, as defined by the various state codes in 1947.

‖ Except in two states.

An interesting comparison of the chlorinated solvents with other solvents is given in Table 8 through the courtesy of the Electrochemicals Department of E. I. du Pont de Nemours & Co.

Nitroparaffin Solvents. The four nitroparaffins listed in Tables 9 and 10 were made available to industry in 1941. The nitroparaffins can be utilized as solvents for vinyl resins, nitrocellulose, cellulose acetate, cellulose acetobutyrate, oils, fats, waxes, and dyestuffs. They are also effective solvents for hydrocarbon-insoluble rubbers, and they have other features such as selective solvent action and entrainers for

Table 9 * Characteristics of the Nitroparaffins

Nitroparaffin	Mol. Wt.	BP, °C	VP at 20°C	FP, °F	Rate of Evap.†	Sp.Gr. at 20/20°C	Pounds per Gallon
Nitromethane, CH_3NO_2	61.04	101.2	27.8	112	180	1.139	9.48
Nitroethane, $CH_3CH_2NO_2$	75.07	114.0	15.6	106	145	1.052	8.75
1-Nitropropane, $CH_3CH_2CH_2NO_2$	89.09	131.6	7.5	120	100	1.003	8.35
2-Nitropropane, $CH_3CHNO_2CH_3$	89.09	120.3	12.9	103	124	0.992	8.24

* Courtesy of Commercial Solvents Corp.
† In comparison with *n*-butyl acetate = 100.

Table 10 * Physical Properties of the Nitroparaffins

Nitroparaffin	Lower Limits of Flammability, % by vol. in air: Observed (1)	Calc.	VP at 25°C	Lethal Dose,† g/kg (2)	Threshold Limit Values, ppm
Nitromethane	7.3	8.0	38	0.75–1.00	100 (4)
Nitroethane	4.0	3.9	25	0.50–0.75	100 (3)
1-Nitropropane	2.6	2.5	13	0.25–0.50	100 (5)
2-Nitropropane	2.6	2.5	20	0.50–0.75	50 (3)

* Courtesy of Commercial Solvents Corp.
† Lethal dose for rabbits by oral administration (2).
(1) Commercial Solvents Corp., unpublished reports.
(2) W. Machle, E. W. Scott, and J. J. Treon, *J. Ind. Hyg. Toxicol.*, *22*, 315 (1940), and *24*, 5 (1942); also W. Machle and E. W. Scott, *Proc. Soc. Exptl. Biol. Ind.*, *53*, 42 (1943).
(3) W. Machle, E. W. Scott, and J. J. Treon, *J. Ind. Hyg. Toxicol.*, *22*, 315 (1940).
(4) E. W. Scott, *J. Ind. Hyg. Toxicol.*, *24*, 226 (1942) and *25*, 20 (1943).
(5) L. Silverman, H. F. Schulte, and M. W. First, *J. Ind. Hyg. Toxicol.*, *28*, 262 (1946).

azeotropic distillations. They find further application in the synthesis of many organic compounds. The nitropropanes are useful in cellulose acetate lacquers because of their good solvent power and medium rate of evaporation. In the latter property they are similar to butyl acetate, and they lie between the fast-evaporating acetone and the slow-evaporating ethyl lactate used previously. Specific combinations of nitropropane with ethyl and butyl alcohols and toluol are given in the discussion of cellulose acetate on page 445.

In addition to being good solvents for the vinyl resins the nitropropanes have a mild odor, and so may replace the ketones where this feature is important. Their good solvent power and medium rate of evaporation will ensure their use in a greater variety of coatings when their prices are competitive with those of the oxygenated solvents. Since the nitroparaffins are derived from natural gas it may be expected that they will be manufactured eventually in adequate quantities at lower cost.

There are no unusual hazards connected with the use of the nitroparaffins in conventional solvent combinations. Unlike nitrobenzene and nitroglycerin, which are hazardous from the standpoint of toxicity and explosive characteristics, the nitroparaffins have the same order of vapor toxicity as other commonly used solvents for coatings. Nitromethane is the only nitroparaffin which can be sensitized to behave in a mildly explosive manner. The ease of detonating nitromethane is reduced considerably by addition of such common solvents as methanol, ethyl acetate, or gasoline. The dry salts of nitromethane formed by the action of alkali metals or amines are explosive, and the chromium and cobalt salts of acetyl acetone will increase the shock sensitivity of nitromethane.

The nitroparaffins can exist in a tautomeric form in the presence of water. The acid form may be responsible for the discoloration which occurs on long storage in metal containers when moisture is present. This can be inhibited by addition of a small percentage of monobutyl phosphate. The weakly acidic nature of the nitroparaffins is also responsible for their reaction with certain basic pigments.

Furan Solvents. The principal furan solvents are furfural, furfuryl alcohol, tetrahydrofuran, and tetrahydrofurfuryl alcohol. Furfural is made by digesting oat hulls, or other pentosan-containing agricultural residues, with steam and dilute acid, followed by distillation. The distillate consists of furfural, water, and organic acids. The furfural is separated in a column still and may be further refined. Furfuryl alcohol is obtained from the catalytic hydrogenation of furfural. Tetrahydrofurfuryl alcohol is obtained by continued hydrogenation so as to saturate the double bonds in the furan nucleus. Tetrahydrofuran may be obtained (12) by a two-step process involving the removal of carbon monoxide from furfural to form furan, followed by the hydrogenation of this compound. Tetrahydrofuran was manufactured in Germany during World War II from formaldehyde, acetylene, and hydrogen. It may also be obtained by the oxidation of petroleum hydrocarbons. The chemistry of the furans is described in the literature of the Quaker Oats Co. and in references 9–12. These also de-

Fig. 4. Furan Solvents

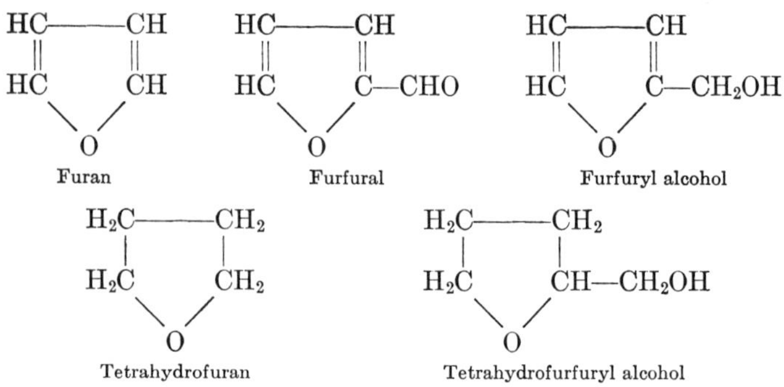

scribe the many chemical derivatives which may be obtained from furfural. The structural formulas for furan and the furan solvents are shown in Fig. 4. Some of the physical characteristics of these solvents are given in Table 11.

Table 11 Physical Characteristics of Furan Solvents

Characteristic	Furfural *	Furfuryl Alcohol *	Tetrahydrofuran †	Tetrahydrofurfuryl Alcohol *
Mol. weight	96.1	98.1	72.1	102.13
Boiling point, °C	158–162	171	65.4	177
Flash point, °C, open-cup	55–57	75	−4	75
Spec. gravity at 20/20°C	1.160	1.134	0.888	1.064
Pounds per gallon	9.6	9.4	7.4	8.8

* The Quaker Oats Co.
† E. I. du Pont de Nemours & Co., Inc.

Furfural. Pure furfural is a colorless liquid with a characteristic odor, but industrial furfural is light yellow to brown in color. Normally, it is shipped and stored in iron and steel equipment without special precautions. When furfural is stored in contact with air there is a gradual darkening in color and an increase in acidity and in polymer formation. Its use as a selective solvent for refining oils is described in Chapter 2, and for the decoloration of wood rosin in Chapter 3. It is a solvent for oleoresinous varnishes, the cellulose esters and ethers, and some of the vinyl resins. It should not be used for white or pastel finishes because of its color characteristics. It has proved satisfactory in varnish removers because of its strong penetrating action. It is used in shoe dyeing both for its solvent action on the dyes and for its good penetrating properties.

Furfuryl Alcohol. Furfuryl alcohol is water white when freshly distilled, but changes color on exposure to air so that the commercial material is amber in color. It is completely miscible with water at temperatures above 21°C. It is a good solvent for many resinous materials, but its largest use is in the manufacture of thermosetting resins. It resinifies readily with acid catalysts (13). It may be polymerized with acid catalysts to the liquid resin stage, and then neutralized so that it will be stable on storage. The addition of acid catalyst before using converts it to the infusible stage. The converted resin is quite dark in color but has good chemical resistance. Furfuryl alcohol also reacts with formaldehyde, urea, and melamine to form resinous materials. Bulletins describing these materials are available from the Quaker Oats Co.

Tetrahydrofurfuryl Alcohol. Tetrahydrofurfuryl alcohol is pale in color as shipped, but it discolors slowly on aging in contact with air and iron or copper containers. It does not discolor in aluminum containers under a cover of inert gas. It is miscible with water and is a solvent for dyestuffs, nitrocellulose, cellulose acetate, ethyl cellulose, vinyl resins, and the usual varnish resins, alkyds, and drying oils. It is a chemical intermediate used in the preparation of tetrahydrofurfuryl oleate, a secondary plasticizer for vinyl resins and other chemical compounds. An industrial grade (#1001, the Quaker Oats Co.) is available which is quite satisfactory for many solvent applications.

Tetrahydrofuran. Tetrahydrofuran, as manufactured by du Pont, is a colorless mobile liquid, which is miscible with water and most organic solvents. It is a solvent for a wide variety of resinous materials, including the polyvinyl chloride and polyvinylidene chloride homopolymers and copolymers. Frequently, lower-viscosity solutions may may be obtained with mixtures of tetrahydrofuran and ketones such as methyl ethyl ketone as the solvent. Clear films can be cast from 10% to 15% solutions of high-molecular-weight polyvinyl chloride, using a solvent mixture of 65% tetrahydrofuran and 35% methyl ethyl ketone. The resin should be dispersed first in the tetrahydrofuran with good agitation; then the methyl ethyl ketone should be added. Solvent mixtures with aromatic hydrocarbons may also be used, such as 75% tetrahydrofuran, 15% methyl ethyl ketone, 10% toluene.

Since tetrahydrofuran is a low-boiling solvent, with low flash point, necessary precautions should be taken to avoid fire. It is recommended that the maximum allowable concentration in air be kept below 100 ppm; therefore good ventilation must be provided where tetrahydrofuran is used. It should also be stored in a cool place and in tightly closed containers because of its tendency to form explosive peroxides when exposed to air without adequate stabilization. The commercial product is stabilized before shipping, but it should be tested for the presence of peroxides if it is to be distilled or evaporated.

BIBLIOGRAPHY

1. Philadelphia Club, *Official Digest Federation Paint & Varnish Production Clubs, No. 241,* 579 (1944).
2. B. H. Shoemaker and J. A. Bolt, *Ind. Eng. Chem., Anal. Ed., 14,* 200 (1945).
3. E. H. McArdle, *Chem. & Met. Eng., 44,* 601 (1937).
4. V. W. Ware and W. O. Teeters, *Ind. Eng. Chem., 738,* 1118 (1939).
5. C. D. Bogin, *Protective and Decorative Coatings,* Volume I, p. 644, J. J. Mattiello, ed., John Wiley & Sons.
6. H. A. Gardner, *Physical and Chemical Examination of Paints, Varnishes, and Lacquers,* tenth ed., 1946.
7. A. K. Doolittle, *Ind. Eng. Chem., 36,* 239 (1944).
8. A. K. Doolittle, *Ind. Eng. Chem., 38,* 535 (1946).
9. A. P. Dunlop, *Ind. Eng. Chem., 40,* 204 (1948).
10. H. Gilman and G. F. Wright, *Ind. Eng. Chem., 40,* 1517 (1948).
11. B. H. Wojcik, *Ind. Eng. Chem., 40,* 210 (1948).
12. O. W. Cass., *Ind. Eng. Chem., 40,* 216 (1948).
13. Dunlop and Peters, *Ind. Eng. Chem., 34,* 814 (1942).

OTHER BOOKS

A. D. Brandt, *Industrial Health Engineering,* John Wiley & Sons, 1947.

K. B. Lehmann and F. Flury, *Toxicity and Hygiene of Industrial Solvents,* The Williams & Wilkins Co., Baltimore, Md., 1943.

National Fire Codes, National Fire Protection Assoc., Boston, Mass.

Properties of Flammable Liquids, Solids, and Gases, Associated Factory Mutual Fire Insurance Co., Boston, Mass.

T. H. Durrans, *Solvents,* D. Van Nostrand Co., New York, 1950.

Most men develop ideas within the limits of recorded experience;
Genius expands those limits beyond the present horizon.

Chapter 7. ALKYD RESINS

INTRODUCTION

The name "alkyd" is a euphonious rendition of alcid, and was coined by Kienle to indicate that alkyd resins are made from alcohols and acids. From the previous discussion of functionality we realize that the alcohols and acids must have functionalities of 2 or more if the products are to be resinous. Hence the classical definition of an alkyd is "the resinous product of polyhydric alcohols and polybasic acids." However, in a great many alkyd resins the product of the polyhydric alcohol and polybasic acid is modified with drying or non-drying oils and both natural and synthetic resins. In addition, more than one polyhydric alcohol or polybasic acid may be used in the same alkyd resin to obtain special properties. Therefore, it will be seen that the term "alkyd resins" includes a very wide variety of resinous products.

The important group of resins which are generally indicated by the term "alkyd resin" is the oil-modified, glyceryl phthalate resins. They are the reaction products of the trihydric alcohol, glycerol, and the dibasic acid, phthalic acid, modified with either the drying or the non-drying oils. The formulation, production, and properties of glyceryl phthalate alkyds will constitute the main part of this chapter. The resins known as resinous plasticizers, typified by glycol or glycerol sebacate, also are alkyd resins. Another group of alkyd resins used extensively in plastics is more generally known as polyester resins. The so-called "reactive polyesters" are based on the linear-type alkyd resins, but they have the C=C type of unsaturation in the resin molecule through the use of maleic or fumaric acid. This type of polyester may be dissolved in a reactive monomer, such as styrene or diallyl phthalate, and the mixture converted to an insoluble and infusible stage by addition polymerization. An extreme type of alkyd is the super polyester Terylene (England), or Dacron (United States), which is a synthetic fiber used extensively in the textile industry. It is the reaction product of glycol and terephthalic acid. Another extreme case is the maleic-modified resins described in Chapter 4; they are the reaction products of glycerol and a mixture of the monobasic acid rosin and tribasic acid adduct of rosin and maleic acid. These hard, brittle resins, used in varnishes and lacquers, are very different in properties from the other alkyds described above. It will be apparent that the classical definition of an alkyd resin may be applied to a very wide variety of products.

Fig. 1. Polyfunctional Alcohols

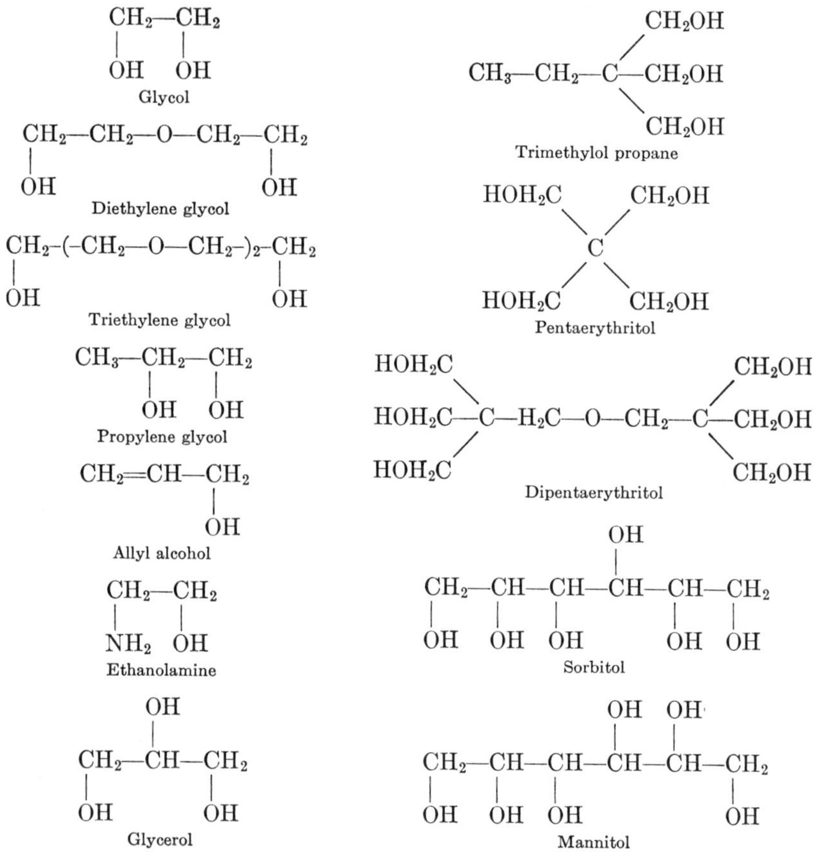

Before discussing any particular class of alkyd resins, it will be of interest to consider the types of polyhydric alcohols, polybasic acids, and modifiers which are available. From the functionalities of these raw materials, and the size and shape of their molecules, we should be able to predict the general properties of the resulting resins. However, we shall see that the ratio in which the various constituents are used, and the type and extent of processing, are also important factors in determining resin properties. Figure 1 shows some of the polyhydric alcohols available for alkyd resin manufacture.

POLYFUNCTIONAL ALCOHOLS

The polyhydric alcohols shown in Fig. 1 have functionalities ranging from 2 to 6. The glycols, or dihydric alcohols, are used in the production of alkyds having a linear structure. In general, they are the non-convertible or thermoplastic type of resin used as plasticizers. The resins are softer as the distance between the hydroxyl groups increases. A resin made with triethylene glycol is softer than one made with glycol, other factors being equal.

Propylene glycol, with a methyl group as a branch, should produce resins having better solubility in hydrocarbons than similar resins made with glycol. However, non-polar-branch chains usually decrease the hardness of linear resins because they do not permit as close packing.

The trihydric alcohol, glycerol, is used in the well-known "glyceryl phthalate" or "phthalic alkyd" resins referred to previously. Its trifunctionality makes possible the cross-linked type of resin structure. Such resins will be convertible or thermosetting, but the type of convertibility will depend on the acid and modifier used with the glycerol, as discussed later. Trimethylol propane is also trihydric, but its hydroxyls are all primary in comparison with the two primary and one secondary in glycerol. Primary hydroxyls are esterified more readily than secondary hydroxyls, which is an advantage in resin processing. The non-polar side chain would increase solubility in hydrocarbons, giving another cost advantage. However, the side chain would produce a softer or slower-hardening resin, which is usually a disadvantage in this class. The material cost is also higher than that of glycerol.

Pentaerythritol, with four primary hydroxyls, is more reactive and permits a greater degree of cross-linking, thereby producing harder resins than can be obtained with glycerol. In fact, pentaerythritol is so reactive that it is generally used in combination with glycerol or glycol. This was referred to in Chapter 3 in the discussion of "penta" resins. The di- and tripentaerythritols are hexahydric and octahydric respectively, and are therefore more reactive. However, the increased distance between the hydroxyl groupings tends to offset the possible increase in hardness of the resins, similarly to the situation with the three glycols.

Sorbitol and mannitol are hexahydric with relatively short distances between the hydroxyls. On this basis one would expect a high degree of reactivity and hard, fast-drying resins. However, they have four secondary hydroxyls, which increase the difficulty of esterification, and they tend to form interethers. Although sorbitol finds use in alkyd manufacture, it is not as reactive as might be expected from its potential functionality of 6.

Fig. 2. Polyfunctional Acids

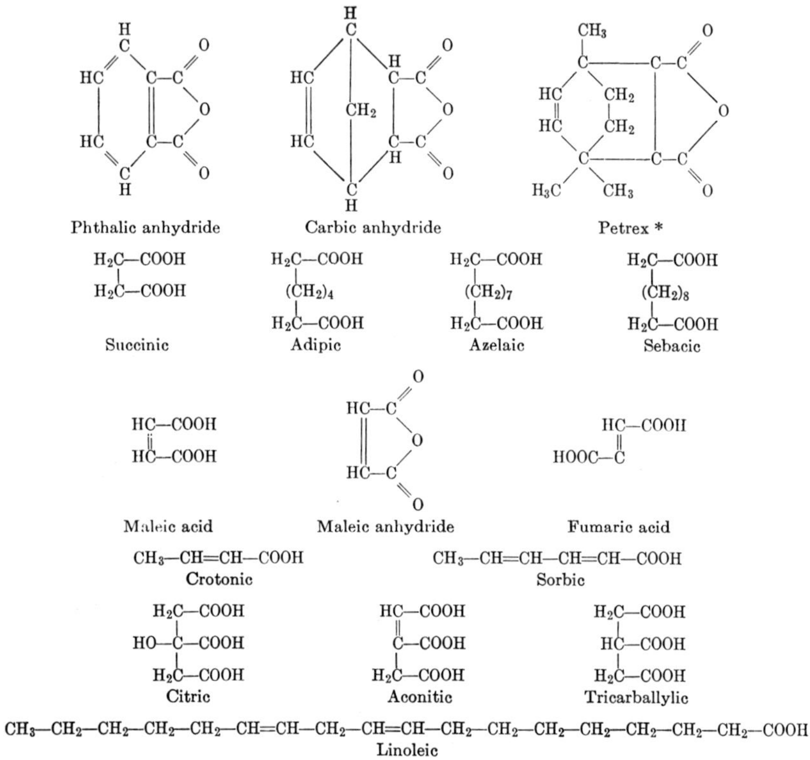

* Trade name, Hercules Powder Co.

Allyl alcohol is only monohydric, but it is potentially difunctional because of its allyl unsaturation. It may be esterified with a dibasic acid such as phthalic, under conditions that will not affect the allyl unsaturation. The diallyl phthalate may then be polymerized with a reactive polyester as mentioned earlier. Ethanolamine contains both a primary hydroxyl and a primary amine group. These may be reacted to form resins with both ester and amide linkages. Although the amide linkage has greater resistance to alkali it is usually quite dark in color. Figure 1 indicates some of the many interesting possibilities in alkyd resin formulation which can be made by variation of the alcohol, and Fig. 2 shows a similar situation with respect to variation of the polybasic acids.

POLYFUNCTIONAL ACIDS

The acids in Fig. 2 may be compared with phthalic as a standard because of its wide use in alkyd resins. It is produced commercially as the anhydride and is utilized as such because there is less water of reaction to be removed. Phthalic anhydride is difunctional and may be used with difunctional alcohols to produce resins having a linear structure. It may also be used with higher functional alcohols to produce resins having cross-linked structures. The resin properties may be modified by mixtures of alcohols or addition of other acids. If phthalic anhydride were used in the correct molar ratios with glycerol, the resulting resin would be hard and brittle because of the highly cross-linked structure obtained through condensation polymerization. If some of the phthalic anhydride is replaced with a fatty acid, such as lauric with a 12-carbon chain, or linoleic with an 18-carbon chain, the resin will be softer because the distance between the cross-linkages will be greater. This is the basis for formulation of the oil-modified alkyds described later. The formula for linoleic acid in Fig. 2 is written in full rather than in the usual abbreviated form so as to emphasize the significance of molecular shape on resin properties. Of course, the acid is not perfectly straight but kinked, as shown in Chapter 2, Fig. 3.

Carbic anhydride is similar in shape to phthalic, but the type of ring structure and its unsaturation are different. Carbic anhydride is the addition product of cyclopentadiene and maleic anhydride, and there is some evidence (1) that at temperatures above 165°C it breaks down into its original components. Some reaction may also be expected from the unsaturation in the carbic ring structure, whereas none would be expected from the benzene nucleus of phthalic. It will also be seen that Petrex, which is a dibasic acid anhydride, is the maleic adduct of a terpene. The branch chains impart somewhat greater solubility in hydrocarbons to resins made with Petrex by comparison with similar resins made with phthalic. However, the branching tends to reduce the hardness of the resins.

The four acids, succinic, adipic, azelaic, and sebacic, offer the interesting possibility of increasing the flexibility of the resin without changing its other components. The chain of eight methylene groups between the carboxyls in sebacic acid acts as an internal plasticizer.

This acid is widely used in the resinous type of plasticizers described later. As the number of methylene groups is reduced, the resin becomes harder, but the other features of the resin structure must always be kept in mind. For instance, adipic acid is used with hexamethylenetriamine to produce the very flexible fiber, Nylon 66.

Maleic anhydride is trifunctional since it can react by condensation as a dibasic acid, and also by addition reactions through its unsaturation. Because of its great reactivity, it is frequently used in small amounts with phthalic to increase the rate of resin formation and its complexity. It also imparts other desirable properties as described later. Fumaric acid is the *trans* isomer of maleic acid and cannot be produced as the anhydride because of this configuration. In some cases, it can replace maleic without significant difference in resin properties. Solubility and hardness are the properties most noticeably affected.

Crotonic and sorbic acids are short-chain monobasic acids having non-terminal unsaturation. If used as modifiers instead of the long-chain unsaturated fatty acids they would produce harder resins, but they could not be expected to be as soluble in aliphatic hydrocarbons. They offer the possibility of producing special resins to meet some specific requirements. The three tribasic acids, citric, aconitic, and tricarballylic, are in a similar category. They would be highly reactive if used with glycerol, but they could be used in small amounts with less reactive acids or with the lower functional glycols.

Another dibasic acid which is available commercially is tetrachlorophthalic anhydride. Its properties and reactions are described by

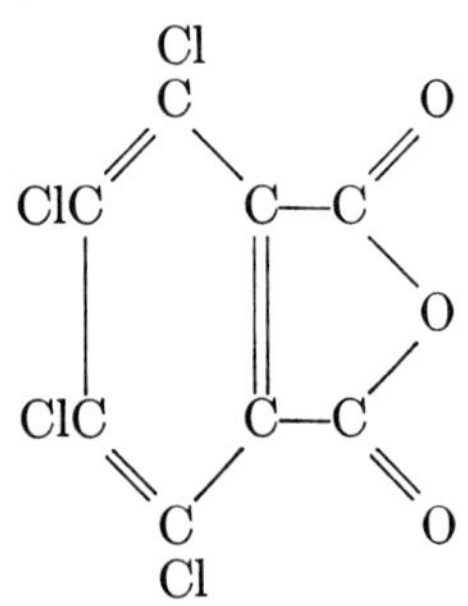

Lawlor (19). Mehta and Payne (20) investigated its use in alkyd resins and plasticizers. Three alkyds were made containing linseed, soybean, and cocoanut oils respectively, with a ratio of 40% oil to 60% glyceryl tetrachlorophthalate. All resins were darker in color than comparable phthalic alkyds. The addition of 1.5% maleic anhydride and 0.5% oxalic acid improved the color considerably, but it was not quite equal to that of phthalic resins. There was no significant difference in air-drying time, but the water and alkali resistances were slightly better with the tetrachlorophthalic anhydride.

Comparison was made (20) between the linseed-phthalic and linseed-tetrachlorophthalic resins for fire resistance, using the method in the National Military Establishment Specification JAN-P-702, November 23, 1948, "Paint, Inside, White Semi-Gloss, Fire Retardant." The paints made with the tetrachlorophthalic alkyd were somewhat more fire retardant than the phthalic alkyd paint, but the improvement was not large. Calculations showed that the resin binder contained only 26.1% chlorine, which is not sufficient for significant fire retardance; 50% chlorine is usually considered the minimum required.

A more recent commercial development is the dibasic acid known as Emery 955 dimer acid, produced by Emery Industries, Inc. It is manufactured by polymerizing linoleic acid by a special patented process (26, 27). The polymerization is carried out by means of heat and pressure in the presence of water. The water prevents the degradation usually associated with thermal polymerization. The product consists essentially of dimers, with small percentages of monomers and trimers. Most of the monomers are removed by distillation. The theoretical concepts regarding thermal polymerization are discussed in Chapter 2, page 117. The mechanism of dimer formation from several types of conjugated drying oils is shown in Chapter 2, Fig. 20, together with the structure of the resulting dimers. It is believed that the same type of structure is obtained from the non-conjugated oils because of isomerization during heat processing. For the dimerization of the acids a mechanism and final structure are postulated similar to those for the oils. Therefore the dimer acid would have the following type of structure. However, it should be realized that the position of the fatty acids with respect to each other could be reversed, which would place the carboxyl groups in the *trans* position instead of the *cis* position shown.

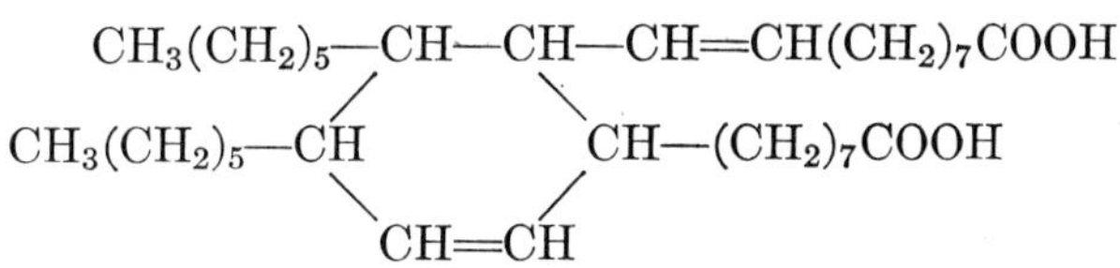

Typical characteristics of the Emery 955 dimer acid are shown in the table on page 276. The dimer acid may be esterified with glycerol to produce a bodied oil, or with other polyhydric alcohols to pro-

Characteristics of Dimer Acid *

Iodine value (modified Wijs)	80–95
Acid value	180–192
Saponification value	185–195
Unsaponifiable, %	2.0 max.
Neutralization equivalent	290–310
Specific gravity at 15/15.5°C	0.95
Refractive index at 25°C	1.4919
Viscosity at 25°C, Gardner-Holdt	Z4
Viscosity at 25°C, centistokes	10,000
Viscosity at 100°C, centistokes	100
Color, Gardner 1933	12 max.
Flash point, °F	530
Fire point, °F	600
Dimer content, by molecular distillation, %	85 app.
Trimer content, by molecular distillation, %	12 app.
Monomer content, distillate below 270°C, 5 mm, %	3 app.

* Courtesy of Emery Industries, Inc.

duce the modified type of bodied oil. It may also be used in the manufacture of varnishes and alkyd resins. The low iodine value indicates that it would be relatively non-oxidizable, which would affect its air-drying properties. The dimer acid may be used to advantage in alkyd resin manufacture to replace some of the monomeric acids or oils, or some of the phthalic anhydride, thereby reducing the time required to obtain the desired viscosity. This is illustrated by the data obtained from processing three medium-oil-length, soybean alkyds, which were reduced to 60% solids with xylol. Alkyd *A* is the blank containing no dimer acid, Alkyd *B* has 10% phthalic anhydride replaced with dimer acid, and Alkyd *C* has 20% replaced.

Effect of Replacement of Phthalic Anhydride with Dimer Acid on Processing Time *

Composition and Characteristics	Parts by Weight		
	A	*B*	*C*
Soybean fatty acids	47.5	43.2	39.5
Glycerol	13.4	12.2	11.2
Pentaerythritol	7.7	7.0	6.4
Phthalic anhydride	31.4	25.8	20.9
Emery 955 dimer acid	....	11.8	22.0
Acid number	1.0	1.6	0.8
Viscosity, Gardner-Holdt	U	Z2	Gel
Processing time, hr	5½	4	3

* Courtesy of Emery Industries, Inc.

Many variations of this procedure are described in the literature of Emery Industries, Inc. Their publications also refer to the use of dimer acid in the manufacture of polyamide resins.

The development of efficient fractionation processes for tall oil has produced tall oil acids suitable for alkyd resin manufacture. Typical formulas and procedures are given on page 103.

It should be apparent that a great variety of alcohols and acids could be prepared for use in alkyd resins. However, availability and cost are limiting factors in the use of such alcohols and acids unless they contribute outstanding properties to the resin. An interesting disclosure is made in U. S. Patent 2,522,586 (September 19, 1950) regarding "alkyd resins prepared from tricarboxylic acid having the structural formula of the addition product of 1-pimaric acid and fumaric acid." This is the same as the maleic adduct shown in Chapter 3, Fig. 3, except that it is made by an aqueous process which is described in Chapter 3, page 161. This gives the pure tricarboxylic acid instead of the mixture with rosin; consequently, the alkyds made with it have good exterior durability.

New raw materials are being made available by chemical manufacturers, and the reader should check the advertisements in trade journals for these possibilities. For example, the monobasic benzoic acid, when combined with more reactive acids, is very useful to control the rate of reaction, particularly in resins having only small amounts of oil modification. A material of this type is sold by the Monsanto Chemical Co. under the trade name of Benthal. A review of the trade magazines will reveal other materials than those listed in Figs. 1 and 2 for resin manufacture.

PHTHALIC ALKYD RESINS

The tremendous popularity of phthalic alkyd resins and phthalic ester plasticizers is based on two factors: the good properties of both the anhydride and its resins, and the relatively low cost of the anhydride. The low cost was made possible by the discovery of Gibbs and Conover during World War I that naphthalene in the vapor phase could be oxidized by air with high yields of phthalic anhydride.

Formerly, naphthalene was obtained from the coke-oven by-product recovery process; therefore, its supply was limited. However, other raw materials have been investigated with the result that phthalic anhydride is now being produced commercially from *o*-xylene. Xylene would suffer from the same limitation as naphthalene except that it is now being produced by the hydroforming of petroleum oils. The differences in boiling points of the three xylenes, *ortho, meta,* and *para,* are very small, making fractional distillation expensive but necessary when the mixed xylenes are produced. *p*-Xylene is of interest in the production of terephthalic acid for textile fiber production, as mentioned earlier.

The good feature of phthalic anhydride for the manufacture of resins is that it is very stable to heat. It will sublime without decomposition, indicating its good heat stability in resin processing. It also resists decomposition by ultraviolet radiation, and it transmits this quality to the resins made from it. This feature is partly responsible for the good exterior durability of alkyd resin finishes, and for the improvement in nitrocellulose lacquers for exterior application when alkyd resin modifiers are used. West (6) gives a good survey of the development and use of phthalic anhydride. About 300 million pounds, solid basis, of phthalic alkyd resins were used in 1950, and the rate of consumption is increasing. An excellent history of the development of alkyd resins is given by Kienle (2) in the April, 1949, issue of *Industrial and Engineering Chemistry.* This issue also contains an article by Hovey (3) on equipment for alkyd manufacture, and one on the theory of alkyds by Earhart (4).

Composition of Phthalic Alkyds. An alkyd resin made from phthalic and glycerol without modifiers is heat convertible to the insoluble, infusible stage because it is the product of a 2:3 condensation reaction. Before the resin is condensed as far as the infusible or gel stage, it is soluble in water or alcohol, depending on the degree of condensation. This type of resin finds limited use in electrical insulation, textiles, and plastics, but is not suitable for organic finishes. When the resin is modified by replacing some of the phthalic with a fatty acid, a so-called oil-modified alkyd is produced. These alkyds are soluble in hydrocarbons, but the solubility varies with the extent of oil modification. With relatively small proportions of oil the resins require aromatic hydrocarbons, but with medium to large proportions of oil the resins become soluble in the aliphatic hydrocarbons.

The nomenclature used for alkyds is similar to that applied to varnishes, namely short-, medium-, and long-oil alkyds. However, the proportion of oil to resin is less in the alkyds because the resin portion is tougher and more durable. Alkyds are also specified by the percentage of phthalic they contain; the general relationship between oil length and phthalic content is shown in Fig. 3. An attempt is made to show the type of structure in these resins; but, like all such at-

Fig. 3. Composition of Various Oil Length Alkyds

	A	*B*	*C*		*D*
Oil length	Zero	Short	Medium	Long	Very long
Range % PAA	77.5	40–50	30–40	20–30	
Actual % PAA	77.5	48.0	30.6		10.8

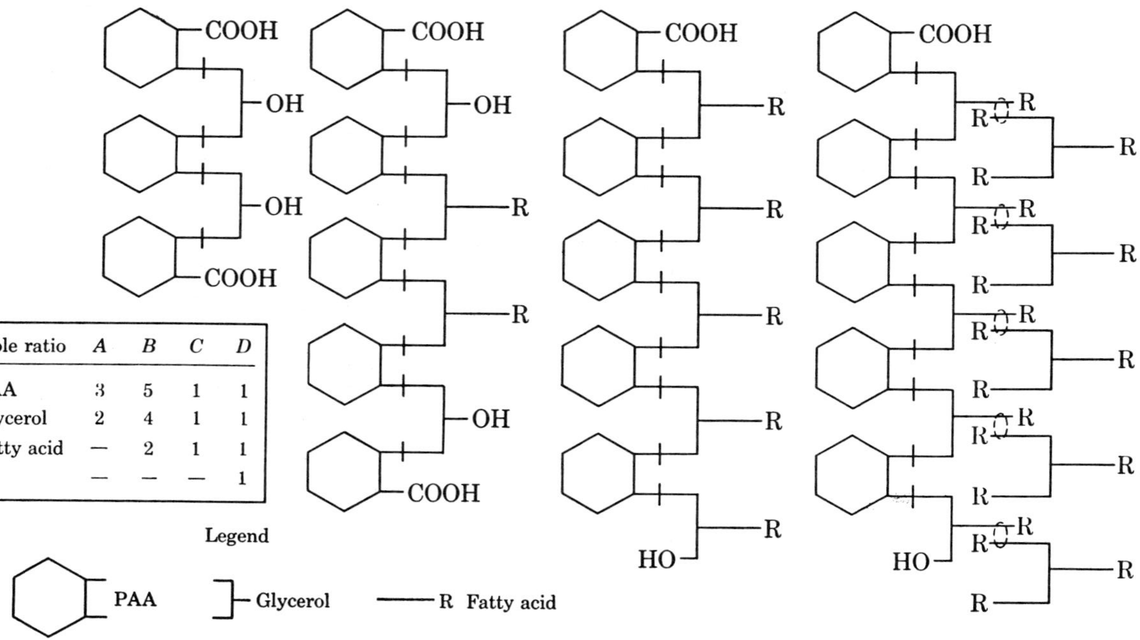

Mole ratio	*A*	*B*	*C*	*D*
PAA	3	5	1	1
Glycerol	2	4	1	1
Fatty acid	—	2	1	1
Oil	—	—	—	1

tempts, these drawings are apt to be misleading unless properly interpreted. In the first place, it is very difficult to show a three-dimensional structure in a two-dimensional drawing; in the second place, the diagrams are not drawn to scale. However, if these limitations are kept in mind, the diagrams help us to visualize the resin structure. The generally accepted designation for oil length of alkyds is:

Oil Length	% Phthalic Anhydride
Short	40–50
Medium	30–40
Long	20–30

The ratio of fatty acids to glyceryl phthalate resin corresponding to these oil lengths is indicated by resins *B*, *C*, and *D* in Fig. 3. Further discussion of the effect of increasing oil length follows the calculations.

Calculations for Phthalic Alkyds. Resin *A* in Fig. 3 is intended to portray an unmodified glyceryl phthalate resin. When the materials are combined in a ratio of 3 moles of phthalic anhydride and 2 moles of glycerol a completely neutral ester should be the product. Resin *A* is shown with two free carboxyl groups which could be neutralized with the two free hydroxyl groups. It should be understood that resin *A* as shown is only a small section of the theoretical, highly cross-linked, three-dimensional structure in which the carboxyls and hydroxyls would be reacted completely. Figure 3 shows resin *A* to have 77.5% phthalic anhydride, and this was calculated as follows:

Calculations for Resin *A*

3 moles phthalic anhydride	$= 3 \times 148$	$= 444$
2 moles glycerol	$= 2 \times 92$	$= 184$
		628
Less 3 moles water of reaction	$= 3 \times 18$	$= 54$
Resin solids		$= 574$

$$\frac{\text{Original phthalic input} = 444}{\text{Total resin solids} = 574} = 77.5\% \text{ phthalic}$$

$100 - 77.5 = 22.5\%$ glyceryl residue

The above is based on theory; in actual practise, an excess of glycerol is used as discussed later. When fatty acid modifier is used to produce the oil-modified alkyds, additional glycerol must be added to esterify the acid. It is convenient to remember the relationship between fatty acid, glycerol, and oil in such calculations.

$$
\begin{array}{lllll}
\text{3 moles fatty acid} & = 3 \times 280 & = & 840 \\
\text{1 mole glycerol} & = 1 \times 92 & = & 92 \\
\hline
 & & & 932 \\
\text{Less 3 moles water} & = 3 \times 18 & = & 54 \\
\hline
\text{1 mole oil} & & = & 878
\end{array}
$$

$$\text{Oil} = \frac{840}{878} = 96\% \text{ fatty acid}$$
$$4\% \text{ glyceryl residue}$$

This ratio may be employed to convert the fatty acid used in resin manufacture to oil in the finished resin, or to calculate the amount of fatty acid required to produce a resin containing a given percentage of oil. The theoretical calculations for resin *B*, containing a ratio of 5 moles of phthalic anhydride, 4 moles of glycerol, and 2 moles of fatty acid, are as follows:

Calculations for Resin *B*

$$
\begin{array}{lllll}
\text{5 moles phthalic} & = 5 \times 148 & = & 740 \\
\text{4 moles glycerol} & = 4 \times 92 & = & 368 \\
\text{2 moles fatty acid} & = 2 \times 280 & = & 560 \\
\hline
 & & & 1668 \\
\text{Less 7 moles water} & = 7 \times 18 & = & 126 \\
\hline
\text{Resin solids} & & = & 1542
\end{array}
$$

The composition of the resin may be expressed in several ways as shown below:

$$\frac{\text{Phthalic anhydride} = 740}{\text{Resin solids} = 1542} = 48\% \text{ phthalic anhydride}$$

$$\frac{\text{Fatty acids} = 560}{\text{Resin solids} = 1542} = 36.3\% \text{ fatty acids}$$

$$\text{Resin} - (\text{phthalic} + \text{fatty acid}) = \text{glyceryl residue}$$
$$100 - (48.0 + 36.3) = 100 - 84.3 = 15.7\% \text{ glyceryl residue}$$

$$\frac{\text{Fatty acid}}{96} = \text{oil} = \frac{36.3}{96} = 37.8\% \text{ oil}$$

$$\frac{\text{Phthalic anhydride}}{77.5} = \text{glyceryl phthalate resin} = \frac{48}{77.5} = 62.0\%$$

Resin *B* Composition

		Per Cent by Weight
1.	Phthalic anhydride	48.0
	Fatty acid	36.3
	Glyceryl residue	15.7
		100.0
2.	Phthalic anhydride	48.0
	Oil	37.8
	Glyceryl residue	14.2
		100.0
3.	Glyceryl phthalate	62.0
	Oil	37.8
		99.8

The reader should make similar calculations for resins *C* and *D* in order to become familiar with the method and with the different ways of expressing the composition. However, the reader may be confused if he tries to calculate the exact composition of a commercial alkyd from the manufacturer's specifications. For example, a certain alkyd is described as having 58% oil and 27.5% phthalic anhydride. If the phthalic is converted to per cent glyceryl phthalate, and the per cent oil added, the percentages should total 100%.

$$\frac{27.5}{77.5} = 35.5\% \text{ glyceryl phthalate}$$

$$35.5 + 58 \text{ oil} = 93.5\% \text{ total resin solids}$$

Obviously something is missing in the above specification. Another manufacturer specifies one of his resins as:

39% minimum phthalic anhydride
40% minimum oil acids

Again, if the phthalic anhydride is converted to glyceryl phthalate, and the oil acids converted to oil, it would seem that the sum of the two should total 100%.

$$\frac{39.0}{77.5} = 50.3\% \text{ glyceryl phthalate}$$

$$\frac{40.0}{96} = 42.0\% \text{ oil}$$

$$50.3 + 42.0 = 92.3\% \text{ total resin solids}$$

The key to this enigma was provided by the second manufacturer in the word "minimum." Since it is not practical to produce commercial resins meeting the various specifications of viscosity, acid number, etc., and maintain an exact composition, some tolerance must be allowed. The manufacturer gave the minimum percentage of phthalic and oil acids, but these values are not necessarily the actual values. For example, in the last resin, if the value for the phthalic were increased from the minimum of 39% to 41.9%, and the oil acids from the minimum of 40% to 44.2%, the sum of the two would total 100%:

$$\frac{41.9}{77.5} = 54\% \text{ glyceryl phthalate}$$

$$\frac{44.2}{96} = 46\% \text{ oil}$$

$$54 + 46 = 100\% \text{ total resin solids}$$

Of course, this is only one of many sets of actual values which would satisfy the requirement of 39% minimum phthalic anhydride and 40% minimum oil acids. It should be understood that these calculations apply only to the oil-modified alkyds. When other resins, such as phenolic modifications, are used in the alkyds, the exact percentage of the modification must be known. It should also be remembered that an excess of glycerol or other alcohol is necessary in commercial resins to prevent gelation before the desired acid value is reached. The actual excess varies with different resins, but a convenient figure for general use is 2.3 moles of glycerol with 3 moles of phthalic anhydride, instead of the theoretical 2 moles of glycerol.

Acid Number—Viscosity Relationship. It is important to have the acid number as low as possible to prevent reactivity of the resin with basic pigments and to improve water resistance. It is also important to obtain the correct viscosity so that the resin will meet the consistency and application requirements of the coatings in which it may be used. The acid number drops quite rapidly in the early stages of the esterification, but the rate of reaction diminishes considerably as the concentration of carboxyl and hydroxyl groups is reduced.

A general form of the acid number-time of reaction curve is shown in Fig. 4. It will be apparent that an excess of hydroxyl groups, introduced by excess glycerol, will tend to lower the acid number more quickly. This will enable the resin processor to obtain the required low acid number without an undesirably high viscosity from prolonged esterification time. However, there is a definite limit to which this procedure may be carried since excess hydroxyl groups in the resin also have a deleterious effect on its properties. They retard the drying of the resin, lower its resistance to water, and detract from its exterior durability. Therefore the excess polyhydric alcohol should be kept to the minimum which will give a reasonably low acid number and still maintain good film properties.

The rate of viscosity increase is much greater for short-oil alkyds than for the long-oil type; therefore a larger excess of polyhydric alcohol is necessary in the short-oil alkyds if a low acid number is re-

Fig. 4. Change of Acid Number and Viscosity with Processing Time

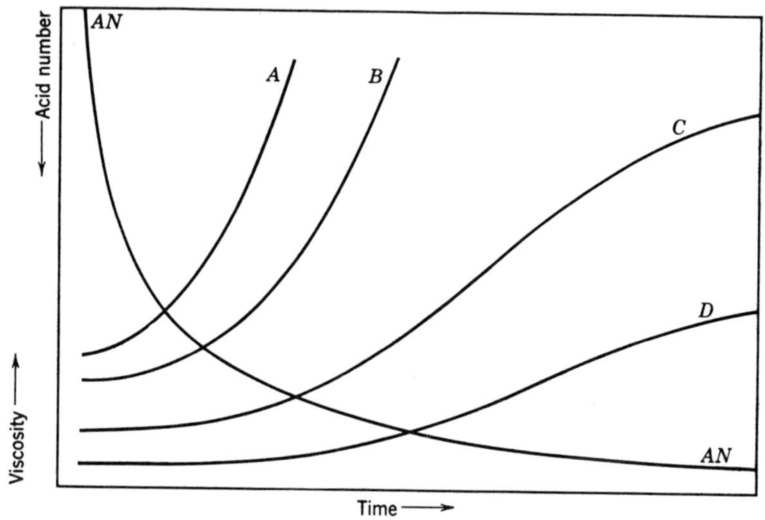

AN = Acid number-time curve. Viscosity-time curves for resins *A*, *B*, *C*, and *D* in Fig. 3.

quired. However, the excess must not be so much that it will detract from the film properties as indicated above. Figure 4 shows the general form of viscosity-time curves obtained in alkyd resin processing. In a general way they may be related to resins *A*, *B*, *C*, and *D* in Fig. 3. However, it should be apparent that the type of oil used in the resin, and the temperature at which the resin is processed, are important factors in the actual viscosity obtained in a given time period. Kienle (21, 22) and others found that the straight glyceryl phthalate increases in viscosity quite rapidly, and it reaches the gelation point when only 75–80% of its components have reacted. A short-oil alkyd, such as resin *B*, Fig. 3, also reaches the gelation stage quite rapidly; but, as the oil content is increased, the rate of viscosity rise is slower. The curve for resin *C* indicates that a high viscosity may be reached, but there is not the sharp tendency to gelation shown in the curves for resins A and B. By contrast, it will be seen that a high viscosity may be difficult to attain in the very long-oil-type alkyds, such as resin D, without extremely long processing time. Again, it must be remembered that the type of oil (drying or non-drying, etc.), the functionality of the polybasic acids and polyhydric alcohols, and the temperature, all affect the rate of viscosity increase and the possibility of gelation.

Functionality and Degree of Polymerization. Carothers (23), from his thorough investigations and clear comprehension of these phenomena, derived a general equation relating the functionality to the extent of the reaction and the degree of polymerization. This equation explains the shape of the viscosity-time curves in Fig. 4 and enables one to visualize the rate of reaction and possibilities of gelation in alkyd resin processing.

Functionality, Extent of Reaction, and Degree of Polymerization (23)

Let

f = functionality (i.e., number of functional groups per monomer molecule).

No = number of monomer molecules present initially.

Then

Nof = number of functional groups present initially.

N = number of molecules present after reaction.

$2(No - N)$ = number of functional groups lost.

$$\frac{2(No - N)}{Nof} = \text{fraction of functional groups lost} = p = \text{extent of reaction}$$

$$\frac{No}{N} = \text{average degree of polymerization} = X$$

$$\text{Hence, } \frac{2}{f} - \frac{2}{xf} = p = \text{extent of reaction}$$

When the average degree of polymerization is very high, x is large and the term $2/xf$ is very small, so that it may be discarded. At gelation x = infinity.

A few calculations, based on the resins in Fig. 3, will serve to illustrate Carothers' equation.

Calculations for Resin *A*

	No	f	Nof
Phthalic anhydride	3	2	6
Glycerol	2	3	6
	5		12

Average $f = 12/5 = 2.4$

p at gelation $= 2/f = 2/2.4 = 83.3\%$

Gelation occurs when 83.3% of the functions have reacted.

Calculations for Resin *B*

	No	*f*	*Nof*
Phthalic anhydride	5	2	10
Glycerol	4	3	12
Fatty acid	2	1	2
	—		—
	11		24

Average $f = 24/11 = 2.18$
p at gelation $= 2/f = 2/2.18 = 92\%$

Gelation occurs when 92% of the functions have reacted.

Calculations for Resin *C*

	No	*f*	*Nof*
Phthalic anhydride	1	2	2
Glycerol	1	3	3
Fatty acid	1	1	1
	—		—
	3		6

Average $f = 6/3 = 2$
p at gelation $= 2/f = 2/2 = 100\%$

Gelation does not occur unless 100% of the functions present have reacted. This extent of reaction is not likely to occur in actual resins; therefore there is little danger of gelation. However, high viscosity may be anticipated, as indicated in Fig. 4.

In the calculation for resin C, the fatty acid was given a functionality of 1 because the condensation reaction only was considered. This would be the situation in actual resin manufacture if the fatty acid were the non-oxidizing, or saturated, type. If the fatty acid were unsaturated, and the conditions of the resin processing were such that polymerization could occur through the fatty acid unsaturation, then the fatty acid should be assigned a functionality of 2. The average f for the reaction would then be $f = 2.33$, and gelation would occur at $p = 2/2.33$, or when 86% of the functions had reacted.

It should be evident that this situation would occur in actual resin processing only when a very reactive fatty acid was used, such as tung or oiticica oil acid. The temperature of alkyd processing is sufficient to polymerize these reactive acids, but it would have considerably less effect on linseed or dehydrated castor oil acids, and the effect would be practically zero on soybean oil acids.

The effect on the extent of reaction at gelation of replacing glycerol with pentaerythritol in resin C may be calculated readily. This will necessitate increasing the mole ratio of phthalic anhydride to 1.5, with 1 mole of pentaerythritol and 1 mole of fatty acid.

Calculation for Pentaerythritol Alkyd

	No	*f*	*Nof*
Phthalic anhydride	1.5	2	3
Pentaerythritol	1	4	4
Fatty acid	1	1	1
	3.5		8

Average $f = 8/3.5 = 2.282$
p at gelation $= 2/f = 2/2.282 = 87.5\%$

Gelation occurs when 87.5% of the functions have reacted.

Comparison of the above $p = 87.5\%$ with the $p = 100\%$ obtained for resin C shows clearly why a pentaerythritol alkyd increases in viscosity more rapidly than a similar glycerol alkyd. It should also be evident that it will be more difficult to obtain a low acid number with a pentaerythritol alkyd processed to the same viscosity as a glycerol alkyd.

Several investigators have attempted to derive equations which may be applied to alkyd resin processing so as to permit the calculation of acid number at the required viscosity, or the reverse. The difficulty of such attempts will be evident from the number of factors which affect the alkyd during processing, some of which are listed below:

1. Temperature of reaction.
2. Degree of agitation.
3. Efficiency of exclusion of oxygen by inert gas.
4. Catalytic effect of metal of kettle, or traces of impurities.
5. Size of kettle.
6. Competition of esterification reactions when more than one acid or alcohol is present.

It will be apparent that, despite the most accurate calculations and the most modern equipment and instrumentation, a good deal of personal experience is necessary to decide just when a particular resin reaction is complete.

Manufacture of Phthalic Alkyds. *Fatty Acid Process.* Alkyd resins may be made by two general processes: (*a*) the fatty acid process; (*b*) the monoglyceride process. These processes may be carried out by direct fusion of the materials (fusion method), or by fusion in the presence of a solvent (solvent method). In the fatty acid process, the ingredients are placed in the reaction kettle and heated at temperatures ranging from 410°F to 450°F or higher. The batch is held at reaction temperature until the required acid value and viscosity have been reached. It is then pumped or blown by inert gas pressure into the thinning kettle and reduced to the required solid content.

Iron kettles produce dark-colored resins; aluminum is satisfactory for color but cannot be cleaned with alkali; stainless steel is satisfactory in both these respects. The kettle should be equipped with an agitator and foam breaker and should be covered to permit the use of a blanket of inert gas above the resin. The heating may be done by oil, gas, or electricity, depending on relative cost in the particular locality. Good descriptions of alkyd resin processing equipment are available from a number of manufacturers.

The temperature of processing is determined somewhat by the oil length and the various modifiers. Short-oil resins react more rapidly than long-oil resins; therefore the top temperature is generally lower. Long-oil resins may be heated as high as 500°F to accelerate the reaction since they are not so sensitive to gelation. The effects of the various types of fatty acid will be discussed later.

It was mentioned previously that maleic anhydride may be used as a modifier to accelerate the reaction and to improve the color of the resin. From 2% to 10% of the phthalic anhydride may be replaced with maleic anhydride; with higher percentage the reaction becomes difficult to control. If an ester-gum-modified alkyd is required, the rosin with the necessary glycerol may be added at the beginning of the reaction, or the ester gum may be added near the end. If modification with congo ester is desired, the ester must be made separately and added about an hour before the end of the alkyd processing. Phenolic-resin-modified alkyds, which have improved water and alkali resistance, may be made by adding a rosin-modified phenolic resin about 1 hr before the end of the alkyd processing, or a phenolic condensate at the beginning. The latter procedure causes considerable foaming.

Monoglyceride Process. It may seem uneconomical to split oil into fatty acids and glycerol, and then to recombine the fatty acids with glycerol to make oil again in the fatty acid process. However, if oil is used instead of fatty acid the glyceryl phthalate will precipitate as it is formed because it is insoluble in oil. With the fatty acids the reaction proceeds smoothly because the glyceryl phthalate is soluble in these acids. However, oil may be used if it is first converted to the monoglyceride. This may be accomplished by heating a mixture of oil and glycerol with catalyst until the product makes a clear solution in methyl alcohol. The formulas and details for making monoglycerides are given in Chapter 2, Formulas 4, 5, and 6. A discussion of monoglyceride theory and practice is given by Petit (10).

The monoglyceride may be prepared separately and used as desired, or it may be made in the alkyd resin kettle and the necessary phthalic anhydride and modifiers added. A procedure for making a simple linseed oil-modified alkyd resin would be as follows:

Formula 1 Linseed Alkyd, Monoglyceride Process

	Parts by Weight
Linseed oil	2000
Glycerol	500
Lime	2
Phthalic anhydride	1400
Glycerol	80

Procedure

Heat the linseed oil and glycerol to 425°F in about 2 hr, add the lime and continue heating to 450°F, hold till clear solution in methyl alcohol is obtained at 3:1 dilution (this requires about 30 min), add the phthalic anhydride slowly to prevent rapid cooling, add the glycerol and maintain at 450°F for acid value of 6–10 and the required viscosity in solution (from 6 to 10 hr), dilute with solvent to the required solid content. It is better to wet the lime with a little linseed oil and add it to the batch in the paste form.

It should be kept in mind that the above formula is a very simple one and that many variations may be made in the ratio of components and in the type of oil, polyhydric alcohol, or acid modification. These changes would necessitate variations in cooking temperature and time schedule. The above resin contains about 55% oil and is therefore fairly long oil. As such, it would be thinned with mineral spirits to a solid resin content of 60%. Short-oil resins are not soluble in mineral spirits but require a stronger solvent such as xylol. They are usually thinned to 50% solids as discussed in a later section.

Solvent Method. The designation "solvent method" for alkyd resins simply means that the resin is processed in the presence of a solvent. In general, about 10% of solvent is sufficient. The method is described in U. S. Patent 2,308,498, dated November 11, 1944. The solvent should be immiscible with water so that the water of reaction may be distilled off as a binary mixture. By means of a suitable separator, water is removed and the solvent returned to the batch. Obviously, the kettle must be equipped with a condenser; and, since the temperature is above the subliming point of phthalic anhydride, some of this material will be found in the condenser. It is important that the solvent be efficient for phthalic; otherwise, the condenser will become choked with the solid phthalic. Mineral spirits is not satisfactory, but a combination of 85% mineral spirits and 15% xylol will prevent the phthalic from collecting in the condenser. All xylol is quite satisfactory where the presence of this material in the final resin solution is not objectionable.

One of the principal advantages of the solvent method is that it permits better control of the resin reaction. In this respect, it is similar to solvent polymerization as compared with mass polymerization. The reactants are diluted; therefore the rate of reaction is slower and the resin or polymer size is more uniform. It will be seen that the solvent method is of greater advantage with the shorter-oil resins since their rate of reaction is faster and more difficult to control than that of the longer-oil resins. In all cases, there is a distribution in the molecular weights of resinous materials, but extremes should be avoided. The high-molecular-weight fraction may not be very soluble in the solvent; it may not be soluble even in the lower-molecular-weight fraction. This condition may show as a haze or turbidity in solution, but it may not be evident at all in the clear resin. However, it can cause serious difficulty if the resin is used in a pigmented coating. The high-molecular-weight fraction, or incipient gel particles, may cause flocculation of the pigment and produce a rough surface on the coating.

Another advantage of the solvent method is that it permits a record to be made of the extent of the reaction and consequently the rate of the reaction. The amount of water obtained can be calculated to the extent of esterification or the condensation reaction. Since all the ingredients may be kept in the kettle except the water of reaction, it is evident that the final resin must have the composition of the original ingredients less the water of reaction. This is especially useful in research where one may wish to have comparable conditions in a series of resins and to know the rates of reaction and also the final compositions without having to analyze them. In open kettles, there is a considerable loss of phthalic anhydride from sublimation, and this occurs to some extent in closed kettles. However, this loss is not serious in a properly closed kettle and does not occur at all in the solvent method.

Types of Oil Modification. Oil-modified alkyds are made with drying, semi-drying, and non-drying oils. The type of oil selected depends on the conditions under which the film will be dried and the color-retention properties required in the film. Air-drying alkyds are made with the good drying oils, such as linseed and dehydrated castor, but the semi-drying soybean oil may also be used in medium and short-oil alkyds. It will be recalled that soybean oil is not generally satisfactory for air-drying varnishes. The ability to use soybean oil in air-drying alkyds makes it possible to produce better non-yellowing white enamels than can be made with varnishes.

The soybean-oil-modified alkyds are also utilized in baking finishes. They may serve as the only film-forming component, or they may be mixed with urea- or melamine-formaldehyde resins to shorten the baking time and improve color retention and surface hardness. When the palest color and maximum color retention are required the non-drying-oil-modified alkyds are preferred. Since they do not form good films alone, they are used in combination with the urea- or melamine-formaldehyde resins. The usual ratios are from 60% to 70% alkyd with 30–40% amino resin. The non-drying alkyds are also used in nitrocellulose lacquers for furniture, automobiles, and other articles.

A wide range of selected fatty acids is available at present for the manufacture of alkyd resins, as referred to in Chapter 2. This permits the alkyd manufacturer to produce resins with special properties, such as improved drying, better water resistance, or maximum color retention in finishes subjected to above-normal temperatures for long periods. This degree of specialization is not possible when the monoglyceride process is used, because the oils are mixed glycerides. For example, if alkali-refined linseed oil were used to make the monoglyceride, the product would contain the usual percentages of oleic, linoleic, and linolenic acid radicals found in the original oil. However, if the oil were split and the fatty acids separated, they could be used to make three types of alkyd. The alkyd made with linolenic acid would be very fast drying, the one with linoleic acid would be medium fast, and the oleic acid alkyd would be quite slow drying. It will also be seen that the degree of yellowing of these three alkyds would be quite different. The first one would have severe yellowing with age since linolenic acid is believed to be the major cause of yellowing, as discussed in Chapter 2.

It will be apparent that the flexibility of an alkyd resin film will be increased as the oil content of the resin is increased, since the phthalic glyceride is a hard and somewhat brittle resin. Of course, the type of oil is also a factor since the non-drying oils act as permanent plasticizers and do not harden on aging like the oxidizing oils. The structure of resin *C* in Fig. 3 indicates that each fatty acid radical is combined with the phthalic glyceride. This is an idealized structure; an actual resin would be expected to be more heterogeneous. The structure of resin *C* also indicates that, when a larger ratio of fatty acid is used, it will combine with glycerol and exist as free oil. This condition is shown in the very long-oil resin *D*. Such resins would be made with oxidizing oils, and some cross-linking may occur between the unsaturated sections of the acid radicals if the reaction temperature is sufficiently high.

The compatibility of alkyd resins with oils increases as the oil content of the resin is increased. Short-oil alkyds are compatible with only small amounts of unbodied oil, but when the oil content is increased to 60–65% (about 25% phthalic anhydride) the alkyd is freely compatible with unbodied and medium-bodied oils. Therefore, there is no need to make alkyds longer in oil than 65%, or about 25% phthalic, since more oil may be added by simple mixing. It will be noted in Table 2 that the resin with the lowest phthalic content in the oxidizing oil-modified alkyds contains 23% phthalic. This resin is listed as containing 60% oil acids, or 62.6% oil, but this value is a minimum; the actual value may be somewhat higher.

The recovery of tall oil from the sulfate process for paper making is described in Chapter 2. The efficient fractionation of distilled tall oil has produced fatty acids with very low rosin content suitable for use in alkyd resins. The characteristics of these acids are described on page 103. The following formulas and procedures serve to illustrate their use in alkyds, and also the difference in procedures for a straight pentaerythritol alkyd and one containing a mixture of glycerol and pentaerythritol. The use of straight pentaerythritol in the short-oil alkyd would make the processing too difficult to control accurately.

Formula 2 Long-Oil Alkyd Resin

	Parts by Weight
Acintol * FA #2 tall oil acid	570
Pentaerythritol	190
Phthalic anhydride	240

Procedure

Heat charge to 390–400°F under inert gas, hold about 1 hr, raise temperature to 480–490°F, hold for acid number of 5–10 and viscosity of W-Z at 70% solids (about 7 hr), cool, and reduce with mineral spirits.

Characteristics

% solids	70
Viscosity, Gardner-Holdt, at 25°C	Y
Color, Gardner 1933	7
Acid number (solid basis)	6.5

* Trade name, Arizona Chemical Co.

This type of alkyd is suitable for use in white and colored architectural paints as described in Volume II. A suggested drier combination is 0.5% Pb, 0.05% Co, and 0.05% Ca as metal to resin solids. Films of the resin dry in 6–8 hr and have good hardness overnight.

Formula 3 Short-Oil Alkyd Resin

	Parts by Weight
Acintol * FA #2 tall oil acid	225
Glycerol	48
Pentaerythritol	58
Phthalic anhydride	148

Procedure

Charge Acintol FA #2 and glycerol, heat to 390–400°F under inert gas, add pentaerythritol slowly, then add phthalic anhydride, raise temperature to 470–475°F hold (3–4 hr) for acid number of 5–10 and viscosity of Z2 at 60% solids in xylene, cool, and reduce with xylene.

Characteristics

% solids	60
Viscosity, Gardner-Holdt, at 25°C	Z2
Color, Gardner 1933	6–7
Acid number (solid basis)	6.5

* Trade name, Arizona Chemical Co.

This type of alkyd resin, with the usual driers, may be used in air-drying or baking finishes. The driers are usually omitted when the alkyd is used in combination with urea-formaldehyde or melamine-formaldehyde resins for baking finishes.

Commercial Alkyd Resins. The physical characteristics of the majority of commercial alkyd resins are listed in the Resin Index of 1950, issued as *Circ. 738* by the Scientific Section of the National Paint, Varnish, and Lacquer Association, Inc. The alkyd resin manufacturers, together with their trade names, are listed in Table 1.

A wide variety of alkyd resins is available from the manufacturers listed in Table 1. A typical series is shown in Table 2 to illustrate the range in phthalic anhydride and oil content, type of solvents, and physical characteristics. The series is divided into three groups: oxidizing oil-modified alkyds, non-oxidizing oil-modified alkyds, and the resin-and-oil-modified alkyds. A consideration of the physical properties of these resins will indicate why they are used in certain applications. However, a few general considerations should be given first.

Solvents. It will be noted that the type of solvent changes from xylol to mineral spirits when the minimum phthalic content changes

Table * 1

Alkyd Resin Manufacturers	Trade Names
Alkydol Laboratories, Inc.	Alkydol
American Alkyd Industries	Amalite
American Cyanamid Co.	Rezyl
Boston Varnish Co.	Kyanol
California Flaxseed Products Co.	Calflax
Falk and Co.	Falkyd
Farnow Varnish Works	FA #
France, Campbell and Darling, Inc.	FCD #
General Electric Co.	Glyptal
Jones Dabney Co.	Syntex
Onyx Oils and Resins, Inc.	Alkydox
C. J. Osborn Co.	Mirasol
Plaskon Division, Libbey-Owens Ford Co.	PCC #
Price Varnish Co.	PVC #
Reichhold Chemicals, Inc.	Beckosol
Rohm and Haas Co.	Duraplex
Schenectady Varnish Co., Inc.	SA #
Sherwin-Williams Co.	XAC #
U. S. Industrial Chemicals, Inc.	Aroplaz
A. J. Wittenberg Corp.	VBR #

* Resin Index of 1950, *Circ. 738*, May 1950, Scientific Section, National Paint, Varnish, and Lacquer Association, Inc.

from 39% to 35%. This indicates the range of solubility in aromatic and aliphatic solvents; resins which are shorter in oil than 35% phthalic require aromatic solvents. The percentage of resin solids in solution is increased from 50% in the medium-oil alkyds to 70% in the long-oil alkyds. The reason for this increase is that the base resin in the medium-oil alkyds is more viscous than the base resin in the long-oil types and requires more solvent to reduce it to shipping viscosity. The short-oil alkyds can be marketed at 50% solids even though the base resin is more viscous than the others, because xylol is a stronger solvent. It should be noted that one of the long-oil alkyds, Rezyl 869, is sold at 70% solids and also at 100% solids. The 70% solution is convenient for general paint purposes, and the 100% resin is used in printing inks without solvent, and in combination with chlorinated rubber and other resins which require special solvents.

Further comparison shows that considerable difference in viscosity may be obtained with resins having practically the same phthalic content. This may be due to a difference in the degree of polymerization of resins of the same composition, or to a difference in the composition of the resins. Replacement of some phthalic anhydride with the more

Table 2 * Characteristics of Commercial Alkyd Resins

	Resin Solution					Solid Resin			
Rezyl † Alkyd Resins	Solids, ±1%	Solvent	Color, Gardner 1933	Viscosity, Gardner-Holdt, at 25°C	Approx. Pounds per Gallon	Min. % Phthalic Anhydride	Acid Number	Min. % Oil Acids	Type Oil ‡
		Oxidizing Oil-Modified Alkyds							
1102-5	40	Xylol	5–9	Y–Z3	8.1	50	40–60	28	A
310-5 §	50	Xylol	4–8	Z2–Z5	8.3	41	4–10	34	C
330-5	50	Xylol	5–9	X–Z1	8.2	39	4–10	40	B
387-5	55	Xylol	3–6	Z2–Z5	8.4	41	6–12	33	C
412-1 §	50	MS	4–8	U–W	7.8	32	3–8	50	C
414-1	60	MS	4–8	Z2–Z5	7.8	32	3–8	50	C
435-1	50	MS	5–9	Z2–Z5	7.7	34	6–12	48	C
807-1 §	50	MS	5–9	W–Z	7.8	30	3–8	51	A
811-1	50	MS	5–9	Z–Z3	7.8	35	4–10	46	A
823-1	50	MS	7–10	Z1–Z4	7.8	35	6–12	46	A
869-1	70	MS	7–10	W–Y	8.1	27	4–8	56	A
869	100	None	See Rezyl 869-1		9.1	27	4–8	56	A
872-1	70	MS	5–9	Y–Z1	8.0	25	4–9	56	C
873-1-70	70	MS	4–8	Z1–Z3	8.0	23	4–8	60	C
		Resin-and-Oil-Modified Alkyds							
728-5	50	Xylol	13–18	U–X	8.2	35	15–25	38	A
729-5	50	Xylol	8–12	U–X	8.3	39	10–20	33	A
775-4	70	Toluol	12–16	Z–Z4	8.4	13	17–24	44	E
		Non-Oxidizing Alkyds							
50-5	60	Xylol	6–12	S–V	8.3	35	6–13	45	ND
92-5	60	Xylol	2–6	Z2–Z5	8.6	45	4–10	30	ND
99-4-60	60	Toluol	3–7	Z–Z3	8.6	43	4–8	34	ND
99-5	60	Xylol	3–7	Z2–Z5	8.6	43	4–8	34	ND

* Courtesy of American Cyanamid Co.
† Trade name, American Cyanamid Co.
‡ Type of oil: A—linseed; B—castor; C—soybean; E—tung; ND—non-drying.
§ Contains castor oil.

reactive maleic anhydride would give greater viscosity if the resins were cooked on the same schedule. Higher viscosity would also be achieved by replacement of some glycerol with the more reactive pentaerythritol.

Acid Number and Reactivity. With the exception of the very short-oil alkyd, Rezyl 1102-5, the resins in Table 2 have low acid numbers. These resins would not be reactive with basic pigments unless the acid numbers of some of them were on the high end of the range, say from 10 to 12. The high acid number is more likely to cause pigment reactivity if it is combined with high viscosity in the resin. Frequently, high viscosity is desirable for fast-drying finishes because it gives faster setting and better solvent release. However, the combination of high viscosity and high acid number may produce instability in pigmented coatings if basic pigments are present. The basic pigment may react with the high acidity to form soaps which increase the viscosity of the paint. The soaps may also form "seeds" because they are not very soluble in the solvent and may separate as the paint ages.

High acidity acts as a dispersing agent for incipient gels; therefore, if the acidity is neutralized by reaction with basic pigments, gelation of the resin may take place with a consequent large increase in viscosity of the pigmented coating. The reactivity of basic pigments with high-acid resins is accelerated by an increase in temperature. Therefore care should be taken when grinding the pigments with the resin to prevent the development of above-normal temperatures. Excellent results may be obtained by grinding the pigments with low-acid, low-viscosity resins, such as Rezyl resins 412-1, 807-1, and 811-1, and adding the higher-viscosity resins to the paste, to complete the formulation. Further details regarding pigmentation are given in Volume II.

Color and Type of Oil. A definite relation may be seen between the color of the resins and the type of oil they contain. The palest color is obtained with the non-oxidizing oils, which also give better color retention with baking and aging conditions. In the oxidizing resins, the best color is obtained with soybean oil and the darkest color with linseed oil. Some exceptions may be found, as in the case of Rezyl 435-1 with soybean oil and Rezyl 811-1 with linseed oil, since both have the same original color. However, Rezyl 811-1 would darken or "yellow" more than Rezyl 435-1 under baking conditions or from prolonged aging at room temperatures. The strong yellowing of linseed oil was discussed in Chapter 2; this yellowing is related to its high content of linolenic acid. Soybean oil contains only about 2% of this acid; therefore it does not yellow appreciably on aging.

Resin-and-Oil-Modified Alkyds. Alkyds may be modified with natural resins such as congo ester or rosin ester, or with phenolic resins, as mentioned previously. Rezyl resins 728-5 and 729-5 are natural resin and oil modified, and Rezyl 775-4 is phenolic resin and oil modified. The resin modification produces faster setting and drying times, and the phenolic modification also improves water and alkali resistance. It will be noted that the oil in Rezyl 775-4 is the tung type, which, of course, further improves the water and alkali resistance. As may be anticipated, these alkyds are darker in color than the straight oil-modified alkyds, but this is of little consequence since they are normally used in colored finishes.

The presence of the resin modification makes it impossible to estimate the alkyd composition from the values for phthalic anhydride and oil acids, as was possible with the straight oil-modified alkyds.

The resin-modified alkyds are dissolved in aromatic solvents to obtain fast setting time. Addition of a phenolic dispersion resin will further reduce the setting time and improve the water resistance. Rezyl 775-4 is dissolved in toluol for use in certain types of nitrocellulose lacquers and in wrinkle finishes. Phenolic-modified alkyds are also available in a mineral spirits solvent for use in fast air-drying enamels. In general, the resin-modified alkyds are not as durable for exterior finishes as the straight, oil-modified alkyds.

General Uses. A comparison of the general characteristics of alkyd resins and of varnishes is given in Chapter 4. It will be noticed that the alkyds are rated better in color and exterior durability than varnishes. These two features contribute greatly to the extensive use of alkyds despite the fact that they are generally less resistant to water and alkali than some varnishes. In marine specifications, it is common to use alkyds in finishes applied above the water line for their good durability, and phenolic varnish finishes below the water because of their better water resistance. Of course, the non-drying alkyds used in white baking finishes have no counterpart in varnishes. There is no varnish resin-oil combination which can compare with them for pale original color and color retention with continued heating.

Detailed formulas for finishes containing alkyd resins are given in Volume II; only a general outline of their use is given here. The short- to medium-oil alkyds in the oxidizing group, Table 2, are suitable for both air-drying and baking finishes. The longer-oil alkyds in this group are generally used in air-drying finishes only. It is a good procedure for the coating manufacturer to make the largest variety of finishes from the smallest number of resins. This reduces inventory, lowers costs, and frequently reduces errors. In order to comply with this procedure, the resin manufacturer makes a minimum number of resins, but they are so designed that they may be used in a wide variety of finishes. Some resins may be blended with others to obtain the desired properties and cost requirements.

As may be expected, the very short-oil resin, Rezyl 1102-5, is generally combined with other resins to shorten the air-drying or baking time. Rezyl resins 310-5 and 330-5 are used primarily in baking finishes in combination with urea- or melamine-formaldehyde resins. Rezyl 310-5 finishes bake somewhat harder than Rezyl 330-5, and they would therefore require less of the more expensive amino resin. White baking enamels for kitchen cabinets and other appliances contain from 70% to 80% Rezyl 310-5 and 20–30% of urea or melamine resin, calculated on a solid resin basis. For maximum whiteness required in refrigerator finishes, the non-oxidizing alkyds are used entirely or in combination with Rezyl 310-5 as described later. Rezyl 387-5 is suitable for either air-drying or baking finishes. A reduction in baking time and a harder surface on the finish may be obtained by blending Rezyl 387-5 with 10–25% of urea or melamine resin. The blend is also better for original color and color retention.

Rezyl 412-1 may also be used for either air-drying or baking finishes. In comparison with Rezyl 387-5 it is slower air drying and requires longer baking because it has higher oil content, lower base resin viscosity, and the slower-evaporating mineral spirits as solvent. However, these features make it a desirable type of resin for finishes which must have good flowing or leveling properties immediately after application. This is a particular requirement of "tin decorating finishes" which are applied to sheets of tin-plated iron by a roller coating machine. The coating must level out rapidly to a smooth finish before it enters the baking oven. Usually, from 10% to 15% of amino resin is added to improve the surface hardness, color, and alkali resistance of the coating. The amino resins which are used with alkyds having mineral spirits as the solvent should be the type which have high mineral spirits tolerance, as described in the next chapter.

Rezyl resins 435-1, 811-1, and 823-1 are typical of the higher-viscosity type, having fast air-drying and baking properties, and used in primers and finish coats. These coatings are used on tools and other implements and on metal furniture, and for refinishing cars and trucks where fast setting and drying are important. The longer-oil Rezyl resin 869-1 is more flexible and has very good exterior durability; it is therefore suitable for architectural trim paint, garden furniture finishes, and porch and deck enamels. It is compatible with unbodied and medium-bodied oils so that its oil length may be increased if desired for use on wood surfaces. Since it is a linseed-type alkyd, it yellows somewhat with aging and is therefore not suitable for white enamels. The soybean-type alkyds, Rezyl 872-1 and 873-1-70, are designed for air-drying white enamels, for either interior or exterior use. Rezyl 873-1-70 is the pentaerythritol-type alkyd which has fast drying characteristics despite its long oil length. The pentaerythritol also improves the durability, which makes this type of resin desirable in white trim paints and marine enamels.

In general, the resin-and-oil-modified alkyds are designed for very fast air-drying or baking colored finishes, for toys, tools, and other low-cost items. However, Rezyl 728-5 was developed for the corrosion-resistant primer for aluminum aircraft, using zinc yellow as the pigment. The phenolic-modified type is used for interior colored finishes where improved resistance to soap is required. Since this type of alkyd usually contains some conjugated oil, tung or oiticica, it is suitable for wrinkle finishes.

The non-oxidizing oil-modified alkyds are used in white baking finishes because of their excellent color retention, and in nitrocellulose lacquers because of their plasticizing action combined with higher gloss retention and good exterior durability. It will be noted that Rezyl 92-5 has very pale original color and excellent color retention; it is therefore combined with about 33% of melamine resin for heat-resistant finishes for lighting fixtures and certain parts of stoves. Rezyl 99-5 is not as good for heat resistance, but it has better toughness. The solution in toluol, Rezyl 99-4-60, is the type used in nitrocellulose furniture finishes and automotive lacquers.

It will be seen from this very brief survey of the uses for phthalic alkyd resins that a very wide range of finishes is included. This subject is covered in greater detail in Volume II. The reader should obtain the current literature from resin manufacturers and become familiar with the relationship between resin characteristics and potential uses for the wide range of general and special alkyds which are available today.

IMIDE-MODIFIED ALKYD RESINS

It was pointed out previously that a great many types of alkyd resins may be produced. The phthalic alkyds are the most important at present, although the sebacic type is widely used as plasticizers and is described in a later section. The styrenated alkyds are growing in importance, and the reactive polyesters are used chiefly in plastics. These are described on pages 304 and 302 respectively. It may be of interest to point out a few possibilities which have not yet assumed commercial importance.

A series of imide-modified alkyds is described by Wright and Dupuis (11). They replaced part of the ester linkages in alkyds with amidic linkages by substituting an amide of glycerylmonoamine for part or all of the fatty acid normally used. The imide-modified alkyds shown in Table 3 were among those investigated. The glyceryl phthalimide was prepared by heating equal molar ratios of glycerylamine and phthalic anhydride under an atmosphere of inert gas, with agitation. The source of heat was removed when the reaction began since it is

Table 3 * Composition of Imide-Modified Alkyds

	Parts by Weight					
Phthalic anhydride	100	90.6	86.8	95.3	124	93.0
Soybean fatty acids	67	67.0	57.4	47.8	38.3	38.3
Glyceryl phthalimide	0	14.9	32.4	31.2	0	46.3
Glycerol †	59.3	50.7	43.2	58.1	65.5	39.8
% oil	35	35	30	25	20	20
Acid number	14	9.2	5.6	18	50.2	10.1

* Courtesy of Wright and Dupuis (11).
† 5% excess glycerol used in all resins.

very exothermic. The reaction was finished at 150–170°C when no further water was evolved. After purification the glyceryl phthalimide is a white crystalline product.

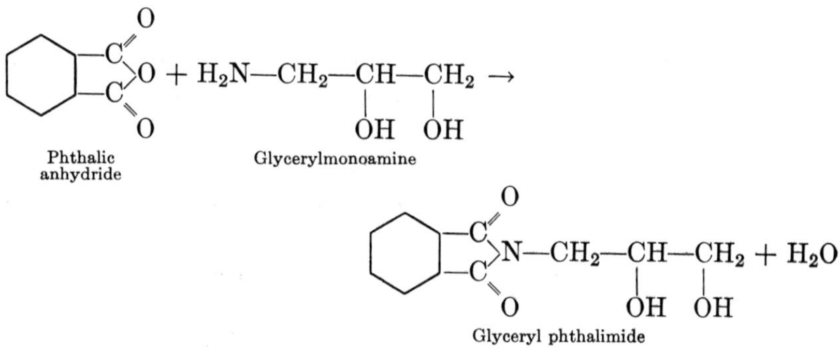

The reactants were placed in the flask and brought to 180°C in 1 hr. The temperature was increased to 235°C in the next hour and held there until the resins developed the required cure time at 200°C on a cure plate. The resins were soluble in aromatic hydrocarbons except the 20% oil resin, which required ester solvents.

The results in Table 3 show that low acid numbers may be obtained in short-oil alkyds by glyceryl phthalimide substitution. Film tests were made by adding 0.01% cobalt drier and baking at 181°F for 20 min. The imide-modified alkyd films were harder, had better water-vapor resistance, and were about the same color as the unmodified films. The film properties may be modified by replacing part of the phthalic anhydride with the succinic-sebacic type of acids. Other modifications may be made by using a fatty acid amide of glycerylamine. The amide is made by adding, dropwise, 1 mole of amine to 1 mole of fatty acid at about 200°C in an atmosphere of nitrogen.

$$\underset{\text{Fatty acid}}{R{-}COOH} + \underset{\text{Glycerylmonoamine}}{H_2N{-}CH_2{-}\underset{\displaystyle OH}{\underset{|}{CH}}{-}\underset{\displaystyle OH}{\underset{|}{CH_2}}} \rightarrow$$

$$\underset{\text{Fatty acid amide}}{R{-}CONH{-}CH_2{-}\underset{\displaystyle OH}{\underset{|}{CH}}{-}\underset{\displaystyle OH}{\underset{|}{CH_2}}} + H_2O$$

RESINS FROM GLYCEROL α-ALLYL ETHER

The preparation of unique alkyd resins from glycerol α-allyl ether is described by Bradley *et al.* (12).

$$\underset{\text{Glycerol }\alpha\text{-allyl ether}}{CH_2{=}CH{-}CH_2{-}O{-}CH_2{-}CHOH{-}CH_2OH}$$

The resins are prepared in the absence of oxygen so that the allyl unsaturation is not affected and is available for subsequent reaction under air-drying or baking conditions. Since the unsaturation is terminal, the resins convert at room temperature relatively slowly. However, their films bake satisfactorily at temperatures ranging from 150° to 300°F. Formula 4 is a representative example of the preparation of this type of resin.

Formula 4 Glycerylol α-Allyl Ether Phthalate

	Moles	Grams
Phthalic anhydride	2.0	296
Glycerol α-allyl ether	2.1	277
Calcium acetate	...	2.15
Barium acetate	...	0.072

Procedure

The reactants were heated with agitation, and with inert gas through the charge, to 240°C at the rate of 20°C increase per hour. The temperature was maintained about 4 hr, and the batch had an acid number of about 30. A vacuum of 3 mm Hg was applied for about 30 min to remove unreacted phthalic anhydride. The batch was cooled to 150°C and thinned to 75% solids with xylene.

A resin was prepared in a similar manner using succinic acid, but the vacuum treatment was unnecessary and the maximum temperature was 230°C. When direct esterification is difficult, the ester-interchange method may be used as indicated in Formula 5.

Formula 5 Ester-Interchange Method

	Mole	Grams
Dimethyl phthalate	0.52	101
Glycerol α-allyl ether	0.78	103
Zinc oxide		0.4
Calcium oxide		0.4

Procedure

Heat the reactants between 200° and 220°C for 7 hr and between 220° and 230°C for 12 hr; apply vacuum during the last 4 hr. Cool, and thin with xylene to 75% solids.

Evaluation of the resin in Formula 4 showed that it had properties comparable to or better than a similar resin made with fatty acids. The air-drying finishes, with 0.025% cobalt and 1% *tert*-butyl hydroperoxide, set to touch within 6 hr and were fully hardened and water resistant within 7 days. The resin was better suited for baking finishes, and with 0.025% manganese it could be cured on wooden furniture in 2 hr at 150°F and 16 hours at 120°F. The finish had good water and solvent resistance, and it withstood a 25-cycle cold-check test without failure. Finishes on metal could be baked at 300°F for about 30 min, and no catalyst was required. However, about 0.05% calcium in the form of octoate was suggested as a wetting agent. The resin showed better color retention on heat aging than a soybean oil alkyd-urea formaldehyde combination. The replacement of the long-chain fatty acid with the short ether chain made no significant difference in flexibility. This rather surprising result was attributed to the relative ease of molecular rotation of the ether oxygen linkages.

These allyl-type alkyds and the previous imide-modified alkyds were included chiefly to illustrate the vast possibilities in the condensation polymers. A great many chemical combinations may be used to prepare interesting resinous materials with valuable properties; unfortunately, availability and cost are also important considerations.

POLYESTER RESINS

"Polyester resins" is the term applied in the plastics industry to those materials made from polybasic acids and polyhydric alcohols, one of which contains C=C unsaturation. This permits cross-linking with other unsaturated materials to form a partially thermosetting product. At present, these products are used almost entirely in the plastics industry in the so-called "low-pressure" or "contact" molding and laminating operations. They are usually based on the glycols in combination with an unsaturated dibasic acid such as maleic or fumaric. The extent of reactivity may be reduced by replacing some of the unsaturated acid with phthalic, adipic, sebacic, etc. It will be realized that considerable variation in the hardness and flexibility of the final product may be obtained from such changes. Similar modification is possible by changing from ethylene glycol to triethylene glycol, or other alcohols.

Originally, the principal unsaturated material used with the polyester resin was styrene. It served the dual purpose of solvent and reactive material. Since monomeric styrene is volatile, it was necessary to put a cover sheet on the laminates to prevent loss of styrene. With insufficient styrene the laminate was soft and sticky and lacked strength. Less volatile reactive materials, such as diallyl phthalate, may be used, but they are more costly.

The polyester resins are made by esterifying equal molar ratios of glycol and dibasic acid at temperatures from 350°F to about 425°F. The time and temperature vary with the proportion of saturated to unsaturated components, but the final acid number is in the range of 20–50. Obviously, a very wide range in composition is possible, but the following two examples illustrate a relatively hard and a relatively soft but more reactive polyester.

Formula 6 Polyester Resin *A*

	Mole Ratio
Ethylene glycol	5.0
Phthalic anhydride	3.5
Fumaric acid	1.5

Formula 7 Polyester Resin *B*

	Mole Ratio
Diethylene glycol	5.0
Sebacic acid	1.0
Fumaric acid	4.0

Laminating materials are made from such polyesters by adding the calculated amount of styrene for reaction with the unsaturation. A catalyst is added, such as benzoyl or lauroyl peroxide, or *tert*-butyl hydroperoxide, in amounts ranging from 0.1% to 0.5%. Selected paper or glass cloth is impregnated with such combinations, and several layers are assembled and pressed between polished platens. Relatively low pressures of about 50 psi are required, and about 10 min at 200–225°F is sufficient.

A completely opposite type of polyester is described in U. S. Patent 2,497,968 (February, 1950). It is soluble in aliphatic hydrocarbons and is intended for an additive in lubricating oils. It increases the viscosity of the lubricating oil, but the oil does not thin out excessively at elevated temperatures, and it will flow freely at very low temperatures. Many variations in composition are given, but a preferred one is equal molar ratios of dimerized fatty acids and a glycol. Dimerized soybean acids in combination with decamethylene glycol are mentioned, with 0.3–1% zinc dilinoleate as catalyst. The reaction temperature is from about 200°F to 225°F, and the reaction should be carried out in the absence of oxygen.

Obviously, the above polyester is an extremely unreactive material which, conceivably, could be used as a plasticizer for surface coatings. These illustrations of polyester resins are given to point out again the wide variation which is possible in resinous materials. The possibility of obtaining a film-forming material with low enough viscosity to be used without volatile solvent would seem to lie within the realm of the polyesters.

STYRENATED ALKYDS

Styrenated alkyds were made available commercially in 1948 (13), and they have established a definite position in the field of fast air-drying and baking finishes. The large production capacity for styrene and its relatively low cost were inducements for its use in coatings. In some respects the styrenated alkyds may be considered resin-modified alkyds, the resin being polystyrene. Polystyrene is tougher and more durable than the natural resin modifiers generally used, and the styrenated alkyds show similar advantages of better toughness and durability. However, they are not as durable on exterior exposure as the pure oil-modified alkyds.

Polystyrene is thermoplastic and may be softened by oils and some hydrocarbon solvents. The styrenated alkyds show this sensitivity to some extent, particularly if the degree of polymerization of the polystyrenes they contain is relatively low and the coatings are air-dried. The sensitivity is decreased as the percentage of styrene is decreased or as the percentage of good drying oil in the alkyd component is increased. The addition of urea- or melamine-formaldehyde resins to the styrenated alkyd when used in baking finishes also makes the film more resistant to solvents and surface marring.

Polystyrene has good resistance to water and alkali, and this quality is imparted to the styrenated alkyds. They also have good color retention; therefore a combination of styrenated alkyd and melamine-formaldehyde resin has been proposed for the baking finish on washing machines. This is discussed further in Chapter 8 on urea- and melamine-formaldehyde resins.

Styrenated alkyds are available in both mineral spirits and xylene solutions. For fastest setting and air drying, the xylene solution is better; but for grinding with pigments and best flow in high-gloss finishes the slower-evaporating mineral spirits is preferred. Of course, it is possible to grind the pigment in the mineral spirits solution and to add the xylene solution to complete the enamel formulation. The styrenated alkyds are practically incompatible with the pure oil-modified alkyds; therefore blends of these resins cannot be made.

The usual lead and cobalt combination of driers is added to air-drying styrenated alkyd finishes. The percentage of metal varies with the type of pigment as in the normal varnish or alkyd finishes. In finishes baked at or below 250°F, the styrenated alkyds require a small percentage of cobalt, but above 250°F driers are not generally used. The type of pigment and the ratio of pigment to resin solids affect the rate of all air-drying finishes. High pigment content reduces the drying time and produces a dull or flat finish. The drying requirements for the Army Ordnance Corps olive drab lusterless finish are:

Set to touch: 3–6 min
Dry hard: 10 min max.
Dry through: 20 min max.
Full hardness: 72 hr max.

These requirements are being met by the xylene solution type of styrenated alkyd. For the usual white and colored enamel finishes made with styrenated alkyds, the air-drying schedules fall within the following times:

Dust free: 15–30 min
Tack free: 1–2 hr
Dry hard: 3–6 hr

The time required before the finished objects may be packed for shipment varies considerably with the color of the finish and the thickness of the coating. From the above discussion, it should be evident that the styrenated alkyds are suitable for fast air-drying finishes on such objects as toys, implements, and wood and metal furniture, and for interior applications which require resistance to alkaline cleaning solutions but are not subject to contact with oils and greases. Greater resistance is developed in baking finishes, particularly in combination with amino resins. The present styrenated alkyds are not as durable for exterior applications as the pure oil-modified alkyds.

Manufacture of Styrenated Alkyds. The methods for manufacturing styrenated alkyds are in process of development, and some changes may be anticipated which may result in new and better resins. Four general methods may be visualized:

1. Styrenate the fatty acid, then combine the product with phthalic anhydride and glycerol.

Fig. 5. Flow Sheet for Styrenated Alkyd (14)

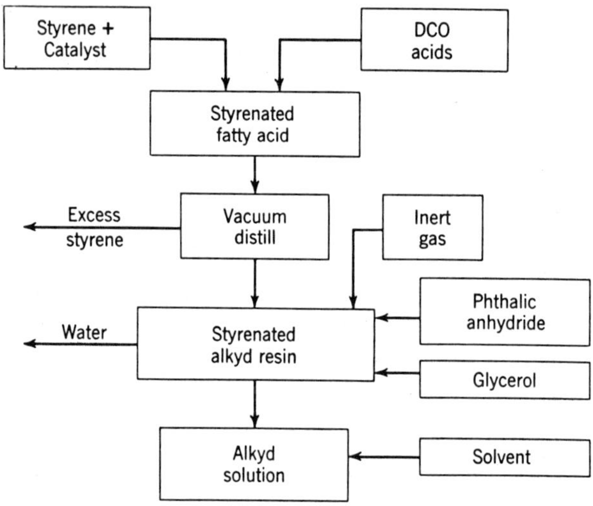

2. Styrenate the oil, react with glycerol to form the monoglyceride, then combine with phthalic anhydride and additional glycerol if necessary.
3. Styrenate the monoglyceride, then combine with phthalic anhydride and more glycerol if necessary.
4. Styrenate the alkyd resin directly.

A comparison of the relative efficiency of these methods is not available at present. The first process is described by Bhow and Payne (14) and was referred to in Chapter 2. The process may be visualized from the flow sheet in Fig. 5.

The difference in weight of styrene input, 416 gm, and that removed, 215 gm, indicated that 201 gm of styrene had reacted. Therefore, the styrenated acids were considered to have 41% styrene; 288 gm fatty acids plus 201 gm styrene = 489 gm styrenated acids. The styrenated

Formula 8 Styrenated Dehydrated Castor Oil Acids

	Moles	Grams
Dehydrated castor oil acids	1	288
Styrene	4	416
Benzoyl peroxide		12.5

Procedure

Heat the DCO acids in a flask with stirrer and condenser to about 240°F, add the styrene-peroxide mixture slowly over a period of 1 hr, with agitation, raise the temperature during this period to about 290°F, hold there for 30 min, and remove unreacted styrene by vacuum.

alkyd was made by adding the required amount of phthalic anhydride and glycerol and raising the temperature to 450°F in 30 min, then holding at this temperature for 50 min. Inert gas was passed through the batch during processing. Two resins were made having the compositions and properties shown in Table 4. Resins *A* and *B* were com-

Table 4 Compositions of Styrenated Alkyds

	A	*B*
Glyceryl phthalate, %	26.8	39.2
Styrenated oil, %	73.2	60.8
Phthalic anhydride, %	20.8	28.4
Styrene, %	30.4	22.4
Solvent	MS	50% MS, 50% xylene
% solids	60	50
Viscosity, Gardner-Holdt	J	Y
Color, Gardner 1933	14	12
Acid number	14	9.2

pared with commercial styrenated alkyds for air-drying and baking schedules and for film properties and were found to be about equal.

In the styrenation of the DCO acids, it was found that, if all the reactants were heated together, a rapid, exothermic reaction occurred when the temperature reached 245°F. In addition, the product was turbid and evidently heterogeneous. It was probably an incompatible mixture of unreacted fatty acids, some styrenated acids, and a considerable amount of polystyrene. The product from the slow addition of styrene-catalyst to the fatty acids was clear and apparently homogeneous. No tests were made to determine if any free polystyrene was present. It was found that the rate of reaction of styrene with the fatty acids increased as the amount of styrene present was increased. The ratio of 4:1, styrene:fatty acid, shown in Formula 8 was considered an optimum. The rates for tung, oiticica, and DCO acids are shown in Fig. 6.

Figure 6 indicates that tung oil acids react more slowly than oiticica oil acids; the fastest rate of reaction was obtained with dehydrated

Fig. 6. Rate of Reaction (14)

Moles of Styrene to 1 Mole of Fatty Acid

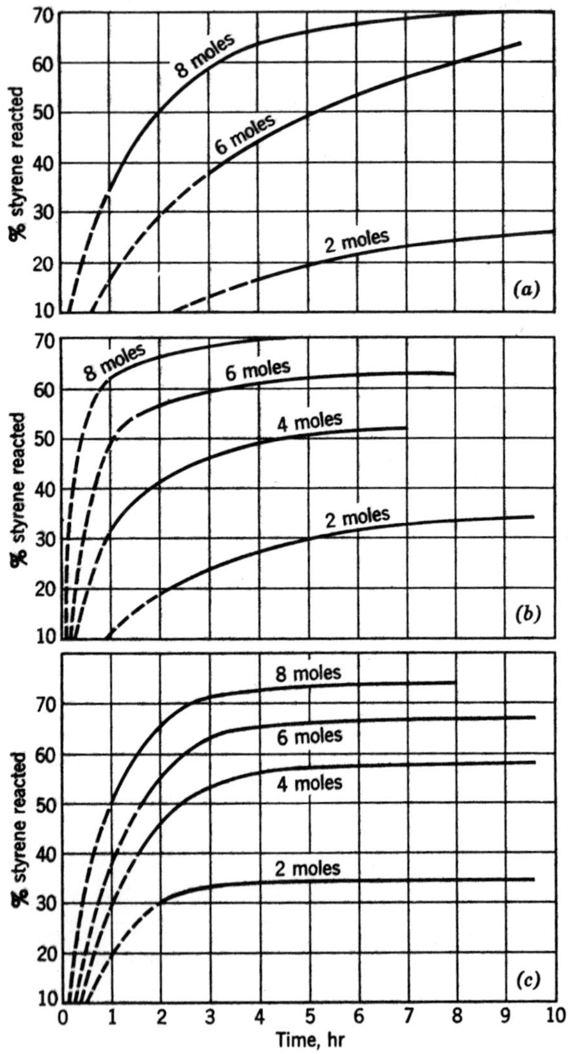

(a) Styrene with tung fatty acid; (b) styrene with oiticica fatty acid; (c) styrene with DCO fatty acid.

castor oil acids. This seems contrary to expectations, since tung oil has the greatest amount and degree of conjugated unsaturation. A possible reason suggested is that styrene copolymerizes almost completely with tung oil acids; therefore there is less polymerization of styrene with its fast reaction rate. With the DCO acids the copolymerization is less and some polystyrene is formed. Since the polymerization of styrene is a faster reaction than the copolymerization reaction, the amount of styrene reacted in a given time would be greater with the DCO acids.

The second method of styrenating the oil, followed by alcoholysis to form the monoglyceride, and then reacting with phthalic anhydride to form the styrenated alkyd is described by Young (18).

Formula 9 * Styrenated Blown Soybean Oil Alkyd

a. Blown soybean oil
Blow soybean oil at 100°C to a viscosity of E and a peroxide value of 334.

b. Styrenation of blown soybean oil

	Parts by Weight
Blown soybean oil	58
Styrene	29.4
α-Methyl styrene	12.6
c. Monoglyceride formation	
Styrenated soybean oil	325.0
Glycerol	54.6
Calcium oxide	0.1
d. Esterification to styrenated alkyd	
Styrenated monoglyceride	Above
Phthalic anhydride	120.4
Xylene	20.0

Procedure

a. Blow soybean oil as indicated above.
b. Heat the reactants to 122°C over a 2-hr period and hold at this temperature for about 8 hr. The product contained 3% non-volatile and was Z4 viscosity.
c. React the components at 230°C for about 1 hr.
d. Esterify at 230°C for about 3½ hr.

Characteristics of resulting styrenated alkyd

Viscosity at 50% solids	H
Color at 50% solids	12
Acid number	16

* Courtesy of A. E. Young, Dow Chemical Co.

Air-dried films were dry to touch in 22 min, were dry hard in 6–8 hr, and attained a Sward hardness of 21 in 24 hr. Films baked at 245°F for 30 min had a Sward hardness of 32 and good flexibility. The films also had good resistance to water, alkali, and solvents.

The third method of producing styrenated alkyds by styrenating the monoglyceride, followed with esterification, is described in the informative brochure, "Styrenated Oils and Alkyds," of the Dow Chem-

Formula 10 Styrenated Linseed-Dehydrated Castor Oil Alkyd

	Per Cent
Linseed oil	19.0
Dehydrated castor oil	19.0
Glycerol	8.7
Styrene N-99	35.0
Phthalic anhydride	18.3
	100.0

Procedure

a. Alcoholysis of oils

	Parts by Weight
Linseed oil, alkali refined	114
Dehydrated castor oil, viscosity G-H	114
Glycerol	51.8
Calcium oxide	0.14

Heat to 450°F and hold about 1 hr to a methanol solubility of at least 3 to 1.

b. Styrenation of monoglyceride

Monoglyceride	Above
Styrene N-99	210
Cumene hydroperoxide (73% solution)	6.3

Add half the styrene and catalyst solution to the monoglyceride and heat to 320°F. Maintain reflux and add remainder of styrene over a 3-hr period. Raise temperature to 410°F over the next 4–5 hr.

c. Esterification to styrenated alkyd

Styrenated monoglyceride	Above
Phthalic anhydride	110
Xylene	30

React at 420–440°F for about 3 hr to the required viscosity and acid number, reduce, and filter.

Characteristics of Styrenated Alkyd

% solids	50
Solvent	Xylene
Viscosity, Gardner-Holdt	N
Color, Gardner 1933	6
Acid number	7

ical Co. Five formulations are given, showing variations in type of oil and polyhydric alcohol and use of styrene and a mixture of styrene with α-methyl styrene. The resin in Formula 10 is Dow Chemical Co. Formulation V168-29; it has the composition given on page 310.

The air-dried films were set to touch in 10 min and had Sward hardness of 25 in 24 hr and 38 in 7 days. The films baked at 250° or 300°F for 20 min had a Sward of 43. Cobalt drier was used in these tests in the ratio of 0.03% cobalt as metal to resin solids.

A wide range of styrenated alkyds may be made by the above general procedure by varying the ratio of reactants, type of oil, and/or polyhydric alcohol, catalyst, and solvent. The use of terpene solvents as chain terminators was discussed previously. The resins should be clear, since turbidity indicates incompatible materials which may cause instability on aging or when pigmented.

The direct styrenation of a completed alkyd might offer some economic advantages in processing, but information has not been made available on this procedure. Good control of the reaction should be possible if the solvent process were used. Obviously the alkyd should be reacted to a viscosity lower than usual before styrenation because the styrenation would increase the viscosity considerably. It would seem necessary to use an oil having some conjugated unsaturation to assure the copolymerization reaction with the styrene. However, it is conceivable that another reactive component, such as maleic anhydride or a reactive polyester, might be incorporated in the alkyd to produce copolymerization.

The general mechanism of the styrenation of an alkyd probably is the same as that for oils, which is discussed in Chapter 2.

EPON * RESINS

The condensation products of epichlorohydrin and *bis*-phenol, known as Epon resins, were made available commercially in 1949 by the Shell Chemical Co. These resins have a chain structure which contains reactive hydroxy and epoxy groups. They may be used for coatings without further chemical modification, or they may be esterified with fatty acids. The fatty acids are monobasic, but they are polyfunctional because of their unsaturation. The functionality may be increased by the use of small amounts of maleic or other polybasic acids. However, these higher functional acids are rarely used. When unsaturated fatty acids are used, the products are suitable for

* Trade name, Shell Chemical Co.

Fig. 7. Reactions Related to Epon Resins

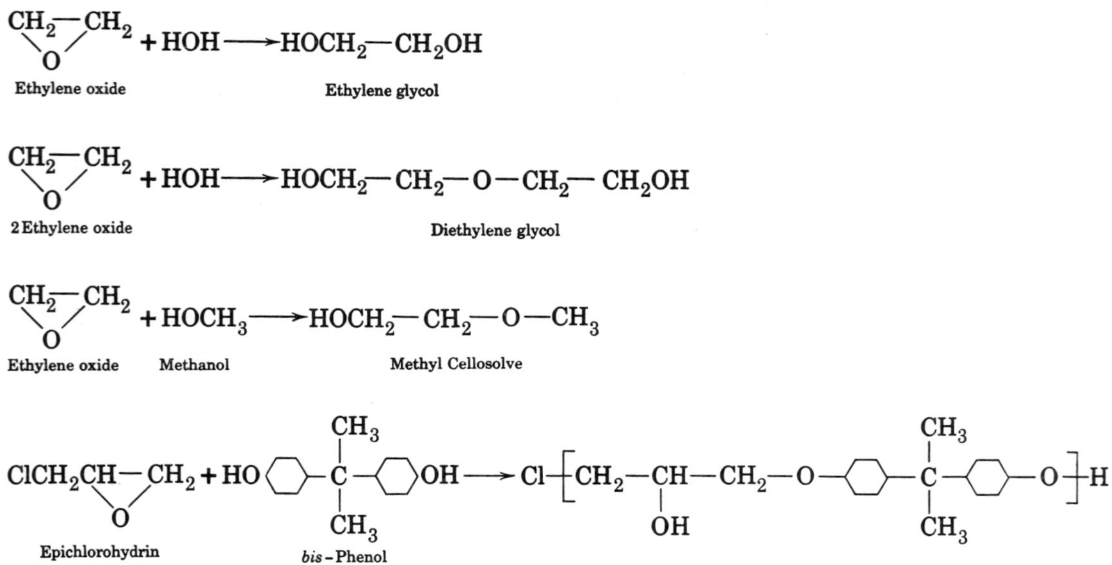

(1)

(2)

(3)

(4)

Let enclosure equal R; then product of 4 is Cl–R–H.

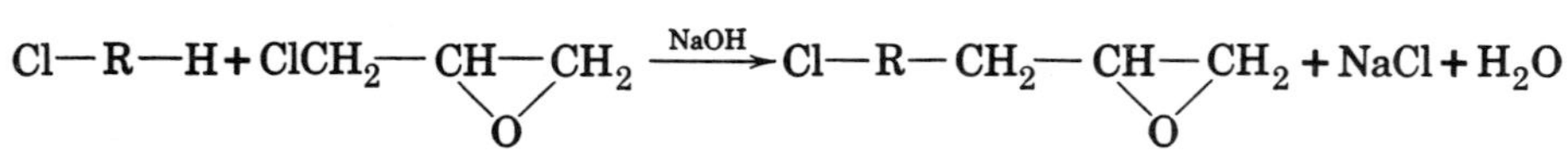

(5)

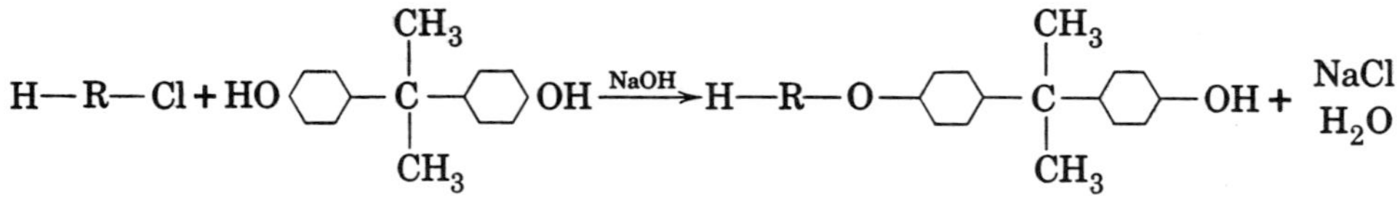

(6)

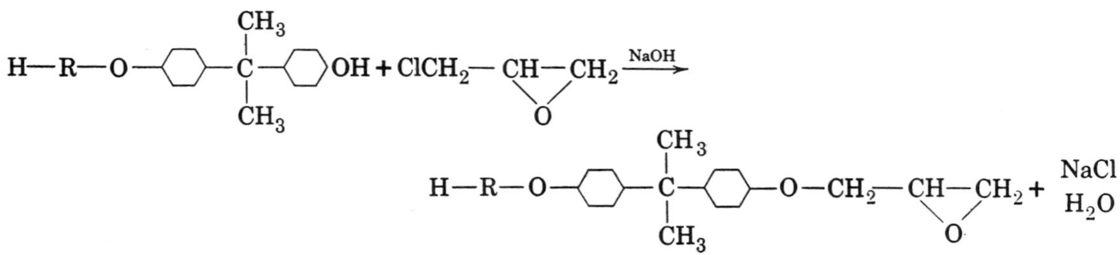

(7)

Combine reactions 6, 7, and 8 in the ratio of:

$(n+2)$ Epichlorohydrin, $(n+1)$ *bis*-Phenol

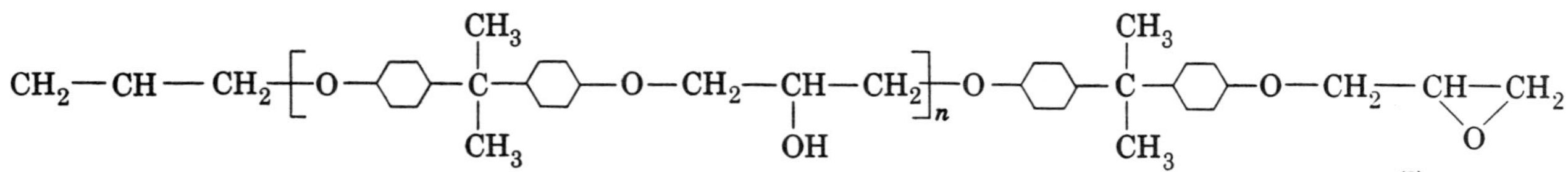

(8)

air-drying coatings. Coatings which will either air-dry or bake may be made with combinations of Epon resins and the reactive Polyamide resin #100 of General Mills, Inc. Baking coatings may be made by the addition of urea- or melamine-formaldehyde resins, or some of the reactive phenolic resins. The Epon resin chains may be cross-linked at room temperature by the addition of polyfunctional amines, such as triethylamine. However, this type of curing has not been used extensively at present.

The Epon resin chains are based on carbon-carbon and ether linkages, and they do not contain the more readily saponified ester linkage. The absence of ester linkages is partly responsible for the very good alkali resistance of coatings made with Epon resins. These coatings also have very good adhesion and flexibility. The good adhesion may be attributed to the presence of the highly polar hydroxyl groups. The toughness and flexibility are derived from the cross-linked structure in which the cross-linkages are not spaced too closely.

The reaction between epichlorohydrin and *bis*-phenol involves both the chlorine and the epoxide of the epichlorohydrin, and both hydroxyls on the *bis*-phenol. The product obtained depends on the ratio of the reactants, the use of chain breakers, and the operating conditions of temperature and pressure. It will be recalled that the epoxide ring is fairly easily hydrolyzed. The reaction is used for the production of ethylene glycol from ethylene oxide as shown in Fig. 7, equation 1. The ratio of water to ethylene oxide determines the product, as shown in equation 2, for the production of diethylene glycol. The reaction of ethylene oxide with alcohols is used to produce the series of ether-alcohol solvents known as the Cellosolves; see equation 3. A similar reaction takes places with *bis*-phenol, as shown in equation 4. It will be noted that the structure within the brackets in equation 4 is the repeating unit in equation 8. This unit may be changed by varying the ratio of epichlorohydrin to *bis*-phenol. An increase in the ratio of epichlorohydrin would increase the number of hydroxyl groups in the unit. The number of hydroxyl groups may also be increased by increasing the degree of polymerization, i.e., by increasing the value of n in equation 8. Equations 5 and 7 show the reactions necessary to terminate the polymer chain with epoxy groups. This type of termination is desirable since it permits further cross-linking reactions with amines.

However, the chains may be terminated with non-reactive groups by the addition of chain breakers. For example, if some of the *bis*-phenol were replaced with phenol a chain would be terminated each time a phenol molecule reacted with the epichlorohydrin because phenol has only one hydroxyl group. This would produce lower-molecular-weight materials with better solubility and compatibility. Conversely, a greater degree of complexity might be obtained by introducing an alcohol such as glycerol, which has more reactive hydroxyls than *bis*-phenol. It will be evident that the possible variations are many, as is typical of resinous materials in general. Some of these variations are described by Greenlee (24) and in the technical bulletins of the Shell Chemical Co. (25).

In one of the processes used by Greenlee (24), the *bis*-phenol is dissolved in a sodium hydroxide solution and about 10% of phenol added to control the polymer size. The epichlorohydrin is then added slowly, with agitation. After the exothermic reaction has subsided, additional caustic is added, and the batch is heated to about 100°C for 1 hr. The upper layer of salt water is removed by decantation, and the product is washed several times with hot water. The last two or three washings should be slightly acid with acetic acid. The product is dried by heating to about 150°C. Higher polymers may be obtained by heating the reaction mass under pressure and to higher temperatures, and also by controlling the type and amount of chain breaker.

At present a series of eight resins is available from the Shell Chemical Co., four of which have only limited use in coatings. The four coating resins shown in Table 5, are hard, brittle solids and vary in

Table 5 * Epon † Resin Characteristics

	Epon Resin			
Characteristic	1001	1004	1007	1009
Melting point,‡ °C	64–76	97–103	127–133	145–155
Viscosity,§ Gardner-Holdt	C–G	R–T	Y–Z1	Z2–Z5
Color,§ Gardner 1933, max.	8	6	8	11
Hydroxyl equivalent ‖	130	175	190	200
Epoxide equivalent ¶	450–525	905–985	1600–1900	2400–4000
Pounds per gallon	10.05	9.63	9.56	9.92

* Courtesy of Shell Chemical Co.
† Trade name, Shell Chemical Co.
‡ Duran's mercury method.
§ 40% solution in butyl carbitol.
‖ Grams of resin to esterify 1 mole monobasic acid.
¶ Grams of resin containing one equivalent of epoxide.

the degree of polymerization and other physical properties as indicated in the table. The increase in molecular weight from Epon 1001 to Epon 1009 is indicated by the increase in melting point of the solid resins, and by the increase in the viscosity of their solutions in butyl carbitol. The actual molecular weight of these materials ranges from about 1000 to about 5000. The unmodified Epon resins are not soluble in aliphatic or aromatic hydrocarbons or in alcohols. They are soluble in esters, ketones, ether-alcohols, and chlorinated solvents. Mixed solvents may be used, which include aromatic hydrocarbons and some alcohols as the latent solvents in combination with true solvents.

In general, the Epon resins are not compatible with vegetable oils, ester gum, maleic- and phenolic-modified rosin esters, alkyd resins, and the cellulosic polymers. They are compatible with an unmodified sebacic-type alkyd, certain pure phenolic resins, the sulfonamide resins, and the urea-formaldehyde resins. Since the Epon resins contain reactive hydroxy and epoxy groups, they may be reacted with fatty acids and polyfunctional amines. The fairly low-molecular-weight Epon 1004 is generally used for esterification with fatty acids, because it is not too reactive, and products with low acid numbers may be obtained. The medium-molecular-weight Epon 1007 is used with urea-formaldehyde resins and heat-reactive phenolic resins for baking coatings. The high-molecular-weight Epon 1009 is also used for this purpose; it has somewhat better chemical resistance and faster baking time but is less compatible. The lowest-molecular-weight Epon 1001 is blended with the others to improve compatibility, and with alkyds and urea-formaldehyde resins.

The structure of the Epon polymer in equation 8, Fig. 7, shows that it is not suitable for air-drying coatings as is. However, if the hydroxyls are esterified with unsaturated fatty acids, the products should air-dry in a manner similar to the drying oils, varnishes, and alkyds. Since the Epon polymer has a higher functionality than glycerol, and a considerably higher molecular weight, it would be expected that the Epon-fatty acid ester will dry very much faster than an oil, and similarly to a fast-drying varnish or oil-modified alkyd. The Epon-fatty acid ester would have better alkali resistance than drying oils, varnishes, or alkyds because of the ether linkages and phenolic nuclei. It would not discolor badly because the color-forming phenolic hydroxyls have been replaced with unreactive ether linkages. Obviously, the type of fatty acid used will be a factor in the rate of drying and the film properties. Typical examples of Epon 1004-fatty acid esters are given in Formulas 11 and 13. It should be understood that these may be modified considerably with respect to oil length and inclusion of maleic or phenolic bodies. Formula 11 is designed primarily for air-drying coatings, and Formula 13 for baking coatings.

Formula 11 * Epon † 1004-Fatty Acid Air-Drying Ester

	Per Cent by Weight
Epon 1004	43
Linseed acids	47
Rosin	10
	100

Procedure

Either the solvent or the fusion method may be used as described previously. In the solvent method the reactants, with from 10% to 15% of aromatic solvent, are placed in a kettle equipped with a reflux condenser and an agitator. The batch is then heated to 500°F. The agitator may be started about 250°F or as soon as the solids become molten. The water is removed from the reflux trap, and the solvent returned continuously. Samples are withdrawn periodically; when the batch has a viscosity of H-J as a 50% solution in a 1:1 mixture of minerals spirits and high-solvency naphtha, and an acid number of less than 10, the heat is shut off and the batch cooled to 375°F for thinning to 60% solids with mineral spirits.

Characteristics

Solids, %	60
Solvent	
Mineral spirits, %	40
Viscosity, Gardner-Holdt tubes, at 25°C	Y–Z
Color, Gardner 1933	7–8
Acid number (solid basis)	7
Pounds per gallon	7.92

* Courtesy of Shell Chemical Co.
† Trade name, Shell Chemical Co.

The above ester may be used as the vehicle for a white air-drying brushing enamel as shown in Formula 12, but it will not have good color retention because of its rosin content. The suggested drier for Formula 11 is 0.4% cobalt as metal to vehicle solids, using cobalt naphthenate. The ratio of the ingredients may be changed to produce clear finishes for floors, furniture, etc. Some of the rosin may be replaced with a maleic adduct if care is taken to prevent premature gelation. The linseed fatty acids may be replaced with dehydrated castor or other drying oil acids.

If Formula 11 is compared with a conventional varnish for oil length it will be seen that it corresponds to a short-oil varnish. Formula 11 contains 47% fatty acids which give 49% oil by the conventional method given on page 280. Approximately equal parts by weight of oil and a varnish resin correspond to a 12½ gallon varnish which

Formula 12 Epon Ester White Air-Drying Enamel

	Per Cent by Weight
Pigment	
Titanium dioxide	100
Binder	
Epon ester (Formula 11)	100
Solvent	
High-solvency naphtha *	50
Mineral spirits	50

Enamel Characteristics

Pigment/binder by weight	1/1
Total solids, %	50.8
Viscosity, sec, #4 Ford Cup, at 25°C	25
Pounds per gallon	9.1
Drier, % Co as metal to vehicle solids	0.05

* Shell TS-28R.

Formula 13 * Epon 1004-Fatty Acid Baking Ester

	Per Cent by Weight
Epon 1004	60
Dehydrated castor oils	40
	100

Procedure

Either the solvent method or the fusion method for processing may be used. The ingredients are placed in a closed kettle and heated to about 250°F; then the agitator is started. Heating is continued to about 500°F and maintained until an acid number of 1.0 or less is obtained at the required viscosity. The batch is then cooled to 400°F and thinned with xylene. Inert gas should be passed through the batch during processing to obtain the best color in the resin.

Characteristics

% solids	50
Xylene, %	50
Viscosity, Gardner-Holdt tubes	X–Z
Color, Gardner 1933	7
Acid number (solid basis)	1.0
Pounds per gallon	7.96

* Courtesy of Shell Chemical Co.

produces a fairly brittle film. However, the Epon ester of this same percentage of "oil" is tough and flexible because of the structure of the "resin" component. It will be recalled that the oil length of pure phenolic resin varnishes must be reduced to obtain maximum alkali resistance (see Formula 18, Chapter 4). It will also be recalled that the oil-modified alkyds are shorter in oil than varnishes of comparable toughness because of the inherent properties of the phthalic-glyceride component.

Formula 13 illustrates an Epon ester which may be used in baking coatings. This ester contains 40% oil acids and may be compared with the short- to medium-oil-length alkyds used in baking finishes in conjunction with urea- or melamine-formaldehyde resins. In general, the color of the finishes containing the Epon esters is not quite as good as that of finishes containing the phthalic alkyds, but the chemical resistance is better. A typical white baking finish is given in Formula 14.

Formula 14 * Epon Ester, Melamine-Formaldehyde White Baking Enamel

	Per Cent by Weight
Pigment	
Titanium dioxide	100

	Per Cent by Weight, Solid Basis
Binder	
Epon ester (Formula 10)	70
Melamine-formaldehyde resin †	30

	Per Cent by Weight
Solvent	
Xylene	84.5
Butanol	8.7
VM & P naphtha	6.8

Enamel Characteristics

Pigment/binder by weight	1/1
Total solids, %	49.2
Viscosity, sec, #4 Ford Cup, at 25°C	25
Pounds per gallon	9.47
Baking schedule: 30 min at 300°F	

* Courtesy of Shell Chemical Co.

† Melmac 245–8, American Cyanamid Co.

Further details regarding the use and performance of the above enamel are given in Volume II. Epon 1001 may be blended directly with non-oxidizing alkyds and amino resins for white baking enamel formulations. Epon 1001 improves adhesion, flexibility, and chemical resistance. A typical ratio of ingredients by weight on a solid resin basis is:

Epon 1001	30
Non-oxidizing alkyd	45
Amino resin	25

Epon resins 1007 and 1009 may also be esterified with fatty acids, but the bodying rates are quite rapid and the esters have high acid numbers. Epon 1007 and 1009 are well suited for baking coatings in combination with amino or phenolic resins for thermosetting. The Epon resin confers good flexibility and adhesion without detracting from chemical resistance. A typical formulation for a white baking enamel is shown in Formula 15.

Formula 15 * Epon 1007, Urea-Formaldehyde White Baking Enamel

	Per Cent by Weight
Pigment	
Titanium dioxide	100
	Per Cent by Weight, Solid Basis
Binder	
Epon 1007	70
Urea-formaldehyde resin	30
Solvent	
Cellosolve acetate	43.7
Toluene	43.7
Xylene	5.0
Butanol	7.6

Enamel Characteristics

Pigment/binder by weight	1/1
Total solids, %	45.7
Viscosity, sec, #4 Ford Cup, at 25°C	25
Pounds per gallon	9.88
Baking schedule: 20 min at 400°F	

* Courtesy of Shell Chemical Co.

A variety of amino resins may be used with the Epon resins, and also the organic-soluble reactive phenolic resin R-108 of the General

Electric Co. A combination of 70% Epon 1009 and 30% G. E. R-108, solid resin basis, is recommended for a baking finish in applications where extreme chemical resistance is required. The baking schedule is about 20 min at 375°F, and the solvent is similar to that used in Formula 15. Combinations of Epon resin 1001 or 1007 with the reactive type of polyamide resin, such as Polyamide resin #100 (General Mills, Inc.), have been suggested for both air-drying and baking coatings. The coatings are hard and tough, and they have excellent resistance to solvents and chemicals.

The Epon resins represent the pure synthetic resins having characteristics not found in natural products. Specific characteristics may be built into the resin structure to meet difficult or unusual application requirements. The reader should review the chart in Chapter 1, and relate the fundamentals of the resin structure to the film properties as indicated.

THE URETHANE REACTION

One of the interesting developments in Germany during World War II was the application of the urethane reaction to drying oils. This work is reported in the English *BIOS Report 628,* item 22, and in the American *OTS Report PB66-130.* Although the "urethane oils and resins" have not found general use in the coating industry at the present time, the urethane reaction illustrates another of the many possibilities for producing synthetic materials.

The urethane reaction was discovered by Wurtz in 1848. He reacted an isocyanate with an alcohol to produce a urethane according to equation 1, Fig. 8. He observed that the urethane was difficult to saponify and was also resistant to acids. This work came to the attention of the I. G. Farbenindustrie, which investigated the urethane for use in coatings, plastics, and fibers. Igamid U of I. G. Farbenindustrie was a polyurethane based on hexamethylene diisocyanate and butyl glycol. As may be expected, the diisocyanates are preferred to the isocyanates for the production of polymeric materials. *Technical Bulletin P-125* of the Monsanto Chemical Co. describes the isocyanates and their reactions. Typical diisocyanates are shown in Fig. 8; as may be expected, the products obtained from the use of hexamethylene diisocyanate are more flexible than those obtained from the tolylene or chlorophenylene diisocyanates. It is claimed that the 1 position must be occupied by a negative group to prevent severe darkening and poor drying. In addition to reacting with alcohols the isocyanates also react with other groups having a replaceable hydrogen atom, such as —COOH or —NH_2.

During the war the Germans extended their limited supply of drying oils by converting them to mono- or diglycerides, and reacting these polyhydric materials with a diisocyanate. Since glycerol was in short supply they used trimethylol propane as the alcohol to react

Fig. 8. The Urethane Reaction

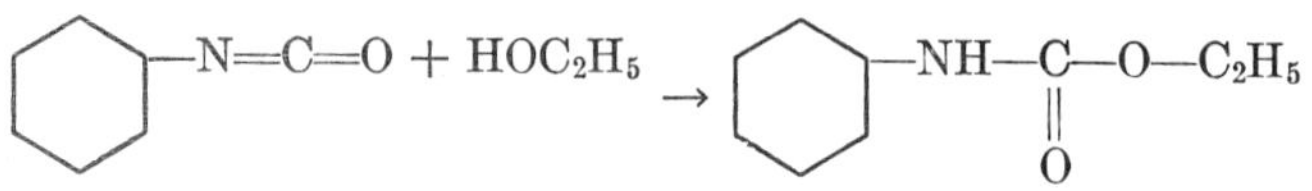

(1)

Phenyl isocyanate — Ethyl alcohol — Urethane

Diisocyanates

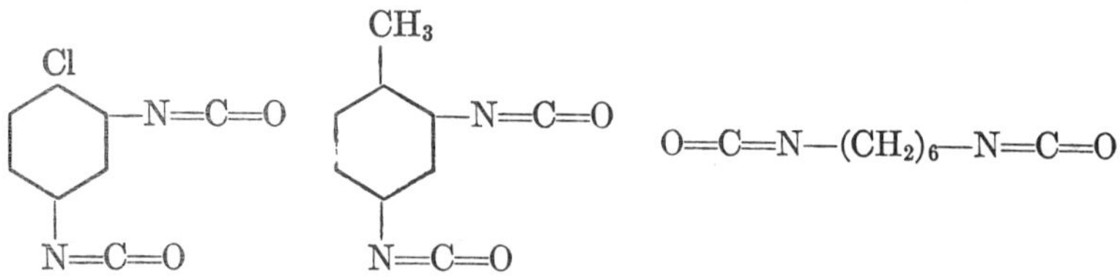

(2)

2,4-Chlorophenylene diisocyanate — 2,4-Tolylene diisocyanate — Hexamethylene diisocyanate

Possible Structure of a Urethane Oil

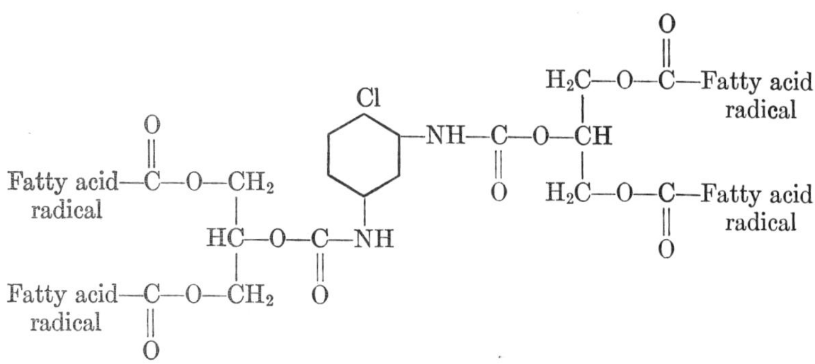

(3)

with the oil to make the mixed glycerides. If 2 moles of a normal diglyceride is reacted with 1 mole of 2,4-chlorophenylene diisocyanate the product will be similar to structure 3 in Fig. 8. This is a considerably larger molecule than the original oil, and it has a higher functionality since it has four instead of three unsaturated fatty acid radicals. This imparts faster drying, and the urethane linkages impart better alkali resistance. There is no loss of unsaturation, since the urethane reaction takes place in a short period at temperatures as low as 100°C.

The urethane oils were investigated by Robinson and Waters (28). They prepared a variety of glycerides in the regular manner and reacted them with the diisocyanates as follows: The glyceride was placed in a flask equipped with a mechanical stirrer, a thermometer, and a device for adding the diisocyanate to the glyceride slowly. Water was excluded by means of a calcium chloride tube fitted to this device. The glyceride was heated to 90°C with agitation, and the chlorophenylene diisocyanate added over a 1- to 2-min period. The reaction is exothermic, and the temperature rose to 100°C. It was maintained at 100°C for 10 min by cooling in a water bath. Different products may be obtained by variation in (*a*) the extent of alcoholysis of the oil; (*b*) the type of oil; (*c*) the type of diisocyanate; (*d*) the ratio of reactants; (*e*) the processing conditions.

The products from hexamethylene diisocyanate gelled on cooling, and those made from chlorophenylene diisocyanate and DCO glycerides gelled on storage. Those made from chlorophenylene diisocyanate and linseed oil glycerides were stable and ranged in viscosity from 13 to 179 poises at 20°C, depending on the extent of alcoholysis of the oil, the ratio of diisocyanate, and the reaction conditions. The acid numbers ranged from 1.9 to 4.2.

Comparison was made between a linseed-chlorophenylene diisocyanate urethane and a linseed-pentaerythritol alkyd for film properties. A mixed drier was used, giving 0.75% lead, 0.03% cobalt, and 0.05% manganese as metal to vehicle solids. The urethane was somewhat better in rate of drying, film hardness, and water resistance. The urethane was darker in color with a pronounced tendency to after-yellow, and it was inferior in flow properties. Exterior durability tests were made on wooden panels which received three coats of standard white undercoats and two coats of the materials under test. Comparison of the clear films showed poor durability for the urethane. It lost gloss quickly and failed by cracking when the alkyd was still in good condition; it also discolored badly. Comparison of white pigmented films showed about equal durability for the urethane and the alkyd, but the urethane was inferior in gloss retention, chalk resistance, and color retention.

The preliminary results obtained from the use of the urethane reaction in coating materials are quite promising. There are many possibilities in this field, some of which are disclosed by Pratt and Rothrock (29). They added diisocyanates to oils and varnishes for improved properties, to linseed oil and rosin for linoxyn in floor coverings, and to castor oil for nitrocellulose plasticizers. They also prepared a low-bake furniture finish by first reacting 40 parts of hexamethylene diisocyanate with 1000 parts of heavy blown castor oil at 80°C for 2 hr in the presence of 1000 parts of xylene. This urethane solution was used in Formula 16.

Formula 16 Low-Bake Furniture Finish

	Parts by Weight
Urethane solution	972
Urea-formaldehyde resin (55% solids)	1720
Maleic resin (50% solids)	956
Isobutanol	432
Hi-Flash naphtha	268

The above finish was sprayed on filled and sealed mahogany panels and baked for 1½ hr at 140°F. The finish was hard and had good print resistance.

It has been reported (28) that some polyisocyanates are being used in commercial coatings at present, and evidently the urethane reaction has interesting possibilities for future exploitation. Care should be taken in handling the isocyanates since it is believed that they are toxic. A good outline of their reactions may be found in the *Monsanto Bulletin P-125.*

BIBLIOGRAPHY

1. C. P. A. Kappelmeier, *Paint, Oil, Chem. Rev., 11* (Aug. 31, 1950).
2. R. H. Kienle, *Ind. Eng. Chem., 41,* 726 (1949).
3. A. C. Hovey, *Ind. Eng. Chem., 41,* 730 (1949).
4. K. A. Earhart, *Ind. Eng. Chem., 41,* 716 (1949).
5. H. F. Payne, L. E. Cadwell, R. H. Brailey, and A. F. Schmutzler, *Am. Ink Maker,* December, 1947, and January and February, 1948.
6. H. J. West, *Am. Paint J., 76* (July 9, 1951).
7. C. E. H. Bawn, *The Chemistry of High Polymers,* Interscience Publications, 1948.
8. R. H. Kienle, P. A. Van der Meulen, and F. E. Petke, *J. Am. Chem. Soc., 61,* 2258 (1939).
9. R. H. Kienle and F. E. Petke, *J. Am. Chem. Soc., 62,* 1053 (1940); *63,* 481 (1941).
10. J. Petit, *Official Digest Federation Paint & Varnish Production Clubs, No. 997,* December, 1948.
11. H. J. Wright and R. N. Dupuis, *Ind. Eng. Chem., 38,* 1303 (1946).
12. H. Dannenberg, T. F. Bradley, and T. W. Evans, *Ind. Eng. Chem., 41,* 1709 (1949).
13. W. C. Norris, *Paint, Oil, Chem. Rev., 16* (March 31, 1949).
14. N. R. Bhow and H. F. Payne, *Ind. Eng. Chem., 42,* 700 (1950).
15. T. H. C. Michael, *Am. Paint J., 60* (Aug. 28, 1950).
16. F. Armitage, D. H. Hewitt, and J. J. Sleightholme, *J. Oil & Colour Chemists' Assoc., 31,* 437 (1948).

17. C. P. A. Kappelmeier, *Paint, Oil, Chem. Rev., 16* (Feb. 1, 1951).
18. A. E. Young, *Official Digest Federation Paint & Varnish Production Club, No. 610,* September, 1949.
19. F. E. Lawlor, *Ind. Eng. Chem., 39,* 1419, 1424 (1947).
20. D. J. Mehta and H. F. Payne, *Paint Varnish Production, 30,* 6 (1950).
21. R. H. Kienle, *Ind. Eng. Chem., 22,* 590 (1930).
22. R. H. Kienle, *J. Soc. Chem. Ind., 55,* 229 (1936).
23. W. H. Carothers, *Trans. Faraday Soc., 32,* 39 (1936).
24. S. O. Greenlee, U. S. Patent 2,456,408 (Sept. 14, 1948) to Devoe & Raynolds Co.
25. Shell Chemical Co., Technical Bulletins on Epon resins.
26. C. G. Goebel, U. S. Patent 2,482,761 (Sept. 27, 1949) to Emery Industries, Inc.
27. C. G. Goebel, *J. Am. Oil Chemists' Soc., 24,* 65 (1947).
28. E. B. Robinson and R. B. Waters, *J. Oil & Colour Chemists' Assoc., 34,* 361 (1951).
29. B. C. Pratt and H. S. Rothrock, U. S. Patent 2,358,475 (1944) to E. I. du Pont de Nemours and Co.

Cherish individuality, but exercise cooperation. He is a rare individual in whom all essential characteristics are conjoined.

Chapter 8. UREA- AND MELAMINE-FORMALDEHYDE RESINS

INTRODUCTION

The term "amino resins" is a short and convenient one to use for the class of resins based chiefly on urea-formaldehyde and melamine-formaldehyde condensation products. The urea-formaldehyde resins were sold to the organic coatings industry in 1936, and the melamine-formaldehyde resins in 1940. An idea of the importance of these resins may be obtained from the volume of their commercial production.

Commercial Production of Urea- and Melamine-Formaldehyde Coating Resins

Year	Pounds
1941	6,600,000
1949	134,000,000
1950	219,000,000

Amino resins are the heat-convertible type and are used in baking coatings and occasionally in some air-drying coatings. When baked at temperatures from 200°F to 350°F, they produce very hard, almost colorless films, but they are quite brittle and have poor adhesion to metal surfaces. It is fortunate that amino resins are compatible with a wide range of alkyd resins because the combination produces films having excellent properties. The short- to medium-oil phthalic alkyds are the types most frequently used with amino resins. The proportions of amino resins vary with the type of coating, but the general range is from 15% to 30%. However, it may drop as low as 5–10% in automobile finishes, or as high as 35–40% in washing machine finishes and up to 50% in low-bake furniture finish.

Amino resins are used with the oxidizing-type alkyds to reduce the baking time and to improve the surface hardness and alkali resistance of the films. In white baking finishes where maximum whiteness and retention of color with heat or age are required, the amino resins are used with non-oxidizing alkyds. There are differences between the urea-formaldehyde and the melamine-formaldehyde resins in baking time and temperature required and also in film properties. These differences are given in greater detail on page 342; but, in general, the melamine resins bake faster and over a wider temperature range than the urea resins. The melamine resins also have better heat stability and alkali resistance, and they are preferred for applications requiring exterior exposure, such as automobile finishes.

This chapter will consider only the amino resins used in organic coatings. These resins are soluble in combinations of organic solvents, such as butanol, xylol, and mineral spirits. They are also compatible with alkyd resins and some varnishes and oils. However, there are other important types of amino resins, some of which are soluble in water and the lower alcohols and are not generally compatible with the usual type of coating materials. These are used in a wide variety of products which include plastics, molding compounds, and adhesives for plywood, etc., and for imparting wet-strength to paper and producing water-repellent, crease-resistant, and shrink-resistant textiles. They have also been used as special tannins for leather. These non-coating applications are described by Payne (1), Dudley (2), and Patterson *et al.* (3), and these authors provide extensive bibliographies on these non-coating applications for amino resins.

CHEMISTRY OF AMINO RESINS

In the description of the resinous state, page 132, one of the conditions given is that "resins are mixtures of mutually soluble materials." Since the amino resins conform to this general rule, any attempt to describe the chemical reactions required for their formation, and the chemistry and structure of the baked films, must necessarily be somewhat simplified. However, the presentation of these simplified chemical formulas is justified because of their assistance in a better understanding of the development and application of amino resins. The resins are the condensation products of amino compounds with aldehydes, which are etherified with various alcohols and polymerized to the required degree. Therefore, competing reactions are taking place simultaneously: condensation, etherification, and polymerization. Control of these reactions is obtained by adjusting the type and ratio of components, and the reaction conditions of catalyst, time, and temperature.

UREA-FORMALDEHYDE RESINS

Although the urea and melamine resin reactions have certain similarities, they also have definite differences; therefore it will be better to describe them separately. One mole of urea may be reacted with 1 or 2 moles of formaldehyde to produce different products. The reactions may be carried out under either acidic or basic conditions, and, again, different products will be obtained. In addition, the dimethylol urea may be produced from urea and formaldehyde and etherified with butanol separately, or the etherification may be carried out simultaneously with the condensation by reacting the urea, formaldehyde, and butanol together. Figure 1 illustrates the condensation reactions between 1 mole of urea and both 1 and 2 moles of formaldehyde under acidic and basic conditions.

Fig. 1. Urea-Formaldehyde Reactions

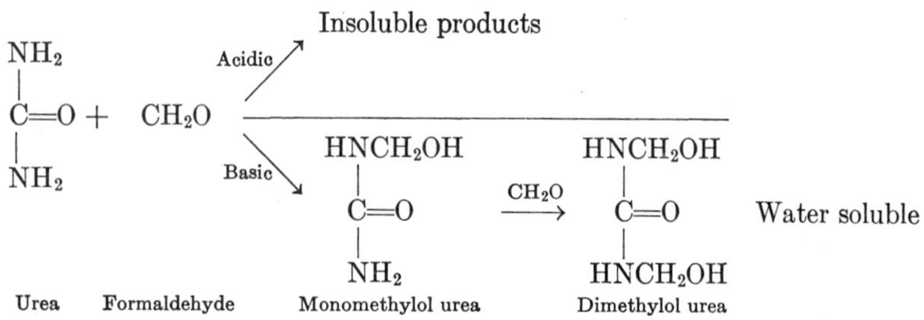

Under acidic conditions, insoluble compounds are formed which cannot be used for coating resins. Under basic conditions the mono- or dimethylol ureas are produced which can be used as intermediate products for coating resins and other purposes. The dimethylol urea is known commercially as DMU and is available as a white solid, containing 88–90% DMU, and 10–12% water. It is soluble in water and alcohol, but it polymerizes slowly at room temperature to the insoluble stage.

The usual source of formaldehyde is formalin, which is a 37% solution of formaldehyde in water. The urea is a white crystalline solid, having a melting point of 133°C and soluble in water to the extent of 80 gm/100 ml; it has a molecular weight of 60. For the production of DMU, the correct amounts of urea and formalin are adjusted to a *p*H of 7–8.5 and reacted at about 50°C. Sodium hydroxide is the usual basic catalyst, but others may be used, including various amines. When the reaction is complete the mass is concentrated under vacuum to the desired solid content and may be tray-dried, spray-dried, or crystallized. The DMU is used in some of the non-coating applications referred to earlier, since it may be polymerized to the insoluble stage by heating. The structure of the insoluble polymer has not been proved, but there can be no doubt that it is highly complex. From the ideas of Thurston (5), Marvel *et al.* (6, 7), Wohnsiedler (4), and others it is probable that the structure contains cross-linked linear

Fig. 2. Possible Structures of Urea-Formaldehyde Polymers

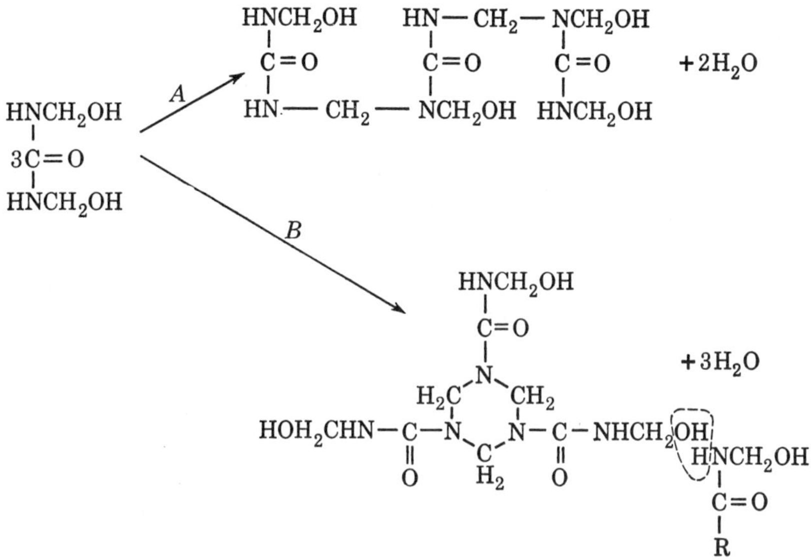

polymers and six-membered rings, as indicated in Fig. 2, *A* and *B* respectively.

Marvel postulated that in order to form the six-membered ring structure (*B*, Fig. 2) the two —NH_2 of urea may react differently with formaldehyde. One may react as a primary amine to form the Schiff's base, followed by trimerization to the ring structure. The other may then react as an amide, followed by elimination of water and formation of methylene linkages connecting the ring structures, as indicated in Fig. 2, *B*. Until more positive evidence is obtained regarding the structure of these insoluble polymers, the above theories afford some idea of their possibilities and their complexity.

Alkylation or Etherification. In order to change the DMU from a water-soluble material to an organic-solvent soluble material it must be made less polar. This is accomplished by alkylation with various alcohols. Obviously, the lower alcohols with very short carbon chains, or small non-polar groups, are less effective than the higher alcohols with longer carbon chains. For example, the methoxy methylol ureas are water soluble, the ethoxy products are soluble in ethanol, but good solubility in organic solvents is not obtained until butyl alcohol is used. It will be shown later that better solubility and compatibility with other resins are obtained if the higher alcohols are used, such as capryl or octyl. However, these are more expensive, and they retard the curing rate of the resin.

Butylated Urea Resin. A typical urea-formaldehyde resin suitable for use in baking coatings may be prepared by dispersing the DMU in butanol which has been slightly acidified. The dispersion is heated, and both etherification and polymerization reactions occur. It is essential that sufficient etherification take place before excessive polymerization occurs so that the product will have good solubility and stability. Conversely, if a high degree of etherification occurs and relatively low polymerization, the resin will have low viscosity and will be slower curing. These factors are controlled by the amount and type of acidic catalyst, the temperature, and the ratio of the components. A variety of acids may be used, including phosphoric, formic, oxalic, and phthalic. In general, the ratio of combined butanol in the final resin is from 0.5 to 1.0 mole per mole of DMU; but, of course, a considerable excess of butanol is used during the resin manufacture.

The water eliminated in the etherification and polymerization reactions, together with any water with the original DMU, is removed either by straight azeotropic distillation or by a continuous decantation procedure. When the desired degree of etherification and polymerization is reached, as indicated by solubility and compatibility tests, the resin is neutralized and concentrated. For a resin solution which is marketed as 50% resin, 30% butanol, and 20% xylol, the original butanol solution would need to be concentrated to 62.5% resin and 37.5% butanol. When 100 parts of this solution are thinned with 25 parts of xylol the resulting product would meet the requirements indicated above. Every effort is made to remove as much water and free formaldehyde as possible because these detract from stability, curing speed, and gloss in the finished enamel. In general, the final resin solution does not contain more than 0.5% water and somewhat less of free formaldehyde. Resins of this type may be prepared from the original ingredients without first preparing the DMU as an intermediate. In such cases the ingredients are reacted first under alkaline conditions, to permit the necessary amount of condensation, then finished under slightly acidic conditions, as indicated above.

The simplified formulas for a butylated urea-formaldehyde resin are shown in Fig. 3. This resin is based on a mole ratio of 1 mole urea, 2 moles formaldehyde, and 1 mole butanol.

The partially polymerized product in Fig. 3 is a highly simplified and idealized representation. The actual product would be much more complex and cross-linked and would probably contain ring structures as indicated in Fig. 2. However, the diagram will serve to illustrate the effect of the type of alcohol and the degree of etherification on the solubility and rate of cure of the resin.

Fig. 3.

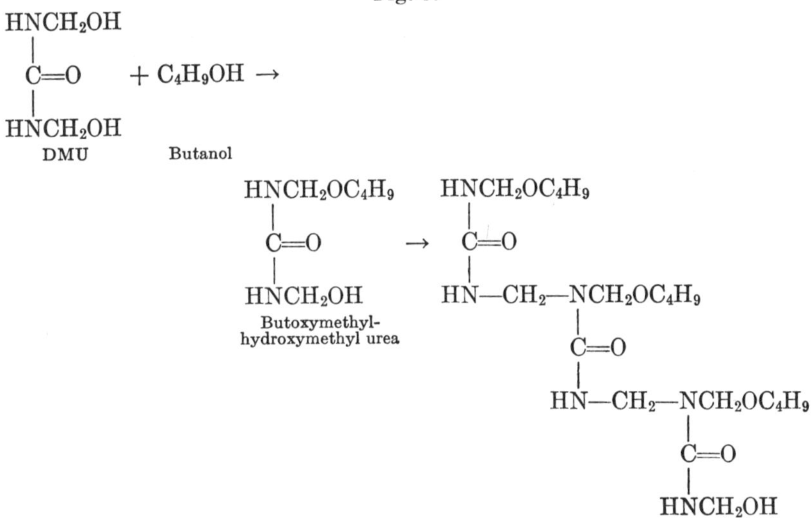

Solubility and Compatibility. It should be apparent that the 4-carbon-chain alkyl group in the butoxymethylol urea serves three purposes: (1) it decreases the amount of cross-linking; (2) it confers hydrocarbon solubility on the resin; (3) it increases compatibility with alkyds and other resins. Decreasing the possible cross-linkages retards the curing rate, and hydrocarbon solubility permits the use of xylol to replace part of the more expensive butanol. In the manufacture of the resin, isobutanol may be used, but the secondary and tertiary butanols react too slowly. If the ratio of butanol were reduced from the 1 mole shown in Fig. 3 to ⅔ mole, there would be fewer butoxyl groups in the resin. This would reduce the hydrocarbon solubility and would provide another point for cross-linking in the trimer illustrated. Higher ratios than 1 mole of butanol would increase the solubility, but these are seldom used since they would retard the curing rate and amount of polymerization excessively.

It should be apparent that increasing the carbon chain from a 4- to a 10- or 12-carbon chain would provide another method for increasing hydrocarbon solubility without reducing the number of possible cross-linkages. This may be done by using capryl, octyl, or other alcohols instead of butanol. However, these longer-chain alcohols are not good solvents for the intermediate DMU. Therefore the DMU polymerizes excessively before any appreciable amount of etherification takes place, and a heterogeneous product is obtained which is not suitable for coating resins. However, the higher alcohols may be incorporated into the resin by the transetherification procedure. The resin is prepared first with one of the lower alcohols, such as methanol, and the methylated methylol urea reacted with capryl or octyl alcohol. The transetherification takes place because the liberated methanol may be removed by distillation at a temperature low enough not to affect the higher-boiling alcohol.

Mineral Spirits Tolerance. Amino resins are used frequently with the medium-oil-length alkyd resins which are thinned with mineral spirits instead of xylol. Since mineral spirits is not as strong a solvent as xylol, it will be necessary for the amino resins to have better hydrocarbon solubility. One method for accomplishing this is the use of the higher alcohols referred to above. From the discussion of solubility in Chapter 6 it will be apparent that the long carbon chain on the resin makes it much more non-polar and therefore more soluble in aliphatic hydrocarbons. The degree of solubility of the resin is referred to as its mineral spirits tolerance. This is measured by adding mineral spirits to the resin solution slowly until turbidity develops. The mineral spirits tolerance is usually expressed as the pounds of mineral spirits tolerated by 100 lb of resin solution before turbidity develops.

The mineral spirits tolerance may also be increased by (*a*) increasing the ratio of formaldehyde to urea; (*b*) reducing the degree of polymerization of the resin; (*c*) increasing the amount of alkylation. It will be evident that all these methods tend to reduce the curing rate of the resin, which means a longer baking time for the finish in order to obtain the same hardness. The values for mineral spirits tolerance of the resins shown in Table 2 indicate the general range for commercial resins.

COMPOSITION VARIABLES

The variation in type of alcohol was the only composition variable considered in the preceding discussion of urea resins. However, both the amino and the aldehyde components may also be changed. The only other amino resin which has achieved commercial importance to date is the melamine-formaldehyde type described in the following section. However, mention should be made of thiourea, since it was one of the early amino compounds investigated for use in resins.

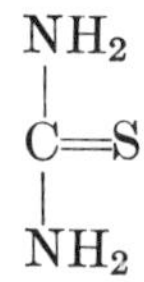

Thiourea

The formula shows that the oxygen of urea has been replaced with sulfur. Resins made with thiourea have slightly better water and alkali resistance than comparable urea resins, but they are not as pale in color, are somewhat odorous, and are inferior in exterior durability.

Formaldehyde is the most useful aldehyde for amino resins utilized in surface coatings. An aldehyde with more carbons, such as acetaldehyde, may be expected to increase the solubility of the resin, but the curing rate, color retention, and film properties have been reported to be inferior (8). Parker (8) also points out that mixtures of formaldehyde and acetaldehyde are impractical because very little of the higher aldehyde is combined under such conditions. It can be removed quite readily by distillation of the resinous material. Furfural has not been used extensively in coating resins to date, but it is employed in amino resins for adhesives and molding compounds.

MELAMINE

In 1834 Justus von Liebig (9) produced a new chemical which he believed was the amine of melam, and which he called melamine. Subsequent investigations have shown that his analysis was not entirely correct, but the chemical has retained its original name. It is a member of the class of compounds known as triazines and may be designated 2,4,6-triamino-1,3,5-triazine. It may also be considered a trimer of cyanamide. It may be prepared from dicyandiamide by heating under pressure in the presence of a diluent such as alcohol or ammonia. The relation of melamine to cyanamide and dicyandiamide is shown in Fig. 4.

The chemistry, historical development, and methods of production of melamine are given in considerable detail by McClellan (11). Hughes (12) studied crystalline melamine and found it to be a resonance hybrid with the position of the atoms as shown in Fig. 5(*a*). Ostrogovich (13) suggested both amino and imino structures in view of the possibility of tautomerism, Fig. 5(*b*). The amino form is generally used in the discussion of melamine in coating resins because its high melting point and heat stability suggest this benzenoid structure.

Fig. 4

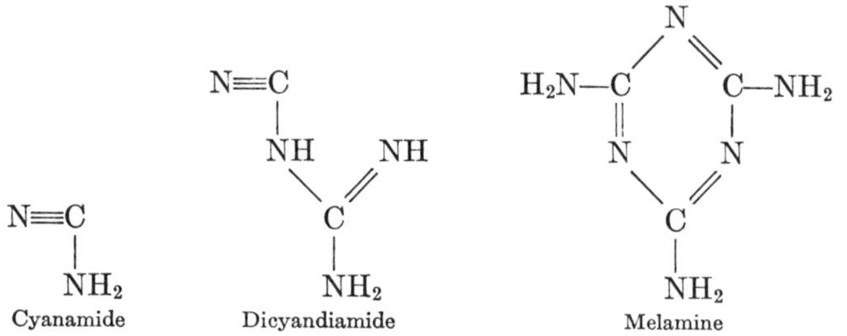

Melamine is a white, crystalline powder, with a melting point of 354°C. It has a molecular weight of 126 and specific gravity of 1.57 at 25°C. Its solubility in water has been reported by Chapman (14) to be 0.5% at 25°C, 1.0% at 50°C, 2.5% at 75°C, and 4.0% at 90°C.

Fig. 5

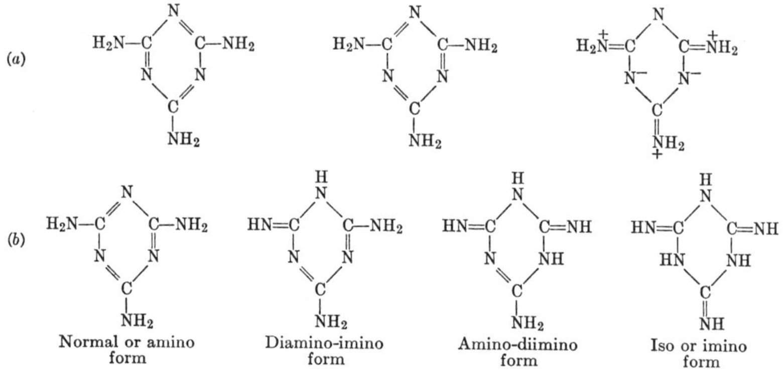

MELAMINE-FORMALDEHYDE RESINS

Melamine may be reacted with formaldehyde to produce a series of methylol melamines. These will vary with the ratio of formaldehyde, the *p*H, and the time and temperature of reaction. In this respect the reactions are similar to the reactions of urea with formaldehyde.

However, melamine differs from urea in that both hydrogens on the amino groups of melamine may be replaced readily with methylol groups. Since there are three amino groups in the melamine structure, it is possible to produce a series of methylol melamines ranging from the monomethylol to the hexamethylol melamine. For coating resins a high formaldehyde ratio is used, but for the water-soluble type a ratio of 3 to 4 of formaldehyde is more general.

Wohnsiedler and Thomas (15) and Norris and Bacon (10) describe the preparation of the water-soluble type of methylol melamines for use in textile treatments, imparting wet-strength to paper, and other applications. Methylol melamines may be prepared under alkaline conditions, but they are less soluble and less stable than the methylol ureas. In the *p*H range of 4–6 clear solutions are produced first, which slowly convert to hydrophilic colloids. These acid colloids continue to increase in viscosity; they finally become hydrophobic and form insoluble, amorphous solids. These polymers are possibly connected with methylene linkages similar to the methylene ureas described previously. In general, when 1–3 moles of formaldehyde are used per mole of melamine, the products resinify readily. When 3–6 moles of formaldehyde are used, resinification is slower; therefore the products are easier to alkylate or etherify.

At the higher formaldehyde ratios a series of alkyl methylol melamines may be produced in a manner similar to that described for the urea resins. The methylated methylol melamines are soluble in water, and the ethylated products are alcohol soluble. Solubility in butanol-xylol mixtures is obtained with the butylated methylol melamines, and increased solubility in hydrocarbons is obtained from alkylation with the higher alcohols, similarly to the situation with urea resins.

Butylated Melamine Resins. In the preparation of melamine coating resins, the methylol melamine may be prepared first, followed by etherification or "butylation." However, it is more general to react all the ingredients together under slightly acid conditions. The characteristics of the finished resin will depend on a number of variables, including the formaldehyde ratio, degree of etherification, type and amount of acid catalyst, and time and temperature of the reaction. In general, from 5 to 6 moles of formaldehyde are used with 1 mole of melamine and an excess of butanol slightly acidified. The batch is heated to reflux, and the reaction continued until the required viscosity and hydrocarbon tolerance are obtained. The water is removed by continuous decantation, and the acidity is neutralized before concentration of the resin. The water and free-formaldehyde contents of the finished resin should be less than 0.5% of each.

The general ratio of combined butanol to melamine is from 1 to 2 moles of butanol per mole of melamine. The structures of a possible melamine-formaldehyde resin unit and a melamine-formaldehyde resin are shown in Fig. 6. The structures are based on a composition of

Fig. 6. Typical Structure of Melamine-Formaldehyde Resin

Composition: 1 mole melamine
5 moles formaldehyde
2 moles combined butanol

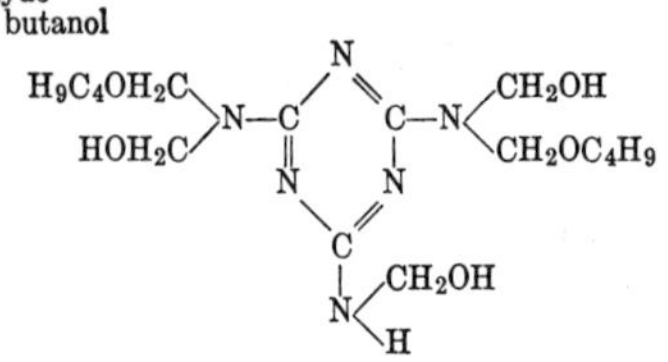

A melamine-formaldehyde resin unit: mol. wt. 320

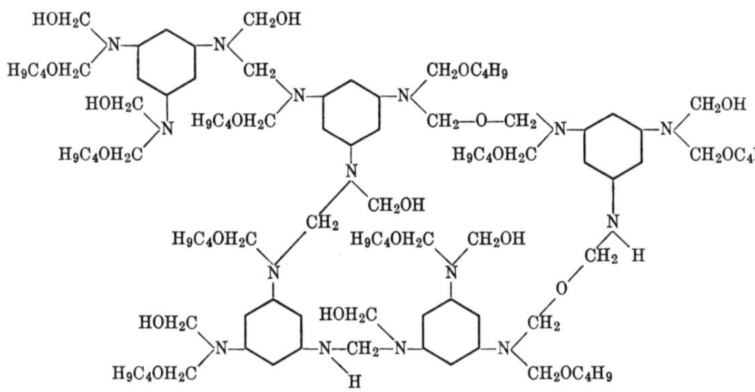

A melamine-formaldehyde resin: mol. wt. 1510

1 mole melamine, 5 moles formaldehyde, and 2 moles combined butanol. The polymer, as shown, contains both ether linkages and methylene linkages. The actual disposition of the resin units and linkages may be expected to be entirely random; therefore the polymer structure may not be precisely as shown in Fig. 6. It should also be remembered that the actual structure is three dimensional instead of two dimensional as in the figure. The possibility of further polymerization is evident from the plurality of methylol groups which are available for additional condensation reactions. The ether linkages are formed from the condensation of two methylol groups with the elimination of 1 mole of water. The methylene linkages are formed from the condensation of a methylol group with the hydrogen attached to a nitrogen atom.

NEW NITROGEN COMPOUNDS FOR AMINO RESINS

A great deal of research has been done on compounds related to urea and melamine for use in coating resins. Urea and melamine resins still dominate the field because of their comparatively low costs

Fig. 7

Triazines

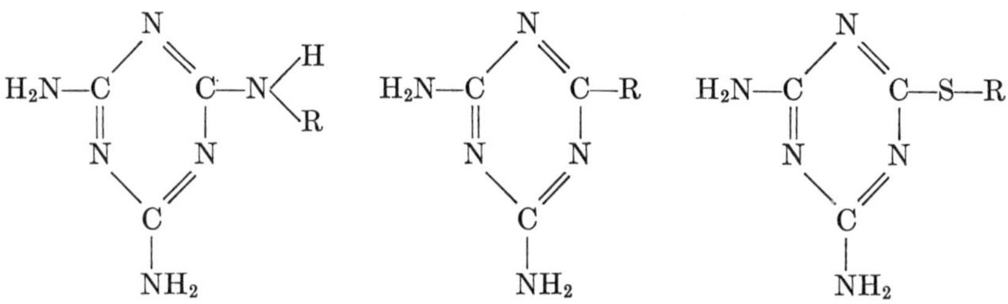

Substituted melamine — Guanamines — Substituted thioameline

Diazines

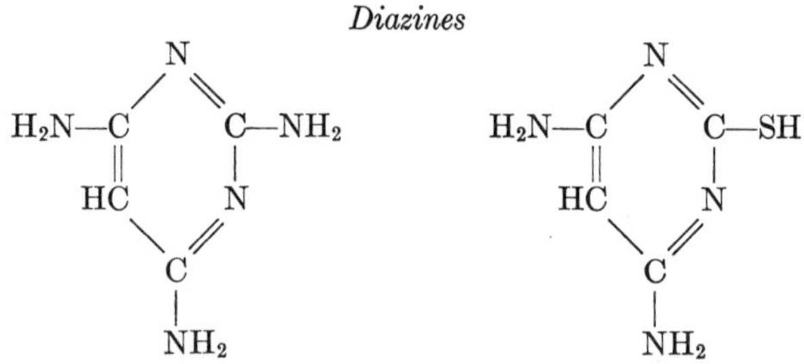

Triaminopyrimidine — 2-Mercapto-4,6-diaminopyrimidine

Triazoles

Guanazole

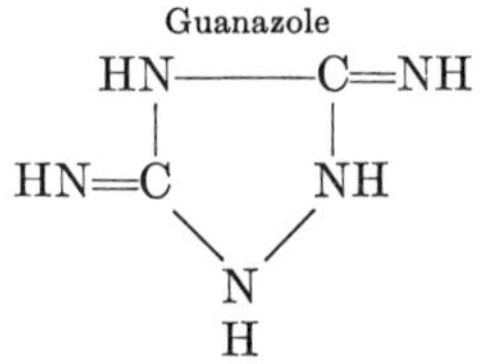

3,5-Diaminotriazole

Carbamyl guanazole

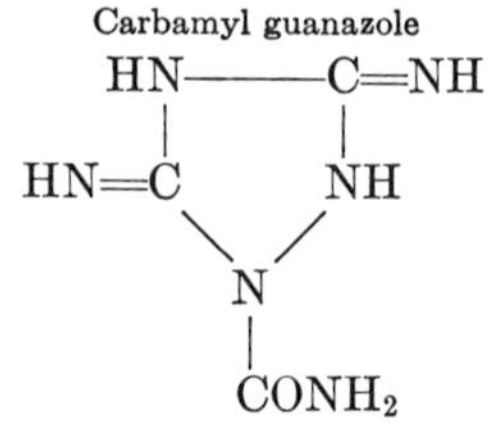

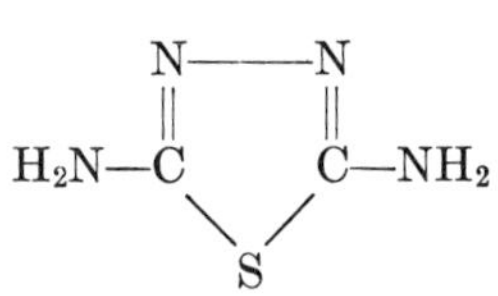

2,4-Diaminothiodiazole

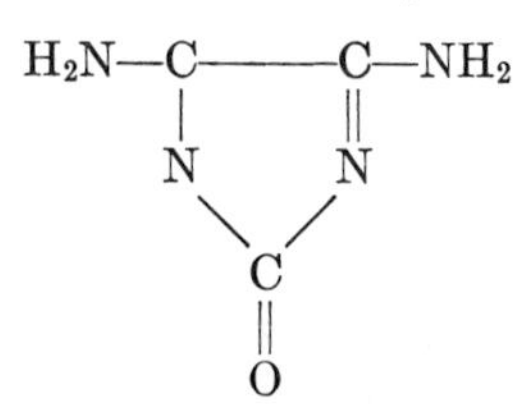

2-Oxo-4,5-diaminoparabanic acid

and generally good curing rates and film properties. However, a new triazine resin has been made available which has alkali resistance superior to that of melamine resins. It also has better compatibility and solubility characteristics but is somewhat slower curing. The general properties of this resin (Uformite MX-61, Rohm and Haas Co.) are given in Table 2. Its use in washing-machine finishes is indicated in Table 3, and further details regarding formulation and use are given in Volume II.

A few of the many possible new compounds for amino resin research are illustrated in Fig. 7. Auten (16) and Simons (17) have discussed the new possibilities in some detail, and many more may be found in the patent literature. A few references are cited in this field: substituted melamine (18), guanamine (19), substituted thioameline (20), diazines (21), triazoles (22).

The substitutent R, shown in the triazines, may be either an alkyl or an aryl group. The size and shape of this group will affect the solubility, compatibility, and curing rate of the resin in a manner similar to the type of alcohol used in the alkylation of these resins. Since this substitutent replaces a reactive amine group in the guanamines, and a reactive hydrogen in melamine, there will be a decrease in reactivity with a corresponding reduction in the curing rate. Considerable reduction in curing rate may be expected from large substituent groups, but the effect may be only slight with small groups. For example, benzoguanamine was used satisfactorily in the German amino resin, Meprenal BC, during World War II (23). Simons (17) also reported good results with a benzoguanamine resin in combination with a dehydrated castor-alkyd resin in white baking enamels. New developments may be expected in this field, but the limited availability and high cost of the amino compounds are restricting factors at present.

COMMERCIAL AMINO RESINS

A list of the manufacturers of amino resins and their trade names is given in Table 1. This was compiled from the Resin Index of 1950, issued as *Circ. 738,* May 1950, by the Scientific Section, National Paint, Varnish, and Lacquer Association.

The characteristics of a few of the commercial amino resins are shown in Table 2 simply to illustrate the types of amino resin solutions which are available. The various manufacturers should be contacted for data regarding newer resins which may have been made available or changes in the characteristics of those listed.

Table 1 * Amino Resin Manufacturers and Trade Names

Manufacturers	Trade Names	
	Urea Resins	Melamine Resins
American Cyanamid Co.	Beetle	Melmac
Jones Dabney Co.	Syn-U-Tex	....
D. H. Litter Co., Inc.	Uralite	....
Monsanto Chemical Co.	Resimene U	Resimene
Onyx Oils and Resins, Inc.	Uforesol	....
Plaskon Div., Libbey-Owens Ford	PCC #	....
Reichhold Chemicals, Inc.	Beckamine	....
Rohm and Haas Co.	Uformite	Uformite

* Resin Index of 1950, *Circ. 738,* May 1950, Scientific Section, National Paint, Varnish, and Lacquer Association, Inc.

The resins in Table 2 were selected to illustrate the variety of solvents used, the wide range in viscosities of the solutions, and the differences in mineral spirits tolerance. It will be noted that the acid numbers of the melamine resins are generally less than 1, but the acid numbers of the urea resins are higher. The color of the resin solutions is not given since they are all practically water white.

SOLVENTS

The regular-type amino resins are thinned with a solvent consisting of approximately equal parts by weight of butanol and xylol. The type of amino resin which has been alkylated with the higher alcohols for improved mineral spirits tolerance usually has some of the higher alcohol in the solvent. This type is illustrated by resins *D* and *E*, which contain capryl and octyl alcohols respectively. It will be noted that these resins have high mineral spirits tolerance. It will be noted that resin *I* also has high mineral spirits tolerance but does not contain the higher alcohols. In this resin the good compatibility and mineral spirits tolerance may be due partly to the lower degree of polymerization, indicated by the lower viscosity, and to the absence of xylol in the solvent. It may also be due partly to variation in the ratio of the resin components and in the processing conditions referred to previously.

Special solvents are used in some resins to meet the requirements of specific applications. For example, resin *C* is thinned with Hi-Flash naphtha, which is slower evaporating than the butanol-xylol combination, and permits better flow in finishes which are roller-

Table 2 Typical Commercial Amino Resins

Characteristic	Uformite * Resins				Beetle † Resins		Melmac † Resins			Resimene ‡ Resins		Uformite * Triazine
	F-200E A	*F-240* B	*F-240N* C	*F-226E* D	*212-9* E	*230-8* F	*245-8* G	*248-8* H	*247-10* I	*882* J	*883* K	*MX-61* L
Solids, % by weight, ±2%	50	60	60	50	60	50	50	55	60	65	60	60
Solvent, % by weight												
Butanol	25	20	..	..	24	27	30	25	40	..	..	25
Xylol	25	30	..	..	..	23	20	20	..	35	..	25
Hi-Flash naphtha	..	..	40	..	..	..	..	..	..	..	..	..
Capryl alcohol	..	..	..	50	..	..	..	..	..	..	..	..
Octyl alcohol	..	..	..	..	6	..	..	..	..	..	..	..
Petroleum aromatics	..	..	..	..	10	..	..	..	..	..	..	..
Butyl Cellosolve	..	..	..	..	..	..	..	..	..	..	20	..
Mineral spirits	..	..	..	..	..	..	..	..	..	..	20	..
Viscosity, Gardner-Holdt, at 25°C	W–Z	L–Q	Z2–Z6	W–Z1	Y–Z2	R–U	L–O	O–R	U–X	Z–Z4	S–W	F–N
Mineral spirits tolerance, min.	10 §	50 §	20 §	45 §	1000 ‖	130 ‖	180 ‖	160 ‖	600 ‖	4 §	10 §	50 §
Acid number, solids	3–8	3–8	3–8	4–7	1–4	0.5–2	1 max.	1 max.	1 max.	1	1	5–9
Pounds per gallon	8.5	8.5	8.7	8.2	8.4	8.3	8.2	8.5	8.6	8.9	8.4	8.7

* Trade name, Rohm and Haas Co.
† Trade name, American Cyanamid Co.
‡ Trade name, Monsanto Chemical Co.
§ Cubic centimeters of mineral spirits tolerated by 10 gm of resin solution.
‖ Pounds of mineral spirits tolerated by 100 lb of resin solution.

coated. These finishes are used in the tin decorating industry and for collapsible tubes. Resin *K* may also be used for these finishes, but it is thinned with another slow-evaporating combination solvent, butyl Cellosolve and mineral spirits.

In some applications, the presence of butanol is objectionable from an odor standpoint and in others because of its incompatibility with the resin used with the amino resin. The three resins *D, J,* and *K* are thinned with solvents other than butanol; resins *D* and *K* have slow-evaporating solvents and resin *J* has the faster-evaporating xylol. Resin *J* is recommended for use with resins which have low tolerance for butanol, such as the condensation of polymers of epichlorohydrin and *bis*-phenol, known as Epon * resins, described in Chapter 7.

VISCOSITY

Resins *B* and *C* have the same base resin, but the 60% solutions are considerably different in viscosity because of the use of different solvents. The high viscosity of resin *C* is due to the relatively poor solvent power of Hi-Flash naphtha compared with the butanol-xylol combination. High viscosity may be an advantage in applications such as roller coating and knife coating where a high consistency in the enamel is desirable. However, high consistency in an enamel which is to be applied by spraying would necessitate a larger amount of thinner to reduce the enamels to spraying consistency. This means that more of such a material must be sprayed to give the same film thickness as would be obtained from a lower-consistency enamel. Obviously, it is not economical to use a high-consistency enamel for spray application if the high consistency is obtained from a high-viscosity resin rather than from a high solids content.

If a resin solution, or an enamel made with it, tends to gel or be unstable in storage, the condition is made worse if these materials have a high original viscosity. Of course, the quality of the solvent is also an important factor in such cases. Frequently, a high percentage of butanol in the solvent combination improves stability considerably. However, in all cases, there is less tendency to gelation on storage if the original consistency is low rather than high.

The advantages of melamine resins over the lower-cost urea resins are discussed in the following section. In some formulations the optimum properties at lowest cost are obtained with a mixture of urea and melamine resins. Resin manufacturers provide combination urea

* Trade name, Shell Chemical Co.

and melamine resins to meet such requirements. Resin *F* in Table 2 is a typical example of such combination resins.

UREA VERSUS MELAMINE RESINS

Urea-alkyd combinations are quite satisfactory for many baking finishes. However, films of melamine-alkyd combinations have better resistance to water, solvents, and chemicals, better color retention at above normal temperatures, and better exterior durability. They also bake to the same hardness in less time than urea-alkyd combinations. A general comparison of the performance of urea and melamine resins in baking finishes is given below.

Baking Temperatures. Urea resin finishes are normally baked at 250–300°F. A special case is the acid-catalyzed furniture finish, which is force-dried at 140–160°F.

Melamine resin finishes are normally baked over the range of 200–350°F. Some roller-coated finishes on strip metal for venetian blinds bake at 400–500°F for short periods.

Baking Time. With equal resin concentrations, melamine resin finishes bake to the same hardness in 25–40% less time than urea resin finishes at the same baking temperature. Lower concentrations of melamine resin may be used to obtain the same hardness at equal baking schedules.

Hardness and Mar Resistance. When baked to the same film hardness, there is no difference in the mar resistance of urea and melamine resin finishes. Both are markedly superior to straight alkyd resin finishes for mar or scratch resistance.

Adhesion to Metal Surfaces. Amino-alkyd resin combinations do not have as good adhesion as straight alkyd resin finishes. In many cases, however, the adhesion is satisfactory, particularly when the finish is applied over phosphate-treated metal. When 40% or more of amino resin is used it may be necessary to apply a primer under the finish coat.

Color and Color Retention. At baking temperatures up to 300°F, urea resin finishes are as good as or slightly better in color than melamine resin finishes. At baking temperatures above 300°F, melamine resin should be used. Melamine resin finishes have better color retention on stoves or lighting fixtures which are subjected to above normal temperature for long periods.

Water, Alkali, and Solvent Resistance. Melamine resin finishes are superior to urea resin finishes in resistive properties. The superiority is related to the fact that melamine does not have the polar oxygen atom of urea, and it is quite probable that the melamine resin structure is more highly cross-linked than that of urea resin.

Exterior Durability. The excellent durability of straight alkyd resin finishes is well known. Alkyd-urea resin combinations are not quite as good, but alkyd-melamine resin finishes are as durable as straight alkyds with the added advantage of greater hardness and chemical resistance.

GENERAL USE OF AMINO RESINS

Films of straight amino resins do not bake as rapidly as the combination of alkyd and amino resin. This characteristic, which applies particularly to the melamine type, is due in part to the low acidity of the amino resins. Addition of acidic catalysts, such as phosphoric acid or the alkyl phosphates, accelerates the cure, but the films are too brittle for normal use. Alkyd resins with high acid numbers bake faster with amino resins than alkyds with low acid numbers. Basic pigments also have a retarding effect on the cure of the films. For example, 10–15% zinc oxide in a pigment combination for white baking enamels will increase the baking time or lower the hardness at equal baking time, in comparison with 100% titanium dioxide as pigment. In general, acid catalysts are not satisfactory with pigmented finishes because they tend to react with pigments and reduce the gloss of pigmented enamels. However, acid catalysts may be used in clear finishes as described later.

ACIDIC FUMES IN BAKING OVENS

Since amino resin finishes are susceptible to accelerated cure by acid materials, it may be expected that acidic fumes in the baking oven can be a source of difficulty. The fumes will be in contact with the surface of the finish and will therefore produce a rapid surface cure. In some respects, this is similar to the wrinkle finishes obtained with varnishes and described previously. However, the wrinkle produced on the amino finish is irregular, and varies from a mere dulling of the gloss to a pronounced shrivel, all of which are objectionable. The chief source of the acidic fumes is the trichloroethylene degreasing unit used for cleaning the metal parts which are to be coated. A small amount of the heavy trichloroethylene may be lost from the unit and be drawn into the oven by convection currents. There it is decomposed in the heating elements to liberate hydrochloric acid fumes which attack the amino finish. Attempts have been made to overcome this difficulty by putting ammonium hydroxide in the oven or incorporating an amine such as triethanolamine in the finish. However, the only satisfactory remedy is to prevent the acidic fumes from entering the oven.

PACKAGE STABILITY

Since acidic materials are necessary to accelerate the cure of amino resin finishes, it may be anticipated that there will be an increase in the viscosity of these finishes after storage for prolonged periods of time. This applies particularly to the types which are developed to have the fastest curing speed. Frequently, these finishes have higher initial acid numbers and also higher viscosity. The high viscosity indicates that the resin has been polymerized to a higher degree and is therefore less stable in the presence of alkyd resins and xylol. In such cases, a larger percentage of butanol is necessary in the solvent combination of the finishing material. It has been recommended that the solvent in all amino resin finishes contain at least 25% butanol to assure satisfactory stability in storage. It will be noted that resin *I* in Table 2 is thinned with butanol as the entire solvent to assure maximum stability.

Triethylamine (Sharples Chemicals, Inc.) is recommended as a stabilizer, or antilivering agent, for alkyd-amino resin finishes. The amount required varies with the amount of amino resin in the finish, the general stability of the particular amino resin, and the solvent combination in the finish. It was mentioned previously that finishes with high package consistency tend to be less stable than if they were shipped at lower consistency. However, high viscosity is frequently used because it is more economical to ship a finish in as concentrated a form as possible, and allow the finishers to reduce it to application consistency.

SPECIFIC APPLICATIONS FOR AMINO RESINS

Amino resins are practically colorless and non-yellowing; consequently they are of prime importance in white baking finishes. They are also used in many colored finishes because they reduce the baking time and improve the surface hardness. It would be impractical to illustrate all the applications for amino resins; therefore a limited number will serve as examples to show the general proportions used and the baking schedules. Details of manufacture and use are given in Volume II.

White baking finishes may be classified into four groups: the general appliance finishes, refrigerator finishes, washing-machine finishes, and heat-resistant finishes. Each group has its specific requirements and cost limitations. The first group includes kitchen cabinets and appliances which must conform to relatively low-cost finishing schedules. Some sacrifice in performance and color retention is made in the interest of economy. Whiteness of color and color retention are very important in refrigerator finishes. They must also have good resistance to soap, grease, and general wear. Washing-machine finishes must resist damage from hot soap solutions as a primary requisite. White finishes are used on stove parts which may be subjected to temperatures up to 250°F and on lighting fixtures which may be heated from 150° to 200°F for long periods; therefore such finishes are formulated to resist discoloration under these conditions. These finishes are described in detail in Volume II, but an outline of the types and proportions of the resins used is given in Table 3. Reference is made to the alkyd resins in Table 2, Chapter 7, and to the amino resins in Table 2 of this chapter. However, other resins are available from different manufacturers, some of which may be more suitable than those illustrated, and more recent products may have been developed since this was written.

The short- to medium-oil-length oxidizing alkyd resins produce very durable baking finishes for metal objects. The soybean oil type have fair color retention which is satisfactory for low-cost applications. However, the added cost of amino resins is practically offset by the shorter baking time and superior surface hardness which they impart. The styrenated alkyds are faster baking than the straight alkyds, but the surface hardness may be improved by addition of amino resins as indicated in Table 3. The styrenated alkyds show better results when baked at 275–300°F than at lower temperatures, but the oxidizing alkyds tend to discolor at 300°F. Higher percentages of amino resin are used with refrigerator finishes to assure good color and wear resistance, and satisfactory chemical resistance may be obtained with a mixture of urea and melamine resins. When maximum whiteness is required the non-oxidizing type of alkyd resin is preferred.

It will be noted that the washing-machine finishes are formulated with melamine resins or the special triazine resin, since the urea resins do not have as good resistance to hot soap solutions. In addition, these finishes have a high percentage of amino resin because the alkyds alone have relatively poor alkali resistance. It might be assumed that

Table 3 White Baking Finishes

	General Appliance					Refrigerator			Washing Machine		Heat-Resistant		
	Alkyds,* % by weight, solid resin basis												
Rezyl resin 310-5, 330-5, or 387-5	100	75	85	85	..	70	50	..	60	..	..	..	..
S101-1 †	..	..	..	..	100	..	..	..	..	70	..	..	..
S102-5 †	..	..	..	15	..	..	..	..	..	..	..	..	..
Rezyl resin 92-5	..	..	..	..	..	..	25	65	..	..	70	70	60
	Amino Resins, % by weight, solid resin basis												
A	..	25	..	..	..	..	12.5	..	..	..	..	..	..
G or *H*	..	..	15	..	..	..	12.5	..	40	30	..	30	40
F	..	..	..	..	..	30	..	35	..	..	..	..	..
L	..	..	..	..	..	..	..	..	..	..	30	..	..
	Baking Time, min												
Temperature, °F													
200	90	..	60	..	..	..	..	..	..	..	..	..	..
250	60	45	45	..	..	50	60	..	..	..	..	..	..
275	45	30	30	25	40	40	45	45	..	..	..	45	45
300	..	..	20	15	20	30	30	30	30	30	30	30	30

* See Chapter 7, Table 2, for description.

† Cycopol resins, styrenated alkyds, American Cyanamid Co. Cycopol: Reg. U. S. Patent Office.

these finishes, with high amino resin content, would bake in much shorter time than 30 min at the high temperature of 300°F. In fact, they are hard enough for normal use in 15 min at 300°F, but they have not developed sufficient alkali resistance at the shorter baking schedule; therefore the longer baking schedule is used. For maximum color retention with heat, the non-oxidizing alkyds must be used with a high percentage of melamine resin. High ratios of amino resin detract from adhesion of the finish to metal surfaces; therefore it may be necessary to use a baking primer under such finishes.

Amino Resins in Primers. Baking primers for metal surfaces are formulated to have excellent adhesion to the metal, and they must also be hard so that the finish will not penetrate them severely, thereby causing a loss of gloss. Frequently, they are sanded lightly for greater smoothness; if the surface is too soft, it will show the marks of the sandpaper. The primer may be hardened by using a higher pigment content, but this makes it more porous and allows the finish to sink into it. The required hardness may be obtained with a small percentage of amino resins without increasing the porosity or sacrificing adhesion.

Primers may be formulated with varnishes or alkyd resins in combination with amino resins. The varnishes in Chapter 4, Formulas 5–10 inclusive, are the type generally used with about 15% of urea resin or 10% of melamine resin. Since these varnishes are thinned with mineral spirits it will be necessary to use amino resins having high mineral spirits tolerance. These are illustrated in Table 2 as resins *B, D, E,* and *I*. The type of alkyd resin used in primers is the medium-oil-length, mineral spirits-soluble type such as Rezyl resins 412-1, 435-1, 807-1, and 823-1 in Table 2, Chapter 7. Primers of this type usually include a small percentage of manganese drier for the oxidizing resin. The usual baking schedule is 30–40 min at 275–300°F, but a shorter time at 400°F may be used with the melamine-type primer.

Automotive Finishes. The two major types of finishes on automobiles are the forced-dried nitrocellulose lacquer finishes and the baking alkyd-amino resin finishes. Although the lacquer finishes are only forced-dried at temperatures as low as 160–180°F for about 1 hr, it appears that a small percentage of melamine resin improves the water-spot resistance and gloss retention of such finishes.

The alkyd-amino type finishes are baked from 30 to 45 min at 275–300°F. The fast-drying, medium-oil-type alkyds, similar to Rezyl resins 330-5, 387-5, and 435-1 in Table 2, Chapter 7, are typical of those used with amino resins. If the alkyds are used alone, a small amount of metallic drier is usually added to reduce the baking time, but if the coating is applied a little more heavily than usual, it will wrinkle badly when it is baked. This then requires an expensive clean-off and refinishing operation. In general, amino resins will prevent the wrinkling when used in amounts as low as 5–10% of the total resin content. Urea resins were utilized for this purpose before melamine resins became available; but, where cost will permit, the melamine resins are preferred because of their superior durability. The amino resin also increases the surface hardness so that the finish is considerably more resistant to scratching and surface marring than a straight alkyd resin finish.

Tin Decorating Finishes. Tin decorating finishes are applied to sheets of tin-coated steel by a roller-coat method and generally baked from 10 to 30 min at temperatures from 225° to 410°F. Various shapes are die-punched from the coated sheets and formed into bottle caps, tobacco tins, toys, etc. Obviously, the finish must have good adhesion to the metal and excellent toughness to withstand the stamping and forming operations. The medium-oil-length alkyd resins meet these requirements quite well, but the addition of amino resin improves the surface hardness. It also makes possible the use of the non-oxidizing alkyds where better whiteness and color retention are required. Very flexible finishes may be obtained with combinations of the longer-oil, soybean-type alkyd and the amino resins which have high mineral spirits tolerance. Either urea or melamine resins may be used, depending on requirements, and the ratio of amino resin is usually from 10% to 20%. Resin *B* in Table 2 is internally plasticized and has a high degree of toughness, so that somewhat more of it may be used in finishes of this type.

Low-Bake Furniture Finish. Very good results have been obtained with acid-catalyzed, urea-alkyd resin combinations for furniture finishes. These are forced-dried at temperatures of 140–150°F for 1 or 2 hr. They have good toughness and cold-check resistance and better alcohol and cigarette-burn resistance than nitrocellulose lacquers. Wooden furniture is very bulky, and the finishing room in a furniture factory would require a very large floor space unless the finish were fast drying. Nitrocellulose lacquer meets this requirement better than any other air-drying finish, which accounts for its wide use in furniture finishing. However, it requires expensive solvents and it contains a low percentage of solids at spraying consistency. This means that from three to six coats must be applied to build up the necessary film thickness.

The urea-alkyd resin finish may be sprayed at 40–45% solids so that one coat over a properly sealed surface is usually sufficient. Best results are obtained when a thin coat of the finish material is used as the sealer coat. The wood is stained, and filled if necessary, and a light coat of the finish is applied as the sealer. This coat is baked for 1 hr at 140°F; when cool, it is sanded lightly. A full coat of the finish is then applied and baked for 1–2 hr at 140–150°F. After cooling, it is wet-sanded and rubbed or polished to the finish desired. The formulations fall in the following range:

Low-Bake Furniture Finish

	Parts by Weight *A*	Solid Basis *B*
Standard urea resin	45–50	45–50
Short-oil alkyd	35–40	.....
Resin-modified alkyd	.....	50–55
Hard resin	10–15	.....

Catalyst: an alkyl phosphoric acid, about 10% based on the urea resin solid content.

The urea resin may be similar to resin *A* in Table 2. The short-oil alkyd should be similar to Rezyl resin 387-5, and the resin-modified alkyd similar to Rezyl resin 728-5 in Table 2, Chapter 7. The hard resin may be a modified phenolic or a maleic resin similar to those discussed in Chapter 3. The finish should be thinned with a solvent combination containing a high proportion of butanol or other strong solvent, and the package viscosity should be low, preferably no higher than E on the Gardner-Holdt scale. The catalyzed solution is usually stable for 2 or 3 months at normal temperatures, but the viscosity will increase rapidly and perhaps reach the gelation stage if the temperature rises to 100° or 120°F.

The above discussion illustrates a few of the more important uses for amino resins. Their excellent color and color retention and their thermosetting properties impart desirable characteristics to many specialized finishes. Certain types of amino resins are compatible with thermoplastic materials such as vinyl resins and nitrocellulose. When the amino resin is used in the ratio of 10–20% the thermoplastic coating may be partially thermoset, so that it has better resistance to heat softening and particularly to the softening action of solvents. A similar use for amino resins is described in Chapter 7 under Epon resins.

BIBLIOGRAPHY

1. H. F. Payne, *Paint, Oil, Chem. Rev.*, March 21, 1946.
2. J. R. Dudley, *Chem. Inds.*, *224* (February, 1949).
3. D. C. Patterson, E. B. Detwiler, and T. J. Suen, Paper at Boston Meeting, American Chemical Society, April, 1951.
4. H. P. Wohnsiedler, Paper at Boston Meeting, American Chemical Society, April, 1951.
5. J. T. Thurston, Gibson Island Conference, July, 1941.

6. C. S. Marvel, J. R. Elliot, F. E. Boettner and H. Yuska, *J. Am. Chem. Soc., 68,* 1681 (1946).
7. C. S. Marvel and E. C. Horning, in *Organic Chemistry,* Volume I, H. Gilman, ed., John Wiley & Sons.
8. C. H. Parker, Paper at Boston Meeting, American Chemical Society, April, 1951.
9. J. von Liebig, *Ann., 10,* 18 (1834).
10. W. C. Norris and J. C. Bacon, *Official Digest Federation Paint & Varnish Production Clubs,* October, 1948.
11. P. P. McClellan, *Ind. Eng. Chem., 32,* 1181 (1940).
12. E. W. Hughes, *J. Am. Chem. Soc., 63,* 1737 (1941).
13. A. Ostrogovich, *Gazz. chim. ital., 65,* 566 (1935).
14. R. P. Chapman *et al., Ind. Eng. Chem., 35,* 137 (1943).
15. H. P. Wohnsiedler and W. M. Thomas, U. S. Patents 2,345,543; 2,485,079; 2,485,080.
16. R. W. Auten, *Resinous Reporter, 8,* No. 3, p. 2 (May, 1947); *Resinous Reporter, 8,* No. 4, p. 12 (July, 1947).
17. J. K. Simons, Paper at Boston Meeting, American Chemical Society, April, 1951.
18. U. S. Patents 2,228,161; 2,328,961; 2,524,727.
19. U. S. Patents 2,321,052; 2,309,681; 2,510,761.
20. U. S. Patents 2,217,667; 2,295,561; 2,258,130.
21. U. S. Patents 2,295,560; 2,295,563; 2,295,564; 2,211,270.
22. U. S. Patents 2,218,077; 2,456,090; 2,510,935; 2,514,313.
23. deBell, Groggin, and Gloor, *German Plastics Practice,* pp. 266–268.

Well-chosen words stimulate ambition and initiate great achievements, but a careless word can shatter hope and destroy confidence. Use only those words which generate constructive thought, and impel men to noble action.

Chapter 9. RUBBER RESINS

INTRODUCTION

Not long ago the term "rubber resins" applied only to such modifications of natural rubber as cyclized rubber, rubber hydrochloride, and chlorinated rubber. Today the term also includes the resins and latexes which are modifications of synthetic rubbers. Therefore, a short review of natural and synthetic rubbers may assist in visualizing the general chemistry and structural characteristics of these materials. It will be noted that they are related to the vinyl resins, discussed in Chapter 12, and to the styrenated oils and alkyds in Chapters 2 and 7 respectively. They are also somewhat related to the petroleum resins discussed in Chapter 3.

The natural and synthetic rubbers are very high-polymeric materials designed especially for toughness, high tensile strength, and flexibility. Their high molecular weight produces relatively low solubility in organic solvents. One of the principal objects in modifying rubbers for rubber resins is to obtain better solubility, with the resulting higher solids at application consistency. Another object is to improve the surface hardness of the materials. Many rubber goods for household use are coated with shellac, nitrocellulose lacquer, etc., to improve the surface characteristics. Of course, it should be realized that "rubber resins," some of which contain as much as 67% of chlorine, cannot be expected to have the same properties as natural or synthetic rubber.

In this discussion of the properties and structure of rubbers and rubber resins we shall use the customary formulas for the structural units to assist in visualizing the differences and relationships which exist among them. However, it must be kept in mind that the actual polymeric materials are much more complicated than the simple structural units might imply. The polymers will contain geometric isomers, and the units may have a random distribution of head-to-head or head-to-tail arrangement. The arrangement is of particular importance in the copolymeric materials. We are also apt to think of these polymers as long, straight chains of carbon atoms with or without occasional side groups attached to the chain. However, we must realize that in most cases the polymer chains are not straight but are "coiled" and intertwined. They can be straightened somewhat by a stretching process, but this can be applied only to fibers or sheeting. Natural and synthetic rubbers contain some unsaturation; therefore they can be changed from thermoplastic to thermosetting by cross-linking. In the rubber industry, this process is known as "vulcanization." Obviously, the degree of vulcanization or cross-linking is an important factor in the relative hardness of rubber.

NATURAL RUBBER

Rubber exists in certain trees and plants as a colloidal dispersion in the milky liquid known as "rubber latex." The chief commercial source is the rubber tree, *Hevea brasiliensis*. The latex is obtained by tapping the tree in a manner similar to that described for rosin in Chapter 3. Rubber latex has the following approximate composition:

	Per Cent
Rubber	35
Proteins	2
Fats	1
Miscellaneous	2
Water	60

The latex is coagulated by smoke to produce "smoked sheet" rubber or by dilute acids to produce "crepe sheet" rubber. The crude rubber is milled with a variety of chemicals and then vulcanized to form the rubber goods of commerce. The chemicals include plasticizers or softeners which reduce the power and time required for milling. Carbon black and a few white pigments are used as reinforcing agents, and other pigments are added as "fillers" to improve hardness and processing properties. Sulfur is the chief vulcanizing agent, but accelerators are included to reduce the time or temperature of vulcanization, and activators are added to assist the accelerators and to obtain specific properties in the product. Since vulcanized rubber still contains considerable unsaturation, it is subject to oxidation on aging with consequent hardening. Therefore antioxidants are added to reduce the aging effect as much as possible. The rubber goods may be colored by addition of pigments or dyes.

In view of the very wide variety of rubber goods produced, no single formula can cover adequately all the compounding mixtures. However, the general range of materials used is as follows:

	Parts by Weight
Rubber	100
Sulfur	2–6
Accelerators	0.5–1.0
Activators	2–5
Antioxidants	1–2
Stearic acid	1–2

The compound must be milled thoroughly on a rubber mill or Banbury mixer before shaping and vulcanizing. The carbon black and other pigments are added on the mill; from 30 to 50 parts of carbon

Fig. 1. Natural and Synthetic Rubbers

Type	Monomers	Polymer Units
Natural rubber	Isoprene	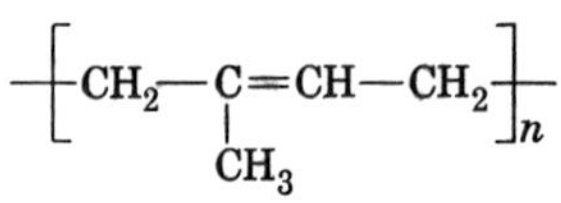
GR-S	Butadiene 25% styrene	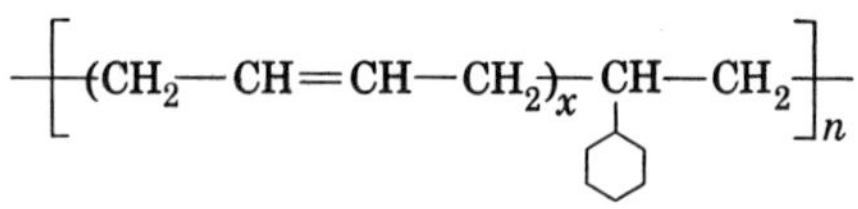
Nitrile rubbers	Butadiene 18-42% acrylonitrile	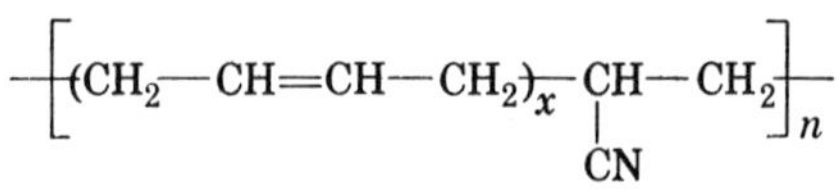
GR-I	Isobutylene 5% isoprene	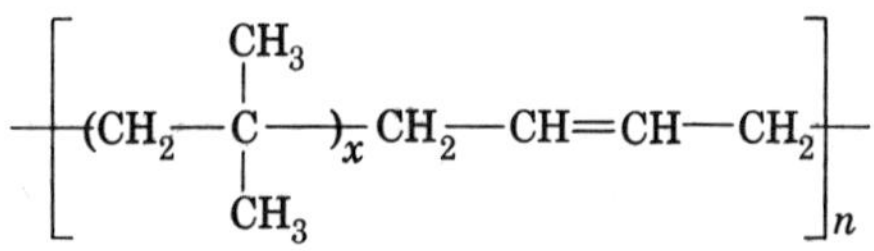
GR-M	Chloroprene	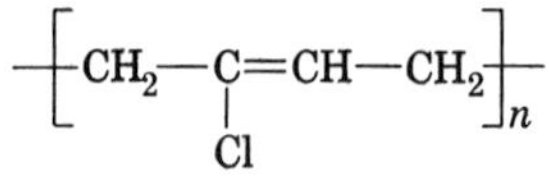
Polysulfide rubbers	Several types	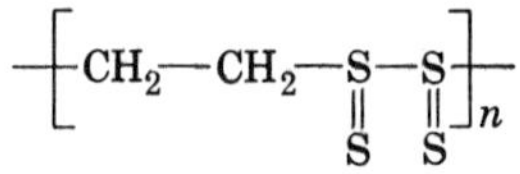

black are used with 100 parts of rubber for such articles as tire treads. The vulcanization time and temperature vary with the type of compound and the shape of the article. An automobile tire may require from 30 to 50 min at 275°F, whereas a bathing cap may not require more than 5–10 min at this temperature. The general temperature range for vulcanization is from 260°F to about 320°F. The synthetic rubbers are milled and processed in the same general manner as natural rubber. They are more difficult to break down on the mill, and frequently different softeners are used with them. Their specific properties are outlined in the next section on synthetic rubbers.

Natural rubber is a hydrocarbon corresponding to the empirical formula $(C_5H_8)_n$. The degree of polymerization, n, is generally considered to be approximately 2000, indicating an average molecular weight of about 136,000. This high value accounts for the high viscosity of natural rubber solutions, and the need for modifying rubber for use in coatings which must be applied from solution. It has been established that the chemical corresponding to C_5H_8, from which the rubber structure is formed, is the conjugated diene, isoprene, $CH_2{=}\underset{\displaystyle CH_3}{\underset{|}{C}}{-}CH{=}CH_2$. The activated form is shown in Fig. 1 together with the fundamental structural units of some of the synthetic rubbers.

SYNTHETIC RUBBERS

Five general types of synthetic rubber have been developed, and their government designations, starting monomers, and polymer units are shown in Fig. 1. It will be realized that variation in the physical properties of the synthetic rubbers may be made by changing the ratio of reactants, or by varying the processing conditions, type of catalyst, etc. However, the general differences among the five types and their comparison with natural rubber are summarized conveniently by the B. F. Goodrich Chemical Co. in Table 1.

Several structural requirements are essential in synthetic polymers if they are to have physical characteristics similar to those of natural rubber. The most important of these requirements are:

(*a*) High molecular weight.

(*b*) Essentially linear hydrocarbon chains.

(*c*) No highly polar groups to produce strong secondary valence forces between the chains.

(*d*) Not too many reactive points along the chain for cross-linking during vulcanization.

It will be seen from Fig. 1 that the five synthetic rubbers meet these requirements. However, it must be remembered that the actual polymers are not so well ordered as the structural units imply. The fact that the synthetic rubbers are more difficult to break down on the mill suggests that some cross-linking does occur. It is well known that conjugated dienes, such as butadiene, polymerize chiefly by 1:4 addi-

Table 1 * Property Relation of Natural and Synthetic Rubber

Abbreviations: E: Excellent F: Fair
G: Good P: Poor

Property	Natural Rubber	GR-M Neoprene	Thiokol	Ameripol D (Nitril Type)	Ameripol D (GR-S Type)	GR-I (Butyl)
	Comparison to Rubber					
Abrasion resistance †	E	G	P	E	E	F
Abrasion resistance—soaked in oil	P	F	P	E	P	P
Adaptability for contact with food †	G	F	P	F	G	G
Adhesion to fabrics	E	E	F	G	G	G
Adhesion to metals	E	E	P	E	E	G
App. specific gravity, basic material	0.93	1.23	1.34	1.00	0.94	0.92
Cold resistance †	E	G	F	G	E	G
Color range	G	G	P	G	G	G
Dielectric strength †	E	F	F	F	E	E
Electrical conductivity †	F	F ‡	F	F ‡	F	F
Elongation	E	E	F	E	G	E
Flame resistance	P	G	P	P	P	P
Freedom from odor †	E	F	P	F	G	G
Gas diffusion (air, ammonia, carbon dioxide, hydrogen, oxygen, and nitrogen)	F	G	E	G	F	E
Hardness Durometer A (100 is bone hard)	20–100	20–90	35–80	20–100	35–100	15–90
Heat resistance †	G	G	P	G	G	G
Rebound elasticity (snap)	E	G	G	F	G	P
Resistance to aging	G	E	E	E	E	E
Resistance to alkali solutions † sodium and potassium hydroxide)	G	G	F	G	G	G
Resistance to aromatic hydrocarbons [benzene (benzol), toluene (toluol), xylene (xylol), etc.]	P	P	G	F	P	F
Resistance to carbon tetrachloride	P	P	G	G	P	P
Resistance to other chlorinated hydrocarbons (ethylene dichloride, trichloroethylene, methyl chloride, chloroform)	P	P	P–F	P–F	P	P
Resistance to cold flow †	E	G	P	E	E	F
Resistance to deterioration in petroleum oil	P	E	F	E	P	G
Resistance to diffusion of petroleum products	P	F	E	E	P	P
Resistance to hydroxy compounds † (water, alcohols, glycerine, glycol)	G	G	F	G	G	G
Resistance to ketones (acetone, methyl ethyl ketone, methyl isobutyl ketone, etc.)	G	F	G	P	G	G
Resistance to lacquer solvents (ethyl, butyl, and amyl acetates)	P	P	G	F	P	P
Resistance to "Naval Stores" (wood rosin, turpentine, rosin oil)	P	F	G	G	P	P
Resistance to ozone †	F	E	E	F	F	E
Resistance to paint and ink driers	P	E	E	E	E	E
Resistance to saturated inorganic salt solutions † (sodium chloride, sodium sulfate, alums, ferrous and ferric sulfate, etc.)	E	E	F	E	E	E
Resistance to strong oxidizing acid † (nitric, sulfuric, chromic)	P	P	P	P	P	F
Resistance to other inorganic acids † (hydrochloric, phosphoric)	G	F	P	F	F	F
Resistance to sunlight †	F	E	E	F	F	E
Resistance to swelling in lubrication oil	P	G	E	E	P	P
Resistance to vegetable oils (corn, soybean, linseed, rapeseed, cotton seed, sunflower seed, etc.)	P	F	G	E	P	G
Resistance to water absorption †	F	G	P	G	G	F
Tear resistance †	E	G	P	F	F	G
Tensile strength †	E	G	P	G	F	F
Workability	E	G	F	G	G	G

All synthetic rubbers can be worked on rubber machinery, but in some products they are more difficult and expensive to fabricate than natural rubber due to lack of tack.

* Courtesy of B. F. Goodrich Chemical Co.

† These properties available only in specific compounds.

‡ Electrically conductive compounds with more "rubbery" characteristics can be made of these synthetics than is the case with natural rubber, which has to be very heavily "loaded" to attain the same degree of conductivity.

tion. This was illustrated in Chapters 1, 2, and 3 with regard to polymerization of drying oils and the production of maleic-treated oils and resins. However, it is also known that some 1:2 addition can take place in the polymerization of butadiene. The 1:2 addition would produce some cross-linking between chains as illustrated in Fig. 2.

In addition to the possibility of cross-linking between chains, as indicated in Fig. 2, some branched chains may be formed which would affect the physical properties of the rubber. Various modifiers are

Fig. 2. Butadiene Polymerization Reactions

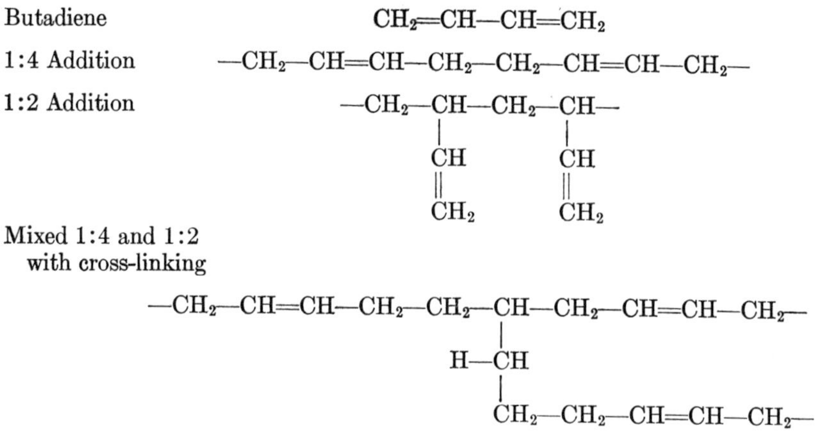

added to assist in controlling the polymer size and shape, and extreme care must be taken to maintain uniform conditions during the polymerization reaction. For this reason, most of the polymers are made by the emulsion polymerization process because it provides better control of the reaction. A general outline of the ingredients and conditions for the emulsion polymerization of the butadiene-styrene copolymer for synthetic rubber is given in Formula 1. The ratio of the monomers can be varied widely to produce products ranging from GR-S rubber (3:1, butadiene:styrene) to some of the "reverse rubbers" or rubber resins which have not less than 85% styrene. In the production of the rubber and solid rubber resins the emulsion is "broken" when the copolymerization is finished, and the products are coagulated, washed, and dried. However, it will be realized that the emulsion is similar to the natural rubber latex, and may be used to impregnate paper or coat textiles, or for the production of latex paints. Special care must be taken in the selection of emulsifying agents when the latex is to be used directly since they are not removed and will appear in the finished coating.

Formula 1 Emulsion Copolymerization of Butadiene-Styrene for Synthetic Rubber

	Per Cent by Weight
Butadiene	26.3
Styrene	8.7
Soap solution (3% soap)	65.0
Modifiers, based on monomers	0.5
Catalyst, based on monomers	0.1–0.4
Antioxidants, based on finished product	1.0–2.0

Procedure

The monomers, modifiers, and catalyst are added to the soap solution in the reactor slowly with agitation. Temperatures from 45° to 60°C are used, and the reactor is under pressure since the butadiene is a gas at these temperatures. When the desired degree of polymerization has been attained, the batch is moved into a tank under lower pressure where the unreacted butadiene is "flashed off." The unreacted styrene is removed by steam stripping, and the antioxidants or stabilizers are added. The emulsion is then ready for use as a latex. The solid product may be recovered by flocculating with brine, coagulating with acid, and filtering. The product is washed free of acid, dried in a tunnel oven, and compressed into blocks.

Many variations in processing are possible to impart specific properties. For example, the so-called "cold rubber" is polymerized at lower temperatures, which produces a more uniform polymer size and better physical properties. Of course, different processing conditions will be required for the other types of synthetic rubbers, but the butadiene-styrene type illustrated above is the one most widely used in rubber resins at the present time. The polysulfide type is used to a limited extent in special corrosion-resistant coatings. These are usually applied by the flame-spraying process in a semi-molten condition as described in Volume II.

RUBBER RESINS AND LATEXES

The types of rubber resins and latexes used at present are based on:

1. Chlorinated rubber.
2. Cyclized rubber.
3. High styrene-butadiene copolymers.
4. Butadiene-acrylonitrile copolymers.

5. Plasticized polystyrene.
6. Vinylidene chloride-acrylonitrile.

A typical example of type 1 is Parlon, of the Hercules Powder Co. Type 2 is represented by Pliolite NR, and type 3 by Pliolite S-5 of the Goodyear Tire and Rubber Co. The latter is also available in latex form as Chemigum Latex 101. The Dow Chemical Co. produces a series of plasticized polystyrene and styrene-butadiene copolymer latexes. The B. F. Goodrich Chemical Co. produces under the trade name of Hycar a series of latexes based on different ratios of butadiene and acrylonitrile. The Bakelite Co., a division of the Union Carbide and Carbon Corp., the Dow Chemical Co., and the Monsanto Chemical Co. produce the plasticized polystyrene type of latex. Other manufacturers are entering this field because of the suitability of these latexes as vehicles for coatings on porous surfaces such as concrete, plaster, and paper.

There is another group of chlorinated resins and plasticizers which are frequently used with the rubber resins. The group includes the Aroclors *, chlorinated biphenyl and polyphenyls of the Monsanto Chemical Co.; the Clorafins *, chlorinated paraffins of the Hercules Powder Co.; and the Chlorowaxes, chlorinated paraffins of the Diamond Alkali Co. These materials range from liquids to brittle solids, and they are non-flammable because of their high chlorine content.

CHLORINATED RUBBER RESINS

Parlon is produced by chlorinating natural rubber; it contains about 67% chlorine. When natural rubber was not available during World War II a chlorinated synthetic rubber was marketed under the trade name Parlon X. Although it was similar in many respects to Parlon, it did not have as good compatibility with other film-forming materials.

The crude rubber used in the manufacture of chlorinated rubber is milled on rubber rolls to break down its high molecular weight, and then dissolved in carbon tetrachloride. Chlorine gas is passed through the rubber solution in glass-lined kettles equipped with condensers. The carbon tetrachloride is maintained in the kettle by reflux, and excess chlorine and liberated hydrochloric acid are removed. The hydrochloric acid is absorbed in tantulum absorbers. After chlorination is completed the solution flows into storage tanks lined with acid-

* Reg. U. S. Patent Office.

resistant brick. The solution is then pumped into tanks containing hot water, and the chlorinated rubber is obtained by precipitation. The precipitate is washed thoroughly, then dried; it appears on the market as a granular white powder with a specific gravity of 1.64, or 13.65 lb per gal. It is available in four viscosity ranges as described later.

Fig. 3. Chlorinated Rubber

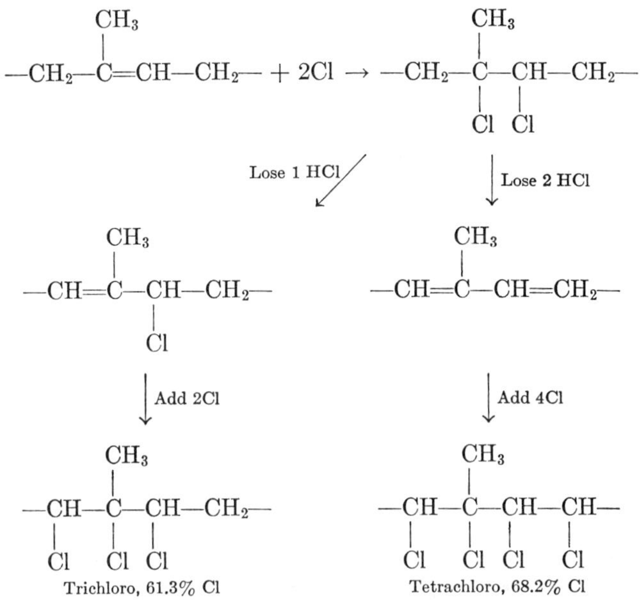

It appears that both addition and substitution reactions take place during chlorination. It is important that a high chlorine content be obtained for maximum stability and resistance to acids and alkalies, and for non-flammability. The addition and substitution reactions are indicated in Fig. 3.

Since the commercial chlorinated rubber has an average chlorine content of 67%, it is believed to be a mixture of the trichloro and tetrachloro polymeric compounds. The manufacturing process must be controlled very carefully, and all traces of free hydrochloric acid must be removed to ensure stability. Chlorinated compounds have a tendency to decompose by splitting off hydrochloric acid, and the decomposition is auto-catalytic. Since free hydrochloric acid accelerates the decomposition it is desirable to add a stabilizer which will react with the acid and neutralize its effect. A similar situation exists with the chlorinated paraffins, and the vinyl chloride resins discussed in Chapter 12. Effective stabilizers are lead carbonate and several organic compounds of lead. Epichlorohydrin is recommended for clear solutions which are stored in tin containers. A further discussion of these stabilizers is given in Chapter 12.

The decomposition is accelerated by contact with iron or tin surfaces, and also by ultraviolet radiation in sunshine, and by heat. Chlorinated rubber is used chiefly in air-drying coatings, but it is also satisfactory in coatings which are baked no higher than 250°F. It shows a slight decomposition at 125°C (257°F), and considerable at 135°C (275°F). Clear films are apt to have considerable discoloration and embrittlement when exposed to strong sunlight, but pigmented films are affected only slightly by sunlight.

Chlorinated rubber films have excellent resistance to both strong and weak acid and alkali and to water and salt solutions. They are also resistant to mineral oils, but are softened by animal and vegetable oils. The films are readily affected by solvents for chlorinated rubber because chlorinated rubber is a thermoplastic material. In this respect it is different from the thermosetting and oxidizing film-forming materials discussed in previous chapters, and is similar to the cellulosic polymers described in Chapter 11. Chlorinated rubber films dry by evaporation of the solvent; there are no cross-linkages formed by oxidation or polymerization. Consequently, the films remain soluble in solvents, but, of course, they are unaffected by non-solvents. In general, the aliphatic hydrocarbons and the lower alcohols are essentially non-solvents for chlorinated rubber; therefore the films are resistant to these liquids.

We shall find that the high-polymer, thermoplastic materials have excellent cohesive strength and form very strong films. A rough estimation of the relative strength or toughness of these materials may be obtained from the values for tensile strength and per cent elongation measurements made on free films. For any particular material the values vary somewhat with conditions of measurement, thickness of film, etc. Average values for the ½-sec type of nitrocellulose, and the 20- and 1000-cp types of chlorinated rubber, are as follows:

	Tensile Strength, psi	% Elongation
Nitrocellulose, ½ sec	8000	3.5
Chlorinated rubber, 20 cp	4300	3.6
Chlorinated rubber, 1000 cp	4900	3.3

The values show that the nitrocellulose produces stronger films than the chlorinated rubber. This is due to the two factors discussed in Chapter 1, namely polarity and molecular weight. The secondary valence forces are considerably stronger between the highly polar molecules of nitrocellulose. However, the polar compounds are much less resistant to water, acid, and alkali. The cohesive strength of all high-polymeric materials is so great that they tend to pull away from the surface on which they are applied. This is generally referred to as poor adhesion. It may be overcome by blending with other film-forming materials; therefore the degree of compatibility of high polymers with other materials is one of their important features.

The low values given above for per cent elongation of free films show the need for plasticizers. Since these materials soften the films and always reduce the tensile strength or cohesiveness, they improve the adhesion. However, it is generally desirable to add also certain resinous materials to improve the gloss and film-building properties. These have lower molecular weight than the high polymers; therefore they also improve the adhesion. It will be realized that the compatibility of the high-polymeric materials with oils, resins, varnishes, etc., is of considerable importance.

LIFTING

Practically all high-polymeric coating materials are non-convertible and form films by evaporation of the solvent. When more than one coat is applied, the subsequent coats soften or dissolve the undercoats and the entire film dries to a normal, smooth finish. However, if oxidizing oils or resins are used with the non-convertible polymers they may become sufficiently insoluble in the drying period between coats to prevent the formation of smooth films on redrying. This phenomenon is known as "lifting." It does not occur when oxidizing oils or resins are used as the only film-formers because, in these cases, the solvents are relatively weak and they do not soften the dried film appreciably. However, strong solvents must be used with high-polymeric materials, and they attack and swell any partially converted film-formers in the dried film. These converted materials are sufficiently cross-linked to prevent true solution even in the strong solvents, but they become considerably swollen and they do not dry down to smooth films when the solvent evaporates. Of course, if one coat only is applied there is no difficulty from lifting with subsequent coats, and oxidizing materials may be used to advantage.

PARLON

Types. The four types of Parlon which are available all have the same fundamental composition, but they vary in degree of polymerization. Since this characteristic is measured readily by viscosity they are classified by the viscosity obtained from 20% solutions in toluene. The viscosity is measured at 25°C with a capillary-type viscometer which is calibrated to read directly in centipoises. The classifying viscosities for the five types (one no longer available), and the viscosity range for each commercial type, are shown in Table 2.

Table 2 Types of Parlon * Available

Designation	Viscosity Range, cp
5	4–7
10	8–13
20	15–23
125	115–180
1000 †	800–2000

* Trade name, Hercules Powder Co.
† Discontinued.

The various viscosity types may be blended to obtain any specific viscosity which is desired. The viscosity which will be obtained from a blend of two solutions of known viscosity may be read directly from a logarithmic chart available from Hercules Powder Co. Obviously, the 5-cp type will give higher solids than the 125-cp type at equal viscosities in the same solvent. However, the high-viscosity types are not soluble in some of the solvents which may be used for the low-viscosity types, and the high-viscosity types are also less compatible with other film-forming materials.

Solvents. Parlon is soluble in most organic solvents; the principal exceptions are aliphatic hydrocarbons and alcohols. The solvent power of the various solvents is indicated by the viscosity obtained from their solutions. Some typical values from 20% solutions of the 20-cp type are given in Table 3.

Solutions in strong solvents will tolerate considerable mineral spirits before actual precipitation occurs, but an increase in viscosity is obtained when the lower-cost volatile material is used as a diluent.

Compatibility. As mentioned earlier, it is necessary to use plasticizers with Parlon to obtain the desired degree of flexibility. Parlon is compatible with a wide range of chemical, oil, and resinous plasti-

Table 3 Parlon, 20% in Various Solvents

Solvent	Viscosity, cp, at 25°C
Toluene	26
Ethyl acetate	32
Xylene	32.5
Hi-Flash naphtha	47
Butyl acetate	53
30% Cellosolve, 70% turpentine	138
Carbon tetrachloride	586

cizers. Since Parlon is used specifically for its resistance to acid, alkali, water, and mineral oil, the plasticizer should be selected which will not detract appreciably from this resistance. Also, when nonflammability is important, the chlorinated plasticizers such as Clorafin, Chlorowaxes, and the Aroclors should be used.

For improved film-building properties, gloss, and application characteristics it is necessary to blend Parlon with alkyd resins, varnishes, etc. It has good compatibility with long-oil, low-viscosity alkyds such as Rezyl 807 and Rezyl 869, described on page 295. Combinations with alkyds have good application characteristics and may be applied without difficulty by brushing. The combinations also dry considerably faster than the straight alkyds and have better chemical and water resistance. However, alkyds are relatively poor in alkali resistance; therefore for maximum chemical resistance, combinations with Clorafin, Chlorowaxes, or the Aroclors are preferred. The use of Parlon with alkyds may be considered from two viewpoints: (1) to shorten the drying time and improve the chemical resistance of the alkyds; (2) to improve the adhesion and application characteristics of Parlon. The limitations of lifting with oxidizing materials, described earlier, should be kept in mind. A slight improvement in the drying or film properties of alkyds is obtained from additions of 10–20% of Parlon, but very good results are obtained from 35–50%. A typical combination which has good chemical resistance and freedom from lifting is as follows: 60% Parlon, 20% Rezyl 869, 20% Aroclor 1254.

The 10- to 20-cp types of Parlon are compatible with medium- to short-oil varnishes. The varnish-Parlon combinations are not as fast drying or as durable as the alkyd-Parlon combinations. In general, from 10% to 20% of Parlon is used on a solid basis with varnishes. The modified phenolic resin varnishes, with either tung, dehydrated castor, or linseed oil, are satisfactory. Parlon improves the resistance to chemicals and mineral oils, but these combinations should not be used as clear films for exterior exposure because of the sensitivity of Parlon to ultraviolet breakdown.

General Uses. A few typical combinations of Parlon with resins and plasticizers are given in Table 4 to indicate the types and ratios for general use. Details regarding solvents and pigmentation are given in Volume II.

Table 4 Parlon-Resin-Plasticizer Combinations

	Per Cent by Weight, Solid Basis					
	a	*b*	*c*	*d*	*e*	*f*
Parlon, 20 cp	50	60	50	55	52	52
Rezyl resin 869	50	20	25	..	..	..
Aroclor 1254	..	20	25	25	26	..
Aroclor 1260	..	..	..	20	..	..
Aroclor 5460	..	..	..	..	22	..
Clorafin 42	..	..	..	..	..	26
Chlorowax 70	..	..	..	..	..	22

Combination *a* has good brushing characteristics, is fast drying, and has good resistance to mineral oils and greases and cleaning solutions. It is suitable for finishes for garage floors, pumps, and other metal work, or for marine enamels, and it may be used as is or modified with linseed oil for inclusion in traffic paints. It may show lifting under some drying and recoating schedules.

Combinations *b* and *c* have practically no tendency to lift, and they have better alkali resistance than *a*. Combinations *d*, *e*, and *f* have maximum chemical resistance and are designed for finishes for interior applications. These combinations also have maximum fire retardance because of their high content of chlorine.

For fire-retardant coatings on awnings, canvas duck, and similar textiles a higher ratio of plasticizer is required, together with about 0.5% of paraffin wax for increased water repellency. It is also desirable to use a higher-viscosity Parlon, such as the 125 cp, for increased film toughness. Ratios of 2–4 parts of plasticizer to 1 part of Parlon, 125 cp, give good results. The plasticizers include Aroclor 1254, Chlorowax 40, Clorafin 42, and tricresyl phosphate. Other fire-retardant compounds are shown in Formulas 2 and 3. Fire-retardant paints will be discussed in Volume II.

Combinations of Parlon with hard resins and plasticizers are used as moisture-resistant coatings for paper and paperboard. These coatings may be heat-sealed, and are non-blocking at the lower plasticizer content. Typical ratios, resins, and plasticizers are:

	Parts by Weight, Solid Basis
Parlon, 125 cp	55–60
Teglac resin Z152	15–20
Diamyl phthalate	15–25
Wax	4–5

Pigmented Parlon finishes have good salt-spray resistance, and a combination such as *b* in Table 4 is satisfactory for marine work. Parlon, in combination with rosin or coumerone-indene resin, is also specified in antifouling ship-bottom paints. A typical combination is:

	Per Cent by Weight
Parlon, 20 cp	1.4
Rosin	21.1
Hercolyn	10.6
Cuprous oxide	42.2
Diatomaceous silica	7.1
Coal-tar naphtha	17.6

CYCLIZED RUBBER RESINS

A typical example of cyclized rubber is Natural Rubber Pliolite, or Pliolite NR, of the Goodyear Tire & Rubber Co. This type of material is made by treating a benzene solution of natural rubber with a metallic halide such as stannic chloride or titanium tetrachloride. A rubber isomer is formed which is considered to be a cyclized product. However, the structure has not been established definitely, and the suggested one contains the highly improbable cyclobutane ring. The cyclized rubber contains less unsaturation than the natural rubber and gives solutions of lower viscosity at equal concentration. The material is harder than rubber and is available as a white powder. It is soluble in aliphatic and aromatic hydrocarbons and has limited compatibility with other resins, oils, and plasticizers. It is widely used in combination with paraffin wax as a hot-melt coating for paper. Commercial mixtures of Pliolite NR and wax are available and are known as Pliowax. Pliolite NR may also be applied from solution in toluene, xylene, VM & P naphtha, etc. A typical combination is the following:

	Parts by Weight
Pliolite NR, P-1230	85
Paraffin wax, mp 134°F	15
Toluene	400

This combination may be modified somewhat with coumarone-indene resins and rosin esters. The coatings may be heat-sealed at 275–300°F; they have excellent resistance to water vapor. They can be formulated to be non-blocking at temperatures up to 130°F. Although the cyclized Pliolite NR has given good results for paper coatings it is not generally used in chemical-resistant paints. Pliolite S-5, described in the next section, is preferred for this purpose.

HIGH STYRENE-BUTADIENE RUBBER RESINS

The high styrene-butadiene resins are copolymers which usually contain 85% or more of styrene. Pliolite S-5 of the Goodyear Tire & Rubber Co. is a typical example. These resins have a structure similar to that shown for GR-S rubber in Fig. 1, except that the ratio of styrene is very much higher. They may be made by the emulsion polymerization process described in Formula 1, and the product may be obtained as the solid resin, or it may be used as the latex. A careful selection of emulsifying agents and stabilizers is necessary when the latex is to serve as the vehicle for pigmented coatings.

Pliolite S-5 is available as light amber granules, with a specific gravity of 1.03, or 8.58 lb per gal. It produces clear films, free from taste or odor, and non-toxic. These films are resistant to acids, alkalies, alcohol, water, and mineral and vegetable oils. The tensile strength has been reported as 1450 psi with zero per cent elongation; this is lower than the values for chlorinated rubber shown on page 360. The resin is soluble in aromatic hydrocarbons, ketones, and esters. A mixture of aromatic and aliphatic hydrocarbons having the same boiling range, and adjusted for solvency to a kauri-butanol value of 60, is recommended for brushing enamels. This mixture is lean enough so that it will minimize the tendency of a second-coat application to soften the undercoat. The aliphatic component has a lower heat of vaporization than the aromatic; therefore it will evaporate more readily and the solvent will become richer as the film dries.

Pliolite S-5 is not generally compatible with alkyd resins or maleic resins and has only limited compatibility with drying oils. However, it has good compatibility with rosin and many of its modifications. It is also compatible with coumarone-indene resins, the Aroclors, chlorinated paraffins, and many of the chemical plasticizers. It is more stable to heat than the chlorinated rubber resins; it shows no apparent breakdown at 300°F for extended periods. However, it will support combustion, since it is a hydrocarbon; therefore it does not have the fire-retardant properties of the chlorinated materials. There is a considerable difference in the weight per gallon of these resins; the chlorinated rubber resin weighs 13.65 lb, and the Pliolite S-5 is 8.58 lb per gal.

General Uses. The Pliolite S-5 type of rubber resins is used in fast air-drying finishes which are required to have good water and chemical resistance and good abrasion resistance. A series of color bases is available containing various pigments dispersed in the solid resin. These are described in Volume II, together with details of the pigmented paints. The general range of the binder composition may be seen from the data in Table 5.

Table 5. Typical Binder Composition

Per Cent by Weight, Solid Basis

Binders	Wall Sealer	Flat Paint	Stucco Traffic	Concrete Enamel	Aluminum Paint	Metal Primers	
Pliolite S-5	75	40	50	87	50	70	77
Aroclor 1254	..	20	..	...	..	..	..
Aroclor 5460	25	40	..	...	50	..	..
Clorafin 42	..	..	25	8.7	..	..	..
Chlorowax 70	..	..	25	...	..	..	..
Tung oil	..	..	..	4.3	..	..	..
Varnish base *	..	..	..	...	..	30	23

* 21 gal tung-oil-Bakelite BR-254 resin varnish in xylene.

CHLORINATED BIPHENYLS

A series of chlorinated biphenyls and polyphenyls is available from the Monsanto Chemical Co. under the trade name of Aroclors. They range from mobile liquids to brittle solids and are used as plasticizers and resins. The consistency increases with the chlorine content, as indicated in Table 6 for the pure chlorinated biphenyls. A partial list of the available Aroclors is given in Table 7.

Benzene may be polymerized in the vapor phase by passing it over a catalyst at about 800°C. A mixture of biphenyl, biphenyl benzene, and polyphenyls is formed which may be separated by distillation. For many commercial purposes it is not necessary to separate so closely as to obtain the individual compounds so long as uniform physical properties are obtained. The biphenyl and related compounds are chlorinated by passing chlorine through the molten materials. The chlorine atoms replace hydrogen atoms on the benzene

Table 6 * Pure Chlorinated Biphenyls

	% Chlorine	Condition
Monochlorobiphenyl	18.8	Mobile liquid
Dichlorobiphenyl	36.8	Mobile liquid
Trichlorobiphenyl	41.4	Mobile liquid
Tetrachlorobiphenyl	48.6	Viscous liquid
Pentachlorobiphenyl	54.4	Viscous liquid
Hexachlorobiphenyl	59.0	Soft resin
Heptachlorobiphenyl	62.8	Soft resin
Octachlorobiphenyl	65.9	Semi-crystalline
Nonachlorobiphenyl	68.8	Crystalline
Decachlorobiphenyl	71.1	Crystalline

* *Protective and Decorative Coatings,* J. J. Mattiello, Ed., Volume I, page 486.

Table 7 * A Partial List of the Aroclors

Properties	Aroclor						
	1242	1248	1254	1260	1270	4465	5460
Form	Colorless mobile liquid	Yellow mobile liquid	Yellow viscous liquid	Yellow sticky resin	White cryst. powder	Yellow brittle resin	Yellow brittle resin
Color, APHA (max.)	100	100	150	150	1.5 †	2.0 †	2.0 †
Acid number	0.015	0.015	0.015	0.015	0.175	0.05	0.07
Pounds per gallon at 25°C	11.50	12.08	12.83	13.50	16.24	14.28	14.50
Distillation range, °C	325–360	340–375	365–390	385–420	450–460	230–320 ‡	280–335 §
Pour point, °C	−19	−7	10	31	....	....	....
Softening point, °C	....	....	....	....	249–300	60–66	100–105.5
Refractive index (av.)	1.628	1.630	1.640	1.648	....	1.665	1.663

* Courtesy of the Monsanto Chemical Co.
† Color, NPA.
‡ At 4 mm Hg.
§ At 5 mm Hg.

rings, and hydrochloric acid is obtained as a by-product. The operating conditions determine the extent of chlorination.

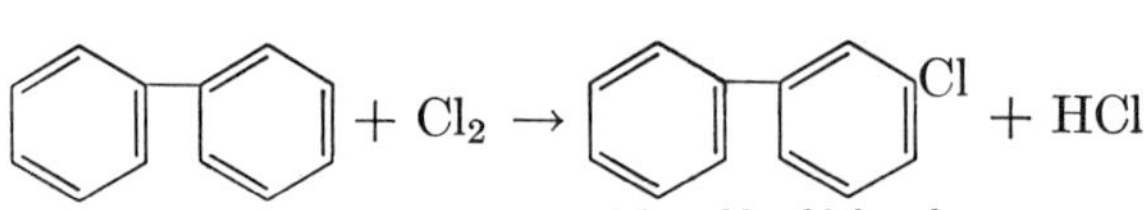

Monochlorobiphenyl

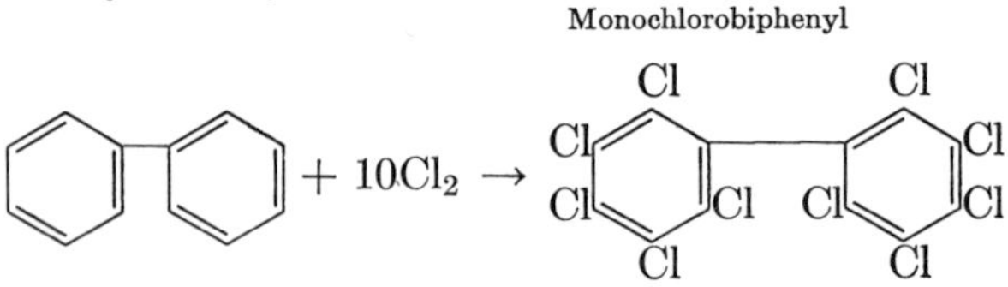

Decachlorobiphenyl

The Aroclors are soluble in most organic solvents, partially soluble in the lower alcohols, and insoluble in glycerol, glycols, and water. They are compatible with drying oils and some varnish resins, the cellulosic polymers (except the acetate), chlorinated rubber, and the rubber resins and vinyl resins. They are very resistant to oxidation, are heat stable up to 600°F, and are permanently thermoplastic. They have excellent chemical resistance, being practically unaffected by sodium hydroxide solutions, mineral acids, and water. They are non-flammable, in that they do not support combustion, and they decrease the flammability of mixtures in which they are used. They have low vapor pressures, and show practically no evaporation loss at 100°C and only very slight loss at 163°C for 6 hr. The liquid members are used as plasticizers, and the solid Aroclors as resins in a variety of coatings. Their use with rubber resins is indicated in this chapter and with other film-formers in other sections of this book. Further information is available from the Monsanto Chemical Co.

CHLORINATED PARAFFINS

The paraffinic hydrocarbons may be chlorinated by the catalytic vapor-phase process, or by passing chlorine through a molten bath of the solid members of the series. The chlorinated paraffins are available commercially in two degrees of chlorination, containing approximately 40% and 70% chlorine respectively. At 40% chlorination the chlorinated paraffins are viscous liquids, and at 70% they are brittle solids. They are similar in most respects to the chlorinated biphenyls discussed in the previous section, but they are not as stable to heat or ultraviolet radiation. At elevated temperatures dehydrochlorination takes place with the formation of unsaturated compounds and the liberation of hydrogen chloride. The polyvinyl chloride resins are also susceptible to this type of degradation. The liberated hydrogen chloride catalyzes the reaction; therefore it must be removed by the acid-acceptor type of stabilizer. The types of stabilizers which are available are discussed on page 495. In general the 70% chlorinated paraffins are more stable than the 40% type, but stabilizers should be incorporated if these materials are used at temperatures of 275°F or higher.

The chlorinated paraffins are soluble in most organic solvents, but are insoluble in glycerol, the glycols, lower alcohols, and water. They have good compatibility with the standard film-forming materials; the liquid type may be used as a plasticizer, and the solid type as a hard resin. They do not support combustion, and consequently they reduce the flammability of compositions in which they are used. They are non-oxidizing and remain permanently thermoplastic. The vapor pressure of the 40% type is comparable to that of tricresyl phosphate, and considerably lower than that of dibutyl phthalate. The chlorinated paraffins have excellent resistance to acids, alkalies, and alcohols and impart these characteristics to formulations in which they are used. They do not detract from exterior durability in pigmented compositions.

Two commercial types of chlorinated paraffins are referred to earlier in this chapter: the Clorafins of Hercules Powder Co. and the Chlorowaxes of Diamond Alkali Co. The general characteristics of these materials are shown in Tables 8 and 9.

Table 8 * Physical Properties of Clorafins 42 and 70

	42	70
Form	Amber liquid	Yellow solid
% chlorine	42–43	69–72
Viscosity, poises, at 25°C	22.5–31.5	....
Softening point,† °C	....	90–100
Color, Gardner 1933	12 max.	....
Pounds per gallon at 25°	9.72–9.82	13.33–13.75
Stability, max. % HCl liberated, 4 hr at 175°C	0.80	Approx. 0.27

Non-flammable—cannot be ignited by direct contact with gas flame.

* Courtesy of Hercules Powder Co.
† Ring and ball method.

Table 9 * Physical Properties of Chlorowaxes 40 and 70

	40	70
Form	Amber liquid	Cream powder
% chlorine	40–42	68–73
Average formula	$C_{24}H_{43}Cl_7$	$C_{24}H_{29}Cl_{21}$
Average molecular weight	570	1060
Viscosity, poises, at 25°C	20–28	....
Softening point,† °C	....	95–110
Pounds per gallon at 25°C	9.6	13.75
Stability, max. % HCl liberated, 4 hr at 175°C	0.50	0.50

* Courtesy of Diamond Alkali Co.
† Ring and ball method.

The use of chlorinated paraffins with the various rubber resins is described earlier in this chapter, and their use with the cellulosic and vinyl polymers will be discussed in later chapters. A solution of the high-chlorine-content chlorinated paraffin will serve as a fire-retardant coating on wood. This type of coating may be used on wood which has been impregnated with water-soluble fire-retardant salts. It reduces the leaching of the salts and presents a satisfactory surface for further coats of paint. Since its pigmented coatings have good exterior durability it may also be used as a fire-retardant material in shingle stains. A suggested formulation is as follows:

Formula 2 * Fire-Retardant Shingle Stain

	Pounds	Gallons
Red iron oxide	140	3.26
Antimony trioxide	100	2.09
Barium sulfate	100	2.79
Asbestine	80	3.36
Aluminum stearate	4	0.40
Chlorowax 40	32	3.33
Chlorowax 70 (60% in MS)	217	22.77
Mineral spirits	403	62.00
	1076	100.00

* Courtesy of Diamond Alkali Co.

A variety of colored stains may be produced by proper pigment variation, and a fungicide may be added if desired. A wide variety of formulations is used as fire-retardant coatings on heavy textiles for awnings, tents, tarpaulins, etc. A combination typical of that used by the Quartermaster Corps during World War II is given in Formula 3.

Formula 3 * Fire-Retardant Textile Compound

	Pounds	Gallons
Chromium oxide	138	3.28
Synthetic yellow iron oxide	102	3.06
Red iron oxide (85%)	71	1.93
Phthalocyanine blue	8	0.58
Antimony trioxide	150	3.25
Calcium carbonate, ppt	300	13.40
Urea-formaldehyde resin (60%)	90	10.40
Chlorowax 40	300	30.10
Chlorowax 70 (70% in xylol)	193	18.00
Copper naphthenate (8%)	135	16.00
	1487	100.00

* Courtesy of Diamond Alkali Co.

This combination is reduced with a high-solvency hydrocarbon solvent for application, and it should be adjusted to provide for a pick-up on the textile of about 40–45%. The urea-formaldehyde resin modifier may be replaced with phenol formaldehyde resin, ethyl cellulose, chlorinated rubber, and other suitable film-formers. The color components were chosen to give the Army's olive drab color, but these may be replaced with others if desired.

SYNTHETIC RUBBER RESIN LATEXES

During the past few years stabilized latexes of a variety of polymeric film-formers have been made available to the paint industry. These latexes are the result of the emulsion polymerization of the polymeric materials. Like all colloidal systems, the latexes must be used with care to prevent breaking the emulsion or coagulating the dispersed phase. However, there need be no difficulty in this respect if the proper precautions are followed. The use of these latexes in interior and exterior architectural paints for plaster and concrete surfaces is expanding rapidly because they have certain inherent advantages over oil- and resin-base paints for these applications. Since the continuous phase in the latexes and subsequent paints is water it will be realized that brush application is very easy. In fact, water-dispersible thickeners, such as casein, methyl cellulose, and alginates, are incorporated in the paints to increase the consistency and stability. Another advantage of latex paints is their very fast drying property. When part of the water has evaporated the emulsion breaks, and the remainder of the water evaporates very rapidly, leaving the dry film of paint. No oxidation or further polymerization is necessary. Latex interior flat paints will dry in 20–30 min and some of them may be recoated the same day. Freedom from odor of solvent or the oxidation of oils is another advantage. This is particularly important in hotels, hospitals, etc., where it is desirable to occupy a room as soon as possible after redecorating. The absence of organic solvents also reduces the fire hazard connected with the use of oil paints. The painter can clean his brushes easily with soap and water and also remove spots of spilled paint in this manner.

The polymeric materials in the latexes have high molecular weight; therefore they are non-penetrating into porous surfaces. They are relatively hard and have high cohesive strength; therefore they must be properly plasticized to assure good adhesion, both originally and after the film has aged. They may be plasticized in three ways: (1) internally plasticized, whereby the plasticizer is a part of the copolymeric molecule; (2) preplasticized, whereby the plasticizer is added during the polymerization and is incorporated in the latex; (3) post-plasticized, whereby the plasticizer is emulsified separately and added to the latex. Examples of each of the three types are available commercially.

From the discussion of synthetic rubbers in the first part of this chapter, of the polyvinyl resins in Chapter 12, and of the acrylic ester resins in Chapter 13, it will be apparent that a wide range of copolymeric materials can be formulated for these latexes. Some of them would be eliminated on an economic basis, because low cost is important for materials used in architectural paints. The large-volume production and price advantage of styrene account for its wide use in these latexes. Since polystyrene is a hard material it requires plasticizing by one of the three methods cited above. It may be plasticized internally by copolymerizing with a reactive monomer such as butadiene. However, it would require the reverse ratio of that shown in Fig. 1 for GR-S rubber; for latex paints the styrene content may be as high as 85% with the butadiene about 15%. The use of butadiene introduces a small amount of unsaturation in the polymer which is subject to oxidation on aging with possible hardening and discoloration in the film. The unsaturation may be avoided by adding a nonoxidizing plasticizer to the polystyrene latex as indicated in methods 2 and 3 above.

Types Available. A series of butadiene-acrylonitrile copolymer latexes is available in which the ratio of butadiene to acrylonitrile is varied. This provides a range in relative hardness or flexibility of the films. Some types are suitable for use in latex paints, and others as

Table 10 Types of Commercial Synthetic Resin Latexes

1. Plasticized polystyrene:
   - BKS-92, BKS-114, and BKS-177; Bakelite Co.
   - Koppers polystyrene emulsion P; Koppers Co.
2. Unplasticized polystyrene:
   - BKS-90, BKSQ-227; Bakelite Co.
   - Dow latexes 579 and 580; Dow Chemical Co.
   - Lustrex X601; Monsanto Chemical Co.
   - Wallpol 9200; Reichhold Chemicals, Inc.
   - Koppers polystyrene emulsions F, M, and R; Koppers Co.
3. Styrene-butadiene copolymers:
   - Dow latex 762-W; Dow Chemical Co.
   - Chemigum 101 and 160; Goodyear Tire and Rubber Co.
   - Kralac 2724; Naugatuck Chemical.
4. Vinyl chloride-vinylidene chloride copolymers:
   - Dow latex 744-B; Dow Chemical Co.
5. Butadiene-acrylonitrile copolymers:
   - Chemigum 235; Goodyear Tire and Rubber Co.
   - Hycar latexes; B. F. Goodrich Chemical Co.
   - Amerite compounds; American Anode, Inc.
6. Plasticized vinylidene chloride-acrylonitrile copolymers:
   - Saran latexes F122 and A20; Dow Chemical Co.
7. Polyvinyl acetate:
   - Elvacet emulsions; du Pont Co.
   - Wallpol 9100; Reichhold Chemicals, Inc.
8. Polyacrylic copolymers:
   - Rhoplex AC-33; Rohm and Haas Co.

impregnants for paper and for leather finishes which require a high degree of flexibility and toughness. A non-flammable copolymer of vinyl chloride and vinylidene chloride is available from the Dow Chemical Co.; it must be blended with plasticizers or other film-forming materials since it does not form a continuous film on drying. It finds use in paper and textile coatings where non-flammability is required. A partial list of the latexes available at present is given in Table 10. However, this field is expanding so rapidly that many others will no doubt be available when this book appears in print.

General Uses. A detailed description of the use of these latexes in architectural and industrial finishes is given in Volume II. However, Formulas 4, 5, and 6 are presented here to show the general principles

Formula 4 * Latex Paint Made with Dow Latex 762-W

Pigment Dispersion	Gallons per Pound	Pounds	Gallons
Titanium dioxide (rutile)	0.0286	250.00	7.14
Lithopone	0.0279	100.00	2.79
Mica, 325 mesh	0.0426	55.62	2.37
Potassium tripolyphosphate (10% solution)	0.1101	12.25	1.35
Water	0.1200	204.00	24.48
Water	0.1200	61.15	7.34
Protein (15% solution, KOH cut)	0.1150	72.70	8.36
Dowicide (50% solution)	0.0952	11.12	1.05
Antifoamer (as required)		1.82	0.22 approx.
Dow latex 762-W (48% solids)	0.1185	379.20	44.90
Synthetic thickener (as required)			
NH_4OH or KOH to *p*H of 9.5			
		1147.86	100.00

Stormer viscosity, 70–80 KU
Per cent solids, 52.5
Pigment volume concentration, 35%

Procedure

The equipment needed is a conventional change can mixer and a suitable means to grind pigment, such as high-speed stone mill, roller mill, or pebble mill.

Pigments are ground together with the phosphate dispersing agent and part of the protein solution. The amount of protein solution and water may vary with the chosen method of milling. The ground pigments and additional water are blended with the protein, Dowicide, antifoamer, and latex in the change can mixer.

Viscosity and *p*H are adjusted as indicated if necessary.

* Courtesy of Dow Chemical Co.

Formula 5 * Latex Paint Made with Alkyd-Modified Dow Latex 762-W

	Gallons per Pound	Pounds	Gallons
Pigment Dispersion			
Titanium dioxide (rutile)	0.0286	250.00	7.14
Lithopone	0.0279	100.00	2.79
Mica, 325 mesh	0.0426	42.90	1.84
Potassium tripolyphosphate (10% solution)	0.1101	11.82	1.30
Water	0.1200	196.60	23.60
Alkyd Emulsion			
Long-oil soya alkyd	0.1200	34.43	4.13
Cobalt naphthenate (6%)	0.1264	1.08	0.14
Lead naphthenate (24%)	0.0996	2.78	0.27
NH_4OH (28%)	0.1350	1.63	0.22
Non-ionic wetting agent	0.0953	3.43	0.33
Dowicide (50% solution)	0.0952	10.94	1.43
Protein (15% solution, KOH cut)	0.1152	83.10	9.56
Water	0.1200	108.60	13.04
Antifoamer (as required)		1.73	0.21 approx.
Dow latex 762-W (48% solids)	0.1185	287.00	34.00
Synthetic thickener (as required)			
		1136.04	100.00

Stormer viscosity, 70–80 KU
*p*H, 9.5
Per cent solids, 52.0
Pigment volume concentration, 35%
Alkyd/latex solids ratio by weight, 20/80

Procedure

The equipment needed is a conventional change can mixer and a suitable means to grind pigment, such as high-speed stone mill, roller mill, or pebble mill.

Pigments are ground together with the phosphate dispersing agent and part of the protein solution. The amount of protein solution and water may vary slightly with the chosen method of grinding.

The alkyd emulsion can be prepared in a change can mixer. The ingredients are added with stirring in the order listed, allowing enough time between additions to obtain thorough mixing.

The ground pigments and additional water are blended with the alkyd emulsion followed by addition of the antifoamer and latex.

Viscosity is adjusted as indicated if necessary. The NH_4OH in the alkyd emulsion is adjusted to obtain the *p*H indicated.

* Courtesy of Dow Chemical Co.

involved in their formulation. It will be understood that many variations of the formulations and procedures are possible.

The wetting agents facilitate the wetting of the pigments, and the sequestering agents act as water softeners. The latter reduce the activity of soluble salts which may insolubilize the casein or react with the emulsifying agents. The complex polyphosphates are suitable sequestering agents, and they include potassium tripolyphosphate and tetrasodium pyrophosphate. In some cases methyl cellulose may be used instead of casein or soya protein as a protective colloid or stabilizer. See Chapter 13 for a brief discussion of the stability of emulsions, and for a formulation based on the acrylic resin latex, Rhoplex AC-33.

Formulas 6(*a*) and 6(*b*) * Semi-Gloss White Finishes

	6(*a*) Pounds	6(*b*) Pounds
Titanium dioxide (rutile)	250	250
Mica, water ground, 200/325 mesh	60	60
Water (soft)	260	260
Lustrex 820	30	20
Ammonium hydroxide (28% NH_3)	9	6
Foam suppressor †	10	10
Santicizer 160	83	..
Aroclor 1248	...	116
Water	56	30
Ethylene glycol	...	40
Thickener ‡	6.5	9
Lustrex 601-40	234	220
Yield, pounds	998.5	1021
Yield, gallons	91.03	89.69

Procedure

1. In a mixing vessel charge 260 lb of water, the pigments, and the foam suppressor and mix to wet out.
2. Add the ammonium hydroxide and continue mixing for a few minutes until viscosity breaks slightly.
3. Add the plasticizer (Santicizer 160 or Aroclor 1248 in the above examples) and mix in.
4. Add the balance of the water, the ethylene glycol (if indicated), the thickener, and finally the Lustrex 601-40 in that order, mixing in each before adding the next one.
5. Run the mix through a colloid mill or high-speed stone mill at the proper setting to yield the desired rate of through-put combined with good dispersion.

* Courtesy of Monsanto Chemical Co.

† Example used here: Glyceryl Monoricinoleate, S-grade, Glyco Products Co.

‡ Example used here: Acrysol GS New, Rohm and Haas Co.

Typical examples of formulations and method of handling the post-plasticized polystyrene system are given in Formulas 6(*a*) and 6(*b*). These are semi-gloss finishes, varying in the type of plasticizer.

Further details regarding the use of latexes in paints are given in Volume II. The latexes may be used as a sealer coat for paper and paperboard or other porous materials. Soft paper sheet which has been lightly beaten may be impregnated by immersion in the latex with subsequent withdrawal between squeeze rolls and drying. The latex pick-up may be adjusted to any desired amount. At about 100% pick-up the impregnated paper develops high tear and bursting strength, but retains its softness and flexibility. In this condition it makes an excellent base for coating with nitrocellulose lacquer or polyvinyl finishes. The latex may also be incorporated in the paper before sheet formation by special processes; see bulletin on Hycar latexes by the B. F. Goodrich Co.

The flexible type of latex also serves as an impregnant for low-cost textile sheeting materials to produce bases for artificial leather and similar compositions. It may be used on leather as a finishing material. Several applications are made by swabbing, and the finish is usually embossed under heat and pressure. The first application may be the straight latex, but subsequent applications usually contain casein, shellac, and other resins. The finish application usually contains a wax such as carnauba or candelilla.

Trace the development of the human body from nebulous elements, through protozoa to its present capacity for performance, and you will pay homage to its Creator, and appreciate more fully the marvelous mechanism provided for your experience on earth.

Chapter 10. PLASTICIZERS

INTRODUCTION

The term "plasticizer" is applied to a variety of materials which are added to high-polymeric film-formers to increase the flexibility of their films. There are three general types of materials used for plasticizers:

1. Vegetable oils (non-drying type).
2. Monomeric chemicals having high boiling points.
3. Polymeric resinous materials.

The principal non-drying oils which are used as plasticizers are castor, rapeseed, and cottonseed oils. These are used both as raw oils and as blown, or otherwise processed oils. The so-called "lacquer oils" are linseed and soybean oils, which have been processed to make them essentially non-drying and soluble in alcohol, so that they can be used as plasticizers for polymers such as nitrocellulose. Typical examples are the ADM lacquer oils of the Archer-Daniels-Midland Co. In general, mineral oils are not satisfactory plasticizers, but they are used in a few low-cost applications.

The monomeric chemical plasticizers include a wide range of materials which have low vapor pressure or high boiling point, and are reasonable in cost and chemically stable. An important group is the alkyl and aryl esters of dibasic acids, such as phthalic and sebacic acids. The esters of phosphoric acid, and ricinoleic and other fatty acids are also widely used as plasticizers. A number of proprietary plasticizers are sold under trade names without disclosure of their composition.

The polymeric resinous plasticizers are non-drying polymers of fairly low molecular weight. They are chiefly alkyd-type resins, since they are based on polybasic·acids esterified with polyhydric alcohols such as glycerol and ethylene glycol. These polyesters may be used as such, or modified with non-drying oils.

A large quantity of plasticizers is used each year in the cellulosic polymers and the vinyl resins. The proportion of plasticizer to polymer varies considerably with the type of polymer and the application for which the coating is intended. Polymers vary a great deal in relative hardness; nitrocellulose is harder than ethyl cellulose, and polyvinyl chloride is harder than polyvinyl acetate. The vinyl chloride-acetate copolymer (see Chapter 12) is a compromise in which the vinyl acetate groups may be considered a "built-in" or "internal" plasticizer. Drying oils may also be considered to have a built-in plasticizer. It will be recalled that the structure of the fatty acids in Chapter 2, Table 3, shows an arrangement of seven methylene groups in each fatty acid radical. This non-reactive, seven-membered chain is, in effect, a built-in plasticizer for drying oils. Of course, the oils used in oleoresinous varnishes and in oil-modified alkyds may be considered internal plasticizers. Similarly, the non-oxidizing alkyd resins which are used with amino resins in baking finishes may be considered as the plasticizers for these coatings. However, the term "plasticizer" is generally reserved for those materials which are added to high-polymeric materials as stated at the beginning of this chapter. Although plasticizers are necessary with high polymers for the desired degree of flexibility, they lower the tensile strength of the films, as discussed in Chapter 1 and illustrated in Tables 3 and 4 of the same chapter. In addition, plasticizers are often more sensitive than the polymers to the action of water, alkalies, acids, solvents, and oils. In addition, plasticizers vary considerably in rate of volatilization; some are essentially non-volatile, such as the resinous type and the vegetable oils, but the monomeric chemical type has a significant volatility, especially at elevated temperatures.

GENERAL USES OF PLASTICIZERS

Since plasticizers are used in a wide variety of finishes it may be well to consider the general proportions of plasticizer in the film-forming components. The three types of finishes shown in Table 1 were chosen to illustrate the range in use of plasticizers. In addition to these three general applications there are many intermediate ones and also many special uses. The furniture and automobile finishes represent the relatively hard type of finish; one is designed for interior and the other for exterior use. The finishes for fabrics and cable insulation represent the very flexible type which may be required to withstand extremes in temperature, to resist soap and hot water or hot lubricating oil, or to be non-toxic for use in food-packaging materials. The two most widely used high polymers, nitrocellulose and vinyl resins, are used for illustration in Table 1. Detailed formulations for specific applications are given in other chapters.

Table 1 General Uses of Plasticizers

Parts by Weight, Solid Basis

	Finishes			
Components	Furniture	Automobile	Fabric	
Nitrocellulose	100	100	100	.....
Vinyl resin	......	......	....	100
Non-oxidizing alkyd	50–100	75–125	....	.....
Hard resin	50–150	......	....	.....
Plasticizer	25–50	30–50	100–200	35–80

The plasticizer in furniture and automobile finishes may be the chemical type entirely, but often it is a 50:50 combination of chemical type and blown castor oil. In these finishes the non-oxidizing alkyd resin provides considerable plasticizing effect, together with good film-building qualities.

In the nitrocellulose and ethyl cellulose lacquers for fabrics, cable lacquers, etc., the plasticizer is usually the resinous type or a processed oil, depending on the service requirements. In the vinyl coatings the chemical type may be used entirely, but combinations of chemical with resinous or oil types give stronger films. Some special resinous plasticizers have been developed for use as the sole plasticizer for vinyl resins.

REQUIREMENTS IN PLASTICIZERS

A comparison of a group of plasticizers for flexibilizing a given high polymer may show that they have about equal efficiency at room temperature, but they may vary widely as the temperature is raised or lowered. Many applications require use at temperatures considerably below zero, so that both the high polymer and the plasticizer must continue to function satisfactorily at the low temperature. Conversely, for use in certain insulation lacquers resistance to heat and hot lubricating oil is required. The correct high polymer may be used for these applications, but if the plasticizer decomposes under heat, or is extracted by the hot oil, or does not function at low temperature, the lacquer film will become hard and brittle under the specific service conditions.

It is fairly obvious that a plasticizer for a particular film-forming material must have good compatibility with the material. However, it is not quite so apparent that the plasticizer should have a solvent action on the film-former. This action promotes better retention of plasticizer in the film, particularly at low temperatures, and reduces exudation, as will be discussed in more detail later. When certain articles fabricated from vinyl resins were placed in contact with furniture finished with nitrocellulose lacquer or oleoresinous varnish, the articles marred the surface of the furniture finish. It was found that certain plasticizers in the vinyl film migrated from the vinyl film into the furniture finish, with a consequent softening and marring of the finish. This phenomenon is characteristic of some plasticizers and is known as plasticizer migration; it is discussed in more detail later in this chapter.

Plasticizers are used in many coatings and sheeting materials which come in contact with human skin. In such cases they should be tested for toxicity, or ability to produce dermatitis. This is a difficult test to make because of the variation in susceptibility of human skin, and it is complicated by the fact that the other components of the coating may affect the relative toxicity of the plasticizer. Therefore the test should be made with the complete coating formulation, rather than with the plasticizer alone. The odor of the plasticizer is also important in many coatings and sheeting materials. Some excellent plasticizing materials cannot be used for such applications because of their objectionable odor. A list of the requirements necessary in plasticizers is given in Table 2.

Table 2 Desirable Characteristics in Plasticizers

1. Compatibility with the film-former.
2. A slight solvent action on the film-former.
3. Low vapor pressure, or volatility from the film.
4. Resistance to degradation by heat.
5. Resistance to ultraviolet radiation.
6. Resistance to extraction by water, hot oil, solvents, etc.
7. Resistance to chemical change by alkalies, acids, foodstuffs, etc.
8. Retention of flexibility at low temperatures.
9. Minimum viscosity change over a wide range of temperatures.
10. Pale color and good color retention with heat and aging.
11. No objectionable odor.
12. Non-toxic and non-irritating to the skin.
13. Flame retardancy or low rate of combustibility.
14. Good electrical properties for insulation lacquers.

It is not likely that all the above properties will be found in any one plasticizer. Therefore it is necessary to select the plasticizer having the most important property or group of properties for a particular application. However, in many cases, more than one plasticizer is used in a coating composition for the best compromise in properties. The types of plasticizers which are available and their characteristics are discussed after a brief outline of the mechanism of plasticization.

MECHANISM OF PLASTICIZATION

The effect of the plasticizer on the tensile strength and per cent elongation of nitrocellulose films is given in Chapter 1, Tables 3 and 4. The data show that these physical properties of films will vary considerably, depending on the type of plasticizer used. Chapter 1 also points out that the molecules in films of high-polymeric, thermoplastic materials are held together by secondary valence forces. These forces are weaker than primary valence forces, or direct chemical bonds, and they may be neutralized or blocked fairly easily. The relatively small plasticizer molecules distributed among the large polymeric molecules effectively neutralize many of the secondary valence forces. This permits greater freedom of movement between the polymeric molecules with a corresponding increase in flexibility, but it will be apparent that the cohesive force between the molecules is also decreased, with a corresponding loss in tensile strength.

This points to the reason for preferring the solvent-type plasticizer in comparison with the non-solvent type. The solvent type necessarily has a stronger affinity for the large polymeric molecules and will be held more firmly to them when the film dries. The non-solvent type will be squeezed out by the strong cohesive forces between the large molecules, and only a relatively small percentage will be retained mechanically. Furthermore, if the temperature of the film is increased, the motion of the molecules will be greater because of increased kinetic energy, and the loosely held non-solvent plasticizer will be more readily exuded from the film.

There is another factor which affects the extent of exudation of plasticizer molecules, and that is their size. Obviously, small molecules would be exuded more readily than larger molecules. However, if the large molecules were the non-solvent plasticizer type, they would dilute the true film-formers to such an extent that the film would be weakened considerably. Conversely, if the relatively large resinous plasticizer molecules were the solvent type, they would impart the required flexibility with a minimum loss in tensile strength.

Plasticizer efficiency is usually expressed in terms of the ratio of strain to stress, i.e., the concentration of plasticizer required to produce a specified elongation of the film when stretched under a given load. Further details regarding the method for measuring these properties are given later.

It appears therefore that plasticizers neutralize some of the secondary valence forces which bind polymeric molecules together. This permits greater freedom of movement of the molecules under stress, which means greater flexibility in the coating, but it poduces some loss of tensile strength. The solvent-type plasticizers are more effective than the non-solvent type. They are also retained in the film more strongly, and they show less exudation on aging. In general, plasticizers with small molecules are more effective softeners than those with large molecules, but they also tend to decrease the strength of the film to a larger extent. Compounds having small molecules usually show a greater vapor pressure than those having large molecules, and, consequently, a greater loss from the film on heating or from aging.

TYPES OF PLASTICIZERS

Plasticizers range from high-boiling-point solvents to non-volatile oils and resinous materials, and their properties are described in detail in the literature of many chemical manufacturers. An outline of the chemical structure and physical properties of the principal types is given in this chapter. Some of the characteristics of the castor oils and alkyl ricinoleates manufactured by the Baker Castor Oil Co. are given in Table 3. A series of resinous plasticizers manufactured by Rohm and Haas Co. is listed in Table 5 with some of their properties, and the characteristics of typical chemical plasticizers are given in Table 6. Reed (1, 2, 3) has described the behavior of an extensive series of plasticizers in the vinyl chloride-acetate resins, and Bogin (4) has discussed considerable data on plasticizers for lacquers. Gearhart *et al.* (12, 13) evaluated a series of plasticizers for cellulose acetate butyrate. The behavior of plasticizers for various rubbers is described (9, 10, 14). The effect of the structure of the plasticizer on the properties of the films is given by Lawrence and McIntyre (5) and the general use of plasticizers by Woodcock (6). Plasticizers from lactic acid are described in a series of articles by Rehberg *et al.* (15, 16), and from tetrachlorophthalic anhydride by Stevenson *et al.* (7) and by Mehta and Payne (8). Alcohols from the new oxo process have made possible new plasticizers (19), and the creep behavior of new diphosphate esters is described by Mesrobian *et al.* (18).

Fig. 1. Structure of Typical Plasticizers

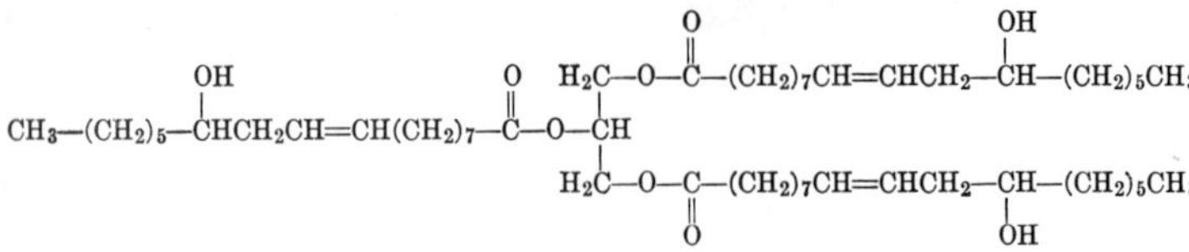

Castor oil

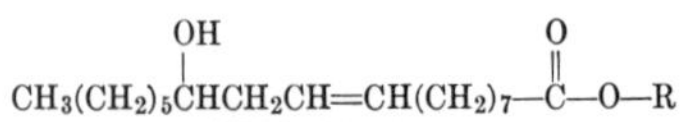

$R = CH_3, C_2H_5, C_4H_9$, etc.

Alkyl ricinoleates

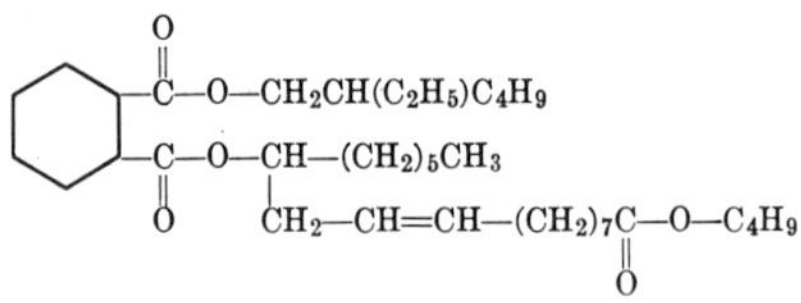

2-Ethyl hexyl phthalyl butyl ricinoleate

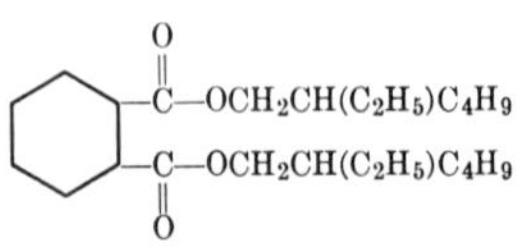

Di-2-ethyl hexyl phthalate

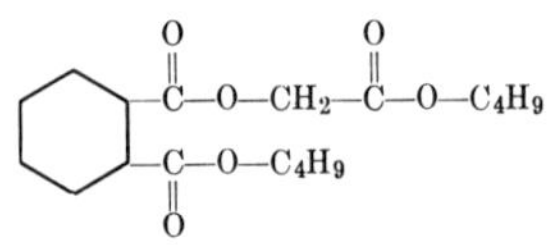

Butyl phthalyl butyl glycollate

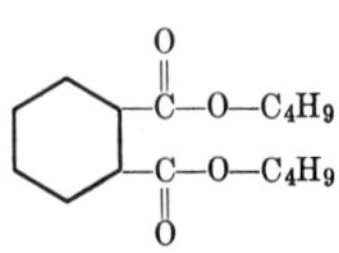

Dibutyl phthalate

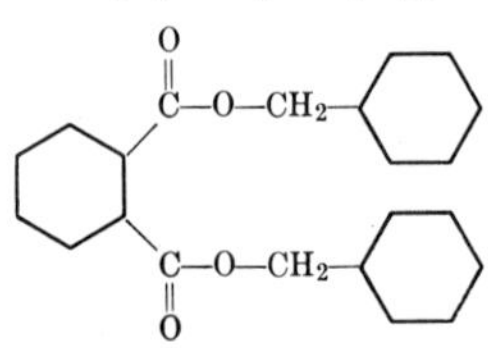

Dibenzyl phthalate

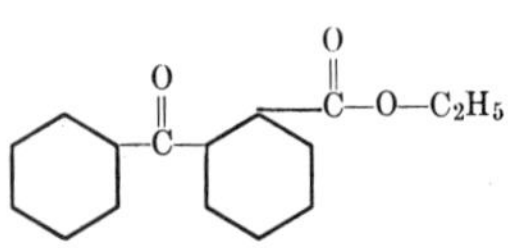

Ethyl-*o*-benzoyl benzoate

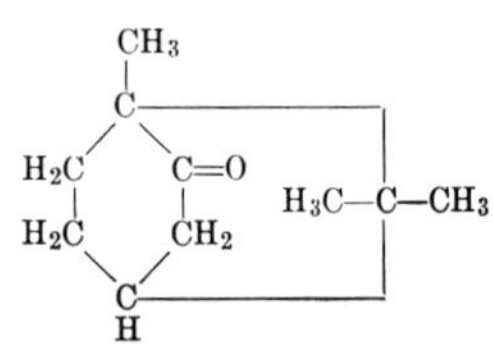

Camphor

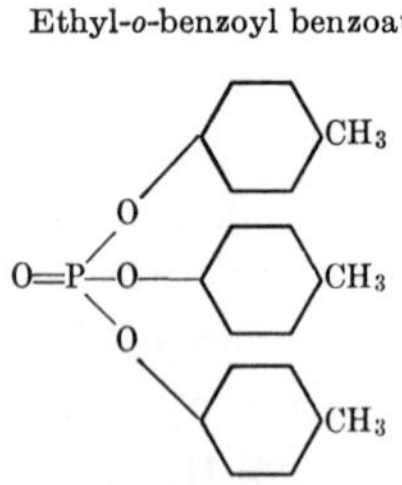

p-Tricresyl phosphate

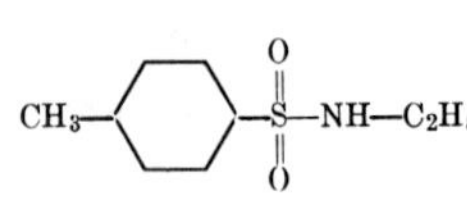

N-ethyl *p*-toluene sulfonamide

Fig. 1 *(Continued)*

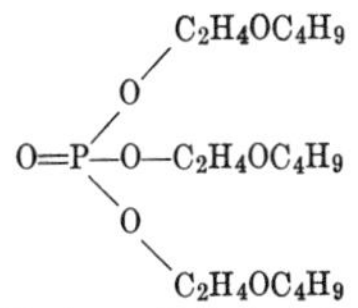

Tributoxy ethyl phosphate

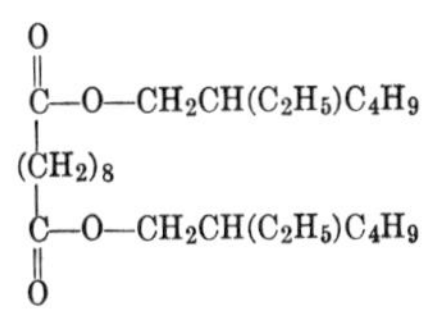

Dioctyl sebacate

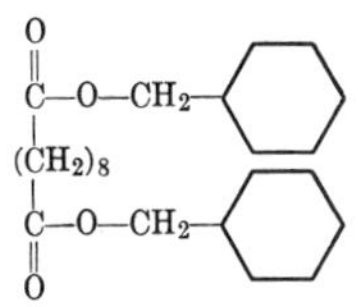

Dibenzyl sebacate

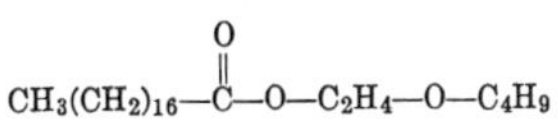

Butoxy ethyl stearate

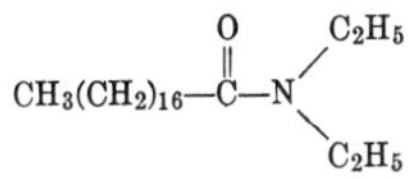

Diethyl stearamide

Typical examples of the various types of plasticizers are given in Fig. 1. This illustrates the chemical types and their structures, from which some of their properties may be deduced.

Plasticizers Based on Castor Oil. The value of castor oil and some of its derivatives as plasticizers for nitrocellulose lacquers was recognized by the Baker Castor Oil Co., and some of the properties of these products are shown in Table 3. These include the refined cold-pressed castor oil, a series of processed castor oils, and a group of the alkyl esters of ricinoleic acid, which is the acid component of castor oil.

Table 3 * Plasticizers Based on Castor Oil

Plasticizers	Pounds Per Gallon	Viscosity, poises, at 25°C	40°C	70°C	100°C	Pour Point, °C	Acid No.
Refined AA castor oil	8.00	6.75	2.40	0.50	0.17	−21	2.0
Processed oils, regular series							
Oil #2AC	8.11	14.00	4.60	0.88	0.28	−15	4.0
Oil #15	8.47	310	80	7.7	2.00	−9	14.5
Oil #30	8.54	1230	250	23	4.3	−9	15.0
Oil #40	8.54	2140	423	36	6.0	−8	15.0
Processed oils, pale series							
Oil #4	8.38	37.50	10.9	1.74	0.48	−11	17.0
Oil #1000	8.55	172	45.4	5.84	1.38	−10	20.8
Oil #16	8.64	310	80.0	7.7	2.0	−9	26.0
Alkyl ricinoleates							max.
P-1: Methyl ricinoleate	7.68	0.32	0.155	0.061	0.027	−30	12
P-4: Methyl acetyl ricinoleate	7.85	0.22	0.200	0.077	0.039	−40	6
P-6: Butyl acetyl ricinoleate	7.75	0.23	0.132	0.055	0.028	−40	4.6
P-8: Glyceryl triacetoricinoleate	8.07	2.3	0.910	0.325	0.128	−35	4.0

* Courtesy of Baker Castor Oil Co.

The oils are practically non-volatile, and the monohydric alkyl ricinoleates have very low vapor pressure at room temperature. The processed oils are oxidized and polymerized, and are generally referred to as "blown castor oils." They are produced in a range of viscosities to meet application requirements. The alkyl ricinoleates are thin liquids having very low viscosities. All liquids increase in viscosity with decreasing temperature, but the ricinoleates change less with temperature than the aromatic esters.

These plasticizers vary in their solubility for nitrocellulose. The AA castor oil is a non-solvent; the processed oils are active solvents in the presence of alcohol, and the alkyl ricinoleates are good solvents for nitrocellulose. They all have good compatibility with nitrocellulose and are soluble in lacquer solvents. The specific applications for these plasticizers in nitrocellulose lacquers are discussed in Chapter 11.

The AA type of castor oil has been used for many years as an efficient, low-cost plasticizer in nitrocellulose coatings for fabrics. Since it is a relatively low-molecular-weight material and a non-solvent for nitrocellulose, it tends to exude more than the blown oils. The #15 blown castor oil is very widely used because its higher viscosity and molecular weight, and its solvent properties, impart toughness and permanent plasticization. The monomeric alkyl ricinoleates have low molecular weight and low viscosity, but their good solvent action and extremely low vapor pressure assure their retention in the film. They also impart better flexibility at low temperatures.

Lacquer Oils. In general, oxidizing materials are not used in lacquers because the application of a second coat frequently "lifts" the first coat. Some oxidation takes place during the time between the application of the first and second coats, so that the first coat becomes partially insoluble in the lacquer solvents. Consequently, when the second coat is applied it does not redissolve the first coat but produces a wrinkled or "lifted" appearance which is quite objectionable. The phenomenon of lifting is discussed in detail in Chapter 11 on cellulosic polymers. However, the Archer-Daniels-Midland Co. produces a plasticizing lacquer oil from linseed oil which has given very good performance for many years. More recently, ADM has produced lacquer oils from soybean oil which have paler color than the linseed type. Some of the characteristics of these lacquer oils are shown in Table 4.

The ADM lacquer oils are processed so that they are practically non-drying, as indicated by their low iodine numbers. They add toughness to lacquer films as well as flexibility because they have fairly high viscosity and are solvents for nitrocellulose. They are used in very flexible lacquers for fabrics and leather where a high proportion of plasticizer is required. They are also satisfactory in

Table 4 * Lacquer Oils

	100 Oil	150 Oil	SL-200
Per cent non-volatile	85 ± 1	85 ± 1	100
Viscosity, Gardner-Holdt	Z–Z1	X–Y	Z–Z1
Color, Gardner 1933	6–7	6–7	5–6
Acid number, max.	10	12	8.5
Iodine number	90–105	60–75	84–94
Pounds per gallon	8.23	8.08	8.12

* Courtesy of Archer-Daniels-Midland Co.

clear and pigmented lacquers for wood and metal surfaces. Further details regarding their use are given in the next chapter.

Resinous Plasticizers. Resinous plasticizers are frequently referred to as polyester resins, but they are not the reactive type which may be copolymerized with reactive monomers as described in Chapter 7. They are non-drying resins and have long linear structures produced from the reaction between dibasic acids and the glycols. However, a considerable range in properties is produced by modification with non-drying oils, particularly castor oil, and by replacing some of the glycol with glycerol. Combinations of various dibasic acids also affect the resin properties. For example, a resin made with both phthalic and sebacic acids would have higher viscosity and greater hardness than a similar resin made from sebacic only.

Resinous plasticizers are used with both the cellulosic and the vinyl polymers. They are essentially non-volatile, and they impart greater toughness than the monomeric chemical plasticizers. The comparison between two resinous plasticizers and dibutyl phthalate in Chapter 1, Table 4, shows that the resinous plasticizers impart greater tensile strength at equal elongation than dibutyl phthalate, particularly at high plasticizer concentration. They have fairly high viscosity and usually a solvent action on the polymeric material. Therefore they show good retention in the film under severe conditions, such as the hot-oil extraction test for insulation materials. Their good retention and toughness permit the use of high percentages which, in turn, produce very flexible coatings free from plasticizer exudation. The Paraplex resins shown in Table 5 illustrate this type of plasticizer.

The series of Paraplex resins was designed chiefly for use in nitrocellulose lacquers and vinyl resin compositions. These resins are all compatible with nitrocellulose, but vary in their degree of compatibility with the other cellulosic polymers and the vinyl resins. Some of them may be used with urea- and melamine-formaldehyde resins in baking finishes. Paraplex RG-2, RG-8, and RG-10 are compatible

Table 5 * Paraplex Plasticizing-Type Resins

Para-plex †	% Solids, ±2%	Approx. Viscosity, Gardner-Holdt	Acid No., Solids	Color, Gardner 1933	Reducing Solvents	Type Resin, All Non-Drying
RG-2	60 ‡	U	22–35	6–8	Aromatics	Oil-modified sebacic alkyd
RG-8	100	Z2	0–3.5	6–8	Aromatics	Oil-modified sebacic alkyd
RG-10	100	Z10	0–8	4–8	Aromatics	Oil-modified dibasic acid alkyd
G-20	60 §	Z10+ ‖	10–20	10–13	Esters Aromatics	Unmodified sebacic alkyd
G-25	100	Z10+	0–2	8–11 ¶	Esters Aromatics	High-mol.-wt. polyester
G-40	100	Z10+	0–2	0–2 ¶	Esters Aromatics	High-mol.-wt. polyester
G-50	100	Z	0–2	4–7	Esters Aromatics	Med.-mol.-wt. polyester
G-60						
5-B	80 ‡	Z1	45–58	5–7	Aromatics	Oil-modified maleic polyester
AL-111	100	Z8	0–15	Dark	Esters Aromatics	Oil-modified polyester

* Courtesy of Rohm and Haas Co.
† Trade name, Rohm and Haas Co.
‡ Solvent, toluol.
§ Solvent, butyl acetate.
‖ At 100% solids.
¶ At 50% solids.

with nitrocellulose, ethyl cellulose, polyvinyl butyral, and urea-formaldehyde resins, but they differ in color and consistency. The RG-10 has paler color and also better solvency for nitrocellulose. The latter feature imparts greater resistance to exudation from the film. The RG-8 has the lowest viscosity and is a very convenient medium in which to grind the pigment for pigmented lacquers. It has good wetting property for pigments, and it may be used at 100% solids without volatile solvent. This eliminates the usual loss of solvent when grinding on a roller mill.

Paraplex G-20 is compatible with nitrocellulose, urea-formaldehyde resins, and the polyvinyl chloride-acetate copolymer resin. The absence of oil modification makes it less soluble in hydrocarbons, but this feature is used to advantage in the formulation of compounds which are required to have high resistance to gasoline. The absence of oil modification detracts somewhat from its flexibility at low temperatures.

Paraplex G-25, G-40, G-50, and G-60 are compatible with nitrocellulose, but they are designed especially for use with polyvinyl chloride and coplymer resins. They have good permanence and do not migrate from the vinyl into other coatings. The G-40 has the palest color and has better resistance to extraction with aromatic solvents or gasoline. The G-50 has lower molecular weight and viscosity and somewhat better compatibility. The G-60 has maximum heat resistance and has a heat-stabilizing action on polyvinyl chloride. From the discussion of alkyds and polyesters it will be apparent that a wide range of properties may be obtained in this type of resin by variation of the components.

An illustration of the above statement is provided in Paraplex 5-B. It is described as a "maleic alkyd polyester" designed especially to impart good adhesion to nitrocellulose lacquers for wood and metal finishing. As will be noted, it has a high acid number which imparts a high degree of polarity. Polar materials have a greater attraction for, and consequently greater adhesion to, wood and metal surfaces than non-polar materials. Increased polarity may be obtained in several ways. We shall see in Chapter 12 on vinyl resins that there is a polyvinyl chloride-acetate copolymer resin which is maleic modified to improve its adhesion. The free carboxyl groups of the maleic acid increase the polarity considerably. Paraplex AL-111 is somewhat similar to RG-2 but was designed with the economic factor of low cost in mind. It is darker in color, but finds use in many applications where pale color is not required.

The polyester plasticizing resins have somewhat better toughness and film-building qualities than the oils, and they are decidedly superior in this respect to the simple chemical plasticizers. The present scarcity of raw materials and higher cost retard their development and use in wider applications. They form one section of a very large group of polymeric materials which range from fibers such as Terylene, or Dacron, to the reactive polyesters which are copolymerized with materials such as styrene, as described in Chapter 7. An interesting use for polyesters in linoleum manufacture is described by Walker and Mackay (20). The scarcity of linseed oil in England during World War II, and its subsequent high cost, prompted the investigation of polyesters as a replacement for the highly oxidized linseed oil used previously. Investigators found that a polyester gel made from adipic acid, pentaerythritol, and ethylene glycol could be used, but it was somewhat more susceptible to hydrolysis than the oxidized linseed oil.

Chemical Plasticizers. It would be impractical to list all the chemical plasticizers which are available commercially. Therefore, the characteristics of only three important groups, the phthalates, the sebacates, and the phosphates, are shown in Table 6. The values shown are general averages and not necessarily those of any particular product or manufacturer. The values serve to illustrate the physical condition of these materials.

The characteristics in Table 6 show that the molecular weights of the chemical plasticizers are within the range of 200–400. This is

Table 6 Typical Characteristics of Chemical Plasticizers

Plasticizers	Molecular Weight	Boiling Point, °C At 760 mm	Boiling Point, °C At 4 mm	Freezing Point, °C	Viscosity, cp, at 25°C	Pounds per Gallon
Phthalates						
Dimethyl	194	282	....	0	11	9.94
Diethyl	222	290	....	−3.5	10	9.34
Dibutyl	278	340	....	−34	15	8.74
Dioctyl (di-2-ethyl hexyl)	390	...	215–235	−55	88	8.21
Dimethoxy ethyl	282	...	190–210	−40	58	9.75
Diethyoxy ethyl	310	...	198–215	31	Solid	9.33
Dibutoxy ethyl	366	...	210–233	−50	45	8.86
Dicarbitol	398	...	200–260	−35	84	9.42
Dicyclohexyl	330	...	200–235	58	Solid	9.58
Diphenyl	318	405	....	69	Solid	10.6
Butyl benzyl	312	370	....	−35	35	9.3
Sebacates						
Dibutyl	314	345	190	−11	7.9	7.8
Dioctyl (di-2-ethyl hexyl)	426	...	256	−48	11.5	7.6
Dibenzyl	382	...	265	26	28	8.8
Phosphates						
Tricresyl	368	420	240–260	−35	120	9.7
Tributoxy ethyl	398	...	200–232	−70	13	8.5
Tri-2-ethyl hexyl	435	...	220	....	14	7.7
Triphenyl	326	370	....	49	Solid	10.5
Cresyl diphenyl	340	390	....	−35	33	10.0
Alkyl aryl	362	375	....		16	9.1

quite low in comparison with oils and resins, since the molecular weight of an unbodied oil is about 880; many resins range from 1000 to 2000, but high polymers range from 10,000 to 100,000. In general, the chemical plasticizers are thin liquids, but it will be noticed that several are solid materials. The freezing point varies considerably, the lowest being −70°C, and the highest being 69°C.

The values for the boiling points at atmospheric pressure, 760 mm, or at a low pressure such as the 4 mm given in Table 6, are intended to indicate the degree of volatility of these materials. However, these values tell us only that the materials have vapor pressures of either 760 mm, or 4 mm, at the temperatures given in the table. In general, we are more interested in knowing the vapor pressure at room temperature, since this is the condition under which most of the finishes containing the plasticizers are aged, and under which the plasticizers will volatilize. Unfortunately, we cannot be sure that the plasticizers will maintain the same order of volatility at room temperature as is indicated by the boiling points at 760 mm or at 4 mm. This situation should be kept in mind when evaluating plasticizers for volatility. The boiling points at various pressures are useful criteria, but they are not necessarily absolute indications of relative volatility at room temperature.

Another complicating factor is the effect of the other components of the coating on the volatility of the plasticizer. If they exert a mutual solvent action, the plasticizer may be retained more strongly than if the components are mutually non-solvent. The best method for determining volatility is to measure the loss in weight of a specimen cut from a sheet of the composition and maintained under specified conditions. Obviously, the test would take too long if held at room temperature; consequently a higher temperature is used. Again this is an accelerated method, which may not give identical results to those which would be obtained at room temperature, but it is the best which can be done under the circumstances.

EVALUATION OF PLASTICIZERS

The efficiency of a plasticizer may be expressed in two ways: (1) the amount of plasticizer required to produce the desired degree of flexibility; (2) the degree of flexibility obtained from a given amount of plasticizer. Both expressions are used in the literature since there is no accepted standard method. Furthermore the degree of flexibility, and also the relative hardness and stiffness of coatings and plastic sheetings, vary with the use of plasticizers. Consider the flexibility requirements for the following four compositions: furniture finish, automobile finish, artificial leather fabric finish, and a vinyl sheeting material. In the case of the furniture finish there will be no bending of the wood, but there will be some dimensional changes due to variation in moisture content and to severe and sudden change in temperature. The temperature change may produce a great many cross-grain checks or cracks in the finish if the lacquer is poorly formulated. This property is measured by the cold-check test described in Chapter 15. Although a furniture lacquer must be quite hard to resist being marred by objects placed on it, a limited degree of flexibility is required of it.

The finish on an automobile is not intended to be subjected to severe strain from the bending of the metal, but it is always tested for this type of flexibility. The test is made by bending finished metal panels over a series of mandrels ranging from ⅛ in. to 1 in. in diameter and noting the size of the mandrel required to produce cracking of the finish. A conical mandrel is also available so that the test may be made on a single panel; see Chapter 15. In both the wood and the metal finishes only a limited degree of flexibility is required, but it is measured by conditions which produce cracking in the finish. However, in the case of the fabric finish, different requirements must be met; the finish is flexed a great many times during its useful life, and it must not distort badly or crack from the flexing. This type of flexibility may be measured on a flex and fold machine, but it is frequently estimated from a combination of tensile strength and per cent elongation measurements made on a free film of the fabric lacquer.

Most sheeting materials are not required to withstand severe flexing, but they must be soft, or have a soft "hand"; they must drape readily and not be stiff or boardy. However, they are required to have considerable tensile strength, and ability to recover their shape after a reasonable amount of deformation. These characteristics are measured by stiffness tests at various temperatures, and also by tensile strength and elongation measurements.

Tensile Strength and Elongation. The tensile strength and elongation measurements are made on strips of lacquer film or plastic sheet-

Fig. 2. Stress-Strain Curves

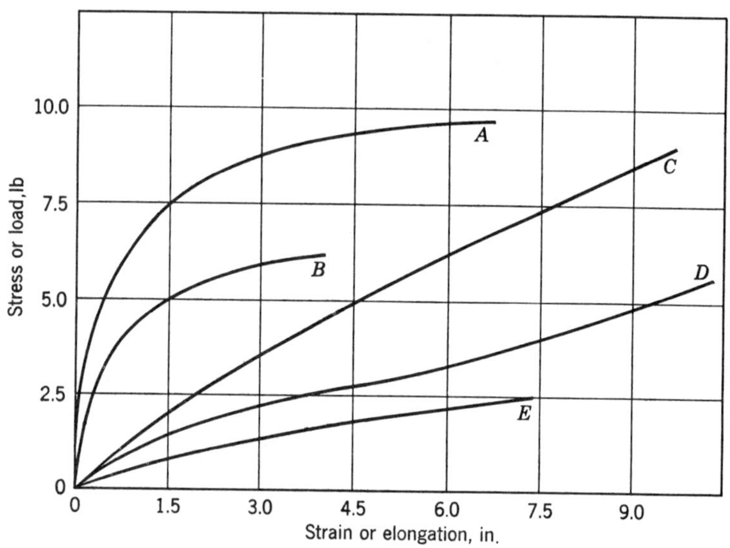

Free films of compositions *A*, *B*, *C*, *D*, and *E*, Table 7.

ing cut to a specific size. A strip is placed in a machine in such a manner that an increasing load may be applied to it. The strip stretches under the load and finally breaks. With a machine such as the Scott Tester IP-4 (21) an automatic recording device draws a curve on a chart which shows the elongation obtained as the load is increased. A series of typical stress-strain curves is shown in Fig. 2.

The end of each curve in Fig. 2 is the point at which the specimen breaks. If we assign arbitrary dimensions to the specimens, and use the values for load in pounds and elongation in inches given in Fig. 2, we can calculate the tensile strength and per cent elongation at the breaking point for the hypothetical compositions *A*, *B*, *C*, *D*, and *E*. Such values are frequently given in the literature for lacquer and other high-polymer films and sheeting. The values for apparent modulus at a specified elongation are also used in connection with plasticizer efficiency. These are known as the apparent modulus since they are not true Young's moduli, which are calculated within the elastic limits of truly elastic materials such as steel.

Calculations

Assume size of test specimens to be:

Length, 6 in.; width, 0.5 in.; thickness, 0.02 in.
Cross-sectional area = 0.5 × 0.02 = 0.001 sq in.

$$\frac{\text{Load, pounds}}{\text{Area, square inches}} = \text{psi}$$

Hence, load of 10 lb = 10,000 psi for specimens.

Therefore, values for tensile strength at break may be read directly from the curve (see Table 7).

$$\text{Per cent elongation at break} = \frac{\text{Final length} - \text{Original length}}{\text{Original length}} \times 100$$

If the specimen is elongated 6 in. before breaking, this would be equal to 100% elongation; other values may be read directly from curve (see Table 7).

Modulus at 75% elongation:

$$\text{Modulus} = \frac{\text{Stress}}{\text{Strain}} = \frac{\text{Load, psi}}{\text{Elongation, inches per inch of original length}}$$

$$\text{Composition } A \text{ at } 75\% \text{ elongation} = \frac{9150}{0.75} \text{ psi} = 12{,}200 \text{ psi}$$

See Table 7 for the other values.

An examination of the values in Table 7 and the curves in Fig. 2 might lead to different conclusions regarding the materials. For example in Table 7 materials *A* and *C* have about the same tensile strength, but *C* is capable of being stretched further than *A* before it breaks. Of course, the curves in Fig. 2 confirm these conclusions, but they give further information. They show that material *A* is quite resistant to deformation with light stress, whereas material *C* is de-

Table 7 Tensile Strength, Per Cent Elongation, and Apparent Modulus

Properties at Breaking Point	Compositions				
	A	*B*	*C*	*D*	*E*
Tensile strength, psi	9,625	6,125	9,000	5,500	2,375
% elongation	112	66	162	172	123
Apparent modulus, psi at 75% elongation	12,200	Failed	6,500	3,675	2,335

formed easily with light stresses. Evidently, material *C* is considerably softer and would drape better than material *A*. An attempt is made to show this condition by use of the apparent modulus at some arbitrarily chosen elongation. The values in Table 7 for apparent modulus at 75% elongation show that of *A* to be almost double that of *C*. However, a better indication would be obtained if a series of values were given at various elongations such as those in Table 8.

Table 8 Apparent Modulus at Various Elongations

% elongation	100	75	50	25
Composition *A*	9,875	12,200	17,500	30,000
Composition *C*	6,025	6,500	7,000	8,000

The values in Table 8 more nearly represent the stress-strain behavior of materials *A* and *C* as shown by the curves in Fig. 2. Of course, for general commercial reference a large series of stress-strain curves would be very inconvenient to use for estimation of plasticizer performance in high-polymeric materials. The values for apparent modulus at a series of elongations are also somewhat cumbersome to use. Therefore, a single condition is usually chosen which experience has shown to be fairly representative. Reed and Harding (3) use the apparent modulus at 75% elongation as measured with the Scott Tester IP-4. Lawrence and McIntyre (5) use an apparent modulus of 1800 psi at 25°C as obtained by the torsional method of Clash and Berg (22). The values in Tables 9 and 10 are taken from these two important contributions and will serve to illustrate the effect of a few plasticizers with polyvinyl chloride-acetate copolymer, and with polyvinyl chloride. The copolymer used by Reed and Harding was the Vinylite Resin VYNW described in Chapter 12. Data showing a comparison of dioctyl phthalate and tricresyl phosphate with the sebacate and polyester types of plasticizers are presented in Table 11.

The data in Tables 9 and 10 show that considerable variation in plasticizer content may be made among various plasticizers to obtain

Table 9 * Physical Properties of Polyvinyl Chloride-Acetate with Various Plasticizers

Plasticizers	% by Weight †	At Break Point and 25°C: Tensile Strength, psi	At Break Point and 25°C: % Elongation	Brittle Temperature,‡ °C
Phthalates				
Dibutyl	30.0	3000	260	−27
Dioctyl	35.5	2800	325	−14
{Dicyclohexyl {Dioctyl phthalate	18.0 20.0	3000	290	−5
Butyl benzyl	34.0	3000	250	−4
Phosphates				
Tricresyl	39.5	2800	250	−3
Trioctyl	33.0	2100	230	<−60
Cresyl diphenyl	38.5	3400	270	−12
Polyesters				
Paraplex G-25	43.0	2100	215	−19
Paraplex G-40	46.0	1850	192	−20
Paraplex G-50	40.0	2700	310	−16
Ricinoleates				
{Methyl acetyl {Dioctyl phthalate	13.0 20.0	2800	300	−32
{Methyl Cellosolve acetyl {Dioctyl phthalate	14.5 20.0	2900	275	−34

* Courtesy of M. C. Reed and J. Harding (3).

† Per cent of plasticizer by weight to obtain elongation of 75% on a Scott Tester IP-4 with constant rate of loading of 15,000 psi per min.

‡ Lowest temperature at which five successive specimens do not fracture when subjected to a sudden shearing action (ASTM Designation D746-44T).

the same degree of flexibility and softness. The data in Table 11 for 100% modulus show that the same percentage of sebacates produces softer films than the other plasticizers but with a slight sacrifice in tensile strength.

Low-Temperature Flexibility. The data show a wide range in behavior of the plasticizers with the polyvinyl type of resins as the temperature is lowered. The trioctyl phosphate and dioctyl and dibutyl sebacates are outstanding in this respect, having brittle points lower than −60°C. Good low-temperature flexibility is obtained with dibutyl phthalate and dioctyl phthalate and the ricinoleates. The polyester plasticizers impart only a fair degree of flexibility as the temperature is lowered, and tricresyl phosphate is the poorest in this re-

Table 10 * Properties of Polyvinyl Chloride with Various Plasticizers

Plasticizers	Parts Plasticizer per 100 Parts Resin †	Flex Temperature,‡ °C	Volatility §	Plasticizer Migration Loss from Plastic,‖ mg		Mol. Weight
				Lacquer	Varnish	
Phthalates						
Dimethyl	46	−5.5	71.3	360	580	194
Diethyl	41	−5.5	55.2	310	400	222
Dibutyl	41	−12	36.0	200	400	278
Di-2-ethyl hexyl	50	−23	4.1	25	20	390
Dibutoxy ethyl	46	−17	9.7	120	80	366
Dicarbitol	52	−17.5	14.6	470	540	398
Butyl benzyl	45	−2	14.4	90	260	312
Sebacates						
Dibutyl	37	−34.5	36.4	320	1150	258
Di-2-ethyl hexyl	54	−57.5	1.0	130	50	370
Dibenzyl	42	−13	9.9	260	840	326
Phosphates						
Tricresyl	57	+1.5	1.5	>600	>600	368
Tri-2-ethyl hexyl	56	−58	5.1	>600	>600	435
Tributoxy ethyl	44	−31	21.4	>600	>600	398
Triphenyl	52	−2.5	5.8	>600	>600	326

* Courtesy of R. R. Lawrence and E. B. McIntyre (5).

† Plasticizer necessary for apparent modulus of 1800 psi at 25°C by torsional method of Clash and Berg (22).

‡ Temperature at which the composition has elastic modulus of 135,000 psi.

§ % plasticizer loss from sheet 3 × 1 × 0.02 in. at 105°C for 24 hr, air flow 200 ft per min.

‖ Loss in milligrams from sheet 2 × 3 × 0.02 in. after 30 days at 25°C in contact with lacquer or varnish.

Table 11 * Properties of Polyvinyl Chloride with Various Plasticizers

Plasticizers 35% by Weight	At Break Point and 25°C		100% Modulus, psi	Bend Brittle Point, °C †
	Tensile Strength, psi	% Elongation		
Dioctyl phthalate	2580	300	1040	−32
Tricresyl phosphate	2960	290	1380	0
Dibutyl sebacate	2050	340	630	−67
Dioctyl sebacate	2310	360	910	−62
Dibenzyl sebacate	2520	320	890	−33
Paraplex G-25	2640	290	1970	−7
Paraplex G-40	2770	270	2490	+3
Paraplex G-50	2750	295	1640	−5

* Courtesy of Rohm and Haas Co.

† By the Kemp Machine (ASTM D746-43T); for the monomeric plasticizers the conditioning period was 15 hr in air at −15°C followed by 2 min in the machine, but the polyesters were allowed 5 min in the machine.

spect. The previous statement that no single plasticizer has the maximum amount of all requirements should be recalled. Therefore, a plasticizer must be selected on the basis of its ability to meet the most important requirement of the specific application for which it is to be used.

Volatility. The rate of volatility of plasticizers from the compositions in which they are used was discussed previously. It was pointed out that the temperature at which a plasticizer boils under atmospheric pressure or at a reduced pressure is not a satisfactory criterion of the volatility of a plasticizer from the composition in which it is used. Actual loss in weight measurements from the composition must be made, but since these would take too long at normal temperatures an elevated temperature is used. Again, it must be realized that elevated temperature constitutes an accelerated test which may not give the same relative values for the behavior of a series of plasticizers aged at room temperature; however, it is the best compromise we have at present.

The conditions used by Lawrence and McIntyre (5) for the results shown in Table 10 are as follows: An oven was constructed in which the air was continually exhausted to the atmosphere at a rate of 200 ft per min. The oven was maintained at 105°C, and the loss from the test specimens was determined by weighing after 24 hr in the oven. The dimensions of the test specimens were as follows: length, 3 in.; width, 1 in.; thickness, 0.020 in.

It will be noted that the degree of volatility of the plasticizers in a related group may be correlated with their molecular weights and boiling points. This is shown clearly by the four phthalates from dimethyl to dioctyl inclusive; the lower-molecular-weight plasticizers volatilize more readily than those having higher molecular weights. This is also verified by comparison of dibutyl and dioctyl sebacate. However, although dibenzyl sebacate is in line with respect to molecular weight and volatility, it is an exception with respect to boiling point at 4 mm. Dibenzyl has a higher boiling point (265°C) than dioctyl (256°C) at the reduced pressure, but it has a greater volatility in this test.

The tricresyl and triphenyl phosphates obey the general rule of increased volatility with lower molecular weight and lower boiling point. However, the high volatility of the tributoxy ethyl phosphate compared with the tri-2-ethyl hexyl phosphate is not accounted for by this rule. It was pointed out earlier that the relative solvent power of the plasticizers for the polymeric material is a factor in their volatility which may have some bearing on the results obtained above.

Of course, the oils and resinous plasticizers are practically non-volatile and show little or no loss on aging at room temperature or in the accelerated tests.

Exudation. It was also pointed out previously that certain plasticizers exude or "sweat out" of the film on aging. This applies particularly to plasticizers which are not solvents for the polymeric materials. The amount of exudation is difficult to measure quantitatively since it is not volatile. Any attempt to remove it from the surface of the film with a solvent might remove more than the actual exudation. Consequently, exudation is usually estimated qualitatively. The degree of greasiness on the surface of the film may be estimated visually by smearing with a finger. A more sensitive test is to place a cigarette paper on the film and note the degree of absorption into the paper of the exuded plasticizer. The paper becomes darkened and partially translucent from absorption of the plasticizer. The paper may be held in contact with the film by a circular weight, and the amount of exudation at various intervals may be estimated from the condition of the circle on the paper. The test may be made at room temperature or at slightly elevated temperatures if necessary.

Migration. The migration of a plasticizer from one polymeric material into another was not a problem until the introduction of the vinyl-type resins, and it should not be confused with exudation as described above. For example, the data in Table 10 show that the phosphate plasticizers migrate severely into lacquer or varnish films, but there would be little or no exudation of these plasticizers on the surface of a polyvinyl chloride film. The precise mechanism of migration has not been determined, but it is presumed to be a form of selective absorption. If the affinity of the lacquer or varnish films for the plasticizer is greater than that of the vinyl compound, then migration of the plasticizer from the vinyl to the lacquer or varnish could occur. In mild cases only a slight marring of the surface occurs, but in severe migration the lacquer or varnish films may be softened considerably, and the plastic material cannot be removed without damage.

The amount of migration may be measured as the loss in weight from the plastic in contact with lacquer or varnish under specific conditions. For the results in Table 10 Lawrence and McIntyre (5) prepared films of clear lacquer and varnish on steel panels. The nitrocellulose lacquer was a typical furniture finish containing dibutyl phthalate and ester gum. The varnish was the linseed oil–*p*-phenylphenol resin type used in high-grade furniture finish. The coated panels were air-dried at least 7 days before being used in the test. A sheet of the plastic was placed on each of the finishes under a pressure of 0.33 psi. The dimensions of the sheet were as follows: length, 3 in.; width, 2 in.; thickness, 0.020 in. The sheets were weighed before the test and again after contact for 30 days at 25°C. The loss in weight was considered as the amount of plasticizer which had migrated into the lacquer or varnish films. The effect was also observed after contact for 7 days.

The data show the retarding effect on migration of increased molecular weight in the phthalate esters from the dimethyl to the di-2-ethyl hexyl, and also in dibutyl and di-2-ethyl hexyl sebacates. However, the difference in migration between di-2-ethyl hexyl phthalate and di-2-ethyl hexyl sebacate is greater than can be attributed to the relatively small difference in molecular weight. This is further emphasized by comparison of these two plasticizers with tri-2-ethyl hexyl phosphate; it has the highest molecular weight, but it migrates much more severely. The differences between migration into the lacquer and into the varnish films do not follow a regular pattern. In some cases the plasticizer migrates into the lacquer more than into the varnish; in others, the reverse is true.

Plasticizer Extraction. A plasticizer may lower the resistance of a polymeric material to hot water, hot lubricating oil, gasoline, or other solvents. Extraction tests are usually made on strips of the plastic sheeting, on free films of the lacquer, or on the coated cable in insulation lacquers. The strips of material are weighed and immersed in the liquid for a specified time and at a given temperature. At the end of the time the strips are removed and conditioned at 50–80°C for periods ranging from 30 min to several hours before the final weighing to determine the weight loss. In the oil extraction tests the specimens are washed with naphtha to remove oil on the surface before conditioning and final weighing. The oil test may be run at room temperature or as high as 120°C, depending on the specific requirements. Typical examples of the rate of plasticizer extraction from cellulosic and vinyl polymers are given in Chapters 11 and 12.

Flammability. There is a vast difference in the degree of flammability of polymeric materials and also of plasticizers. Nitrocellulose is highly flammable, whereas polyvinyl chloride may be considered non-flammable since it stops burning if the source of fire is removed. Most of the oils and resinous plasticizers, and many of the chemical plasticizers, are flammable, but some of the phosphates and chlorinated plasticizers are non-flammable. It would not be practical to add a non-flammable plasticizer to nitrocellulose because of the large excess which would be required. However, slow-burning compositions may be made by dilution such as the following: nitrocellulose, 100 parts; vinyl resin VYCC, 100 parts; and the plasticizer trichloroethyl phosphate, 50–80 parts. Further details on such compositions are given in subsequent chapters.

In a comparative flammability test between dioctyl phthalate and tricresyl phosphate as plasticizers for polyvinyl chloride the Monsanto Chemical Co. determined whether the compositions would continue to flame after the source of heat was removed and, if so, for how many seconds. Of course, the composition containing tricresyl phosphate did not continue to flame, since neither the polyvinyl chloride nor the plasticizer is flammable. However, the composition with 35% dioctyl phthalate flamed for 65 sec, and that with 50% dioctyl phthalate flamed for 107 sec.

The Rohm and Haas Co. compared a series of plasticizers in polyvinyl chloride by the ASTM Method D568-41T. This requires a strip of the composition measuring 12 × 1.0 × 0.010 in. to be ignited by a cellulose nitrate fuse. The time of burning and amount of specimen burned are indications of the flammability. The results shown in Table 12 were obtained with polyvinyl chloride compositions containing 35% plasticizer. The data show that the polyester plasticizers have good resistance to burning.

Table 12 * Relative Flammability of Plasticizers in Polyvinyl Chloride

Plasticizers 35% by weight	% Burned	Time of Burning, sec
Tricresyl phosphate	0	2.4
Dioctyl phthalate	100	19.1
Dibutyl sebacate	100	17.0
Dioctyl sebacate	100	20.5
Dibenzyl sebacate	57	32.0
Paraplex G-25	7	9.0
Paraplex G-40	1	3.5
Paraplex G-50	4	5.0

* Courtesy of Rohm and Haas Co.

BIBLIOGRAPHY

1. M. C. Reed, *Ind. Eng. Chem.*, *35*, 896 (1943).
2. M. C. Reed and L. Connor, *Ind. Eng. Chem.*, *40*, 1414 (1948).
3. M. C. Reed and J. Harding, *Ind. Eng. Chem.*, *41*, 675 (1949).
4. C. Bogin, *Paint, Oil, Chem. Rev.*, *15* (March 2, 1950).
5. R. R. Lawrence and E. B. McIntyre, *Ind. Eng. Chem.*, *41*, 689 (1949).
6. W. A. Woodcock, *Ind. Eng. Chem.*, *41*, 663 (1949).

7. J. K. Stevenson, L. E. Cheyney, and M. M. Baldwin, *Ind. Eng. Chem., 42,* 2170 (1950).
8. D. J. Mehta and H. F. Payne, *Paint Varnish Production, 30,* 6 (1950).
9. L. E. Cheyney, *Ind. Eng. Chem., 41,* 670 (1949).
10. K. K. Fligor and J. K. Sumner, *Ind. Eng. Chem., 37,* 504 (1945).
11. D. K. Rider, J. K. Sumner, and R. J. Myers, *Ind. Eng. Chem., 41,* 709 (1949).
12. L. W. A. Meyer and W. M. Gearhart, *Ind. Eng. Chem., 40,* 1478 (1948).
13. W. M. Gearhart and W. D. Kennedy, *Ind. Eng. Chem., 41,* 695 (1949).
14. E. M. Cunningham, *The Rubber Age,* November, 1947.
15. C. E. Rehberg, M. B. Dixon, T. J. Dietz, and P. E. Meiss, *Ind. Eng. Chem., 42,* 2374 (1950).
16. C. E. Rehberg, M. B. Dixon, T. J. Dietz, and P. E. Meiss, *Ind. Eng. Chem., 42,* 1490 (1950).
17. W. S. Emerson, R. I. Longley, J. R. Darby, and E. E. Cowell, *Ind. Eng. Chem., 42,* 1431 (1950).
18. R. H. Oliver, N. M. Wiederhom, and R. B. Mesrobian, *Ind. Eng. Chem., 42,* 488 (1950).
19. W. J. Sparks and D. W. Young, *Ind. Eng. Chem., 41,* 665 (1949).
20. F. T. Walker and T. Mackay, *J. Oil & Colour Chemists' Assoc., 34,* 311 (1951).
21. Scott Testers, Inc., Providence, R. I., *Bull. Scott Serigraph,* IP-4.
22. R. F. Clash and R. M. Berg, *Modern Plastics, 21,* 119 (July, 1944).

The structure of life is built of accumulated experience. Plan well the experience of each day; then the structure will meet the requirements of the Great Architect, and be an honor to its builder.

Chapter 11. CELLULOSIC POLYMERS

INTRODUCTION

The cellulosic polymers used in organic coatings, plastics, laminates, adhesives, etc., are esters and ethers of cellulose. The esters are cellulose nitrate (usually referred to as nitrocellulose), cellulose acetate, and the mixed esters: cellulose acetobutyrate and acetopropionate. The ethers are ethyl cellulose and benzyl cellulose. Methyl cellulose and carboxymethyl cellulose are also available; but, since they are water soluble, they are not generally included in the cellulosic polymers for organic coatings. However, these materials are described in this chapter.

The cellulosic polymers are extremely useful film-forming materials because of their inherent qualities of fast air drying and good film strength. These qualities are derived from their high molecular weight and their relatively high softening point. They are thermoplastic materials; therefore films are formed from them directly by simple evaporation of the solvent. The speed of drying is related to the rate of evaporation of the solvent and the ability of the film to release the final few per cent of solvent. The polymers vary considerably in the type of solvent required to put them into solution.

The cellulosic polymers are tough, horny solids and are available in a range of molecular weights. In general, the film strength and flexibility increase as the molecular weight increases, but the solubility decreases. A low degree of solubility necessitates the use of more expensive solvents, and results in solutions of low solid content at application consistency. The latter difficulty has been relieved somewhat by the development of efficient methods for heating the lacquers so that they may be applied by the "hot-spray" process described later. The recent use of dry steam as the combined propellant and heater in spray application also is of definite interest in this respect.

A characteristic of high-molecular-weight thermoplastic polymers is high cohesive strength. This imparts toughness, or film strength, but it detracts from adhesion to smooth surfaces. Improved adhesion is obtained from the addition of plasticizers and other resinous materials having lower molecular weight. It will be recalled from discussion in previous chapters that such additions necessarily reduce the film strength somewhat. However, in addition to improving adhesion they increase gloss and the solid content of the lacquer at spraying consistency. Plasticizers are usually necessary to impart sufficient flexibility to the coatings. A good deal of experience is required to formulate lacquers having the optimum combination of polymer, plasticizer, resin, and solvent for best performance at minimum cost.

The individual members of the cellulosic polymers will be described separately because of their different uses and characteristics. However, the general comparison shown in Table 1 will be found useful to differentiate among them.

Table 1 * Characteristics of Cellulosic Polymers

Cellulose	Softening Point, °C	Tensile Strength, psi	% Elongation	% Water Absorption †	Specific Gravity	Pounds per Gallon
Nitrate	Decomposes	8,500–17,000	4–12	3.1	1.65	13.7
Acetate	235–245	8,000–13,000	3–12	8.4	1.38	11.5
Acetobutyrate	210–225	7,000– 9,000	3–11	3.1	1.20	10.0
Acetopropionate	200–210	7,000– 8,000	7–11	3.2	1.25	10.4
Ethyl cellulose	160–180	8,000–11,000	7–30	2.8	1.14	9.5
Benzyl cellulose	149–155	7,000	19	0.6	1.23	10.2

* Courtesy of Hercules Powder Co.
† 24-hour test at 25°C, 80% RH.

Nitrocellulose is the most widely used member of this group of cellulosic polymers for organic coatings. Its disadvantages are its sensitivity to degradation by heat and ultraviolet, and its high degree of flammability. Cellulose acetate is very much more stable to heat, and it transmits ultraviolet so freely that sheets of it are used as windows in solariums and poultry houses. Its disadvantages are poor compatibility with plasticizers and other film-forming materials, and sensitivity to absorption of moisture. Table 1 shows that the mixed esters are as good as nitrocellulose for moisture resistance. However, they are much softer and have lower tensile strength. Ethyl cellulose has good heat and ultraviolet resistance and is frequently used in hot-melt coatings. It also has good water resistance, but its films are softer than those from nitrocellulose. Some grades have a good combination of tensile strength and flexibility and therefore require only small amounts of plasticizer. The benzyl cellulose has excellent moisture resistance and fair tensile strength and flexibility, but its softening point is quite low. The values in Table 1 were obtained on average materials; of course, they will vary somewhat with the range of materials which are available commercially.

The cellulosic polymers all support combustion, but the very high rate of burning of nitrocellulose makes it unsatisfactory for many coatings on fabrics used in public buildings. The burning rate may be reduced by blending with polyvinyl chloride-acetate resins, and using non-combustible plasticizers and resins. The cellulosic polymers constitute a valuable group of film-forming materials because of their good film properties and relatively low cost. They have good non-penetration on porous surfaces and are widely used on paper and fabrics. Pigmented nitrocellulose lacquers have excellent durability and are used in automotive finishes; the pigment protects the nitrocellulose from degradation by ultraviolet.

CELLULOSE

Cellulose, which is the raw material from which the cellulosic polymers are made, is the chief constituent of living plants. It is a high-polymeric material in the natural state, believed to be formed by the condensation of glucose molecules. It is a long-chain polymer of very high molecular weight. It is very insoluble, and attempts to isolate it result in partial degradation and reduction of its molecular weight. However, the natural cellulose polymer is believed (1) to have an average molecular weight of at least 500,000.

The physical structure of cellulose is quite complex, but X-ray analysis shows that it has definite crystallinity. This would indicate that the long-chain molecules are arranged in an orderly manner and probably parallel to each other. The molecule is highly polar; therefore strong secondary valence forces may be expected to exist between the molecules. This would tend to bind the molecules into orderly bundles known as micelles which produce the crystal pattern. The micelles are submicroscopic, but they aggregate into larger bundles known as fibrils, which constitute the fine structure of cellulose fiber.

The important commercial sources for cellulose are trees such as pine, poplar, and spruce, and plants such as cotton, hemp, and flax. The composition of wood varies with the species of tree, and the part of the tree from which it is taken. An average value for the composition may be taken as 40–50% cellulose, 20–30% lignin, and 10–30% hemi-celluloses, and polysaccharides other than cellulose (2). Cotton is the purest natural form of cellulose since it contains about 90% of cellulose. The cellulose in wood is separated from the lignin by either the sulfite process, the sulfate process, or the soda process.

These processes require pressure and high temperature, and are followed by bleaching and other purification treatments to produce wood pulp. These treatments tend to split the cellulose molecule with a consequent reduction in molecular weight. Less drastic treatments are required to purify the cellulose obtained from cotton linters. The linters are the short fibers or fuzz which grows on the cotton seed. The long fibers are called lint; if some of these are broken into short fibers during removal from the seed they are frequently classified as linters. The lint or long fibers are used for spinning cotton textiles.

Cellulose is not soluble in normal solvents, but it may be solubilized by chemical treatments and regenerated by precipitation. Cellophane is produced by precipitation from a cellulose xanthate or viscose solution. The regenerated cellulose is washed, plasticized with glycerol, and sheeted over heated rolls. The Cellophane sheet is moisture-proofed by applying a very thin coat of nitrocellulose lacquer. The production of rayon filaments is another illustration of the methods used for solubilizing cellulose. Three processes are used: (1) the cuprammonium; (2) the viscose; (3) the acetate. In the cuprammonium process cotton is dissolved in cuprammonium hydroxide, $Cu(NH_3)_4(OH)_2$, and spun into filaments which are insolubilized by passing through water and dilute acid baths. A modification of this process gives higher concentration of cellulose in solution. This is accomplished by precipitating copper hydroxide on the cotton fibers, and then treating with ammonium hydroxide to produce the cuprammonium *in situ*.

Viscose rayon is produced by treating cellulose pulp with sodium hydroxide to make alkali cellulose, followed by treatment with carbon disulfide to produce cellulose xanthate. This is soluble in dilute alkali and is spun into filaments from this solution.

Acetate rayon is made by treating cotton linters with acetic anhydride to produce cellulose acetate. This is partially hydrolyzed so that it becomes soluble in acetone, and the filaments are spun from this solution.

Cellulose may be hydrolyzed by acids, and a variety of products will be obtained, depending on the strength of the acid and conditions of hydrolysis. Hydrocellulose is obtained by hydrolysis with dilute mineral acids, but this material is not a homogeneous product. It is a mixture of saccharides produced by splitting the original cellulose molecules into chains of shorter length. Continued hydrolysis with strong acids always produces the monosaccharide, glucose, as the final product. From this result cellulose is considered to be composed of glucose anhydride units, except for the two end groups. The empirical formulas may be written as follows:

Glucose: $C_6H_{12}O_6$
Cellulose: $C_6H_{11}O_6(C_6H_{10}O_5)_nC_6H_{11}O_5$

However, it has been shown from X-ray analysis of cellulose that the length of the unit cell in the direction of the fiber is 10.3 A, which corresponds to the dimensions of anhydro-β-cellobiose. Since cellobiose may be obtained from cellulose by appropriate chemical means it was decided that the unit in the physical structure of cellulose is anhydro cellobiose. Cellobiose is composed of two β-glucose units connected with a glucosidic linkage. The glucose units are considered to be in the pyranose ring form; therefore the structure of cellobiose is usually shown as follows:

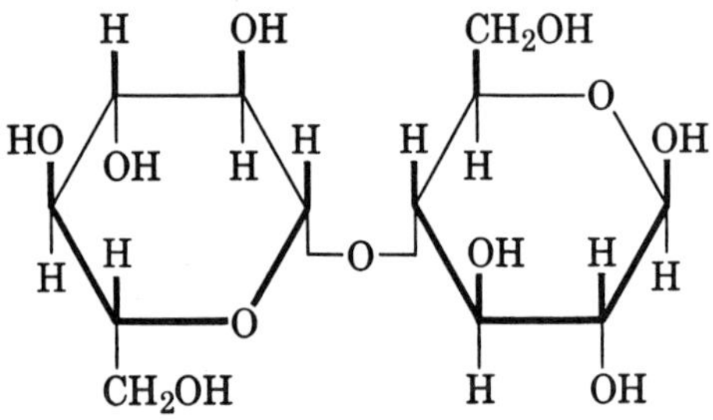

Cellobiose

This structure shows three hydroxyl groups on each glucose unit, one of which is a primary hydroxyl and the other two are secondary hydroxyls. We shall see later that these groups may be nitrated, etc., to produce the organic solvent-soluble polymers used in surface coatings. The above structural formula is written with heavy and light lines to indicate the relative positions of the various groups. The five carbon atoms in the pyranose ring are considered to be in the same plane. If they are visualized as being in the plane of the paper, then those groups on heavy lines are above the plane of the paper, and those on light lines are below this plane. The cellulose molecule is made up of a large number of these units as indicated below, but the above formula enables us to visualize the arrangement of the groups in the structure. As we have seen previously, many of the physical properties of organic coatings are related to the structural arrangement in the film-forming molecules and to the forces between the molecules.

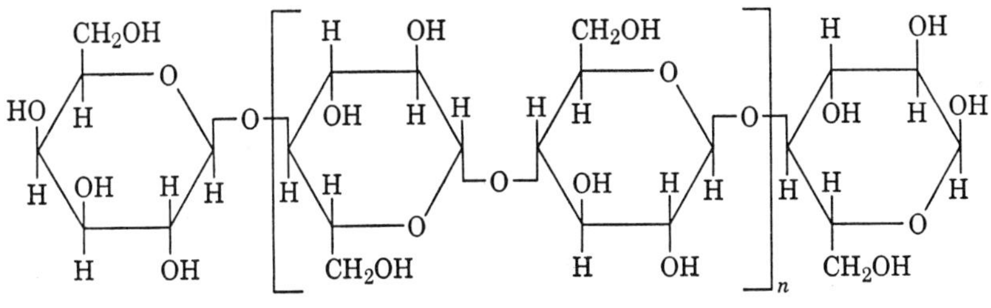

Cellulose

ALKALI CELLULOSE

Alkali cellulose is formed when cellulose is treated with about 18% sodium hydroxide solution. There are differences of opinion regarding the nature of the compound, but it is usually regarded as an addition product. Its composition has been given (1) as either $(C_6H_{10}O_5)_2 \cdot NaOH$ or $C_6H_{10}O_5 \cdot NaOH$, depending on the method of analysis. Alkali cellulose is decomposed easily by water into its components, cellulose and sodium hydroxide. However, the regenerated cellulose is considerably different physically from the original cellulose. The fibers have greater strength and luster, and they absorb water and dyestuffs more readily. Alkali cellulose is used in the manufacture of the cellulose ethers described later.

CELLULOSE NITRATE

Cellulose nitrate, more commonly known as nitrocellulose, is made by reacting highly purified cellulose with nitric acid in the presence of sulfuric acid. The function of the sulfuric acid is to remove the water of reaction so that nitration may be carried to the desired degree more readily. The nitric acid reacts with the hydroxyl groups on the cellulose molecule to form the ester. Since there are three hydroxyls on each glucose unit it is theoretically possible to form the cellulose trinitrate, with a nitrogen content of 14.14%. However, the practical limits for commercial materials are between 10.5% and 13.8% nitrogen. The grades used for surface coatings range from 10.7% to 12.2% nitrogen. See Tables 2 and 3. The theoretical values for the nitrogen content of the mono-, di-, and trinitrates are:

	% Nitrogen
Mononitrate	6.76
Dinitrate	11.11
Trinitrate	14.14

The various grades of nitrocellulose used in surface coatings are not definite compounds. They are probably mixtures of all possible compounds, ranging from unreacted cellulose to the trinitrate. However, the grades are maintained uniform in their properties by careful control of the nitrating conditions and subsequent washing and purification. The structural formula for a dinitrate is given below, from which it will be seen that there are both nitrate and hydroxyl groups in the molecule.

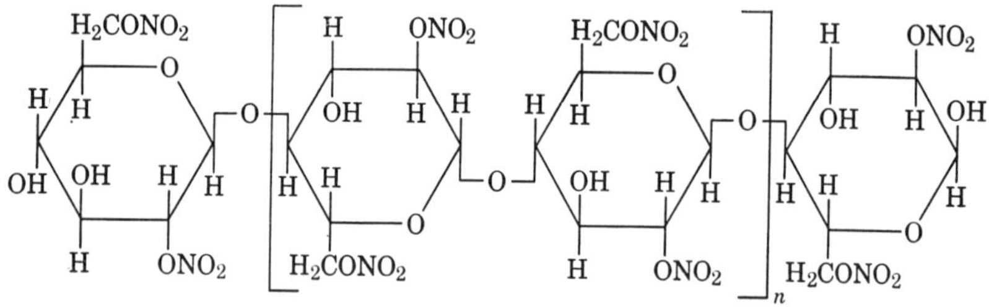

Cellulose dinitrate

Manufacture. The carefully purified cellulose may be obtained from either wood pulp or cotton linters. The cellulose is added to a mixture of nitric and sulfuric acids in a dipping pot equipped with paddle agitators. The ratio of the nitric and sulfuric acids, and the concentration of the solution, are important factors in determining the nitrogen content of the product. The reaction is exothermic, and the temperature during nitration is usually maintained at about 20°C because of instability at higher temperatures. When the desired degree of nitration has been reached the charge is dropped from the dipping pot into a basket-type centrifuge. Here the excess acid is removed and pumped to spent acid storage. The charge is then dropped from the centrifuge into a container of cold water and removed to the purification building. Here it is washed to remove free acids, and boiled with dilute sodium carbonate solution to decompose any sulfates or other compounds which produce unstable products.

The next step is to digest the nitrocellulose with water at temperatures up to 160°C, and under a pressure of about 100 psi to reduce its viscosity. This treatment splits some of the nitrocellulose molecules into shorter chains, or reduces their average molecular weight. This automatically reduces the viscosity of the solution of the nitrocellulose as used in lacquers, and consequently increases the solid content at application consistency. The digestion is carried out in a Miliken (3) digester, which is a 4-in. stainless steel coil almost a mile long. The material from the digester is neutralized and washed to remove traces of decomposition products. The majority of the water is removed by draining or centrifuging. The nitrocellulose is then compressed and alcohol pumped through it to remove the last traces of water. The block of nitrocellulose containing about 30% of alcohol is removed from the press and broken into a fluffy mass for easy dissolution before placing in drums for shipment. The alcohol content is adjusted to the amount specified by the Interstate Commerce Commission for safe transportation of nitrocellulose. The standard RS and SS types contain 30% of ethyl alcohol. The RS nitrocellulose is available containing 35% of ethyl alcohol, and also containing 30% or 35% of either isopropyl or butyl alcohol.

The packaged nitrocellulose is a white fluffy lint which should be handled and stored carefully. Detailed directions for safety in lacquer plants are available from Hercules Powder Co. (5) and other manufacturers. Nitrocellulose burns readily and very rapidly, particularly if it has become "dry" because of the evaporation of its alcohol content. Nitrocellulose has good stability at normal temperatures, 60–70°F (15–20°C), but as the temperature increases it decomposes at a logarithmic rate (4). The average rate of decomposition is 3.45 per 10°C rise in temperature. Therefore the rate of decomposition at 30°C will be 3.45 times greater than that at 20°C, and the rate at 40°C will be 11.9 times that at 20°C. The ignition point is about 200°C (400°F). The drums of nitrocellulose should be kept closed until used, and stored in a cool place away from steampipes or other sources of heat. All batches of nitrocellulose are tested for stability before shipment. The stability is measured by the time required at 135°C for the nitrocellulose to evolve sufficient nitrous oxides to change the color of methyl violet test papers. Properly stabilized nitrocellulose will withstand this test for 20 min. The test is described in detail in *Nitrocellulose Handbook,* Hercules Powder Co.

Types Available. Nitrocellulose is available in different viscosity types and solubility types. We have seen that the viscosities of polymer solutions are related to the molecular weight and shape of the polymer; therefore the low-viscosity types have shorter molecular chains than the high-viscosity types. We have also seen that the secondary valence forces between the chain molecules increase as the length of the chain increases; therefore we would expect the films of high-viscosity ntirocellulose to have greater tensile strength than films of the low-viscosity type. The values in Table 7 show that the tensile strength and flexibility increase with chain length, but not in direct proportion to it. Unfortunately, high viscosity in solution means low solid content at application consistency; therefore some sacrifice in film strength is made in many lacquers in order to obtain a higher solid content. The ½-sec type shown in Table 2 is very widely used

Table 2 Nitrocellulose-Viscosity Specifications

Standard Designations	Actual Viscosity: 12.2% NC Centipoises	12.2% NC Seconds	20% NC Seconds	25% NC Seconds
18–25 cp	18–25	0.0477–0.0663	0.513–0.717	1.59–2.21
30–35 cp	30–35	0.0795–0.0928	0.857–1.0	2.85–3.09
¼ sec	45–56	0.120–0.150	1.29–1.61	4.0–5.0
½ sec	105–140	0.279–0.372	3.0–4.0	9.3–12.4
¾ sec	210–263	0.558–0.698	6.0–7.5	18.6–23.3
5–6 sec	1,885–2,450	5–6.5	53.7–70.0	166–216
15–20 sec	5,600–7,540	15–20	161–215	500–667
30–40 sec	11,300–15,100	30–40	322–430	1,000–1,334
60–80 sec	22,600–30,200	60–80	644–860	2,000–2,667
125–175 sec	47,200–66,000	125–175	1,340–1,880	
250–400 sec	94,400–151,000	250–400		
600–1,000 sec	226,000–377,000	600–1,000		

because of its good compromise of film strength and solid content in solution.

The method of expressing the viscosity of the various types may appear confusing unless its origin and development are understood. Some time ago the 5- to 6-sec nitrocellulose was the lowest-viscosity type available, and its viscosity was measured on a solution containing 16 oz of nitrocellulose in a gallon of solution. This corresponded to a 12.2% nitrocellulose solution which had a viscosity of 5–6 sec by the falling-ball method. When the lower-viscosity ½-sec type was developed the 12.2% solution was too thin to be measured conveniently; therefore the concentration was increased to 20% nitrocellulose. The viscosity of this solution is 3–4 sec as measured on the 20% solution, but it is always known as ½-sec nitrocellulose. Further complications arose when the still lower-viscosity types were developed. The concentration was then increased to 25% nitrocellulose, which gave a viscosity of 4–5 sec for the so called ¼-sec type. However, the two viscosity types lower than the ¼-sec are referred to in centipoises rather than in seconds. The relationships between these various viscosity types and concentrations are shown in Table 2. The values were obtained by the falling-ball method, ASTM D301-33, described in Chapter 15, using the standard solvent:

	Per Cent by Weight
Ethyl acetate	20
Ethyl alcohol No. 1	25
Toluene	55

The viscosity of a nitrocellulose solution of any one of the three standard concentrations may be converted into the viscosity of another concentration by use of the factors in Table 3. Of course, the solvent

Table 3 Viscosity-Concentration Conversion Factors for Nitrocellulose Solutions

12.2% NC		20% NC	25% NC
Second	Centipoises	Seconds	Seconds
1	377	10.75	33.33
0.093	35.06	1.0	3.1
0.030	11.31	0.322	1.0
0.00265	1.0	0.0285	0.0884

must be the same in all cases. The application of these factors will be apparent by considering any set of values in Table 2. For example, if the ½-sec nitrocellulose has a viscosity of 4 sec at 20% concentration, it will have a viscosity of 4 × 0.93 or 0.372 sec at 12.2% concentration. This is equal to 0.372 × 377 or 140 cp. At 25% concentration the solution will have a viscosity of 4 × 3.1 or 12.4 sec, which is equal to 12.4 × 377 or 4680 cp.

Nitrocellulose is also classified according to its type of solubility, which is related to its nitrogen content as shown in Table 3. The volatile solvent used in nitrocellulose lacquers is always a mixture of organic liquids which usually contains true solvents, latent solvents, and diluents, as discussed in a later section. Table 4 shows the types of solvents and diluents for RS, AS, and SS nitrocellulose.

Table 4 Solubility and Nitrogen Content

	% Nitrogen	Solvents	Diluents
RS type	11.8–12.2	Esters, ketones, ether-alcohols	Hydrocarbons
AS type *	11.2–11.7	Esters, ketones, ether-alcohols	Alcohol
SS type	10.7–11.2	Ethanol	Hydrocarbons

* The commercial production of AS type has been discontinued.

From the discussion in Chapters 1 and 6 of the relationship between relative polarity and solubility we can see why the SS type has better solubility in alcohol than the RS type. The structure of the nitrocellulose molecule on page 408 shows that as the percentage of nitrogen is decreased the number of free hydroxyl groups will be increased. Since the hydroxyls are more polar than the nitrate groups, we can expect better solubility in the highly polar alcohol from molecules having more free hydroxyls. Although the SS type in Table 4 is indicated to be soluble in alcohol, it is customary to add a small percentage of ester solvent to obtain somewhat lower viscosity.

Solvent Combinations for Nitrocellulose. Most nitrocellulose lacquers are composed of nitrocellulose, resins, plasticizers, and volatile solvents, with or without pigments. The volatile solvent is referred to as the "lacquer thinner" because it is a mixture of true solvents, latent solvents, and diluents. The diluent is added primarily to reduce the cost, but it is often a good solvent for the resin and plasticizer components. Both aromatic and aliphatic hydrocarbons are used as diluents, and they are rated by their dilution ratio as discussed in Chapter 6. The dilution ratio is the ratio of the amount of diluent which must be added to a true solvent to first produce turbidity in a nitrocellulose solution. It is important that the concentration of nitrocellulose at the end point be 10% or the 8% concentration required by the ASTM Test. The dilution ratios for a number of diluents are given in Tables 4 and 5, Chapter 6.

The dilution ratio for any specific diluent varies with the true solvent with which it is tested. This may be illustrated by the following dilution ratios for toluene, xylene, and VM & P naphtha with ethyl acetate and butyl acetate.

	Dilution Ratio		
	Toluene	Xylene	VM & P
Ethyl acetate	3.3	3.1	1.1
Butyl acetate	2.7	2.6	1.45

The above indicates that toluene and xylene are better diluents when used with ethyl acetate than when used with butyl acetate. However, the reverse is indicated for VM & P naphtha; it is better when used with butyl acetate. It was pointed out in Chapter 6 that the dilution ratio is a useful criterion for hydrocarbon diluents. However, the test is made with nitrocellulose as the only solid component in the solution; therefore the results are not necessarily the same as when the diluent is used in normal lacquers containing resins and plasticizers.

Some of the best true solvents are esters, ketones, and ether-alcohols; the latent solvents are alcohols. Sometimes the combination of a true solvent and a latent solvent is a better solvent than the true solvent alone. The solvent power may be indicated by the relative viscosity of nitrocellulose solutions, or by the dilution ratio when mixed with a standard diluent. A few values showing the change in dilution ratio with solvent-latent solvent combinations will illustrate this point.

	Dilution Ratio for Toluene
Butyl acetate	2.7
50% butyl acetate } 50% ethyl alcohol }	3.0
50% butyl acetate } 50% butyl alcohol }	2.4
Cellosolve	4.8
50% Cellosolve } 50% butyl alcohol }	2.9

The alcohols are usually lower in price than the esters and also more readily available in critical periods. It will be apparent that considerable saving may be realized by skillful blending of the lacquer thinner to obtain optimum solvent performance at the lowest cost. One factor which is frequently overlooked is the possible reduction in volume when two or more solvents are mixed. There are not many data available on this phenomenon, but a mix of 64 parts of ether and 36 parts of ethanol yields 98.5 parts instead of the expected 100 parts. This reduction in anticipated yield would seriously affect the cost calculations.

The physical characteristics of a number of true and latent solvents for nitrocellulose are given in Table 7, Chapter 6, and the dilution ratios for typical diluents are given in Table 4, Chapter 6. It would be impractical to attempt to show all possible combinations of solvents, latent solvents, and diluents suitable for lacquer thinners. However, the six combinations in Table 5 will serve to illustrate the range in

Table 5 Typical Lacquer Thinners

Per Cent by Weight

	Lean			Rich		
	Std.	Fast	Slow	Good	Hi Solv.	Hot Spray
Solvents						
Ethyl acetate	20	15	5	..	10	....
Butyl acetate	..	10	20	35	20	37.5
Methyl ethyl ketone	..	..	..	..	25	....
Latent solvents						
Butanol	..	10	10	15	10	12.5
Ethanol	25	5	..	..	5	10
Diluents						
Toluol	55	60	30	50	30	....
Xylol	..	..	30	..	..	40

diluent used for "rich" and "lean" thinners and also the relatively fast- and slow-evaporating types.

Blushing. It is well known that the rapid evaporation of a highly volatile material will cool the surface from which it is evaporating. This phenomenon occurs during the spray application of lacquers because evaporation is rapid from both the finely atomized lacquer as it leaves the spray gun and from the surface on which the lacquer is sprayed. If the degree of cooling of the surface of the object being sprayed is sufficient to cool it below the dew point of the atmosphere, it can be seen that moisture will condense on the surface of the freshly applied lacquer. The moisture is carried into the wet film by some of the water-miscible solvents such as ethyl alcohol and ethyl acetate. Since these solvents evaporate much faster than the water, the water becomes entrapped in the film and produces the phenomenon known as "blushing." The degree of blushing may be mild, causing a slight turbidity in a clear lacquer film, or it may be severe enough to produce complete opacity and whitening of the film. Blushing produces milkiness and loss of gloss in colored lacquer films. The difficulty may be overcome by using about 5% of a "retarder" in the lacquer thinner. The retarder is a slow-evaporating solvent which is miscible with water. The retarder slows the drying of the film sufficiently to permit the water to leave the film without causing blushing. Typical retarders are butyl lactate and butyl Cellosolve. Retarders are not necessary when spraying lacquers under normal atmospheric conditions. However, they may be required during periods of high humidity in the summer months. It has been found (21) that lacquers applied by the hot-spray process, described on page 434, are practically free from blushing because the temperature of the lacquer is not lower than room temperature when it reaches the object being sprayed.

Resin Blush. Turbidity or haze may also develop in clear lacquer films because of incomplete compatibility of the resin with the nitrocellulose. This is known as a "resin blush" and usually occurs in lacquers having high resin content. If the resin and nitrocellulose are not completely miscible they will tend to form different layers in the film; this will produce turbidity since they do not have the same refractive index. Resin blushing may also be produced by use of a poorly formulated thinner. If the thinner does not contain sufficient solvent for the resin, or if the resin solvent evaporates first, leaving non-solvents in the drying film, the resin will be partially precipitated and will produce haziness in the film.

Rate of Evaporation. The rate of evaporation of the volatile constituents affects the physical properties of the lacquer film considerably. The most apparent property is the appearance of the surface of the film; it will be rough, or contain "orange peel," if the solvent has evaporated too fast to permit proper flow or leveling of the film. Orange peel may also be produced by poor spraying technique as discussed later. A property which is not so readily apparent is the loss of tensile strength due to the inability of the nitrocellulose molecules to align themselves in the film in a manner which produces maximum strength. The correct rate of evaporation may be obtained by incorporating enough "high boilers," or slow-evaporating true solvents, in the lacquer thinner. However, it was pointed out in Chapter 6 that boiling point is not a safe criterion of the rate of evaporation of volatile liquids. Highly polar liquids are associated more strongly than less polar liquids. The association retards evaporation at normal temperatures, but association becomes almost zero at the boiling point. This may be illustrated by comparison of the rate of evaporation and the boiling range of the solvents shown in Table 6. The rate of

Table 6 Effect of Association on Rate of Evaporation

Solvent	Relative Evaporation	Boiling Range, °C
Ethyl ether	1	34–35
Methyl acetate	2.2	56–62
Methyl alcohol	6.3	64–65
Ethyl acetate	2.9	74–77
Ethyl alcohol	8.3	78
Butyl acetate	12.5	110–132
Butyl alcohol	33.0	114–118
Benzol	3.0	80–81
Toluol	6.1	110
Xylol	13.5	138

evaporation is compared with ethyl ether as the standard. It will be noted that the more highly polar alcohols evaporate more slowly than the esters even though their boiling ranges are not significantly higher. However, the rate of evaporation of the non-polar hydrocarbons is more closely related to their boiling points.

The rate of evaporation may be measured by allowing a specified quantity of solvent to evaporate from a container of given dimensions under controlled conditions of temperature and air movement. However, the results obtained will not necessarily be the same as those obtained from a normal lacquer, which is a solution of nitrocellulose, resins, and plasticizers in the solvent. The solid components retard the evaporation of the volatile constituents. In addition, good solvents will be retarded more than poor solvents or diluents because the solid components exert a greater attraction for the good solvents. This phenomenon, which was discussed in Chapter 6 under "solvent release," points to the necessity for careful formulation of lacquer thinners.

The strength of a lacquer film depends on the secondary valence forces which operate between the nitrocellulose molecules. If the drying film is maintained in a fluid condition long enough for these molecules to arrange themselves so that the maximum forces between them are developed, the film will have its greatest possible strength. A loss of strength will result from the film becoming set so rapidly that the best molecular arrangement is not possible. Good solvents tend to straighten out the long chains of high-polymer molecules, but poor solvents leave these chains coiled or kinked, thereby preventing best alignment and film strength. Also the film will have poor strength and reduced gloss if the rate of evaporation of the diluent is slower than that of the true solvent. In this case the diluent will be the last to leave the film, and it will tend to disrupt the film structure and cause partial precipitation of the polymer. However, no matter how carefully a lacquer is formulated the best performance will not be obtained unless it is applied correctly. The lacquer must be thinned sufficiently for spray application so that good atomization is obtained, and the spray operator must apply a wet coat so that it can flow and level out to a smooth surface relatively free from orange peel.

Blends of Viscosity Types. In the formulation of lacquers it may be necessary to blend solutions of various viscosity types of nitrocellulose in order to obtain a solution of required viscosity. The best results are obtained when blends of closely related viscosity types are used; it is not good practice to blend types of widely different viscosity, such as ½ sec and 1000 sec, because of their immiscibility. By means of a blending chart such as Fig. 1, the viscosity of a blend of two solutions of different viscosities may be obtained. Conversely the percentages of two different viscosity solutions necessary to produce a solution of a required viscosity may be obtained.

Facility in the use of the chart can be readily attained through use of an example calculation:

Problem: To get nitrocellulose of 5-sec viscosity by blending nitrocellulose lots of ½-sec and 15- to 20-sec viscosity.

Step 1: Use viscosity values shown on the Hercules nitrocellulose drum labels for the two nitrocellulose lots to be blended, or use your own laboratory-determined viscosity values. For illustration, assume the viscosity value for the low-viscosity type is 4 sec in 20% solution and the value for the high-viscosity type is 16 sec in 12.2% solution.

Fig. 1. Nitrocellulose Blending Chart

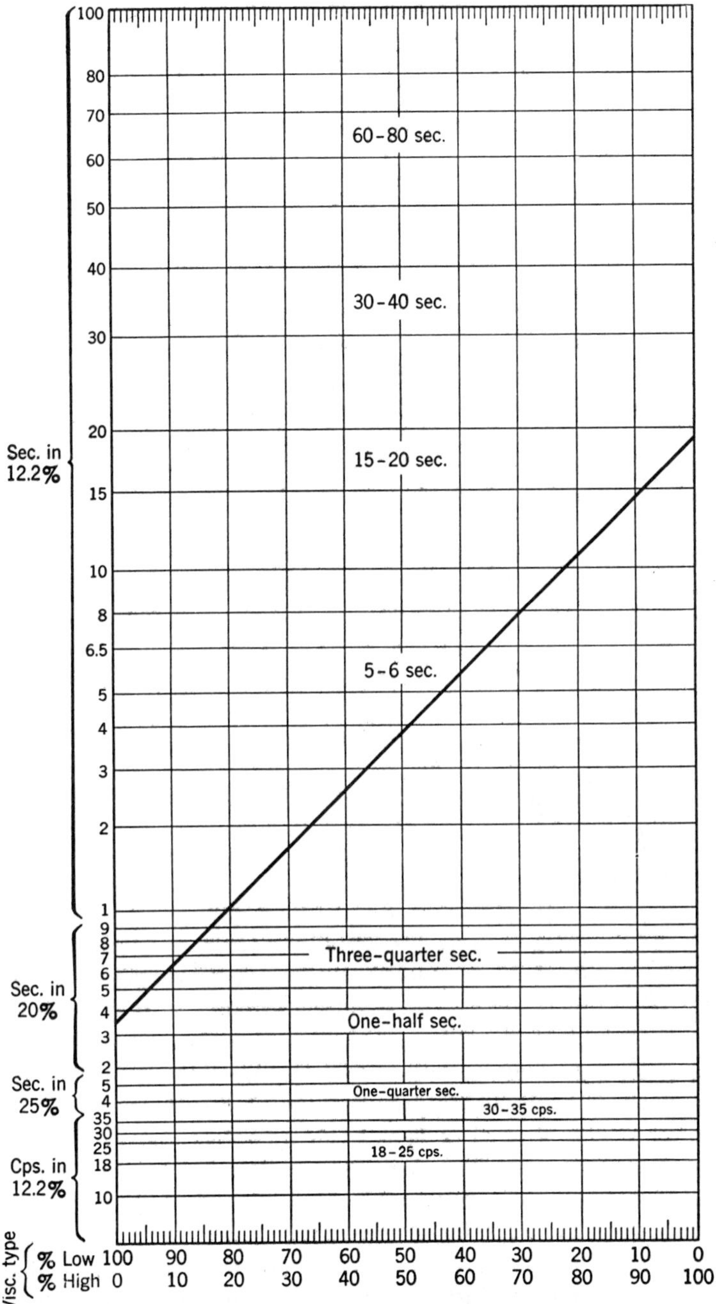

Courtesy of Hercules Powder Co.

Step 2: With a pencil, spot the point for the 4 sec in 20% solution on the left-hand margin line, and the point for the 16 sec in 12.2% solution on the right-hand margin line.

Step 3: Using a straightedge, draw a line between the 4- and 16-sec points spotted in Step 2 and locate the point where this diagonal line crosses the horizontal line representing 5-sec viscosity. The point thus located falls on a vertical percentage line, actual or interpolated. For the example chosen, the point is found to fall on the vertical line representing 40% of the ½-sec (low-viscosity) nitrocellulose and 60% of the 15- to 20-sec (high-viscosity) nitrocellulose.

Step 4: Now, knowing the percentages of the two available nitrocelluloses required to obtain a blended nitrocellulose of the desired viscosity, calculations are made for the poundages necessary for the size charge to be made.

Viscosities for the various types of nitrocellulose should be determined at the standard solution concentrations indicated on the chart.

These values for various blends may also be obtained by use of the Arrhenius equation, which relates viscosity and concentration. In this case all viscosities should be expressed in centipoises on solutions of equal concentrations.

$$\log V_s = \frac{n \log V_1 + (100 - n) \log V_2}{100}$$

where V_s = viscosity sought.
n = % of component of blend having viscosity V_1.
V_2 = viscosity of second component of blend.

Plasticizers. One of the distinct advantages of nitrocellulose is its wide range of compatibility with plasticizers and other film-forming oils and resins. This makes possible its use in coatings as widely different as brittle sanding sealers for furniture finishing, tough, durable finishes for automobiles, and soft, flexible coatings for fabrics. Nitrocellulose is also used in blends with ethyl cellulose to increase flexibility and heat resistance, and with certain polyvinyl resins to obtain maximum toughness and flexibility and also to retard its rate of combustion. Films of nitrocellulose are very strong but they have a low degree of flexibility; therefore they require plasticizers as discussed in Chapter 10. Various types of resins are also used with nitrocellulose to impart higher solids in lacquers and increased gloss and adhesion to the coatings.

The relative strength of lacquer films is indicated by tensile strength and elongation measurements on free films. The strength varies somewhat with the solvent in the lacquer, the thickness of the film, and the conditions of measurement. Although comparison of values in any particular set of tests is justified, it will be found that the values for tensile strength and per cent elongation vary considerably when reported from different laboratories. An indication of the difference in strength of films of nitrocellulose of different viscosity types (or molecular weights) may be obtained from Table 7.

Table 7 Variation in Physical Properties with Viscosity Type

Viscosity Type	Tensile Strength, psi	% Elongation	Softening Point, °C
18–23 cp	9,000	2.0	160
¼ sec	12,000	5.0	182
½ sec	12,800	7.5	190
5 sec	14,200	12.5	202
40 sec	14,600	13.0	210
1000 sec	15,000	13.0	218

It will be seen that the ½-sec and the 5-sec types offer the best compromise of film strength and amount of solids in solutions of equal viscosities. The higher-viscosity types are used in paper and textile coatings where maximum strength is required. These coatings are usually applied by the knife-coating method which requires high-viscosity lacquers. However, plasticizers are necessary in all lacquers, and we have seen in Chapter 1, page 36, and Chapter 10, page 395, that the strength of the film is necessarily reduced when plasticizers are added. The values on page 36 show that there is considerable difference in the strength of lacquer films due to the type of plasticizer used. The values indicate that resinous plasticizers produce stronger films than chemical-type plasticizers. Although this is true generally, it is unsafe to make generalizations where so many variables are involved.

The importance of using a solvent-type plasticizer whenever possible cannot be stressed too strongly. In general, less of this type of plasticizer is needed to produce a given degree of flexibility than is required of the non-solvent type. This automatically increases the strength of the film at equal flexibilities. The solvent type is retained in the film more strongly, and therefore it shows less tendency to exude at either elevated or reduced temperatures. Since nitrocellulose is compatible with almost all the chemical and resinous plasticizers shown in Chapter 10, it would be impractical to list them here. A few examples are given later under Applications for Nitrocellulose.

Resins. Nitrocellulose is compatible with a very wide range of resins and oils. However, the non-oxidizing types are generally preferred because of the danger of lifting when oxidizing materials are used. This phenomenon was discussed previously in connection with resinous materials for use with rubber resins. We shall see later that if only one coat of lacquer is to be applied the oxidizing materials may be used to advantage. However, only small amounts of oxidizing materials can be tolerated in multi-coat lacquers if freedom from lifting is to be assured. Resins improve the gloss or fullness of lacquer films. Resins also have much lower viscosity than nitrocellulose at equal concentrations in solution; therefore they increase the solid content of the lacquer at spray viscosity. This reduces the number of coats which must be applied to obtain a required film thickness. The general types of resins used with nitrocellulose are outlined below.

Natural Resins. The natural resins for lacquers have been largely replaced with synthetic resins, but some dammar, shellac, and elemi are still used. The dammar should be dewaxed as described on page 145. The dewaxed dammar has pale color, and it imparts good adhesion and easy polishing properties. Shellac also contains a natural wax which is not compatible with nitrocellulose. Therefore the shellac is dewaxed and usually bleached for use in lacquers. In clear metal lacquers it imparts adhesion and gloss. A high percentage of alcohol is required in the thinner when shellac is present since it is not soluble in esters or lacquer diluents. Gum elemi is a soft balsam-type resin with pale original color, but it tends to yellow on aging in the film. It is not soluble in alcohol or ketones but dissolves in esters and aromatic hydrocarbons. It acts as a semi-plasticizer and imparts adhesion to lacquer films.

Rosin Resins. The entire range of rosin esters is used in nitrocellulose lacquers and also the maleic-modified rosin resins. The rosin resins do not have good exterior durability; therefore they are suitable for interior lacquers only. They add gloss and ease of sanding and polishing in furniture lacquers, and they reduce the cost. The methyl esters are liquids and are used as plasticizers. The regular methyl ester is known as Abalyn, and the methyl ester of hydrogenated rosin as Hercolyn; both are produced by Hercules Powder Co. The glycerol ester is ester gum and is used in low-cost lacquers. The speed of drying and hardness of the lacquer film are related somewhat to the softening point of the resin. Therefore the pentaerythritol esters and the maleic-modified esters are preferred to ester gum. Special modifications of rosin resins, such as the Cellolyn resins (Hercules Powder Co.), have been developed for use in interior lacquers. In some special resins of this type part of the rosin may be replaced with dibasic acids such as phthalic to improve toughness. Although the dibasic acid usually lowers the softening point, this does not interfere seriously with the drying of the lacquer or the hardness of the film. The phenolic-modified rosin resins are not generally used in nitrocellulose lacquers.

Alkyd Resins. The non-oxidizing, short-oil, glyceryl phthalate type of alkyd resin as shown in Table 2, Chapter 7, is extremely useful for nitrocellulose lacquers. This type has pale color, is non-yellowing, and can be used in white lacquers. It has excellent durability and is the type used in automotive finishing lacquers. Although this type of resin is softer than the rosin resins the lacquer film will withstand the mechanical polishing used on new car finishes. It is also combined with rosin resins for furniture finishes to improve toughness and cold-check resistance. It functions partly as a plasticizer since it does not convert but remains thermoplastic.

The long-oil, oxidizing type of alkyd resin usually is not compatible with nitrocellulose. However, the medium- to short-oil oxidizing type can be used successfully if the phenomenon of lifting is kept in mind. The oxidizing-type resin converts and produces an excellent lacquer film, but it is safe only for one-coat applications. An example showing the use of this type is given later. The so-called non-reactive polyester alkyds (see Table 5, Chapter 10) are excellent resinous plasticizers for the very flexible lacquers used on fabrics, cable insulation, etc. They are based on adipic or sebacic acids esterified with the glycols, and they may or may not be modified with non-drying oils. This type of plasticizer reduces the strength of lacquer films less than the chemical plasticizers for the same degree of flexibility, and it does not volatilize from the film at elevated temperatures. Some of these plasticizers also impart very good low-temperature flexibility to lacquer films.

Lacquer Oils. Raw and blown castor oil and blown rapeseed oil are used as plasticizers in lacquers. However, the lacquer oils, such as those described in Table 4, Chapter 10, serve both as plasticizers and film-formers. They are linseed and soybean oils, processed so that they are soluble in alcohol and compatible with nitrocellulose. Since they are film-forming materials they may be used in higher ratios than the castor oils and still produce fast-drying films. They are suitable as dispersing media for the pigments in colored lacquers. They have good durability, and the soybean type is practically non-yellowing.

Amino Resins. Amino resins are heat-convertible and are not generally used in air-drying finishes. However, it has been found that small percentages of the melamine-formaldehyde type improve the water-spot resistance and gloss retention of automotive lacquer finishes. The general range in which the melamine-formaldehyde resin is used is indicated below:

	Parts by Weight, Solid Basis
Nitrocellulose, RS-½ sec	100
Non-oxidizing alkyd	150–250
Melamine-formaldehyde resins	15–25
Chemical plasticizer	15–25

It has been general practice in the lacquer industry to refer to formulations on the basis of 100 parts of dry cotton, the cotton being nitrocellulose. The term "dry" means without the alcohol used for shipping purposes. This is a very convenient way to express formulations because it enables one to visualize the relationship between the nitrocellulose and the other ingredients more readily than when they are expressed on a percentage basis. However, at present there is a tendency on the part of some manufacturers to use the percentage method; therefore both methods will be shown in this book for some formulations.

Acrylic Ester Resins. Acrylic ester resins are the polymerized ester derivatives of acrylic and α-methyl acrylic acids. Typical examples are ethyl acrylate, which is a soft, rubbery resin, and methyl methacrylate, which is a hard, tough resin. Many other variations are possible, covering a wide variety of physical properties. The Acryloid resins of the Rohm and Haas Co. belong to this class of resin. The flexible type produces lacquers which have good adhesion to fabrics and are therefore used as undercoats or "base" coats. The film-forming materials for a typical fabric base coat are given below.

	Parts by Weight,* Solid Basis	Per Cent
Nitrocellulose, RS-½ sec	100	30
Acryloid † C-10	133	40
Dibutyl sebacate	100	30

* Courtesy of Rohm and Haas Co.
† Trade name, Rohm and Haas Co.

The acrylic ester resins have excellent color retention and may be used in white lacquers for interior applications. A typical combination for white finishes on wooden articles is as follows:

	Parts by Weight,* Solid Basis	Per Cent
Nitrocellulose, RS-½ sec	100	35
Acryloid B-72	143	50
Paraplex G-50	43	15

* Courtesy of Rohm and Haas Co.

Polyvinyl Resins. The standard polyvinyl chloride and polyvinyl chloride-acetate resins are not compatible with nitrocellulose. However, when the ratio of vinyl acetate in the vinyl copolymer is increased the resin becomes compatible with nitrocellulose. A typical example of this type of resin is Vinylite VYCC, Union Carbide and Carbon Corp. This resin contains 62% vinyl chloride and 38% vinyl acetate and is described in more detail on page 500. It may be used with nitrocellulose to produce very flexible lacquers which have good chemical resistance and a slower burning rate than that which is obtained with the polyester resinous plasticizers. The burning rate will be reduced further if the phosphate plasticizers such as tricresyl phosphate and trichloroethyl phosphate are used instead of plasticizers such as dioctyl phthalate. A suggested ratio for the use of these materials is:

	Parts by Weight,* Solid Basis
Nitrocellulose, RS-½ sec	100
Vinylite VYCC	100
Chemical plasticizer	40–80

* Courtesy of Bakelite Co., Div. Union Carbide and Carbon Corp.

Pigments. A wide range of pigments is available for use in colored nitrocellulose lacquer enamels. Pigments impart color and opacity, but another of their important functions is to protect the film-former from degradation by the ultraviolet component of sunshine. In practically all cases a pigmented film is more durable on exterior exposure than a clear film. Pigments vary a good deal in their ability to absorb ultraviolet and thus protect the film-former, as discussed in more detail in Volume II. The outstanding durability of black finishes in comparison with other colors is related to the ability of carbon black to absorb the destructive ultraviolet radiation. Conversely, the relatively poor durability of organic reds and maroons is also related to their relative transparency and consequent inability to protect the film-former.

Pigments having maximum hiding power and lowest oil absorption should be used whenever possible with nitrocellulose to avoid increasing the consistency any more than is absolutely necessary. The extender pigments are not used in lacquer enamels although they may serve to advantage in lacquer primers. Alkaline materials tend to react with the nitrate groups on nitrocellulose to produce gaseous nitrogen oxides. Therefore pigments having an alkaline reaction should not be used in nitrocellulose lacquers if they can possibly be avoided. If alkaline pigments are used the addition of small quantities of acids such as citric, benzoic, or phosphoric may partially neutralize the reaction and reduce the evolution of gas. Not only does the alkaline reaction with nitrocellulose change the solubility and film characteristics of the nitrocellulose, but also the gases which are evolved create considerable pressure in the container during storage. Under some conditions the pressure may be sufficient to blow the lid from the container.

An accelerated test to determine the possibility of gas formation in pigmented lacquers is described in the nitrocellulose literature of Hercules Powder Co. It involves heating the lacquer to boiling (about 80°C) in the absence of air, and collecting any evolved gases over a standardized sodium hydroxide solution. The gases are then oxidized, and their reaction with the sodium hydroxide solution may be determined by back-titration with a standard acid solution. However, it will be apparent that good lacquer formulation would avoid the use of any material which produces an alkaline reaction.

Applications for Nitrocellulose. Nitrocellulose enjoys a unique position among surface coating materials in that finishes made with it air-dry faster than those made with any other film-former. Nitrocellulose finishes dry to a dust-free condition in a few minutes. However, they may require from several hours to overnight air drying before they are hard enough to rub and polish, or to pack for shipment. However, at this time they are usually more print resistant to the paper or other packaging materials than finishes which dry by oxidation. The fast air-drying and print-resistant features of nitrocellulose lacquers make them particularly useful for finishing large, bulky objects which cannot be baked, such as wooden furniture. Fast air-drying schedules reduce the space required in finishing and drying rooms for such objects. However, there are many instances where nitrocellulose lacquers are force-dried at temperatures from 150°F to 180°F in heated rooms or tunnel ovens. For example, lacquer finishes for automobiles are usually force-dried for about 1 hr so that they will be hard enough to withstand the polishing operation. Lacquer finishes applied to rolls of paper or textiles are often force-dried by passage through tunnel ovens in from 2 to 10 min. This removes the final few per cent of solvent, and the coated material may be rolled after cooling without danger of sticking.

Nitrocellulose is used in a very wide variety of coatings which require specialized formulation. However, there are certain fundamentals of formulation related to these coatings which aid in visualizing the materials and proportions used. Obviously any attempt at such simplification will suffer from lack of detail, but it will indicate the general types and proportions of resins and plasticizers used in lacquers. This is illustrated in Table 8, showing three general types of

Table 8 General Proportions for Lacquers

Parts by Weight, Solid Basis

Ingredients	Hard Interior	Durable Exterior	Highly Flexible
Nitrocellulose, ½ sec	100	100	....
Nitrocellulose, 5–80 sec	......	......	100
Hard resin	50–200	......	....
Non-oxidizing alkyd	0–150	75–150	....
Resinous plasticizer	......	......	150–200
Chemical plasticizer	15–25	15–25	0–25
Blown castor oil	15–25	15–25	0–50

lacquers: (1) the hard, brittle type for furniture and interior use; (2) the durable type for automobiles; (3) the flexible type for fabrics. Further details regarding these lacquers are given later and may also be obtained from raw material manufacturers.

There are, of course, many exceptions to the ratios in Table 8, but the information will illustrate certain fundamentals. The ½-sec or lower-viscosity nitrocelluloses are preferred wherever possible because they give higher-solid-content lacquers at application consistency. The ½-sec type has good exterior durability and is to be preferred in this respect to the ¼-sec or lower-viscosity types. The higher-viscosity types are used where maximum toughness and flexibility are required.

The hard resins, such as the maleic-modified rosin esters, are suitable for interior lacquers only, since they do not have sufficient durability for exterior finishes. A lacquer sealer for furniture may contain hard resin as the only modifier in ratios ranging from 100 to 400 resin to nitrocellulose. On the other hand, a finish lacquer for furniture may not contain more than 50 parts of hard resin, but it will contain from 75 to 100 parts of the durable non-oxidizing alkyd resin. A better finish would be obtained if no hard resin were used, but it would be too soft to rub and polish easily in the drying schedules of furniture finishing.

The non-oxidizing alkyds are mandatory in automobile finishing lacquers to provide the necessary durability. Proportions up to 300 of alkyd to 100 of nitrocellulose would be satisfactory for durability, but such lacquers would require longer drying schedules than would be tolerated by the industry. The oxidizing alkyds cannot be used except in one-coat finishes because of the possibility of lifting. The resinous plasticizers are too soft for general use in automotive finishing lacquers.

In both the hard- and the durable-type lacquers it is good practice to have at least 15 parts of a chemical plasticizer to 100 parts of nitrocellulose. The plasticizer should be the solvent type to ensure good retention in the film on aging. The same degree of initial flexibility may often be obtained with 30 parts of blown castor oil as would be produced with 15 parts of chemical plasticizers and 15 parts of blown castor. However, after aging for about 3 months the film with all blown castor will be considerably more brittle than the other. This statement regarding chemical plasticizers does not apply so much to the lacquers containing resinous plasticizers because these often have some solvent action on the nitrocellulose. The service requirements of the flexible lacquers may be such that chemical plasticizers cannot be used. A specific case would be insulation lacquers for electric cables; these are subject to contact with hot oils which might extract the chemical plasticizer and the film would lose flexibility. In such cases, part of the nitrocellulose may be replaced with ethyl cellulose to obtain better heat resistance and increased flexibility so that less plasticizer is necessary. Of course, cost is always an important factor in industrial formulations. Raw castor oil is one of the cheapest plasticizers, and it is used in large proportions in low-cost fabric lacquers despite its tendency to exude on aging. It is retained in the film somewhat better when some ethyl cellulose is included in the formulation.

It should be realized that the proportions and types of materials shown in Table 8 are not limiting proportions. They are intended to indicate only the general characteristics of lacquer materials.

Furniture-Finishing Lacquers. Since nitrocellulose is an important ingredient in air-drying furniture finishes a brief outline of typical finishing schedules is given below. The use of alkyd-amino resin combinations in low-bake furniture finishes is given on page 348, and the type of varnishes used in furniture finishes is given on page 207. The finishing schedule and the type of lacquer vary with the quality of finish required. Unfortunately, the quality of finish on a large percentage of furniture is sacrificed to meet the demands of fast drying schedules and easy sanding and rubbing operations, as referred to previously in connection with furniture varnishes.

Wood is a variable product; therefore the color of boards from different trees of the same species may vary considerably. Stains are used not only to obtain a desired color but also to ensure uniformity of color in the pieces of furniture of a given set, or in the various doors, drawers, etc., of a single piece of furniture. The final color uniformity is obtained by "shade staining," usually after the sealer coat has been sanded. The wood for pale or "blonde" finishes is bleached before staining. Three types of stains are available: oil stains, water stains, and non-grain-raising (NGR) stains. The oil stains may be either the soluble type or the pigment type. The soluble oil stains are solutions of dyestuffs in hydrocarbon solvents, and they may be applied by spraying or dipping. They have good penetration and excellent color, but they are not permanent to light and they tend to "bleed" into the finish coats, producing an uneven color and sometimes turbidity. The pigment type of oil stains is permanent and does not bleed, but these stains are dull in color and not as clear or transparent. The water stains are based on water-soluble dyestuffs which have good color and are permanent to light, and these stains do not bleed into the finish coats. However, the water raises the grain of the wood so that an additional sanding operation is necessary after the stain has dried. The NGR stains were developed to overcome the grain-raising difficulty. They are based on dyestuffs similar to those used in water stains, but the solvent is a combination of alcohols. Many finishers prefer to put a very thin coat of shellac or lacquer over the stain before the wood filler is applied. This is called a "wash coat" and serves chiefly to stiffen the fuzz on the wood so that it can be removed by a very light sanding.

Woods such as oak, mahogany, and walnut, which have large open pores, require the application of "wood filler" to fill the pore depressions before the application of sealer and finish coats. The best wood filler is based on raw linseed oil heavily pigmented with a coarse inert pigment such as silica. The filler may be "natural" or without a color pigment, or it may be colored with burnt sienna, burnt umber, yellow iron oxide, etc., for use on the various woods. It is thinned with VM & P naphtha, and a heavy coat should be applied by brushing or spraying. After it has "set" about 15–20 min, the excess is wiped off across the grain with a piece of burlap. The excess must be wiped off completely to avoid a "muddy" appearance, but care must be taken not to wipe the filler out of the pores. The linseed oil-type filler requires at least overnight drying before applying subsequent coats. Fast-drying fillers, 2–4 hr, have been developed based on varnishes, but it is difficult to wipe such a filler off clean without pulling it out of the pores. The so-called 30-min fillers are based on nondrying vehicles. They wipe off more easily than the varnish type, but they tend to shrink when subsequent coats are applied. Some fillers "pop" or expand when the finish coats are applied, and they do not shrink back uniformly, thereby producing an uneven appearance. It will be apparent that proper filling is essential to good finishing of open-grain wood.

A coat of sealer is applied after the filler has dried. The sealer may be shellac, although more often it is the lacquer type. It is intended to seal the irregularities in porosity of the wood, filler, etc., and to sand easily so that a smooth uniform surface may be obtained on which to apply the finish coats. In the formulation of sealers the greatest emphasis is placed on ease of sanding because sanding is an expensive operation; therefore sealers are usually weak, brittle coatings which are poor foundations for finish coats. Efforts are being made (6) to improve the quality of sealers without severe loss of sanding ability. The difficulty would be reduced if practically all the sealer coat were sanded off, but this is not done usually.

The sealer develops the color of the stain sufficiently to bring out any irregularities in color. These are then shaded out by applying a shade stain where necessary, and the finish coats are applied. Both gloss- and dull-finish lacquers are used as finish coats. The gloss finish may be "dull rubbed" or "rubbed and polished," depending on requirements. These are manual operations; consequently they are expensive. The dull-finish lacquer may be used to avoid this expense, but the appearance of this lacquer does not equal that of the hand-rubbed finish. If the dull-finish lacquer is based on the use of metallic soaps, there is usually some sacrifice in the durability of the finish. A typical finishing schedule is outlined below, but it will be realized that some variation is permissible to meet requirements of cost and performance.

Furniture-Finishing Schedule

1. Stain: Water or NGR type, spray or dip, air-dry 30–60 min.
2. Wash coat: Very dilute shellac or lacquer, spray on, air-dry 30 min, sand lightly.

3. Filler:	For open-pore woods only, brush or spray heavy coat; after it has set wipe off across the grain, air-dry overnight, or force-dry 1–2 hr at 110–130°F.
4. Sealer:	Shellac or lacquer sealer, spray 1 coat, air-dry 1–2 hr or force-dry 1 hr at 110–130°F, sand thoroughly to a smooth surface.
5. Shade stain:	Spray on where necessary to obtain color requirements.
6. First finish coat:	Spray 1 full coat or 2 light coats of gloss lacquer, air-dry 2–3 hr or force-dry 1 hr at 110–120°F, inspect, and sand where necessary.
7. Final finish coat:	Spray 1 full coat or 2 light coats of gloss rubbing lacquer or dull-finish lacquer, depending on requirements. Air-dry overnight or force-dry 2 hr at 110–120°F. Dull-rub or polish the gloss lacquer as required.

Sanding Sealers. A brief discussion of sealer formulations will serve to indicate the range and effect of the various ingredients.

Formula 1 Lacquer Sanding Sealers

	Parts by Weight, Solid Basis			
Ingredients	*A*	*B*	*C*	*D*
Nitrocellulose, ½ sec	100	100	100	100
Hard resin *	100	150	150	250
Chemical plasticizer	10	10	20	...
Blown castor #15	10	...	20	...
Zinc stearate	20	20	20	25

* Maleic-type resins as described on page 160. Typical examples are Amberol 801, Arochem 520, Beckacite 1110, Lewisol 33, and Teglac Z-152. Special modifications have been developed, such as the Cellolyn resins, Hercules Powder Co., for use in furniture finishes.

The use of ½-sec nitrocellulose represents the best compromise between film strength and solid content of the lacquers. If higher solid content is demanded the ¼-sec nitrocellulose may be used, but the film will not be quite as strong. Greater film strength may be obtained from 5- to 6-sec nitrocellulose, but at a sacrifice of solid content. With a good solvent combination the lacquers would range from about 20% solids for *A* to about 25% solids for *D*. The relatively low viscosity of the resin solution compared with the nitrocellulose solution accounts for the higher solid content of lacquer *D*. Increasing the resin content weakens the film so that sanding becomes easier. However, at high resin content the film softens more readily under the friction produced by sanding, so that the sandpaper becomes gummy.

Lacquer *B* represents the best compromise between ease of sanding and film strength. Lacquer *C* has better cold-check resistance because of the increase in plasticizers, but it would not sand as easily as *B*. Lacquer *D* presents a sacrifice of flexibility in order to obtain higher solids and ease of sanding. The best film properties would be obtained from lacquer *A*, but this lacquer would be highest in cost, lowest in solid content, and not the easiest for sanding. The reason for the poor performance of many furniture finishes will be apparent from these remarks. The chemical plasticizer is usually chosen on the basis of price. When dibutyl phthalate is lower in price than dioctyl phthalate, it is used even though dioctyl phthalate has greater permanence in the film. Tricresyl phosphate is also widely used for this type of lacquer. The blown castor is included to lower the plasticizer cost.

Zinc stearate makes the sanding easier, but it detracts from film strength and film clarity. The proportions range from 10 to 40 parts of zinc stearate to 100 of nitrocellulose, the higher ratios being used with lacquers having high resin content. For best results the zinc stearate should be dispersed thoroughly. This may be done by dispersing it in the nitrocellulose solution in a Banbury-type mixer, or in a solution of the resin and plasticizer in a pebble mill, or in the xylol and toluol components of the solvent combination. Good dispersion may be obtained by the last method.

Finish Lacquers. From the extensive work which has been done to increase the solid content of lacquers at application consistency the following fundamentals have been developed:

1. Use lowest possible viscosity type of nitrocellulose.
2. Use highest possible ratio of resin to nitrocellulose.
3. Use resins which have best compatibility and solubility.
4. Use strongest possible solvent combination.
5. Use hot-spray application technique.

These deductions are quite logical, and the reason for each should be readily apparent. However, the limitations of each regarding cost and performance must be kept in mind when formulating lacquer finishes. The effect of nitrocellulose viscosity type on film strength was discussed previously, and also the use of high resin ratios to increase solid content. However, resin manufacturers usually produce a series of resins of any given type the members of which range in hardness, solubility, and other physical properties. The number of resins in any such series should be kept to a minimum for practical reasons, but the formulator frequently may lower cost and improve performance by using the most suitable resin for the purpose in hand. For example, two resins may have the same viscosity when dissolved in toluol to the same concentration, but if one has better compatibility with nitrocellulose than the other, the lacquer made with it will have lower viscosity or higher solid content at equal viscosity.

Frequently, resins having lower viscosity in solution also have lower softening point. If this type is used the lacquer may have a higher solid content, but the resulting lacquer film may not meet the requirements for ease of polishing or for print resistance.

Cost is a severe limitation in choice of solvent combination. Rich thinners produce higher-solid lacquers, but they cost more than lean thinners. However, technical grades of some of the true solvents are available at reduced cost because they have not been fractionated so closely. Frequently such grades may be used to advantage in thinner combinations. There is also available a wide range of low-cost diluents, and the skill of the formulator will be shown by obtaining the best balance between cost and performance.

The range of materials for furniture finish lacquers shown in Formula 2 is intended to illustrate the general principles involved in this type of formulation.

Formula 2 Furniture Finish Lacquers *

	Parts by Weight, Solid Basis				
Ingredients	*A*	*B*	*C*	*D*	*E*
Nitrocellulose, RS-½ sec	100	100	100	100	100
Non-oxidizing alkyd	60	75	100	100	...
Hard resin, maleic type	40	75	50	100	200
Chemical plasticizer	20	20	15	20	20
Blown castor oil #15	20	20	15	20	30

* See Chapter 12, Formula 2, for a combination of nitrocellulose and vinyl chloride-acetate copolymer for furniture finish. See also Chapter 13, Formula 2, for a combination with acrylic ester resin for very pale original color and excellent color retention.

The ½-sec nitrocellulose is given in Formula 2 because of its good film properties, but there is justification for the use of the ¼-sec or the 18- to 35-cp type when higher solid content is demanded. The 5- to 6-sec type is used only in special cases where film quality is the most important consideration. The short-oil, non-oxidizing type of phthalic alkyd described on page 299 imparts fullness, gloss, and adhesion. This type of resin has good toughness and durability and acts as a semi-plasticizer. Normally it is not used in excess of 100 of alkyd to 100 of nitrocellulose because of its tendency to soften the film and lower the print resistance. The hard resin component may be the maleic type, as used in the sealers, Formula 1, or it may be the special modifications such as the Cellolyn resins. The standard chemical plasticizers are satisfactory, and the blown castor lowers the plasticizer cost.

The characteristics of the five lacquers in Formula 2 should be apparent from previous discussions. The solid content increases as the ratio of resin to nitrocellulose increases. Lacquer *A* would be the fastest drying and would have fair rubbing and polishing characteristics. It would be highest in cost, based on thickness of film applied. Lacquer *B* is a good compromise, but it would not have as good cold-check resistance as lacquer *C*. Lacquer *D* would have high solid content and fair film properties, but it might be too slow drying for some applications. Lacquer *E* represents the low-cost type for which there is little or no justification. It dries very fast and has good gloss but practically no toughness or cold-check resistance.

Dull-Finish Lacquers. Dull-finish lacquer is made by dispersing either zinc or aluminum stearate, or diatomaceous or colloidal silica, in the regular lacquer. Any of the five lacquers in Formula 2 is suitable, but generally the lower-cost types are chosen because dull-finish lacquers are used on less expensive furniture. The amount of dulling agent depends on the degree of sheen or dullness required and the ratio of resin to nitrocellulose. High resin ratios require more agent to dull the finish. In general from 15 to 25 parts of dulling agent to 100 of nitrocellulose is satisfactory.

The stearates tend to lower the toughness of the film and to introduce some turbidity. The clarity of a finish on furniture is important because it brings out the natural beauty of the wood. Aluminum stearate is used more than zinc stearate because it is less turbid. It is also much more soluble in aromatic solvents, particularly if it is dispersed in a mill which becomes heated. It has less dulling action if it becomes partially dissolved; therefore proper care must be exercised when dispersing it in the lacquer in order to maintain uniformity in dullness. Since it is not dissolved by esters or alcohols it may be dispersed satisfactorily in a mixture of 2 parts of butyl acetate and 1 part of butyl alcohol. These solvents are usually present in the lacquer thinner. Frequently the tops and fronts of furniture finished with dull lacquer are hand-rubbed to improve appearance. When aluminum stearate is the dulling agent the lacquer tends to rub up more glossy than may be desired. This can be overcome by the use of zinc stearate, but this agent introduces more turbidity in the finish than aluminum stearate, as noted above.

Dullness may also be obtained by dispersing fine-particle-size silica in the film. Since silica has about the same refractive index as the film-formers it does not introduce color or opacity. However, the particle size must be very small to avoid roughness, and the colloidal grade is required. When properly dispersed this type of material does not detract from film properties but may actually improve them slightly. The amount of dullness obtained is related to the degree of dispersion, and since the silica is of colloidal dimensions, it is difficult to disperse uniformly and consistently. Insufficient dispersion gives good dullness but may produce objectionable settling in the container during storage. Maximum dispersion reduces the gloss only slightly but may increase the consistency considerably. The ideal dulling agent might be a resin which has been polymerized to practically the insoluble stage and then precipitated in a finely divided state (13). This type of material has only slightly greater density than the lacquer vehicle, and about the same refractive index.

Hot-Spray Process. The fact that the viscosity of viscous liquids is reduced when the temperature is increased is common knowledge; the extent of viscosity reduction with increase in temperature for typical lacquers is shown in Fig. 2, and this relationship has been described in the literature (8, 9, 12, 21). The advantage of always spraying at a uniform temperature, approximately 160°F, in comparison with the variable temperatures encountered with spraying at normal room conditions is shown graphically in Fig. 3. The finishing industry could not take advantage of this simple means for reducing the viscosity, thus increasing the solid content at application viscosity, until a satisfactory apparatus was developed for heating the coating materials. This problem appears to have been overcome by the use of apparatus such as that shown in Fig. 4, manufactured by the Bede Products Co. (9). The apparatus is well designed and constructed, and appears to be reasonably foolproof. The continuous circulation of the coating prevents cooling in the hose when the coating is not being sprayed.

The hot-spray process permits the application of the same film thickness with fewer coats, thus saving material and manpower, and also the application conditions are more uniform. There is less solvent in the lacquer as applied and therefore less tendency to sag. This condition might appear to increase the "orange peel" or roughness of the coating, but the higher application temperature actually improves leveling and reduces orange peel. Cold-sprayed lacquers are cooled by evaporation of the solvent to about 20° below room temperature when they reach the object being sprayed. In humid weather this

Fig. 2.

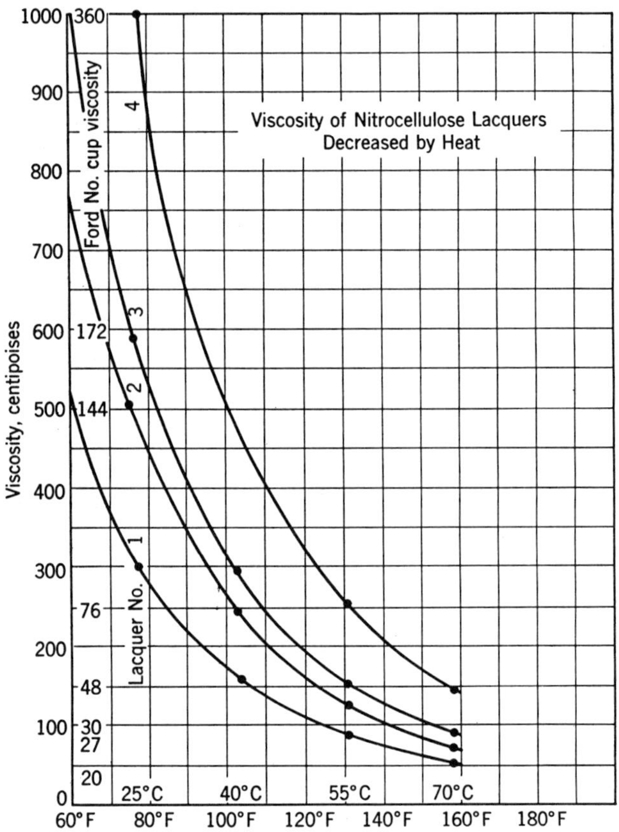

Courtesy of Hercules Powder Co.

temperature drop may be below the dew point, resulting in condensation of moisture on the film and blushing of the lacquer as described on page 414. It has been found (21) that hot-sprayed lacquers are not cooled below room temperature when they reach the object and are therefore practically free from blushing.

Credit for the discovery of the hot-spray process is due to Dr. Charles Bogin of the Commercial Solvents Corp., as disclosed in U. S.

Fig. 3.

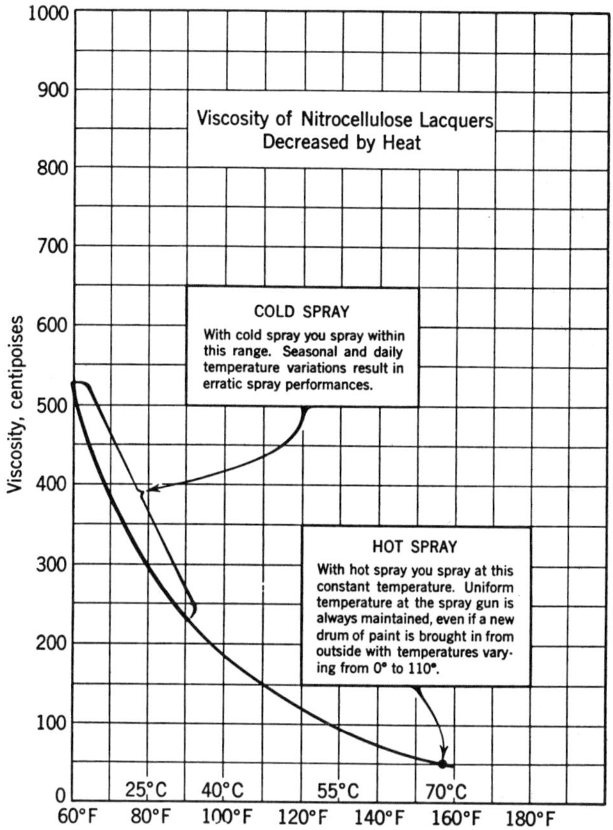

Courtesy of Hercules Powder Co.

Fig. 4. Hot-Spray Equipment

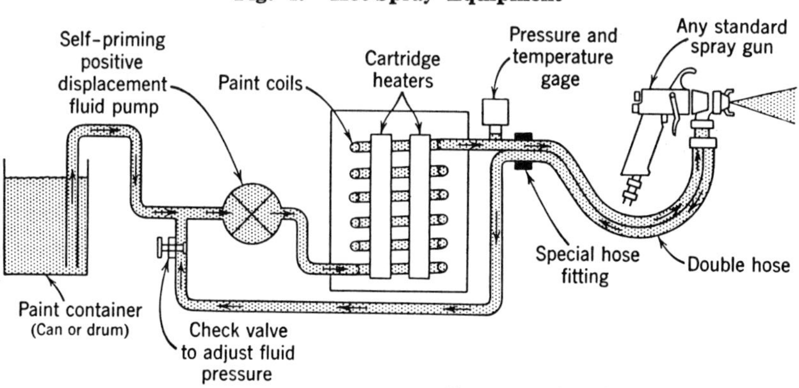

Courtesy of Bede Products Co. (9).

Patent 2,150,096, issued May 7, 1939. This patent has been dedicated to the public by the Corporation, which has published a brochure on the types of solvents to use in the hot-spray lacquers. A recent report of the Hercules Powder Co. (21) describes six different types of heating equipment for hot-spray application of coatings. A summary of the principal features of this equipment together with the addresses of the manufacturers is given in Table 9.

Table 9 * Hot-Spray Units—Comparison Data

	Manufacturer					
	Arco	Beck	Bede	DeVilbiss	M & E	Svenska M.A.B. Greiff
Model	*A*. Heater at gun *B*. Heater in line	Dualheet	U-39	QBH-5001	A	Varmolack VLS-802 Therm-O-Spray SPC-1
Time to heat lacquer from 70° to 160°F	3 min	45 min	15 min	15 min	90 min	20 min
Lacquer hold-up in unit and lines	1 oz (heater only)	0.5 gal	0.5 gal	2.5 gal	1.5 gal	15 cc (0.5 oz) (heater only)
Temperature drop between heater unit and spray gun †	0 †	10°F	10°F	10°F	10°F	0 †
Air temperature for atomization	Cold	160–180°F	Cold to 160°F	Cold	Cold	180–300°F
Heating medium	Steam	Electric, hot water	Electric, metal casting	Steam	Electric, hot water	Electric, hot air
Lacquer circulator	None	Air-motor pump	Air-motor pump	Air-motor pump	Electric-motor pump	None

* Courtesy of Hercules Powder Co.

† Length of fluid hose between heater and spray gun, 8–10 ft. Heater is attached to spray gun on the Arco and Varmolack units; therefore, they do not show a temperature drop.

Addresses:

Arco Company, 7301 Bessemer Avenue, Cleveland 4, Ohio.

Beck Equipment Company, 10118 Detroit Avenue, Cleveland 2, Ohio

Bede Products, Inc., 4311 Ridge Road, Cleveland 9, Ohio

The DeVilbiss Company, 296 Phillips Avenue, Toledo 1, Ohio

M & E Manufacturing Company, 2571 Winthrop Avenue, Indianapolis 5, Ind

Svenska M. A. B. Greiff, Stockholm, Sweden (American distributor—Douglas Wilson and Company, Inc., 101 Park Avenue, New York 17, N. Y.).

Steam-Spray Process. The steam-spray process uses superheated steam for heating and atomizing the coating material. No compressed air is required in the steam-spray process, where pressure on the finishing material is supplied to the gun from pumps or by gravity. Air pressure is used in connection with a standard-pressure paint container when smaller quantities of materials are required. The process is described in U. S. Patent 2,511,797, issued to the E. I. du Pont de Nemours Co. The equipment is available from the American Brake Shoe Co. (10) and is illustrated diagrammatically in Fig. 5. A steam generator is available for plants which do not have steam at a pressure of about 100 psi. The superheater, shown in Fig. 5, dries the steam

Fig. 5.

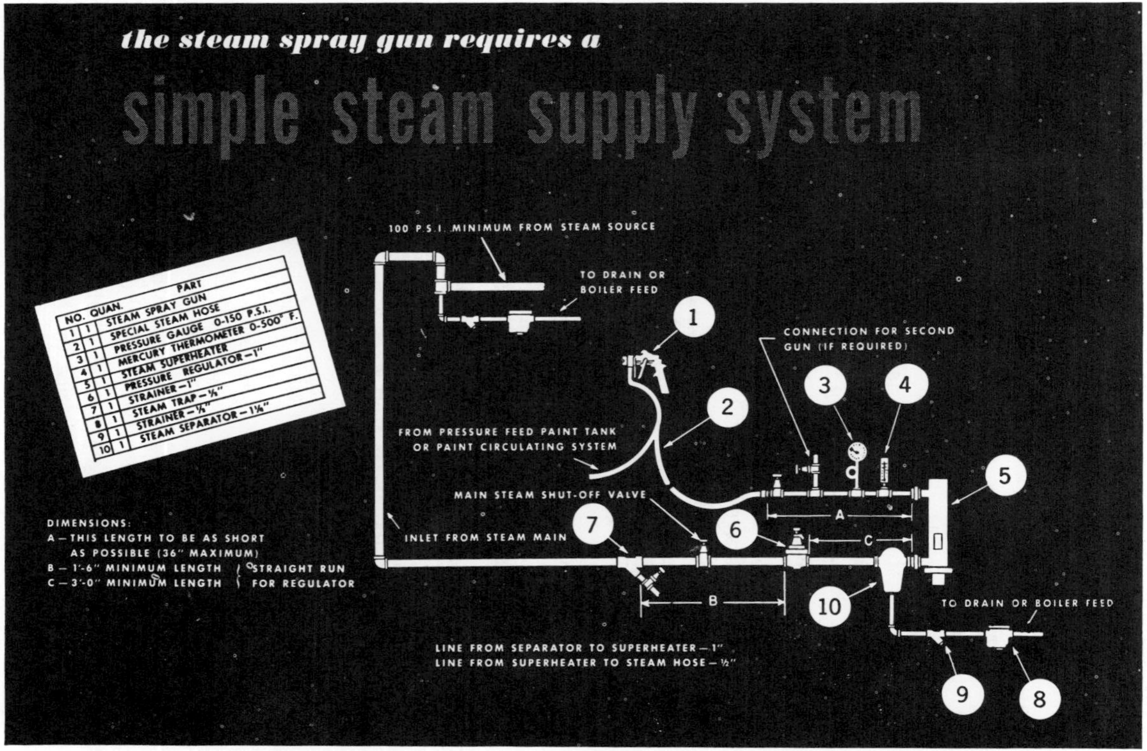

Courtesy of American Brake Shoe Co.

so that no moisture condenses on the objects being coated. The steam dissipates readily when released from the gun; therefore it does not reach the objects with the same force as compressed air. The contracting effect of steam enables the finishing material to fill in deep pockets and cover irregular surfaces more readily than air spray. This reduces the amount of overspray with a consequent saving in material. The fundamentals of the steam-spray process are based on good engineering principles, and the method should be applicable to a variety of coating materials.

Clear Lacquers for Metal. Many metal novelty goods and art objects are given a thin coating of lacquer to prevent the metal from tarnishing. Such lacquers must have good adhesion, pale color, and non-yellowing characteristics, and be non-reactive with the metal surface. Good adhesion is obtained from high resin content, and the pale color requirements from non-oxidizing-type resins. The lacquer components should have low acid value to prevent reaction with the metal. Frequently a blend of oxidizing and non-oxidizing alkyd resins is used to obtain faster-drying and tougher films than those obtained if all the resin is the non-oxidizing type. A typical blend is illustrated in Formula 3, but it should be realized that the ratio of components may be varied considerably.

Formula 3 Clear Lacquer for Metal

	Parts by Weight, Solid Basis
Nitrocellulose, ½ sec	100
Rezyl resin * 387–5	150
Rezyl resin * 99–4–60	150
Dewaxed shellac	25
Dibutyl phthalate	20

* See Table 2, Chapter 7.

Clear lacquers are used by the electroplating industry to prevent the plating metal from depositing on certain parts of an object when it is immersed in the plating bath. These lacquers are usually referred to as "stop-off" lacquers. In addition to good adhesion they are required to have good resistance to the various plating solutions. A lacquer of this type using an acrylic ester resin is shown in Formula 4.

The lacquer should be thinned to about 6% solids, and allowed to air-dry for 48 hr after spraying, or baked for 30 min at 250°F. The

Formula 4 * Stop-Off Lacquer

	Parts by Weight, Solid Basis
Nitrocellulose, ½ sec	100
Acryloid B-7	400
Dibutyl sebacate	100

	Reducing Thinner, % by Weight
Butyl acetate	20
Cellosolve acetate	20
Toluol	60
	100

* Courtesy of Rohm and Haas Co. (see Chapter 13 for description of resin).

coating has good adhesion and is not affected by immersion in the following solutions for 15 min:

5% Sodium hydroxide at 220°F
18% Hydrochloric acid
3% Sodium cyanide at 150°F
0.5% Nitric acid

Lacquers for Fabrics. Formerly, nitrocellulose lacquers were very widely used for fabrics, but they have been replaced somewhat by the fire-resistant coatings based on polyvinyl chloride-type resins. Coatings for fabrics require a very high degree of flexibility combined with film strength, and they should retain these qualities on aging. The resinous-type plasticizers similar to those shown in Table 5, Chapter 10, are most satisfactory, but their cost is usually higher than that of the chemical plasticizers or the oil plasticizers. Therefore, when cost dictates the composition plasticizer blends are made, but this results in some sacrifice of permanence. Since a high ratio of plasticizer is essential the chemical and oil types do not give the best results because they tend to exude or sweat out of the film on aging. This exudation not only produces an objectionable surface greasiness but also reduces the flexibility of the coating. When the coated fabric is intended for cable insulation the resinous plasticizers only should be used, since the chemical and oil types are leached out of the coating when the cable is in contact with hot oil. The Paraplex resins RG-2, RG-8, and RG-10 described on page 388 are typical of those suitable for such lacquers.

The higher-viscosity types of nitrocellulose are used for fabric coatings, the most general range being from the 5–6 sec to the 30–40 sec. The higher-viscosity types are also used when maximum strength is required. Blends with ethyl cellulose may be made for improved heat and light resistance as discussed previously. The range of components in fabric lacquers is shown in Formula 5.

Formula 5 Types of Lacquers for Fabrics

	Parts by Weight, Solid Basis			
	A	*B*	*C*	*D*
Nitrocellulose, 5–40 sec	100	100	100	100
Resinous plasticizer	125	75–150	75	...
Raw or blown castor oil	...	...	50	50
Chemical plasticizer	...	0–25	...	25

Lacquer *A* is the type required for cable insulation, and lacquer *B* is the general-purpose good-quality type. Lacquers *C* and *D* represent the types used for low-cost coatings. The ratios of castor oil may be higher for low-cost flexible coatings.

SS Nitrocellulose. The principal difference between SS and RS nitrocellulose is solubility. The low-viscosity SS nitrocellulose is soluble in ethanol:toluene mixtures, but the higher-viscosity material requires a small percentage of ester added to the ethanol:toluene. Better solubility is obtained with anhydrous ethanol than with the 92.5% grade. Part of the ethanol may be replaced with isopropanol or butanol with only slight change in solubility, but a slower evaporation is obtained with less tendency to blush. Toluene may be replaced with the various toluene substitutes, but the corresponding solubility will be lower in proportion to the relative dilution ratios of these substitutes. A series of three component triangular charts is available from Hercules Powder Co. which show the extent of solubility of SS nitrocellulose in mixtures of various alcohols, toluene, and ethyl acetate. The principal use for SS nitrocellulose is in lacquers which must be free from objectionable solvent odors. Table 4 shows that the nitrogen content of the SS type is from 10.7% to 11.2% and that of the RS type is from 11.8% to 12.2%. The lower nitrogen content of the SS type indicates that it has more free hydroxyl groups on each molecular unit than the RS type. The reason for the better solubility of the SS type in alcohol will be apparent from previous discussions of solubility and relative polarity. Since solubility is one of the factors in plasticizer efficiency some differences in the efficiency of standard plasticizers may be expected, according to whether they are used with the SS or the RS type. However, the differences are relatively small so that the general principles of formulation apply to either the SS or the RS type.

A similar situation exists with resin combinations; there are only slight differences in compatibility between the SS and the RS types. Obviously, alcohol-soluble resins such as shellac will have better compatibility with the SS type. For resins which have low solubility in alcohol it is advisable to increase the toluene content as much as possible. The linseed type of lacquer oils described in Table 4, Chapter 10, is alcohol soluble; therefore these oils lend themselves readily to formulation with SS nitrocellulose. High proportions of the oils may be used with little or no resin to produce specialty lacquers.

CELLULOSE ACETATE

Cellulose acetate is a very versatile cellulosic polymer which is widely used in plastics and textiles but finds only limited application in organic coatings. The chief deterrents to its use in coatings are relatively poor compatibility with resins and plasticizers, and lack of solubility in many organic solvents. The comparison of cellulose acetate with the other cellulosic polymers given in Table 1 indicates that it has about the same film strength and per cent elongation as nitrocellulose. It has a high softening point, and a slow burning rate compared with nitrocellulose. It has excellent resistance to degradation by ultraviolet radiation and, unlike nitrocellulose, does not discolor from light or heat. It absorbs more moisture than any of the other cellulosic polymers.

Manufacture. Cellulose acetate is an organic ester of cellulose made by reacting acetic anhydride and glacial acetic acid with cellulose in the presence of a catalyst such as sulfuric acid. The cellulose pulp is wetted with acetic acid and charged into the acetylators. The acetic anhydride, catalyst, and solvent are added, and acetylation is completed under carefully controlled temperature conditions. This converts the fibrous mass into a solution of cellulose triacetate. The triacetate has only limited solubility characteristics (see Fig. 7); therefore it is partially hydrolyzed with a mixture of dilute acetic and sulfuric acids or other hydrolyzing agents. The degree of hydration is controlled to produce the various grades shown in Table 10. After hydration the charge is precipitated into water and carefully neutralized, washed, and dried.

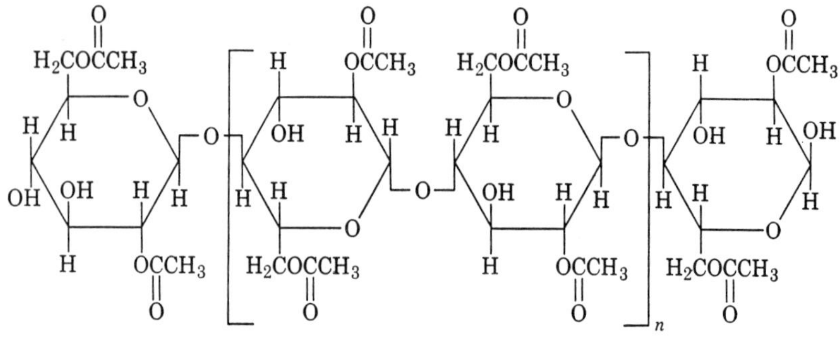

Cellulose diacetate

Types Available. The commercial material is available in the form of acetate flake. The various grades are generally referred to according to their percentage of combined acetic acid. The relationship between combined acetic acid and the number of hydroxyl groups

Fig. 6.

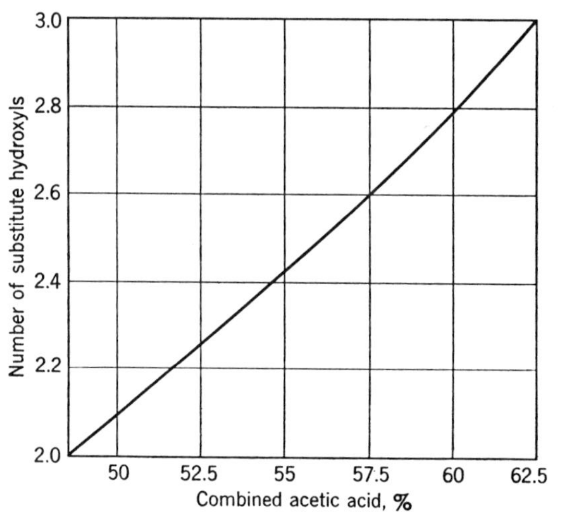

which have been substituted is shown in Fig. 6. Complete substitution gives 62.5% combined acetic acid, and the useful range of substitution extends down to about 2.1 hydroxyl groups or about 51% combined acetic acid. A typical series of commercial cellulose acetates, together with some of their physical properties, is shown in Table 10. The films for the tests were laid down on glass from a 90:10 acetone:ethanol solution to a thickness of 3 mils. After drying they were conditioned at 70°F and 65% relative humidity before testing.

Table 10 * Cellulose Acetates

Standard Grades	Type	General: % Comb. Acetic Acid	General: Vis.‡ Range	Specific Films: % Comb. Acetic Acid	Specific Films: Vis.‡	Specific Films: Tensile Strength, psi	Specific Films: % Elongation	Specific Films: Flex.†	Specific Films: Softening Point, °C
Molding	PM-3	52.5–53.5	30–60	53.2	40	12,400	8.5	25	244
	FM-6	55.5–56.2	35–55						
	WH-1	56.5–57.5	150–250 §						
Lacquer	LL-1	55.0–56.0	2–4	55.4	2.8	9,060	3.0	6	266
	LH-1	54.0–55.0	60–80	54.3	91	13,450	8.2	22	247
Film	FM-1	55.9–56.4	25–60	55.9	41	12,520	6.4	25	234
Special Grades									
Lacquer	FM-3	56.1–56.6	20–25	56.1	22	12,320	5.0	20	239
	TH-2	61.5–62.5	10–25 ‖						

* Courtesy of Hercules Powder Co.

† Flexibility, Schopper double folds.

‡ Viscosity as time of fall in seconds of a 5/16-in. steel ball through 10 in. of a 20% solution in 90:10 acetone:ethanol by weight, at 25°C in a 1-in.-diam. tube. One second of fall equals 3.78 poises.

§ 20% solution in 90:10 methylene chloride:ethanol by weight at 25°C.

‖ 15% solution in 90:10 methylene chloride:ethanol by weight at 25°C.

Table 10 shows that the regular lacquer grades run from 54% to 56% combined acetic acid, and the special grades from 56.1% to 62.5%. If necessary these values may be converted into acetyl content by multiplying by the factor 0.7167. For example, 54% combined acetic acid is equivalent to 38.7% acetyl. A limited range in viscosity types is also available. The effect of viscosity, or molecular weight, on film properties is similar to that noted for nitrocellulose. The low-viscosity type LL-1 has lower tensile strength, per cent elongation, and fold flexibility than the higher-viscosity type LH-1. Blends of different viscosity types may be made if necessary, and a blending chart similar to Fig. 1 may be obtained from the manufacturers to simplify the calculations.

Solvents for Cellulose Acetate. Cellulose acetate is more limited in solubility than the other cellulosic polymers. Frequently the best solvent is a combination of active solvent and cosolvent. The cosolvent alone does not dissolve cellulose acetate, but the combination of cosolvent and an active solvent is better than the active solvent alone. Hydrocarbon diluents may be used, the most important of which is toluene. For special purposes a considerable amount of water may be incorporated in the solvent combination. The following combination was dissolved at 60°C and remained clear after cooling to room temperature:

	Parts by Weight
Cellulose acetate LL-1	10
Ethyl acetate	54
Ethanol (92%)	18
Distilled water	18

A number of solvents, cosolvents, and solvent mixtures are shown in Table 11. The relative viscosities of the cellulose acetate solu-

Table 11 * Solvents Useful with Cellulose Acetate

Solvent	Pounds per Gallon	Boiling Point, °C	Volatility †	Viscosity ‡ at 25°C
Methylene chloride	11.4	42	1.8	Non-solvent alone (see alcohol mixture below)
Methanol	6.6	64.7	6.3	Cosolvent
Acetone	6.6	56.5	2.1	46 (see mixtures below)
Methyl acetate	7.55	57.2	2.2	62
Methyl ethyl ketone	6.7	79.6	2.4	203 §
Ethyl acetate (85%)	7.2	77.1	2.9	151 §
Ethanol	6.6	78.4	7.0	Cosolvent
Ethylene dichloride	10.45	83.7	4.1	Cosolvent
Dioxane	8.5	101.4		308
Nitromethane	9.5	101	5.8	207 (see mixtures below)
Toluene	7.2	109.5–110.5	6.1	Diluent
β-Trichloroethane	12.0	113	4.5	Cosolvent
1-Nitropropane	8.4	132	8.0	Cosolvent
Methyl Cellosolve	8.0	121–126	34.5	360 §
Methyl Cellosolve acetate	8.4	143–145	35	350
Ethyl lactate	8.6	119–176	80	720
Diacetone alcohol	7.8	153–160	147	650
Carbitol acetate	8.4	211–220	150+	Solvent only when hot

Solvent Mixtures	Viscosity	Solvent Mixtures	Viscosity
9 Methylene chloride : 1 Ethanol	84	8 Nitromethane : 2 Ethanol	109
9 Ethylene dichloride : 1 Ethanol	200	7 Nitromethane : 3 Ethanol	102
8 Ethylene dichloride : 2 Ethanol	137	8 1-Nitropropane : 2 Ethanol	185
9 Acetone : 1 Ethanol	42	9 Methyl ethyl ketone : 1 Methanol	62
9 Acetone : 1 Methanol	40	8 Methyl ethyl ketone : 2 Methanol	70
9 Nitromethane : 1 Ethanol	120	86 Methyl ethyl ketone : 14 Methanol	80

* Courtesy of Hercules Powder Co.

† Based on Gardner's ratings, *Paints, Varnishes, Lacquers, and Colors*, page 1082 (1937 ed.).

‡ Rated from centipoise viscosity of 9% LL-1 cellulose acetate solutions in solvents shown.

§ These solutions were shaken overnight, heated 2 hr at 65°C, then cooled overnight to measurement temperature.

tions may be used as criteria of solvent power. Of course, the higher-viscosity types are less soluble than the lower-viscosity types. Also the types containing more combined acetic acid are less soluble than those having lower combined acetic acid. This is shown graphically in Fig. 7. Care should be taken when dissolving cellulose acetate to avoid the formation of lumps of gel which require long agitation to dissolve. Frequently it is better to wet the cellulose acetate with the non-solvent components of the solvent combination and then add the active solvents. For example, the cellulose acetate could be wetted thoroughly with a mixture of the cosolvent and about half of the diluent; the active solvents are then added and stirring is continued until solution is complete. The remainder of the diluent may then be stirred in slowly to prevent local precipitation.

Plasticizers. The low values for per cent elongation of unplasticized cellulose acetate films given in Table 10 indicate the need for plasticizers. However, many of the usual chemical and resinous plasticizers

Fig. 7. Effect of Combined Acetic Acid on Solubility

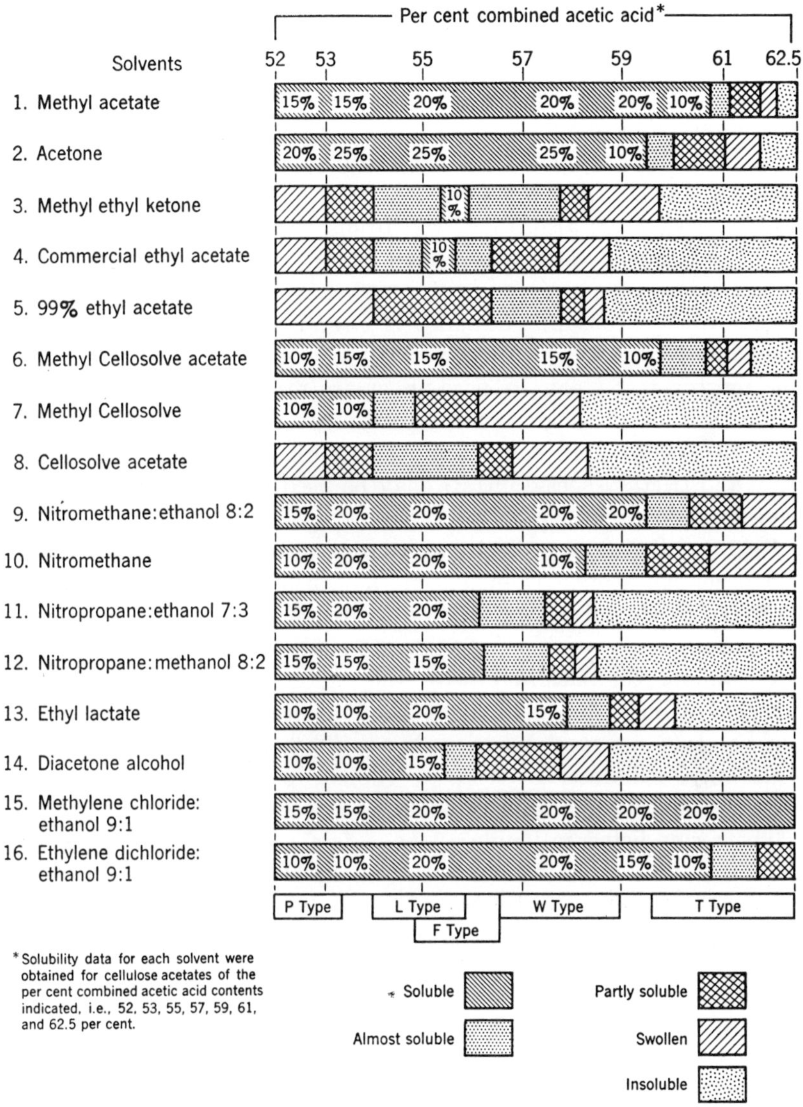

Percentage figures in solid areas show maximum solubility concentrations of cellulose acetate at 25°C

are not compatible with cellulose acetate; for example, only a small percentage of dibutyl phthalate can be used, but almost 50% of diethyl phthalate and about 100% of dimethyl phthalate are compatible. Unfortunately, the diethyl and dimethyl phthalates have fairly high vapor pressures; therefore they do not have good permanence in the film.

Since the tricresyl and triphenyl phosphates are non-solvent-type plasticizers, they should be used in combination with the solvent type. They have fair compatibility with cellulose acetate and also act as fire retardants. They are somewhat toxic and should not be used in contact with food; also, they tend to discolor on aging.

The two mixed ester plasticizers, Santicizer 15 and Santicizer 17, have good solvent action and fair permanence. They are manufactured by the Monsanto Chemical Co. and are ethyl phthalyl ethyl glycollate and methyl phthalyl ethyl glycollate respectively. They are non-toxic and can be used in grease-resistant lacquers for food wrappings. Their degree of compatibility varies somewhat with the combined acetic acid content of the cellulose acetate. Triethyl citrate and triethyl acetyl citrate may be used as plasticizers, but they have relatively poor water resistance.

Resins. Cellulose acetate has very poor compatibility with resins and other polymers. If blends of the various types of acetates are desired they should be limited to mixtures of types which do not vary more than 5% in combined acetic acid content. Cellulose acetate has only limited compatibility with the acetate-butyrate and acetate-propionate, but it has good compatibility with cellulose nitrate if mutual solvents and plasticizers are used. The acetate is not compatible with ethyl cellulose, chlorinated rubber, polyvinyls, polyacrylates, polystyrene, and synthetic rubbers.

The resins listed in Table 12 have fairly good compatibility with cellulose acetate. However, they should be tested carefully before

Table 12 Resins Compatible with Cellulose Acetate

Resin	Type	Manufacturer
Durez 500	Phenolic	Durez Plastics and Chemicals, Inc.
Durez 525	Phenolic	Durez Plastics and Chemicals, Inc.
Bakelite BR-4036	Phenolic	Bakelite Co., Div. Union Carbide and Carbon Corp.
Bakelite XR-4357	Phenolic	Bakelite Co., Div. Union Carbide and Carbon Corp.
Rezyl 14	Alkyd	American Cyanamid Co.
Aroplaz 1015	Alkyd	U. S. Industrial Chemicals
Beckamine P-254	Urea-formaldehyde	Reichhold Chemicals
Vinsol Ester	Ester gun type	Hercules Powder Co.
BR-14590	Carbic resin	Bakelite Co., Div. Union Carbide and Carbon Corp.
Santolite MS	Aryl sulfonamide-formaldehyde	Monsanto Chemical Co.
Santolite MHP	Aryl sulfonamide-formaldehyde	Monsanto Chemical Co.
Santolite K	Aryl sulfonamide-formaldehyde	Monsanto Chemical Co.

using because compatibility varies with the type of cellulose acetate, plasticizer, and solvent combination.

Applications. Cellulose acetate is used much more extensively in plastics and transparent film and sheeting than in lacquers. Its brilliant clarity and resistance to yellowing make it ideal for novelty packaging and food wrappers. It finds use in specialty lacquers for metal, paper, and fabric where resistance to heat and light is important. These lacquers may be formulated to have good resistance to animal and vegetable fats and oils, cheese, and dilute acids. They are somewhat sensitive to alkali and water.

CELLULOSE ACETATE-BUTYRATE

The cellulose acetate-butyrates are mixed esters which vary in the ratio of acetyl to butyryl as shown in Table 13 and Fig. 8. These

Table 13 * Composition of Lacquer-Type Cellulose Acetate-Butyrates

Composition and Properties	AB-161-2	AB-272-3	AB-381-1	AB-500-1	A-48 †	½-Sec NC
Acetyl, %	31.0	20.5	13.0	6.5		
Butyryl, %	16.0	26.5	37.0	48.0		
Hydroxyl, %	1.0	2.5	2.0	0.5		
Group per C_6 unit, %						
Acetyl	2.18	1.45	1.0	0.54		
Butyryl	0.67	1.11	1.67	2.40		
Hydroxyl	0.15	0.44	0.33	0.06		
Melting point, °C	250	205	180	165	235	Dec.
Density, g/cc	1.25	1.23	1.21	1.18	1.31	1.59
Moisture regain						
25% RH	0.8	0.9	0.6	0.4	1.6	0.8
50% RH	1.7	1.8	1.1	0.6	3.5	1.8
75% RH	2.8	3.3	1.8	1.1	6.3	2.7
95% RH	4.4	6.0	3.0	2.0	12.5	5.1

* Courtesy of Malm and Smith (14), Eastman Kodak Co.
† Cellulose acetate.

polymers have somewhat better solubility and compatibility than cellulose acetate, but they are not as good in these respects as nitrocellulose. In general they have slightly lower softening points than the acetates and less film strength. They have lower water absorption than the acetates and about equal to that of nitrocellulose. They have good heat and light resistance and are slow burning in comparison with nitrocellulose. The properties of the lacquer-type cellulose acetate-butyrates are described in detail by Malm and Smith (14). A series of ultraviolet inhibitors for acetate-butyrate plastics was investigated by Meyer and Gearhart (15).

The mixed esters are partially hydrolyzed similarly to the acetates described in the previous section. Figure 8 shows that the four acetate-butyrates lie between the diester and triester lines. The values

Fig. 8. Composition of Commercial Cellulose Acetate-Butyrates

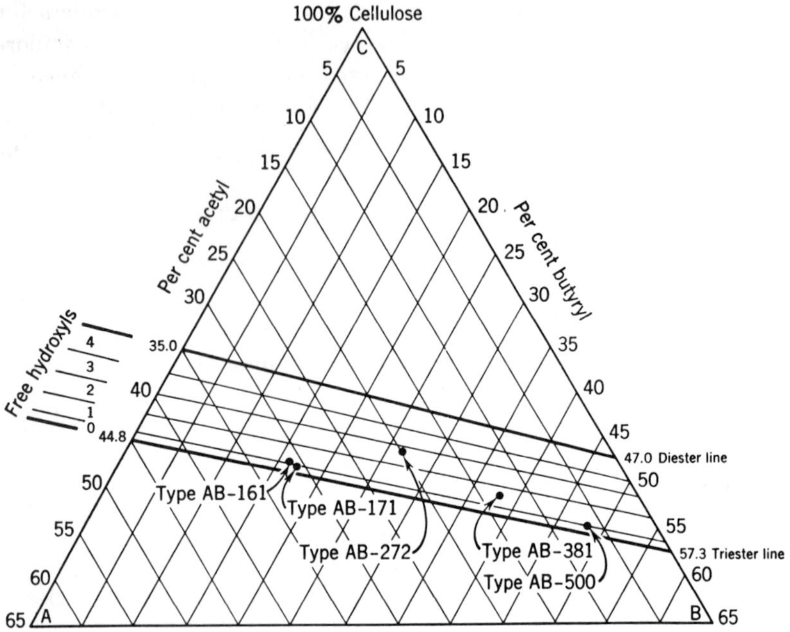

Courtesy of Tennessee Eastman Corp.

for acetyl and butyryl content may be read directly from the chart. The number of free hydroxyls per unit may also be estimated.

The values for per cent acetyl and butyryl are shown in Table 13 together with the average number of these groups per anhydroglucose unit of cellulose. The code for the commercial designations used in Table 13 is as follows: AB indicates acetate-butyrate, the first two

Table 14 * Commercial Cellulose Acetobutyrates

Aceto-Butyrate	Softening Point, °C	Tensile Strength, psi	% Elongation	Moisture Regain, 75% RH	Specific Gravity	Pounds per Gallon
AB-161	208–225	7,000–10,000	3–8	2.8	1.25	10.4
EAB †-381	180–204	7,000– 9,000	3–15	2.2	1.21	10.1
EAB †-500	166–168	6,000– 8,000	3–15	1.1	1.18	9.84

* Courtesy of Tennessee Eastman Co.

† E prefix indicates filtered resins which give clear solutions.

digits represent the approximate butyryl content, the third digit gives the degree of hydrolysis, the digit following the dash represents the average viscosity of the particular type. The physical characteristics of three commercial cellulose acetobutyrates are given in Table 14. The effect of increasing butyryl content on the physical properties and end uses of the acetate-acetobutyrate polymers is shown in Table 15.

Table 15 Effect of Butyryl Content

	Cellulose Acetate	Low-Butyryl Cellulose Acetate-Butyrate		Increasing Butyryl Content →		High-Butyryl Cellulose Acetate-Butyrate
Grade *	E-398-10	AB-161	AB-171	AB-272	AB-381	AB-500
Tolerance for toluene	Very slight	Slight	Slight	Moderate	Considerable	Almost soluble
Compatibility	Very limited	Very limited	Very limited	Limited	Moderately wide	Moderately wide
Tensile strength	Decreasing→					
Hardness	Decreasing→					
Melting point	Decreasing→					
Flexibility	Increasing→					
Moisture resistance	Increasing→					
General usage	Paper coating, flash bulbs, screen coating	Airplane dopes, cable lacquers		Cable lacquers	General lacquer coatings, gel lacquers	Strip and melt coatings

* Tennessee Eastman Co.

Solubility. The physical characteristics of a number of solvents for the cellulose acetate-butyrates are given in Table 16. An indication of the solvent power of these solvents may be obtained from the solution viscosities and relative viscosities shown in Table 17. The solution viscosities are expressed in centipoises for 10% solutions at 25°C and were determined with a Fenske-Cannon modification of the Ostwald pipet. The values for relative viscosities were obtained by dividing the viscosity of the solution by the viscosity of the solvent. The relative viscosity is considered by some authorities to be a better indication of solvent power than the actual viscosity because it takes into account the viscosity of the solvent.

In general, the solubility of the cellulose acetate-butyrates increases as the butyryl content increases. This may have been anticipated from the discussion in Chapter 6 of the factors which affect solubility. The longer carbon chain of the butyryl radical as a side chain on the polymer usually confers greater solubility than the shorter chain of the acetyl radical. Unfortunately the longer side chain usually reduces the melting point (see Table 13) and corresponding hardness of the film.

Table 16 * Physical Characteristics of Solvents

Solvent	Mol. Wt. †	Boiling Point, °C †	Vapor Pressure at 25°C ‡	Density §	Viscosity ‖
Acetone	58	56	220	0.79	0.31
Methyl ethyl ketone	72	80	100	0.80	0.39
Methyl isobutyl ketone	100	116	18	0.80	0.59
Methyl acetate	74	57	210	0.93	0.44
Ethyl acetate	88	77	91	0.88	0.43
Isopropyl acetate	102	89	60	0.87	0.56
n-Butyl acetate	116	126	14	0.87	0.71
Methylene chloride	85	41	410	1.33	0.57
Ethylene chloride	99	83	83	1.25	0.78
Propylene chloride	113	97	52	1.15	0.80
Nitromethane	61	101	35	1.13	0.70
Nitroethane	75	114	20	1.05	0.65
2-Nitropropane	89	120	17	0.99	0.73
Methyl Cellosolve acetate	118	144	4.5	1.00	1.09
Cellosolve acetate	132	156	2.8	0.97	1.49
Methyl Cellosolve	76	124	12	0.96	1.53
Ethyl Cellosolve	90	135	6	0.93	1.66
1,4-Dioxane	88	101	40	1.03	0.85
Tetrachloroethane	168	146	6.4	1.60	1.82
Cyclohexanone	98	157	4.6	0.94	2.52
Dichloroethyl ether	143	178	1.1	1.22	2.08
Ethyl lactate	118	154	2.5	1.03	2.33

* Courtesy of Malm and Smith (14), Eastman Kodak Co.

† Literature values for pure compounds.

‡ Vapor pressure of pure compounds at 25°C in mm Hg, calculated from data of Stull (16) where possible, otherwise experimental values.

§ Density in grams per cubic centimeter at 25°C, experimental values on commercial-grade material.

‖ Viscosity in centipoises at 25°C on commercial-grade material.

The dilution ratios for various non-solvents reflect the solubility characteristics, and generally they increase as the butyryl content of the polymer increases. They are lowest for the AB-161-2, and highest for the AB-500-1. Table 18 shows the values obtained with AB-381-1 at a final concentration of 10%. The values represent the number of milliliters of diluent which can be added to 1 ml of solvent without causing precipitation at 10% concentration.

It will be realized from the method employed to obtain the dilution ratios that the values represent the amount of diluent which can be added to a solution of the polymer in a true solvent. The values are

Table 17 * Solution Viscosities, 10% Solids at 25°C

Solvents	AB-161-2		AB-272-3		AB-381-1		AB-500-1		½-Sec NC	
	Vis., cp	Rel. Vis.	Vis., cp	Rel. Vis.	Vis., cp	Rel. Vis.	Vis., cp	Rel. Vis.	Vis., cp	Rel. Vis.
Acetone	44	142	47	152	19	61	19	61	12	39
Methyl ethyl ketone	64	165	79	205	24	62	25	64	17	44
Methyl isobutyl ketone	In	....	910	1,540	39	66	32	54	33	56
Methyl acetate	98	220	121	275	36	82	37	84	31	71
Ethyl acetate	109	250	155	365	37	86	40	93	37	86
Isopropyl acetate	In	....	290	520	47	84	42	75	37	66
n-Butyl acetate	In	....	1,400	2,000	61	87	50	71	54	76
Methylene chloride	114	200	2,800	4,000	42	74	34	59	In	....
Ethylene chloride	240	305	14,000	18,000	99	127	68	87	In	....
Propylene chloride	In	....	In	....	102	128	48	60	In	....
Nitromethane	330	470	410	590	120	171	118	169	115	165
Nitroethane	190	290	880	740	69	106	111	171	103	160
2-Nitropropane	In	....	7,700	10,500	90	123	60	82	160	220
Methyl Cellosolve acetate	390	360	440	400	133	122	132	121	130	119
Cellosolve acetate	In	....	730	490	155	104	130	88	150	101
Methyl Cellosolve	In	....	470	310	175	115	200	131	170	111
Ethyl Cellosolve	In	....	650	390	160	96	In	....	215	130
1,4-Dioxane	440	520	490	580	150	175	170	200	In	....
Tetrachloroethane	560	310	13,000	7,150	230	126	165	91	In	....
Cyclohexanone	960	380	470	295	300	119	210	83	165	65
Dichloroethyl ether	In	....	23,000	11,000	300	144	260	125	In	....
Ethyl lactate	In	....	1,000	430	320	137	355	152	255	109
Int. vis. in acetone	1.0	....	1.1	....	0.8	....	0.9	....	0.7	....

* Courtesy of Malm and Smith (14), Eastman Kodak Co.
In = Insoluble.

Table 18 * Dilution Ratios for AB-381-1

Solvents	Diluents: Ethanol	*n*-Butanol	Xylene	High-Sol. Naphtha	VM & P Naphtha
Acetone	10+	8.1	8.9	8.8	1.1
Methyl ethyl ketone	10+	4.7	6.4	5.4	1.1
Nitroethane	10+	5.7	10+	9.2	0.9
Nitropropane	10+	5.9	7.2	3.5	0.8
Methyl Cellosolve	Gel	1.0	10+	10+	1.2
Cellosolve	2.2	1.0	9.8	10+	0.3
Ethyl acetate	9.1	3.7	5.1	5.3	1.2
n-Butyl acetate	7.4	2.0	4.0	2.3	0.8
Methyl Cellosolve acetate	10.0	4.0	5.2	5.1	0.7
Ethyl acetate	6.9	3.6	9.6	10.1	0.9
Ethylene chloride	10.0	4.5	5.0	3.5	0.9
Propylene chloride	9.4	4.0	2.4	1.8	0.7

* Courtesy of Malm and Smith (14), Eastman Kodak Co.

not necessarily an indication of the amount of diluent which may be used in an initial solvent mixture. Also, the values will be somewhat different when mixtures of solvents are used as the initial solvent. Although alcohols are not true solvents they frequently lower the viscosity when used with true solvents. The maximum lowering of viscosity will occur at a specific alcohol content for the various acetate-butyrates. It will also vary as the carbon chain length or molecular weight of the alcohol increases. A few solvent combinations which may be used with the AB-381-1 type of cellulose acetate-butyrate are shown in Table 19, together with the viscosity in centipoises at 25°C for four concentrations.

Table 19 * Solvent Combinations for Acetate-Butyrate AB-381-1

	Composition, % by Weight					
Solvents	1	2	3	4	5	6
Acetone	35	...	30	...	...	...
Methyl isobutyl ketone	...	...	...	...	...	20
Ethyl acetate	10	20	15	10	...	...
n-Butyl acetate	...	...	...	20	20	10
Cellosolve acetate	...	15	15	...	12	...
Diacetone alcohol	10	...	...	...	...	...
Ethyl lactate	...	...	5	...	...	5
Ethyl alcohol	...	15	...	20	...	...
Isopropyl alcohol	...	...	...	...	13	...
n-Butyl alcohol	5	5	10	10	...	10
Toluene	30	25	25	40	55	...
Xylene	10	...	...	...	...	45
VM & P naphtha	...	...	...	...	...	...
Hexane	...	20	...	...	...	...
Vis. at % solids						
10	36	38	46	45	46	61
12	65	70	82	81	85	117
14	115	130	145	145	155	225
16	205	240	260	260	280	430

* Courtesy of Tennessee Eastman Co.

An indication of the relative solvent power of these combinations may be obtained from their viscosities. The #6 composition has the lowest solvent power, but it is much slower evaporating than the others. The solvent composition should be adjusted for various applications, such as spray, dip, or knife-coat.

Plasticizers and Resins. The cellulose acetate-butyrates have better compatibility with plasticizers than the cellulose acetate, and the compatibility will vary with the acetyl:butyryl ratio. For coating compositions the plasticizer content is generally 25% or less of the film-former, but for free films it may be as high as equal parts of plasticizer and polymer. The compatibility of some common plasticizers with the four acetate-butyrates is given in Table 20. As may be

Table 20 * Compatibility of Certain Common Plasticizers with Eastman Cellulose Acetate-Butyrate

	AB-161-2		AB-272-3		AB-381-1		AB-500-1	
Plasticizer	25%	100%	25%	100%	25%	100%	25%	100%
Diethyl phthalate	C	C	C	C	C	C	C	SH
Dibutyl phthalate	C	C	C	C	C	C	C	C
Dioctyl phthalate	C	H	H	H	C	C	C	C
Santicizer M-17	C	C	C	T	C	C	C	H
Santicizer E-15	C	C	C	C	C	C	C	SH
Santicizer B-16	C	C	C	C	C	C	C	S
Santicizer 8	C	C	C	T	C	T	C	T
Santicizer 10	C	H	C	H	C	H	S	H
Santicizer 1-H	C	C	C	H	C	C	C	T
Triacetin	C	C	C	C	C	C	C	H
Tripropionin	C	C	C	C	C	C	C	H
Tributyrin	C	C	C	C	C	C	C	C
Tricresyl phosphate	C	C	C	S	C	C	C	C
Triphenyl phosphate	C	H	C	H	C	H	C	SH
Ethyl benzoate	C	H	C	C	C	H	H	H
Dibutyl sebacate	C	C	C	H	C	C	C	S
Dioctyl sebacate	H	HS	H	H	C	S	H	H
Butyl adipate	H	H	C	C	C	C	C	C
Butyl stearate	H	H	H	H	H	H	H	SH
Methyl Cellosolve stearate	C †	S	H	H	C †	S	H	SH
Methox	C	C	C	T	C	T	C	S
Ethox	C	C	C	T	C	C	C	S
Raw castor oil	H	H	H	H	C	H	H	SH
Blown castor oil	H	H	H	H	H	H	H	SH
Camphor	C	H	C	H	C	H	H	H

C—Complete compatibility.
T—Clear film but remains tacky.
S—Plasticizer exudes on standing.
H—Hazy film, plasticizer incompatible.

* Courtesy of Malm and Smith (14).

† This plasticizer is apparently perfectly compatible with the esters indicated at the lower plasticizer concentrations. However, if subjected to cooling to temperatures of around 40°F or lower, it will exude profusely.

expected the compatibility of the acetate-butyrates with resins also varies with the acetyl:butyryl ratio. The details of compatibility with a wide range of commercial resins are available from the manufacturers and are also given by Malm and Smith (14). A general picture of the trend in compatibility of the four acetate-butyrates with various types of resins is given in Table 21. Resins are used to improve adhesion and in some cases to increase surface hardness. A small ratio of the oxidizing type of alkyd resin may be included to advantage in some applications, but the possibility of lifting with such compositions should be kept in mind. Faster drying schedules may

Table 21 * Trends in Compatibility of Synthetic Resins with Eastman Cellulose Acetate-Butyrates

Type of Resin	AB-161-2	AB-272-3	AB-381-1	AB-500-1
Alkyds Phenolics Ureas Maleics Polyesters	Compatibility depends on modifiers used in resins. Many resins of these types designed for lacquer use are compatible, particularly with type AB-381-1 ester.			
Aryl sulfonamide-formaldehyde condensates Modified hydrocarbons	—— Compatibility very good ——			
Chlorinated diphenyls	—— Good ——		—— Very good ——	
Acrylates	Fair	Good	Very good	Fair
Rosin derivatives	—— Fair ——		—— Generally good ——	
Polyvinyl acetates	Fair	Very good	Good	Poor
Polyvinyl chlorides Polyvinyl chloride-acetates Polyvinyl acetals	—— Generally not compatible ——			
Unmodified hydrocarbons Melamines Furfurals	—— Generally not compatible ——			
Alkyd resin plasticizers	Fair	—— Generally good ——		

* Courtesy of Malm and Smith (14).

be obtained from force drying or mild baking to facilitate the removal of the solvent. In general a higher percentage of aromatic hydrocarbon may be used in the solvent combination when high ratios of resin are present in the coating formulation.

Applications. Coatings containing the acetate-butyrates may be applied by the spray, dip, gel-dip, and hot-melt methods. The adhesion of these coatings is good when high ratios of resin and plasticizer are used, but a primer may be necessary for smooth surfaces when the polymer ratio is high. The toughness and water and chemical resistance of the coatings may be varied by choice of particular acetate-butyrate and amount and type of resin and plasticizer. The suggested ratio in Table 22 illustrates the type of formulations for a range of applications.

Formulations 1 and 2 illustrate the type used for spray application. Formulation 3 represents the type for knife-coat application to paper, and formulation 4 the type for cloth coating and cable lacquers. Formulation 5 is the type used by Malm and Smith (17) for gel-dipping lacquers. The gel-dipping process depends on the fact that the particular formulation is a mobile liquid at about 120°F so that it will drain off normally from objects dipped into it. However, the coating sets to a firm gel on cooling to room temperature of about 70–75°F. Coatings up to 25 mils in thickness may be applied by a single dip. The process is well suited to coating small objects of wood, paperboard,

Table 22 * Suggested Formulations for Acetate-Butyrates

	Composition by Weight				
Film-Former	1	2	3	4	5
AB-162-2	100	100	...	...	...
AB-381-1	...	...	...	100	...
EAB-381-20	...	...	100	...	100
Resin and plasticizer	100	75	...	60	...
Tricresyl phosphate	...	...	75	...	...
Dicapryl phthalate	...	...	...	...	20
Solvent by Weight, %					
Acetone	30	12	30	22	...
Ethyl acetate	15	10	...	...	...
Cellosolve acetate	15	...	...	...	...
Ethyl lactate	5	...	...	...	...
n-Butyl alcohol	10	...	...	...	...
Ethylene chloride	...	32	...	...	...
Methyl Cellosolve acetate	...	23	...	...	...
Ethyl alcohol #1	...	...	20	20	...
Isopropyl alcohol (99%)	...	...	...	...	80
Toluene	25	23	50	50	...
Xylene	...	...	...	...	20
Hexane				8	...
% solids	20	14	40	30	20

* Courtesy of Tennessee Eastman Co.

glass, or metal. Since these coatings have relatively poor adhesion it is usually necessary to apply a primer before gel dipping.

The acetate-butyrates (EAB-500 types) are suitable for hot-melt compositions which may be applied to paper or textiles at elevated temperatures without the use of volatile solvents. This technique and compositions suitable for it have been described by Malm *et al.* (18) and are also available from Tennessee Eastman Corp. The properties of such coatings may be varied considerably by choice of acetate-butyrate and variation in plasticizer resins, waxes, etc. The high butyryl acetate-butyrates are also suitable for the so-called strippable coatings. These coatings are applied to metal objects by a hot-melt dip process and are used as water- and rust-proof packaging materials. They are transparent and may be removed readily by cutting the coating with a knife and stripping it off the object. The removed coating may be melted and reused if desired. A slight decomposition of the acetate-butyrate may take place at elevated temperatures with the consequent production of a disagreeable odor. This may be minimized by inclusion of about 0.06% of magnesium carbonate based on the acetate-butyrate or use of 0.25–0.50% of such stabilizers as A-5 (Union Carbide and Carbon Corp.) and Ferro 900 (Ferro Chemical Co.). It is also advisable to include from 0.1% to 0.25% of an antioxidant such as 2,5-di-*tert*-butyl hydroquinone.

CELLULOSE ACETATE-PROPIONATES

The cellulose acetate-propionates are not widely used in coatings at the present time. An extensive article by Roberts and Weissberg (19) describes their use in airplane dopes. Comprehensive data are given on solution and film properties and mechanism of film shrinkage and plasticizer action. The authors found that films obtained from solutions in solvent-diluent mixtures shrink more than those obtained from solutions in a single solvent.

ETHYL CELLULOSE

Ethyl cellulose is an ether of cellulose made by reacting alkali cellulose with ethyl chloride under pressure, followed by precipitation in water and subsequent purification. The reaction may be expressed as follows:

$$\text{Alkali cellulose} + C_2H_5Cl \rightarrow ROC_2H_5 + NaCl$$

R represents the cellulose radical.

Complete substitution on each of the three hydroxyl groups of the anhydroglucose unit of cellulose produces the triethyl ether. This material has very limited solubility and compatibility and is not produced commercially.

The structure of diethyl cellulose may be represented as follows, although it should be recognized that an actual product would be less symmetrical.

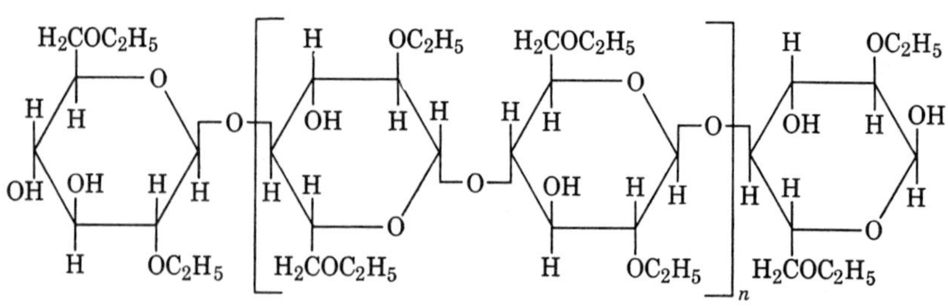

Diethyl cellulose

Types Available. Ethyl cellulose is produced by such manufacturers as Hercules Powder Co. and the Dow Chemical Co. The commercial materials shown in Table 23 vary from 2.21 to 2.53+ ethoxyl

Table 23 * Commercial Ethoxyl and Viscosity Types of Ethyl Cellulose

	Type			
	G	K	N	T
Ethoxyl, %	44.5–45.5	45.5–46.8	47.5–49.0	49.0+
Substitution	2.21–2.28	2.28–2.38	2.42–2.53	2.53+
Vis. range, cp	Viscosity Types Available			
6–8	...	...	7	...
8–11	...	...	10	10
12–16	...	14	14	...
18–24	...	...	22	...
40–52	50	50	50	50
80–105	100	100	100	100
150–250	...	200	200	200

For convenience in specifying, the materials are referred to according to both type and viscosity, such as N-7 or N-50.

* Courtesy of Hercules Powder Co.

groups per anhydroglucose unit. This is usually expressed as per cent ethoxyl content and corresponds to a range of from 44.5% to a value somewhat greater than 49.0%. The N type, 47.5–49.0% ethoxyl, is the one most widely used in surface coatings. The commercial materials are produced in a range of molecular weights which is indicated by the range in viscosity shown in Table 23. The viscosity is measured at 25°C on 5% solutions of the ethyl cellulose in a solvent mixture of 80 parts of toluene and 20 parts of ethanol by weight. The ethanol is the denatured 2B ethyl alcohol of 95% strength. The samples were dried at 100°C for 30 min before making the solution. It will be noted that certain types are not produced in the complete viscosity range because of lack of demand at the present time.

We have seen previously that the tensile strength and per cent elongation at rupture of films of high polymers tend to increase with increase in molecular weight of the polymer. This is also true for ethyl cellulose as shown by the data in Table 24. The films were cast from solution in 80:20 toluene:ethanol, dried 16 hr at 160°F, and conditioned 48 hr at 70°F and 65% relative humidity before testing. The dry films were 3 mils in thickness.

Table 24 * Variation in Physical Properties with Viscosity of N-Type Ethyl Cellulose

Actual Viscosity, cp 5% Solution	Tensile Strength, psi	% Elongation	Softening Point, °C
7.4	7,960	7	151
9.3	7,670	11	151
13.0	8,100	14	156
19.7	8,250	16	156
42.5	8,390	14	161
93.5	9,100	25	167
199.0	9,240	27	171
319.7	9,380	24	170
785.3	9,660	23	173
1,196	10,800	25	172

* Courtesy of Hercules Powder Co.

Ethyl cellulose has a range of good properties which make it a very useful material in surface coatings. Since it is an ether of cellulose it is more resistant to alkali than the cellulose esters. However, it is more sensitive to attack by acids, although it is not seriously affected by dilute acids. Some types are more flexible than the cellulose esters, as shown in Table 1, and ethyl cellulose retains better flexibility at low temperatures. It is not as readily degraded by sunlight or ultraviolet radiation as nitrocellulose and is more resistant to heat. However, antioxidants should be added when it is used in compositions which are heated above its melting point. Ethyl cellulose has good compatibility with a range of plasticizers, resins, oils, and waxes. It also has good solubility in a wide range of solvents, including the lower-cost aromatic hydrocarbons and ethanol.

Solubility. Since the ethyl ether radical is less polar than the nitrate or acetate radical it might be expected that ethyl cellulose would have better solubility than nitrocellulose or cellulose acetate in non-polar solvents such as aromatic hydrocarbons. This would apply more particularly to the N and T types, which have a high degree of substitution. The effect of the degree of substitution on the solubility of ethyl cellulose is shown graphically in Fig. 9. This chart shows at a glance that the N type, 47–49% ethoxyl, has the best solubility in aromatic hydrocarbons and also has the widest solubility in the range of solvents. Since the molecules of the N type have both non-polar $—C_2H_5$ groups and polar —OH groups in their structure it might also be expected that a lower solution viscosity would be obtained with a mixture of hydrocarbons and alcohols than with either solvent alone.

Fig. 9. Effect of Substitution on Solubility

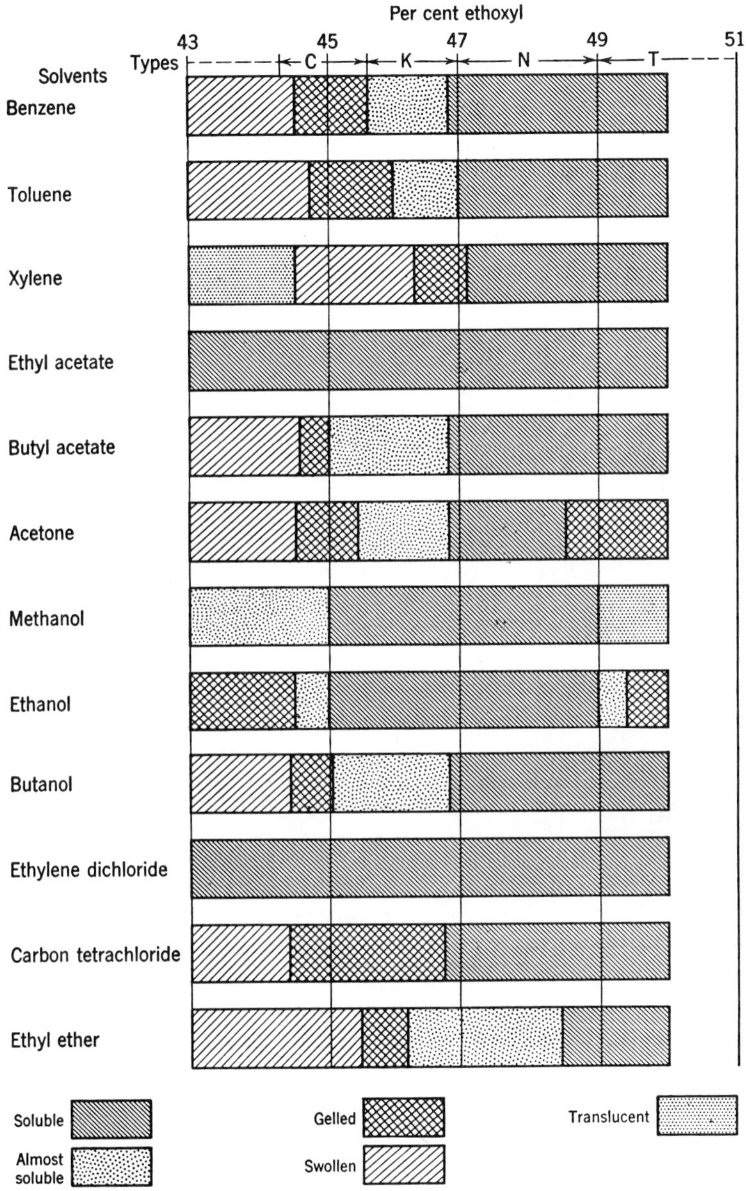

Tests made on 15% solutions

Courtesy of Hercules Powder Co.

The effect of such mixtures on the viscosity of solutions of N-22 type ethyl cellulose is shown in Fig. 10. The solubility phenomenon in a binary solvent is a convenient means for increasing or decreasing the

Fig. 10. Viscosity of N-22 Type in Binary Solvent

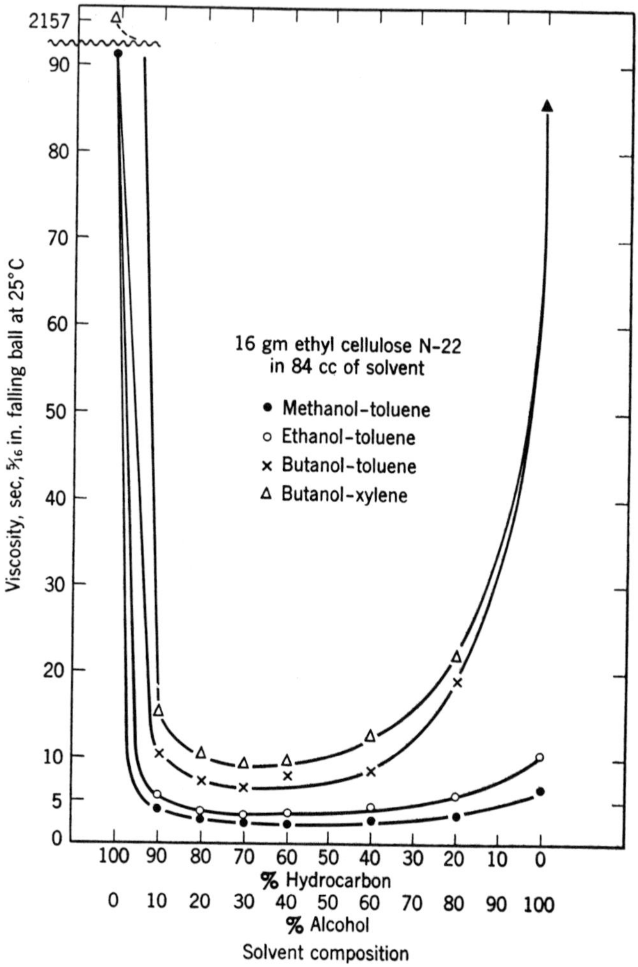

Courtesy of Hercules Powder Co.

viscosity of a solution without changing its solid content or changing the viscosity type of ethyl cellulose used. These binary solvent solutions will tolerate some aliphatic hydrocarbon such as VM & P naphtha to lower the solvent cost. The viscosity of such solutions will depend on the ratio of solvent constituents. The tolerance of ethyl cellulose

Table 25 * Tolerance of Ethyl Cellulose for Petroleum Thinners

Effect of Ethoxyl Substitution

Solvent Toluene:ethanol: VM & P naphtha ratio	Solubility		
	G-Type	N-Type	T-Type
7: 3:90	I	Sw	Sw, G
14: 6:80	W	Sw, G	G, pS
21: 9:70	Sw	G, pS	S
28:12:60	Sw, G	S	S
35:15:50	S	S	S
42:18:40	S	S	S
49:21:30	S	S	S
56:24:20	S	S	S
63:27:10	S	S	S

LEGEND: S—soluble, free of granulation and gel.
pS—almost soluble but contains gel and granulation.
G—gelled.
Sw—swollen, but granules not merged into cohesive gel.
W—particles become translucent but not swollen.
I—insoluble.

* Courtesy of Hercules Powder Co.

for various solvent ratios is shown in Table 25. In some cases it is possible to replace at least one-third of the toluene with a lower-cost aliphatic hydrocarbon without a significant effect on the viscosity of the solution. From this fact it will be realized that ethyl cellulose has much better solubility characteristics than either nitrocellulose or cellulose acetate. This is further emphasized by a comparison of the solubility of these three polymers as given in Table 26.

Blending Viscosity Types. It may be necessary at times to blend two solutions of different viscosity types of ethyl cellulose in order to obtain a solution of the required viscosity. The amount of each solution may be obtained from a chart as described previously for nitrocellulose, or it may be calculated from the equation which was given for this purpose. Care should be taken not to use solutions of widely different viscosities, because a very high-viscosity type may not be compatible with a low-viscosity type. A similar situation exists with the different ethoxyl types; widely different ethoxyl types may not be compatible with each other.

Compatibility. Ethyl cellulose is compatible with many varnish resins, vegetable and mineral oils, long-oil alkyds, chemical and resinous plasticizers, some waxes, and nitrocellulose. It is incompatible

Table 26 * Solubility of N-22 Ethyl Cellulose: Comparison with LL-1 Cellulose Acetate and ½-Sec RS Nitrocellulose

Solvent	N-22 Ethyl Cellulose	LL-1 Cellulose Acetate	½-Sec RS Nitrocellulose
Methanol	S	I †	I
Ethanol	S	I †	I
Butanol	S	I	I
Cyclohexanol	pS	I	I
Diacetone alcohol	Sw	S	S
Methylene chloride	S	I ‡	I
Ethylene dichloride	S	I ‡	I
Chloroform	S	I	I
Carbon tetrachloride	S	I	I
Pentachloroethane	S	I	I
Methyl acetate	S	S	S
Ethyl acetate	S	S	S
Butyl acetate	S	I	S
Ethyl lactate	S	S	S
Butyl lactate	S	I	S
Methyl Cellosolve acetate	S	S	S
Cellosolve acetate	S	I	S
Ethyl ether	S	I	W
Dioxan	S	S	S
Methyl Cellosolve	S	S	S
Cellosolve	S	I	S
Benzene	S	I	I
Toluene	S	I	I
Xylene	S	I	I
Dipentene	pS	I	I
Turpentine	Sw	I	I
Hexane	I	I	I
VM & P naphtha	I	I	I
Acetone	S	S	S
Methyl ethyl ketone	S	S	S
Cyclohexanone	S	pS	S
1:5 Camphor:ethanol	S	I	S
2:1 Ethyl ether:ethanol	S	I	S
80:20 Toluene:ethanol	S	I	I
70:30 Solvesso xylene:ethanol	S	I	I
90:10 Methylene chloride:ethanol	S	S	I
90:10 Ethylene dichloride:ethanol	S	S	I

LEGEND: S—soluble.
pS—almost soluble, but contains gel and granulation.
I—insoluble.
I †—insoluble alone; the solvent acts as a cosolvent with ketones and chlorinated hydrocarbons to form an active solvent mixture.
I ‡—insoluble alone; the solvent acts as a cosolvent with ethanol to form an active solvent mixture.
Sw—swollen, but granules not merged into cohesive gel.
W—particles become translucent but not swollen.

* Courtesy of Hercules Powder Co.

with cellulose acetate, acetate-butyrate and acetate-propionate, benzyl cellulose, polyvinyl chloride, and the polyvinyl chloride copolymers, polyacrylates, polystyrene, and natural and synthetic rubbers. In some cases an incompatible material may be made compatible with ethyl cellulose by the addition of a third component which is a mutual solvent in the solid phase. In view of the large number of materials and combinations which may be used with ethyl cellulose it appears impractical to list them in this volume since this information may be obtained readily from the manufacturers of ethyl cellulose.

In general, the hard resins such as rosin-base resins and the 100% phenolic oil-soluble resins increase the hardness, gloss, and adhesion of ethyl cellulose lacquer films. Ethyl cellulose has been advocated for reducing the drying time of varnishes. It is added to the varnish on the down heat at about 475°F, and the batch should be held at 450°F until the ethyl cellulose is dissolved completely. If the batch is held too long the ethyl cellulose may be heat-degraded considerably with a consequent loss of drying ability. Usually from 10% to 15% of ethyl cellulose on a solid basis is suggested. The ethyl cellulose increases the viscosity of the varnish with a corresponding decrease in solid content at normal viscosity. It also requires a stronger solvent than mineral spirits. Although the idea of using ethyl cellulose to speed the drying time and increase the toughness of varnishes appears of interest it does not seem to be widely used at the present time.

Plasticizers. The data in Table 24 show that ethyl cellulose has better inherent flexibility than nitrocellulose, particularly in the higher-viscosity types. Therefore less plasticizer is required, but, of course, the actual ratio of plasticizer to ethyl cellulose will depend on the type of lacquer being formulated. Both vegetable and mineral oils may be used, but they tend to exude from the composition if the ratios are higher than 30–40 oil to 60–70 ethyl cellulose. Castor oil is an exception, since it may be used as high as 50 oil to 50 ethyl cellulose without exudation.

The large majority of monomeric ester plasticizers are compatible with ethyl cellulose and also many of the polymeric or resinous type. The resinous type is preferred where maximum toughness is required and resistance to mineral oils and solvents, such as in insulating cable lacquers and cloth coatings. Since very flexible films may be made with ethyl cellulose and a relatively small proportion of plasticizer such films may be expected to retain a high percentage of the inherent tensile strength of the polymer. As noted with other polymers the tensile strength of the film varies somewhat with the type of plasticizer. A few examples of this variation are given in Table 27.

Table 27 * Variation in Film Properties with Type of Plasticizer

Ratio—80:20 Ethyl Cellulose: Plasticizer

Plasticizer	N-100 Ethyl Cellulose Tensile Strength, psi	N-100 Ethyl Cellulose % Elongation	G-100 Ethyl Cellulose Tensile Strength, psi	G-100 Ethyl Cellulose % Elongation
Dibutyl phthalate	5500	45	5450	50
Tricresyl phosphate	6400	38	6900	37
Trioctyl phosphate	4200	62	5150	52
Raw castor oil	6700	42	6100	41
Blown castor #16	6700	35	7550	35
Paraplex RG-2	7150	37	7300	26
Butoxyethyl stearate	2700	19	4600	55
Pentaphen	7850	27	8100	21
Dow resin 276-V9	7600	15	7900	34

The films were approximately 5 mils thick.

* Courtesy of Hercules Powder Co.

The values for N-100 type ethyl cellulose shown in Table 24 are as follows: tensile strength, 9100 psi; and per cent elongation at the point of rupture, 25%. The inherent flexibility of ethyl cellulose is one of its important characteristics. It is also significant that considerable flexibility is retained even at temperatures as low as −70°C and after exposures to temperatures as high as 150°C. The necessity for the use of suitable plasticizers under these conditions will be evident.

Stabilization. Ethyl cellulose is sensitive to oxidation when exposed to sunlight and ultraviolet radiation, and when heated above its softening point of about 150°C. However, it may be stabilized by the addition of suitable antioxidants, by the use of certain plasticizers, and by the pigments in pigmented coatings. All clear films, and also compounds for hot-melt application, should include from 0.5% to 1% of such antioxidants as diamylphenol, octylphenol, *p*-benzylcatechol, and phenyl-β-naphthylamine.

For exterior coatings the following plasticizer combinations have been found to exert a stabilizing action:

70:30 mixtures of
- Tricresyl phosphate: pentaphen
- Tricresyl phosphate: triphenyl phosphate

Tricresyl phosphate: diphenyl phosphate
Pentaphen: diphenyl phosphate

In opaque pigmented coatings the pigments provide some protection against degradation by sunlight because of their ability to absorb the harmful ultraviolet radiation. Pigments vary considerably in their ability to absorb ultraviolet as discussed previously and in detail in Volume II.

Hot-Melt Applications. The possibility of applying a coating at 100% solids without solvent has economic advantages. It may be accomplished by heating the coating material to a temperature at which it is a mobile fluid, and applying it by a coating machine or the doctor-knife method to continuous rolls of paper or textiles. Objects may also be dipped into the hot-melt composition; the coating solidifies rapidly after the object is withdrawn. In the latter case the coating can be formulated to have little or no adhesion to the object so that the coating can be removed easily by a simple stripping operation when it has served its purpose. This type of coating is referred to as a "strippable coating" and is applied to protect metal parts and instruments during shipment from the manufacturer to the ultimate user. Such coatings frequently contain a large percentage of refined mineral oil and a small percentage of wax.

Wax is also used as a coating for paper to make it water resistant. Wax films are relatively weak and crack easily when the paper is folded. These films may be strengthened by the addition of high-polymeric materials and certain natural and phenolic resins. The compositions may be varied widely to meet cost and service requirements. Three compositions are given in Formula 6 to indicate the range and

Formula 6 * Hot-Melt Compositions

	Per Cent by Weight		
Composition	(*a*)	(*b*)	(*c*)
Ethyl cellulose, N-14 type	32	23	21.5
Stabelite † ester gum #10	24	52	..
Super-Beckacite ‡ 2000	..	..	55
Carnauba wax	18	7	6.5
Paraffin wax	24	9	8.5
White mineral oil	..	7	6.5
Dibutyl oxalate	1	1	1
Octylphenol	1	1	1

* Courtesy of Hercules Powder Co.
† Trade name, Hercules Power Co.
‡ Trade name, Reichhold Chemicals, Inc.

type of materials which may be used. Formula 6(*a*), having the most ethyl cellulose, would be harder and dryer and have less tendency to "block" or stick when the coated paper is in contact with other surfaces. Formula 6(*b*) has greater flexibility, and Formula 6(*c*) has the highest gloss.

A wide variety of compositions may be formulated for hot-melt strippable coatings; those shown in Formula 7 are indicative only. It

Formula 7 * Hot-Melt Strippable Compositions

Composition	Per Cent by Weight (*a*)	(*b*)
Ethyl cellulose, N-50 type	25	24 †
Lopor 45 (Esso Standard Oil)	61	57
Blown castor #15 (Baker Castor)	10	5
Dow resin 276-V9 (Dow Chemical)	..	7
Paraplex G-60 (Rohm and Haas)	..	3
Paraffin wax	3	3
Octylphenol	1	..
Antioxidant 2246 (American Cyanamid)	..	1

* Courtesy of Hercules Powder Co. and Dow Chemical Co.
† Ethocel Std., 50 cp.

will be noted that a higher-viscosity type of ethyl cellulose is used over that in Formula 6 in order to obtain higher strength. Further strength may be obtained from even higher-viscosity types, but there is a corresponding increase in melting point and viscosity of the melt.

Care should be taken in the manufacture of hot-melt compositions to avoid local overheating in the kettle. All the ingredients except the ethyl cellulose should be heated to about 350–375°F. Then the ethyl cellulose should be added rapidly and the temperature maintained until the ethyl cellulose is melted completely and the foam has subsided. It is preferable to dry the ethyl cellulose before using since it absorbs some water, which increases the amount of foam.

The effect of various waxes, fatty acids, pitches, etc., on the softening point of ethyl cellulose compositions is shown in Table 28.

Solvent Applications. Ethyl cellulose compositions may be applied from solvent solutions to a variety of rigid and non-rigid surfaces such as wood, metal, paper, and textiles. Ethyl cellulose may replace part of the nitrocellulose in furniture finishing lacquers, such as those in Formula 2, to increase flexibility and cold-check resistance. In lacquers for exterior exposure it is about as durable as nitrocellulose. It gives excellent flexibility in fabric lacquers and is much slower burn-

Table 28 * Ethyl Cellulose Wax Mixtures

Effects of Increasing Amounts of Ethyl Cellulose on Softening Points of Mixtures

Compositions					
Wax, %	100	90	75	66.7	50
Ethyl cellulose, N-type	0	10	25	33.3	50
Softening-point data, °C, for mixtures containing:					
Beeswax, refined	62.5	103.5	139	144	182
Candelilla	68	75	130	132	142
Carnauba	82	100.5	122	135	..
Coal tar, refined, 28°C mp	27.5	88	118	132	171
Gilsonite	158.5	172.5	195.5	208.5	218
Japan wax	41	69	121	151	182
Linseed oil fatty acids	Liquid	Liquid	64	83	110
Montan wax, crude	83	87.5	112.5	125	145
Montan wax, refined	69.5	86.5	115	117.5	134.5
Oleic acid	Liquid	Liquid	61	75	106
Spermaceti	45.5	91.5	120	127.5	131.5
Stearic acid	55	56	71	87	114
Stearin pitch	85	106	120	163	182
Tallow	35	62	117	151	165

* Courtesy of Hercules Powder Co.

ing than nitrocellulose. It requires less plasticizer, and for low-cost fabric coatings a large proportion of raw castor oil may be used since this oil does not exude as much from ethyl cellulose as from nitrocellulose films. Ethyl cellulose may be blended with nitrocellulose to reduce the exudation and also to improve the embossability of fabric coatings. Ethyl cellulose compositions may be applied to fabrics by the calender method, thereby eliminating solvents. For lacquers on textiles intended for cable insulation which may come in contact with hot oil it is customary to use the resinous type of plasticizer which is not leached out by the oil.

A great many combinations of polymer-resin-plasticizer are in general use in coatings. The combinations in Formula 8 are intended to indicate various types, but, of course, these may be blended in any proportion. Formula 8(*a*) illustrates the type for rigid surfaces, satisfactory for exterior exposure. Formula 8(*b*) illustrates a strong, flexible coating for fabrics, and Formula 8(*c*) shows the type for cable coatings without chemical plasticizer. Formula 8(*d*) represents a very soft, lower-cost fabric coating which may be hardened if necessary with nitrocellulose as indicated in Formula 8(*e*). Further details regarding the manufacture, application, and performance of such lacquers will be found in Volume II.

Formula 8 Ethyl Cellulose Lacquer Compositions

	Parts by Weight, Solid Basis				
Composition	*(a)*	*(b)*	*(c)*	*(d)*	*(e)*
Ethyl cellulose	100	100	100	100	100
Nitrocellulose	...	...	...	...	100
Hard resin	100	...	...	...	...
Chemical plasticizer	33	25	100	...	...
Resinous plasticizer	...	75	...	...	...
Raw castor oil	...	...	...	50–100	50–100
Antioxidant	1	1	1	...	...

The hard resins include Super-Beckacite 1001, Bakelite 254, Durez 550, Krumbhaar 202, Aroclor 5460. Typical resinous plasticizers are Paraplex RG-20 and Rezyl X-315. The solvent for such lacquers may be the 80:20 toluene:ethanol combination referred to previously. Somewhat better flow may be obtained from a slower-evaporating combination such as:

	Parts by Weight
Toluene	60
Xylene	20
Ethanol	15
Butanol	5

Ethyl cellulose may be used with alcohol-soluble resins to produce shellac substitutes. It is also combined with a variety of resins, oils, and plasticizers for adhesives to laminate films to paper or metal foil and for other adhesive applications. Ethyl cellulose is also used to produce continuous free film and a wide variety of plastics.

Ethyl Hydroxyethyl Cellulose. A bulletin from the Hercules Powder Co. describes a new cellulose derivative which is not available commercially at the time of writing. The new derivative is a mixed ether, ethyl hydroxyethyl cellulose (EHEC-3U). It is soluble in aromatic hydrocarbons, 70:30 mixtures of aliphatic:aromatic hydrocarbons, and aliphatic hydrocarbons with 5–10% of butanol or anhydrous isopropanol. It is also soluble in ethyl and butyl acetate and in methyl ethyl ketone and methyl isobutyl ketone. It is compatible with standard 100% phenolics and modified phenolics, ester gum, and a number of modified rosin esters. In general, it is not compatible with oil-modified alkyds, styrenated alkyds, or amino resins.

The new derivative, EHEC-3U, may be added to short- to medium-oil-length varnishes to improve drying time, toughness, and hold-out on porous surfaces. For example, the addition of 10% EHEC-3U, on a solid basis, to a 20-gal modified rosin resin-tung oil varnish eliminated pinholes when used as a paper coating. The EHEC-3U was dissolved in VM & P naphtha to produce a 20% solution, which was mixed with the varnish and agitated at room temperature for about 1 hr to assure complete dispersion. The mixture was somewhat higher in viscosity than the varnish, but this may be an advantage for knife-coat application to paper. The use of EHEC-3U in combination with other resinous materials is indicated for certain printing inks and silk-screen printing compositions. It has also been suggested for adhesives and for hot-melt coatings.

The new material is available in granular form and in both high- and low-viscosity grades. It shows good tensile strength; free films gave 5000–6000 psi and 8–10% elongation at the break point. It may be plasticized with a variety of standard plasticizers.

METHYL CELLULOSE

Methyl cellulose is the methyl ether of cellulose made by reacting alkali cellulose with methyl chloride under pressure. The reaction may be expressed as follows:

$$\text{Alkali cellulose} + CH_3Cl \rightarrow ROCH_3 + NaCl$$

R represents the cellulose radical.

The commercial methyl cellulose has a degree of substitution of about 2 and a methoxyl content ranging from 27.5% to 32%. The structure of this type of methyl cellulose may be represented as follows, but it should be understood that the commercial product would be less symmetrical.

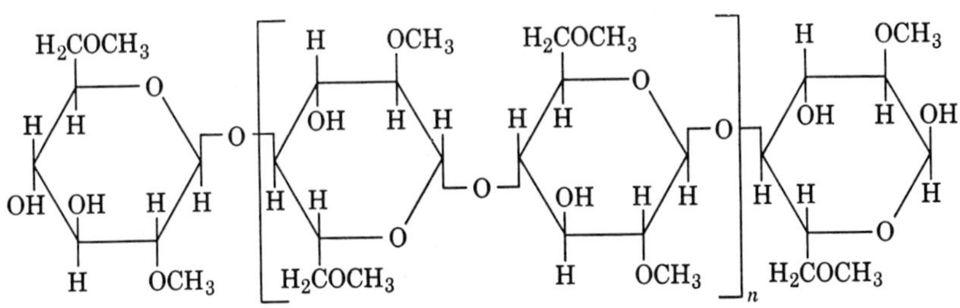

Dimethyl cellulose

Methyl cellulose is soluble in cold water and practically insoluble in most organic solvents. This characteristic differentiates it from the materials usually considered under organic coatings. However, the films of methyl cellulose are unaffected by animal, vegetable, or mineral oils and greases. The films may be rendered insoluble in water by addition of materials which react with the free hydroxyl groups to form cross-linkages between the molecules. Such materials include acids such as citric, dialdehydes such as glyoxal, and resinous materials such as dimethylol urea or the water-soluble urea-formaldehyde and melamine-formaldehyde resins. Methyl cellulose may be used as a paper size either alone or blended with starches, dextrins, glues, and the water-dispersible natural gums. Methyl cellulose is also added to water and emulsion paints as a thickener or protective colloid. Another water-soluble material, sodium carboxymethyl cellulose, has similar uses and is described later.

Types Available. Methyl cellulose is manufactured by the Dow Chemical Co. and marketed under the trade name of Methocel. It is available in several grades and viscosity types. The standard grade only, designated simply as Methocel, will be discussed in this section. A high-gel-point grade, Methocel HG, has a gel point 10–15°C higher than Methocel, which permits somewhat higher solids in solution and better stability in emulsions. The grade Methocel AS is insoluble in water but soluble in caustic solutions. This permits its use as a thickener in strongly alkaline systems, and its alkali-soluble films can be made insoluble by neutralization of the free caustic.

Methocel is available commercially in six viscosity types and in experimental quantities in a 10-cp viscosity type. The viscosity is expressed in centipoises obtained on 2% aqueous solutions. The viscosity-type designations and viscosity ranges are shown in Table 29.

Table 29 * Viscosity Types of Methocel

Viscosity Designation	Viscosity Range
10	9–12
15	13–18
25	20–30
100	80–150
400	350–550
1500	1200–1800
4000	3000–5000

The 10-cp type is available in experimental quantities only.

* Courtesy of Dow Chemical Co.

It is important that the Methocel solution be prepared properly in order to obtain accurate viscosity data. The sample should be dried at 105°C for 1 hr, or to constant weight, to determine the dry Methocel content. A 2.0-gm dried sample should then be added to 98 gm of distilled water at a temperature between 85° and 90°C. The mixture is agitated for 10 min; then the container is placed in an ice bath at 0–5°C until solution is complete. The solution should be free from air bubbles for the viscosity measurement. The viscosity may be determined in a Methocel viscosimeter (Dow Chemical Co.) or other instruments which give absolute viscosity in centipoises. The viscosity is determined at 20° ± 0.1°C.

Solubility. The solubility-temperature relationship for Methocel is the reverse of that found for most materials. In general, the solubility of a material increases with increase in temperature of the solution; the reverse is the case for Methocel. When the temperature of a Methocel solution is raised it reaches a gel point. At this point the Methocel precipitates if the concentration is below specified limits, or it gels if the concentration is high enough. This unusual phenomenon is related to the mechanism whereby Methocel dissolves. In view of this mechanism it may be more accurate, though less convenient, to describe Methocel "solutions" as "colloidal dispersions."

The structure of cellulose fiber, as described previously, indicates that the highly polar, long-chain molecules of cellulose are bound together by strong secondary valence forces into bundles known as micelles. The micelles aggregate further into fibrils which constitute the fine structure of the cellulose fiber. When the fiber is treated with alkali solution the forces between the fibrils are weakened and the fiber swells. The solution then penetrates the micelles and loosens the secondary valence forces, though not to the point of complete separation of the cellulose molecules. In this distended condition the alkali cellulose is reacted with methyl chloride and methyl groups are attached at random along the cellulose chains. Subsequent removal of the alkali solution leaves the cellulose chains kept apart slightly by the methyl groups. In this condition the hydroxyl groups on the molecules in the micelles are more accessible to water and the molecules become highly hydrated. If sufficient hydration is achieved the material becomes water soluble. However, this rather loosely bound water of hydration may be removed if the kinetic energy of the system is increased by raising the temperature. Then the molecular aggregates become insoluble and precipitate if the concentration is low, or form a continuous gel structure at high concentration.

The correct methoxyl content is necessary for maximum solubility in water. If insufficient methyl groups are attached to the chains the hydroxyl groups will not be readily accessible to the water molecules.

If too many methyl groups are attached there will not be enough hydroxyl groups to provide the required degree of hydration for water solubility.

Methocel is available in a powdered form. If the powder is placed in cold water the outside of the particles becomes hydrated and a gel layer is formed which retards further penetration of water, making solution difficult. If hot water, 80–95°C, is used at first, hydration is slower so that the particles become completely wetted and solubility is achieved readily when the mixture is cooled. The solubility phenomenon is reversible, i.e., the Methocel will precipitate if the temperature is raised to the gel point and go into solution again when the mixture is cooled. The gel point depends on the grade of Methocel, the viscosity type, and the concentration.

The gel point is also affected by the presence of soluble salts in the water. A high concentration of these salts may reduce the gel point to normal temperatures so that solution of the Methocel is impossible. In general, salts with ions of higher valence will be tolerated less than those with lower valence. For example, phosphates and sulfates have lower limits of tolerance than acetates or chlorides.

Plasticizers. Free films of Methocel have good strength and fair flexibility as indicated by average values for tensile strength of 8500–11,400 psi, and 10–15% elongation at the point of rupture. These values were obtained at 73°F and 50% RH. Water is an effective plasticizer for Methocel, but the materials listed in Table 30 may also be used.

Methocel coatings on paper are practically impervious to oils, waxes, and greases and thereby protect the paper from discoloration. The coatings are tasteless and non-toxic and may be used for edible oils and fats. Of course, the normal Methocel coatings retain their solubility in water. However, they may be insolubilized by including the materials which react with the free hydroxyl groups as referred to previously.

Use in Emulsion Paints. The thickening action of Methocel increases the viscosity of the water phase of emulsions and emulsion paints. Higher viscosity in the water phase improves application properties and increases stability on aging. The higher viscosity of the continuous phase reduces Brownian mobility of the disperse phase of emulsions, thereby retarding coalescence. Higher viscosity also retards settling of the pigments in emulsion paints.

Frequently it is desirable to use calcium pigments in emulsion paints, but such pigments react with casein to form insoluble caseinates. In such cases the casein must be replaced with other water-soluble pro-

Table 30 * Plasticizers for Methocel

Plasticizer	%	Tensile Strength, psi	% Elongation †	% Volatile ‡	Moisture Sorption 50% RH	Moisture Sorption 90% RH
None	0	10,200	14	....	2.3	21.1
Invert sugar (70% solids)	15	8,230	12	23.6	4.2	27.3
	30	8,090	28	....	8.4	31.0
d-Sorbitol	15	10,000	28	17.0	4.8	46.4
	30	8,860	38	....	6.0	59.4
Glycerol	15	7,100	38	67.0	5.6	45.6
	30	6,620	48	....	5.9	65.7
Triethylene glycol	15	8,010	34	....	14.7	23.1
	30	5,000	50	99.5	....	24.4
Tetraethylene glycol	15	7,080	45	....	12.5	26.4
	30	7,300	56	52.1	16.0	28.4
Triethyl phosphate	15	9,160	19	....	7.3	22.7
	30	9,900	31	100	7.8	21.7
Calcium chloride	15	3,750	52	....	24.1	68.1
	30	2,280	79	....	38.3	103.7
Corpolin	15	8,020	38	39.1	10.5	40.9
	30	7,370	42	....	10.7	44.2
Carbowax 6000	15	8,300	32	3.8	5.9	23.4
	30	6,350	39	....	6.5	23.6
Triethanolamine	15	8,930	40	19.1	9.0	18.7
	30	4,410	44	....	13.2	32.6

* Courtesy of Dow Chemical Co.
† Sample maintained 100 hr at 100°C in open dish.
‡ Per cent water by weight at 73°F.

tective colloids, and Methocel has been found useful in this respect. Further details regarding this use for Methocel are given in Volume II.

Use in Paint Removers. Two types of non-flammable paint removers have been developed based on Methocel as the thickening agent and methylene chloride as the active remover. Both types blister the painted surface, but one type is removed by scraping and the other by flushing with water. The "scrape-off" type is designed chiefly for paint removal from wooden surfaces, and the "flush-off" type for removal from metal surfaces. The materials used in these two types are shown in Formulas 9 and 10.

Formula 9 * Scrape-off Paint Remover

Methylene chloride PRG	90 gal
Methanol	10 gal
Paraffin wax (122–124°F ASTM)	16.5 lb
Methocel PRG-4000 cp	11.0 lb

Procedure

Add melted paraffin to the methylene chloride while agitating vigorously. Add Methocel while agitating. Add methanol while agitating. Stir until uniform.

* Courtesy of Dow Chemical Co.

Formula 10 * Flush-off Paint Remover

Methylene chloride (tech.)	71 gal
Toluene †	3 gal
Methanol	12 gal
Di-triisopropanolamine PRG	9.5 gal
Water	1.5 gal
Methocel PRG–4000 cp	13.0 lb
Areskap 100	33 lb
Potassium oleate	22 lb
Paraffin (122–124°F ASTM)	16.5 lb
Triethyl ammonium phosphate (40% in ethyl alcohol—Monsanto)	0.5 gal

Procedure

Add melted paraffin to the mixture of methylene chloride and paraffin solvent while agitating violently. Stir until dissolved.

Add Methocel while agitating.

Add methanol while agitating.

Add remaining ingredients, one at a time, while agitating.

Stir until uniform.

* Courtesy of Dow Chemical Co.

† Six gallons VM & P naphtha can be substituted for toluene.

A heavy coating of the paint removers should be applied to the painted surfaces and allowed to remain until the paint is blistered or softened. The scrape-off remover is then removed with a putty knife, and the flush-off remover with a stream of water.

SODIUM CARBOXYMETHYL CELLULOSE

Another water-soluble cellulosic polymer which has found uses in surface coatings is the sodium salt of carboxymethyl cellulose. It is manufactured by the du Pont Co., Wyandotte Chemical Co., and Hercules Powder Co. The material made by Hercules is known as Hercules CMC or Hercules Cellulose Gum. It is made by reacting alkali-cellulose with sodium monochloroacetate. The reaction may be expressed as follows:

$$\text{Alkali cellulose} + ClCH_2COONa \rightarrow ROCH_2COONa + NaCl$$

R represents the cellulose radical.

Types Available. Commercial grades of CMC are available having a degree of substitution ranging from 0.3 to 1.2 sodium carboxymethyl groups per anhydroglucose unit. The grade most widely used has a substitution value of 0.7 and is designated as CMC-70.

This is available in a range of viscosity types as shown in Table 31. The viscosity is measured on 2% solutions of the low- and medium-viscosity types and on 1% solutions of the high-viscosity types.

Table 31 * Commercial Types of CMC

Regular	Degree of Substitution	% Solids	Viscosity at 25°C, cp
CMC-70 LL	0.65–0.85	2	18–28
CMC-70 Low	0.65–0.85	2	25–50
CMC-70 Sp. Med.	0.65–0.85	2	150–250
CMC-70 Med.	0.65–0.85	2	300–600
CMC-70 High	0.65–0.85	1	1300–2200
CMC-70 D High	0.65–0.85	1	1300–2200
CMC-CT Low	....	2	15–70
CMC-CT Med.	....	2	100–400
Spec.			
CMC-30	0.3 approx.	.	....
CMC-90 Low	0.9–1.15	2	25–50
CMC-90 Med.	0.9–1.15	2	300–600
CMC-90 High	0.9–1.15	1	200–500
CMC-120 Med.	1.20–1.40	2	400–800
CMC-120 High	1.20–1.40	1	100–300

* Hercules Powder Co.

For food, pharmaceutical, and cosmetic applications the CMC-70 Low and Medium are available in a premium grade. The CMC-70 High is supplied in the premium grade only, but it is also available in a high-density type known as CMC-70 D. The latter type flows better in the dry state, and it is more readily dispersed in water. The CT material is a technical grade for use in detergents.

The sodium carboxymethyl groups are considerably more polar than the methoxy groups on Methocel discussed in the previous section. Therefore, CMC is more readily soluble and does not precipitate or gel when the temperature of the solution is raised to a critical degree. CMC is quite hygroscopic, and absorbs from 14% to 18% water at 50% RH, and about 32% water at 75% RH, both at 25°C. The commercial materials contain about 5% water as shipped; therefore they should be kept in closed containers to prevent moisture pick-up.

Solubility. The various grades of CMC are soluble in either hot or cold water, but care must be taken to obtain clear solutions free from lumps. The CMC should be added slowly to the water with good agitation and maximum possible shear. This breaks the CMC down into its ultimate particle size and prevents the formation of aggregates having a gelled outer skin and dry center. Improved clarity of a solution is obtained by passing it through a homogenizer or colloid mill. Solutions treated in this way are also more stable since they do not increase in viscosity on aging. Since it is difficult to prepare solutions which are free from bacteria the addition of 0.05–0.1% of one of the many available mold inhibitors or preservatives is usually desirable.

The CMC may be precipitated if certain salts are present in solution at sufficient concentration. In some cases the precipitated CMC may be redissolved by addition of alkali, but with certain salts the CMC remains insoluble. The salt solutions which precipitate the CMC do not dissolve dry films of CMC. If the films of CMC remain in contact with these salt solutions for sufficient time for the precipitating metal ions to replace the sodium ions the films become relatively insoluble in water. Water-insoluble films may also be obtained by adding mineral acid to CMC solutions to produce the free acid of CMC, and allowing this solution to evaporate. A similar result may be obtained by passing the CMC solution through an ion exchanger which will replace the sodium ions with hydrogen ions.

Film Properties. Dry films have good strength and fair flexibility and are quite resistant to vegetable, animal, and mineral oils and greases. An increase in flexibility may be obtained by incorporation of suitable plasticizers. The values in Table 32 are typical for tensile strength and per cent elongation of free films of the three viscosity types of CMC-70.

Table 32 * Physical Properties of CMC-70 Films

	Tensile Strength, psi	% Elongation
CMC-70 Low	12,600	10.5
CMC-70 Medium	13,450	9.0
CMC-70 High	13,850	8.5

* Courtesy of Hercules Powder Co.

The films were cast from water solutions and conditioned at 77°F, 50% RH, before testing. The equilibrium moisture content under these conditions is 16%.

Plasticizers. The per cent elongation may be increased up to about 50% by addition of typical plasticizers such as:

Glycerol	Diglycerol
Glycerol α-monomethyl ether	Glycerol monochlorohydrin
1,2,6-Rexanetriol	Ethylene glycols
2,3-Butylene glycol	Ethanolamines
2-Nitro-2-methyl-1,3-propanediol	Corpolin (Heyden Chem. Corp.)

Compatibility. The CMC-70 Medium was found to be compatible with a variety of water-soluble materials, both in solution and in dry film. The following materials were compatible in 3:1, 1:1, and 1:3 ratios on a solid basis when added to a 2% solution of CMC-70 Medium, and films formed by evaporation of the water:

Gum arabic	Hydroxyethyl cellulose
Karaya gum	Polyvinyl alcohol
Locust bean	Sodium alginate
Tragacanth	Starch
Water-soluble urea-formaldehyde resins	

In general the lower-viscosity types of CMC are more compatible with water-soluble colloids than the higher-viscosity types.

General Uses. The sodium carboxymethyl cellulose has found use in many industries because of its good solubility in either hot or cold water, its thickening action in solutions ranging from 0.5% to 2%, its ability to form oil- and grease-resistant films, and its non-toxic properties. Its use as a thickener and protective colloid in emulsion paints is described in Volume II.

BIBLIOGRAPHY

1. Emil Heuser, "Cellulose," Chapter 22 in *Organic Chemistry,* H. Gilman, ed., John Wiley & Sons.
2. Emil Ott, *Cellulose,* Interscience Publishers.
3. M. G. Miliken, *Ind. Eng. Chem., 22,* 326 (1930).
4. Ryan and Lantz, *Ind. Eng. Chem., 20,* 40 (1928).
5. C. L. Jones, *Safety in Lacquer Plants,* Hercules Powder Co.
6. *Lacquer Sanding Sealers,* Hercules Powder Co.
7. W. Koch, *Ind. Eng. Chem., 36,* 756 (1944).
8. C. Bogin and H. L. Wampner, *Ind. Eng. Chem., 29,* 1012 (1937).
9. Bede Products, Inc., *Hot-Spraying Process,* 1110 Brookpark Rd., Cleveland 9, Ohio.
10. American Brake Shoe Co., Kellog Division, *Steam-Spraying Process,* Rochester 9, N. Y.
11. U. S. Patent 2,511,797.
12. Hercules Powder Co., *High Solids Nitrocellulose Lacquers for Furniture Finishing.*
13. U. S. Patent 2,534,717.
14. C. J. Malm and H. L. Smith, *Ind. Eng. Chem., 41,* 2325 (1949).
15. L. W. A. Meyer and W. M. Gearhart, *Ind. Eng. Chem., 43,* 1585 (1951).
16. D. R. Stull, *Ind. Eng. Chem., 39,* 517 (1947).
17. C. J. Malm and H. L. Smith, *Ind. Eng. Chem., 38,* 937 (1946).
18. C. J. Malm, M. Salo, and H. F. Vivian, *Ind. Eng. Chem., 39,* 168 (1947).
19. C. J. Malm, H. B. Nelson, and G. D. Hiatt, *Ind. Eng. Chem., 41,* 1065 (1949).
20. A. G. Roberts and S. G. Weissberg, *Ind. Eng. Chem., 43,* 2088 (1951).
21. Hercules Powder Co., *Report on Hot-Spray Application of Lacquer.*

The material world is for your pleasure and profit,
But beware, lest it become your master!

Chapter 12. VINYL RESINS

INTRODUCTION

Since the vinyl radical is $CH_2{=}CH{-}$, the term "vinyl resins" could be applied to the substituted ethylenes in Table 1 and their many copolymers. This would include polyethylene, polystyrene, polyvinyl chloride resins, some of the synthetic rubbers, the rubber resins in Chapter 9, and the acrylic ester resins in Chapter 13. However, this chapter will cover chiefly the polymers and copolymers of vinyl chloride, vinyl acetate, and vinylidene chloride since these are classified as vinyl resins by the U. S. Department of Commerce in its Census Reports, and are recognized as vinyl resins by the plastics and coating industries. An indication of the growing importance of vinyl resins may be seen from their production volume as given by the U. S. Bureau of Census.

Vinyl Resin Production *

	Pounds
1941	23,000,000
1942	52,000,000
1943	95,000,000
1944	127,000,000
1945	123,000,000
1946	155,600,000
1947	177,450,000
1948	218,237,000
1949	267,686,000
1950	380,986,000

* U. S. Bureau of Census.

The outstanding properties of polyvinyl chloride and the many possibilities which exist for modifying it through copolymerization have developed uses for vinyl resins in packaging, plastics, and coatings. A general indication of the use distribution of vinyl resins is given by Plumb and Hansen (1).

The relatively small side groups on the polyvinyl chloride and polyvinylidene chloride chains, together with their high degree of polarity,

Consumption of Vinyl Resins,* Net Resin Basis, in Millions of Pounds

	1946		1950	
	Pounds	%	Pounds	%
Sheeting and film	26	26.4	118	42.8
Molding and extrusion	46	46.7	88	31.8
Textile and paper coating	11.5	11.7	45	16.3
All other uses	15	15.2	...	.....
Surface coating	....	.....	25	9.1
	98.5	100	276	100

* Courtesy of D. S. Plumb and R. F. Hansen (1).

produce compact structures having great strength, high softening point, and low solubility. Conversely, the large side groups on the polyvinyl acetate chains impart relative softness and solubility in a variety of solvents. Polyvinyl acetate was available as early as 1917 and was used chiefly in high-strength adhesives. Attempts to blend polyvinyl chloride and polyvinyl acetate to obtain a compromise in properties were unsuccessful because of incompatibility. The important discovery that the monomers could be copolymerized to obtain the desired results was made about 1928. In that year at least three companies filed patent applications on copolymers. These included the Reid patent (U. S. Patent 1,935,577, Carbide and Carbon Chemicals Corp.), filed in May; the Voss and Dickhauser patent (U. S. Patent 2,012,177, I. G. Farben.), filed in June; and the du Pont application (U. S. Patent 1,867,014, W. E. Lawson), filed in September.

The first plant for the commercial production of the copolymers was completed by Carbide and Carbon Chemicals Corp. in 1936. One of the first coating uses for the vinyl chloride-vinyl acetate copolymer was the lining of beer cans. The copolymer coating is sufficiently impervious to beer to prevent contamination by the metal, and the coating is non-toxic and does not affect the flavor. It is usually necessary to use a primer under the copolymer coating because the coating does not have sufficient adhesion to the metal. A later development of this type of vinyl resin incorporated about 1% of maleic acid in the copolymer which produced resins having good adhesion to metal. Still another modification improved the compatibility with other resinous materials by introducing some hydroxyl groups along the copolymer chain. To obtain compatibility with nitrocellulose it was necessary to increase the vinyl acetate content from about 13% to about 38%.

A series of polyvinyl acetal resins was produced which finds use in electrical insulation, safety-glass laminates, and strong flexible coatings. One of these, the polyvinyl butyral, has since been used in a special anticorrosive primer known as "wash primer" or "metal conditioner." The development of organosols and plastisols is of special interest in coatings because it permits the application of the high-polymeric vinyl resins without the use of high-price solvents. The resins are dispersed in low-cost dispersant-diluent mixtures for the organosols, or in plasticizers for the plastisols. These will be discussed in detail later. The polyvinyl acetate has now become available in latex form for coatings. New developments in vinyl resins for coatings may be expected because of the wide variety of copolymeric materials which can be produced, and the toughness and good chemical resistance of their films. An extensive description of the preparation, properties, and applications of vinyl and related polymers is given by C. E. Schildknecht in *Vinyl and Related Polymers,* John Wiley & Sons (1952).

ADDITION POLYMERS

Vinyl resins are addition polymers, and they are formed by a different mechanism from that of condensation and autoxidation polymers, as discussed in Chapter 1. The vinyl resins may be considered as substituted ethylenes, as indicated at the beginning of this chapter. When considered as substituted ethylenes the reader will visualize readily their relationship to many other polymeric coating and plastic resins. The names and monomeric units of a variety of substituted ethylenes are given in Table 1 for the purpose of comparison. It should be kept in mind that Table 1 does not show the great many combinations of these materials which can be used in copolymers.

SUBSTITUTED ETHYLENES

A detailed study of the chemical and physical properties of the materials listed in Table 1 and their many copolymers would cover the fundamentals of many of the thermoplastic polymers for coatings, plastics, adhesives, synthetic rubbers, and fibers in use today. The polymers range from the highly extensible polyethylene and synthetic rubbers to the hard though tough materials such as polystyrene, polyvinyl chloride, and some of the polyacrylic esters. They range in solubility from the complete insolubility of polytetrafluoroethylene to fair solubility in such different solvents as aromatic hydrocarbons,

Table 1 Substituted Ethylenes

Monomer	Monomeric Unit	Resin
Ethylene	$-CH_2-CH_2-$	Polyethylene
Vinyl chloride	$-CH_2-CH(Cl)-$	Polyvinyl chloride
Vinyl acetate	$-CH_2-CH(O-C(=O)-CH_3)-$	Polyvinyl acetate
Vinyl alcohol	$-CH_2-CH(OH)-$	Polyvinyl alcohol
Vinylidene chloride	$-CH_2-C(Cl)(Cl)-$	Polyvinylidene chloride
Tetrafluoroethylene	$-CF_2-CF_2-$	Polytetrafluoroethylene
Chlorotrifluoroethylene	$-C(F)(F)-C(F)(Cl)-$	Polychlorotrifluoroethylene
Vinyl ethers	$-CH_2-CH(O-R)-$	Polyvinyl ethers
Vinyl acetals	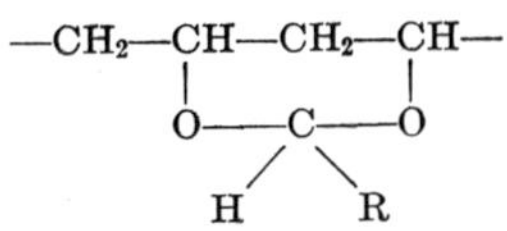	Polyvinyl acetals
Vinyl benzene (styrene)	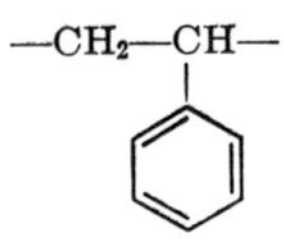	Polystyrene
Acrylonitrile	$-CH_2-CH(CN)-$	Polyacrylonitrile
Acrylic esters	$-CH_2-CH(C(=O)-O-R)-$	Polyacrylates
Methacrylic esters	$-CH_2-C(CH_3)(C(=O)-O-R)-$	Polymethacrylates
Divinyl (butadiene)	$-CH_2-CH=CH_2-CH-$	Synthetic rubber
Isoprene	$-CH_2-C(CH_3)=CH-CH_2-$	Natural rubber

R = short-chain alkyl group.

ketones, esters, alcohols, and even water. This extremely wide range in chemical and physical properties is due principally to three factors:

1. Size and degree of polarity of the side groups along the polymer chains.
2. Length of the polymer chains, or their molecular weights.
3. Relative arrangement of the polymer chains with respect to each other.

These fundamentals are discussed in a general manner in Chapter 1, and, of course, it is not feasible to cover them in detail in a book of this scope. However, even a casual examination will reveal the important features of interest to surface coating formulators. Consider polyethylene; it is a highly non-polar material with no side groups along the chains and is quite sensitive to non-polar solvents, but it is highly extensible. However, when it is stretched it develops higher tensile strength and shows a crystalline pattern in X-ray analysis. Normally the polyethylene chains are coiled or kinked, and not straight or parallel to each other. When straightened by stretching they become more compact, and the symmetry of the molecule permits a symmetrical or crystalline arrangement. The chains are too long to allow a high degree of crystallinity as found in paraffin wax, but there is sufficient crystallinity to increase the tensile strength.

Two other materials in Table 1 represent extremes from polyethylene; these are polytetrafluoroethylene and polyvinyl alcohol. In the former there is the same high degree of symmetry as in polyethylene, but the four fluorine atoms are highly polar, thereby providing strong attractive forces between the chains and a greater degree of crystallinity. In the polyvinyl alcohol, the symmetry is disturbed by the hydroxyl group on each monomeric unit. This group is quite polar and attracts water molecules with sufficient strength to overcome the relatively weak forces of hydrogen bonding between the chains. Consequently polyvinyl alcohol has a high degree of water solubility, but the polytetrafluoroethylene chains are so compact, and held together by such strong secondary valence forces, that no ordinary solvent can separate them.

The other materials in Table 1 may be considered as falling between the extremes of polyethylene, polytetrafluoroethylene, and polyvinyl alcohol. The presence of one chlorine atom in polychlorotrifluoroethylene reduces slightly the symmetry of polytetrafluoroethylene with a consequent slight reduction in fusion temperature and increase in solubility. The presence of one chlorine atom in polyvinyl chloride changes the symmetry considerably from that of polyethylene, and it increases the cohesive or attractive forces between the polymer chains because of its polarity. However, the chlorine shows little attraction for water; therefore polyvinyl chloride is not readily affected by water. A similar situation exists for polyacrylonitrile, but not for polyvinyl acetate. The acetyl group is relatively large and bulky, thereby preventing the development of crystallinity on stretching. The acetyl group has sufficient attraction for water to account for the moisture sensitivity of polyvinyl acetate. The benzyl side group in polystyrene is also large and bulky so that no crystalline pattern can be obtained in X-ray examination, but it is so highly nonpolar that polystyrene shows no affinity for water.

Of course, the molecular weight, or length of the polymer chains, is also a contributing factor. This will be very evident from the fact that the monomers are gases or volatile liquids at room temperature. Therefore, not only the degree of polymerization but also the uniformity of molecular weight is important. A relatively small percentage of low-molecular-weight material may seriously detract from the desirable properties of polymeric resins. Conversely, a relatively small percentage of very high-molecular-weight material decreases solubility and may produce turbidity, instability, loss of gloss, and other undesirable features.

From this very brief outline the reader will realize the many possibilities which exist in homopolymeric, and particularly copolymeric, addition polymers for use in coating materials. A great deal of research in this field is being carried on at the present time, and the scientific literature is replete with excellent articles on the subject. The reader will profit from a comparison of the structures and properties of the rubber resins in Chapter 9 with the structurally related vinyl resins in this chapter, and the structurally unrelated cellulosic polymers in Chapter 11.

COPOLYMERS

Since vinyl coating resins are chiefly copolymers rather than homopolymers it may be well to review the definition and mechanism of copolymerization. Homopolymers are polymers in which the monomeric units are all identical. Typical examples are polyvinyl chloride, polyvinyl acetate, and polystyrene. Of course, the end groups on homopolymers are usually different from the monomeric units because they are derived from free radicals or other terminating groups, but they constitute only a very small percentage of the polymer.

Copolymers are polymers in which the monomeric units are different. Typical examples are polyvinyl chloride-acetate resins, polyvinyl chloride-acrylonitrile resins, and polystyrene-butadiene resins. It is important to remember that, by definition, each of the various monomers in a copolymer must be capable of forming polymers. It will be seen that this requirement is met by the examples cited above. However, certain commercial vinyl resins (see Table 5, page 500) contain a small percentage (about 1%) of maleic acid which is believed to be distributed randomly in the polymer chains. Although maleic acid has an ethylenic type of unsaturation it does not form homopolymers; however, the amount of maleic in these vinyl resins is relatively small so that they may be considered copolymers. The introduction of the highly polar carboxyl groups on maleic acid improves the adhesion of the vinyl resin coating to smooth surfaces. As stated previously, the particular value of the copolymerization technique is that the properties of the copolymer resins may be varied over a wide range by adjusting the type and amount of the various monomers used in their formulation. This range of properties cannot be obtained by blending the various homopolymers since almost invariably they are incompatible.

In some respects the mechanism of copolymerization is the same as that of homopolymerization in that it involves initiation of the monomers and propagation and termination of the polymers. However, the different monomers in a copolymer usually do not polymerize at the same rate. It will be seen that unless something were done to offset this difference in rate of polymerization of the different monomers the finished copolymer would not be a uniform product. The copolymers formed at the beginning of the polymerization would contain a high percentage of the faster-polymerizing monomer, and those formed at the end of the polymerization would contain more of the less active monomer. Usually this difficulty may be overcome by adding the more active monomer continuously or in small amounts at intervals during the polymerization. In this way a fairly uniform copolymer may be obtained.

The difference in rates of polymerization of two or more monomers in a copolymer is not necessarily the same as the difference in their rates of polymerization as homopolymers. The presence of another active monomer may modify the three factors which affect the rate of polymerization of a monomer, namely resonance stabilization of the free-radical initiator, polarity of the active double bonds, and steric hindrance effects.

METHODS OF POLYMERIZATION

Vinyl resins are polymerized and copolymerized commercially by four different processes: bulk, solvent, suspension, and emulsion. The bulk-polymerization process is the simplest, but a high degree of conversion cannot be attained because the monomer is used to keep the polymer in suspension as it is formed. After the bulk polymerization has been carried as far as is feasible the unused monomer is evaporated from the monomer-polymer slurry, and the polymer is washed and dried.

In the solvent or solution process the monomers and catalyst (usually an organic peroxide) are added to a solvent which is capable of dissolving the monomers, and also the polymers as they are formed. When the polymerization has been carried as far as necessary, any residual monomers are stripped off, and the polymer is precipitated from the solution with a suitable non-solvent. The polymer is then filtered, washed, and dried.

In the suspension process the monomers and catalyst are mixed with water which contains a small percentage of a protective colloid, such as polyvinyl alcohol. The mixture is agitated to keep the small droplets of monomer in suspension until the polymerization is essentially complete. In this process some polymers are formed as very small beads; therefore the suspension process is sometimes referred to as "bead polymerization." The polymer is recovered very simply by filtering, washing, and drying.

The emulsion polymerization process is widely used in the manufacture of vinyl resins. In this process the monomer is emulsified in water, using a suitable emulsifying agent. Of course, the monomer must be substantially insoluble in water in order to obtain a good emulsion. After polymerization is essentially complete the emulsion may be broken and the resin coagulated, washed, and dried; or suitable protective colloids, buffers, corrosion inhibitors, fungicides, etc., may be added to produce a stable latex of the polymeric resin.

Each of the four processes has its particular advantages and limitations. The bulk process is the simplest, but the polymerization cannot be carried to a high conversion because of the high consistency which results from conversion of the dispersing monomer. The solvent process gives polymers of good clarity and solubility, but the amount of resin which can be made in an autoclave is limited by the high viscosity of the solution. Both the suspension and the emulsion processes provide a low-viscosity medium, and good heat transfer throughout the process so that the heat of polymerization may be more readily dissipated. However, in these processes there is some unavoidable contamination of the resin with water-dispersible or water-soluble materials. The emulsion process is best when high molecular weight is required, when the ultimate particle size must be controlled closely, and when the resin may be used in latex form.

A great many modifications of the polymerization processes are described in the patent literature; many of them are discussed by Schildknecht (3). It has been found that oxygen inhibits addition polymerization; therefore higher yields in less time may be obtained if the materials and system are free from molecular oxygen. In view of the oxidation effect certain reducing agents may be used with the peroxide catalysts to accelerate the rate of polymerization. Benzoyl peroxide was most widely used in the early work, but many other peroxides are now available. Some are more suitable for high-temperature and others for low-temperature polymerization. In general, low-temperature polymerization produces more uniform polymers with high molecular weight, but the rate of polymerization is slower.

There is a tendency for authors to refer to peroxide catalysts as initiators rather than catalysts. The free radical obtained from decomposition of the peroxide actually reacts with the monomer when it activates it, and usually the free radical becomes the end group on the polymer chain. A true catalyst, by definition, does not enter into the reaction; hence the preference of some writers to designate the peroxides as "activators" or "initiators." However, the peroxides will be refererd to as catalysts in this book because of the wide use of this term in the literature.

RAW MATERIALS AND PROCESSES

Vinyl resins are made from readily available raw materials such as ethylene, acetylene, and chlorine. Ethylene may be reacted with chlorine to produce ethylene dichloride. This may be thermally cracked, at temperatures ranging from 250° to 600°C, depending on catalyst and equipment, to produce vinyl chloride monomer, and hydrogen chloride as a by-product. Under suitable conditions the hydrogen chloride may be combined directly with acetylene to produce the vinyl chloride monomer.

Ethylene may also be oxidized to acetic acid, and this combined directly with acetylene to form the vinyl acetate monomer. The

Fig. 1. Vinyl Resin Flow Chart

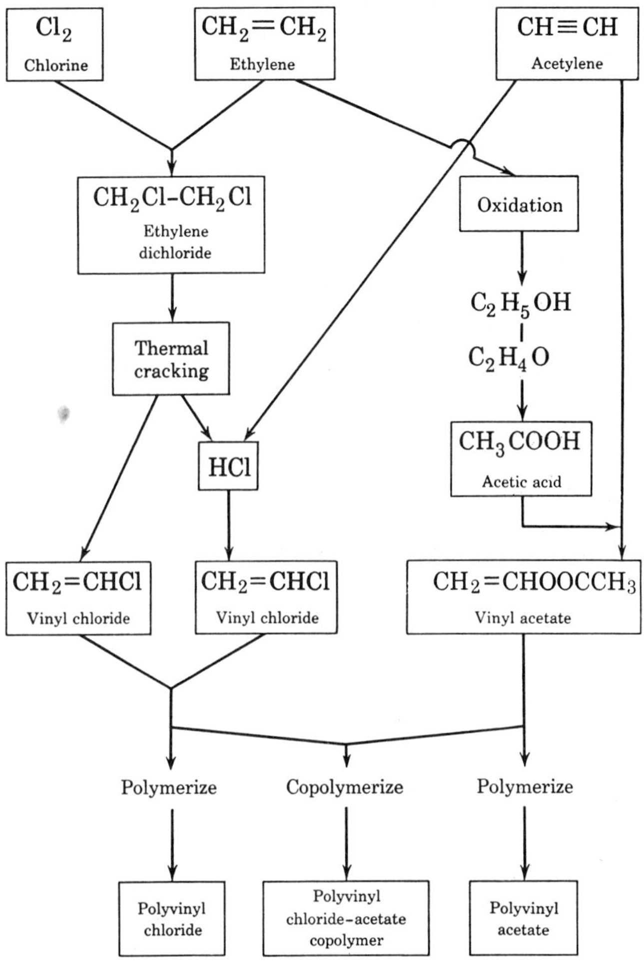

monomers may be polymerized directly or blended to form the copolymers. These general reactions are illustrated in Fig. 1 and in equations 1, 2, and 3.

Equation 1

$$CH_2Cl—CH_2Cl \rightarrow CH_2{=}CHCl + HCl$$

Equation 2

$$CH{\equiv}CH + HCl \rightarrow CH_2{=}CHCl$$

Equation 3

$$CH{\equiv}CH + CH_3COOH \rightarrow CH_2{=}CHOOCCH_3$$

There are two types of vinyl resins which cannot be made directly from the monomers but are obtained by treatment of polyvinyl acetate. These resins are polyvinyl alcohol and the so-called polyvinyl acetals. The polyvinyl alcohol is obtained by hydrolysis of polyvinyl acetate, and the polyvinyl acetals are condensation products of polyvinyl alcohol and aldehydes.

It might appear that monomeric vinyl acetate can be hydrolyzed with water to produce vinyl alcohol and acetic acid as in equation 4. However, monomeric vinyl alcohol is not stable but isomerizes immediately to acetaldehyde as in equation 5. It is generally stated in the literature that polyvinyl alcohol is obtained by "hydrolysis" or "saponification" of polyvinyl acetate. However, if this reaction takes place in the presence of water the resins which are produced are difficult to filter and dry. If anhydrous methanol is used as the source of hydroxyl groups, as in equation 6, the product is satisfactory. It will be seen that this reaction is alcoholysis instead of hydrolysis, but the latter term occurs most frequently in the literature. Either acid or alkaline catalysts may be used for the alcoholysis reaction. The commercial products, described later, vary in the degree of conversion from the polyvinyl acetate to the polyvinyl alcohol with a corresponding variation in physical and chemical properties.

Equation 4

$$CH_2{=}CHOOCCH_3 + H_2O \rightarrow CH_2{=}CHOH + CH_3COOH$$

Equation 5

$$CH_2{=}CHOH \rightarrow CH_3{-}CHO$$

Equation 6

$$[{-}CH_2{-}\underset{\substack{|\\ O-\underset{\substack{\|\\ O}}{C}-CH_3}}{CH}{-}]_n + nCH_3OH \rightarrow [{-}CH_2{-}\underset{\substack{|\\ OH}}{CH}{-}]_n + nCH_3COOCH_3$$

The most widely used polyvinyl acetals are polyvinyl formal, polyvinyl acetal, and polyvinyl butyral. They are prepared by hydrolizing polyvinyl acetate to polyvinyl alcohol and condensing the hydroxyl groups with an aldehyde; formaldehyde produces the formal, acetaldehyde the acetal, and butyraldehyde the butyral. Polyvinyl acetals have been prepared with longer-chain aldehydes for experimental purposes. It will be seen that considerable variation in properties of the acetals may be obtained by controlling the degree of hydrolysis of the polyvinyl acetate and the extent of condensation with the aldehyde.

Also, polymers could be produced with varying ratios of acetyl, hydroxyl, and acetal groups in the chain. The reaction for the hydrolysis of polyvinyl acetate is shown in equation 6, and the condensation with formaldehyde to produce the polyvinyl formal is shown in equation 7.

Equation 7

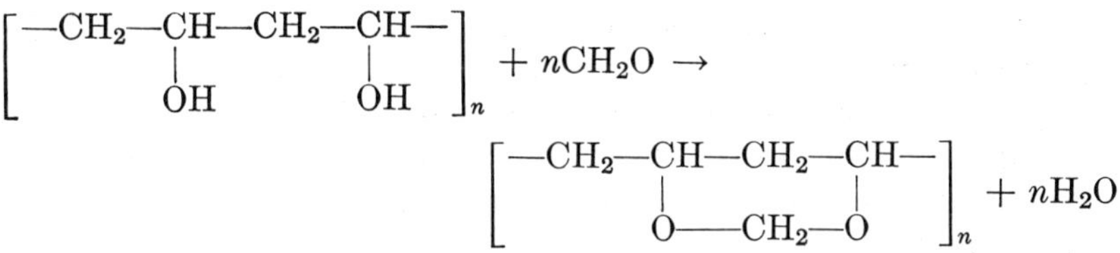

As stated above the extent of the different reactions is controlled easily, and polymers may be produced having the various groups along the chain such as:

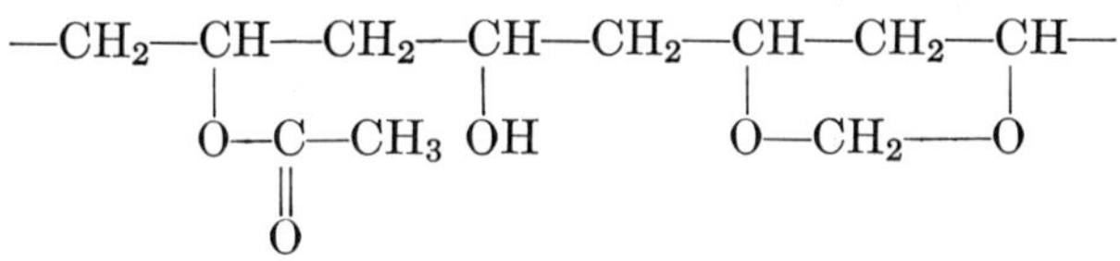

Acetyl Hydroxyl Formal

Replacement of formaldehyde with butyraldehyde would increase the length of the carbon chain in the side group and produce a softer and more soluble resin. The polyvinyl formal is widely used in insulating-wire coating enamels. The polyvinyl butyral serves as the interlayer in windshield safety glass, in coatings for fabrics, and as the vehicle in the so-called "wash primer" for anticorrosive paint systems on metal.

A variety of copolymers of vinylidene chloride is available commercially, but the homopolymer of vinylidene chloride is not widely used because of its difficult solubility and limited stability to heat and sunlight. The monomer may be prepared by thermal cracking of trichloroethane with the elimination of hydrochloric acid as shown in equation 8.

Equation 8

$$\underset{\text{Trichloroethane}}{CH_2Cl{-}CHCl_2} \rightarrow \underset{\text{Vinylidene chloride}}{CH_2{=}CCl_2} + HCl$$

Copolymers of vinylidene chloride and vinyl chloride have been widely used in plastics and some coatings. Those containing high proportions of vinylidene chloride have been marketed as Saran resins

(Dow Chemical Co.) since 1937. In 1943 Dow introduced a copolymer of vinylidene chloride and acrylonitrile. More recently a series of vinylidene-vinyl chloride copolymers and latexes has been made available for coating use which contain a high percentage of vinyl chloride. These are described in more detail later in this chapter.

Acrylonitrile may be converted into an addition-type homopolymer by the various polymerization techniques discussed previously. The high-molecular-weight homopolymers have high softening points and are quite resistant to solvents and chemicals, and they may be formed into oriented fibers and filaments. The copolymers of acrylonitrile and other active monomers are used more widely than the homopolymers in plastics and coatings.

Acrylonitrile may be produced from ethylene by oxidation to ethylene oxide, and reaction with hydrogen cyanide to form ethylene cyanohydrin, followed by dehydration to acrylonitrile according to the following equations:

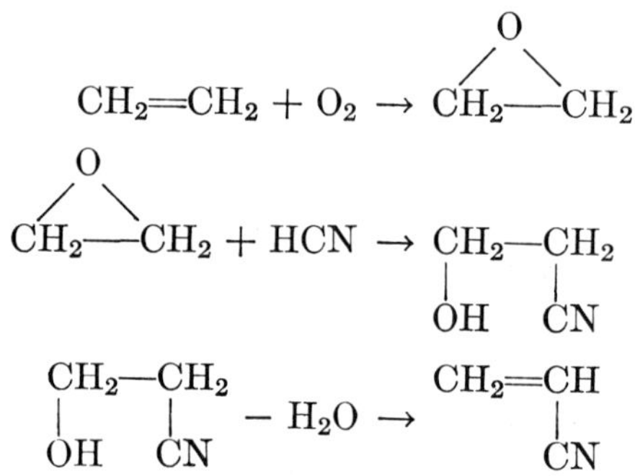

Acrylonitrile may also be produced from the direct addition of hydrogen cyanide to acetylene according to the following equation:

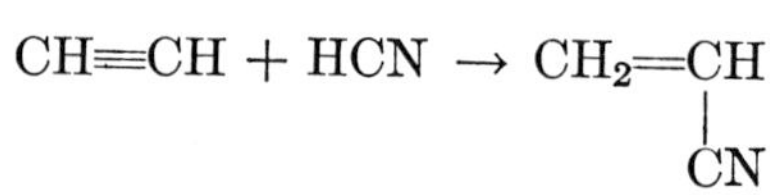

The chemistry of acrylonitrile is discussed in considerable detail in a booklet published by the American Cyanamid Co.

The polyfluorovinyl resins have such excellent resistance to chemicals, solvents, and heat that they have been called the "noble" vinyl resins. Three commercial products are available: Teflon, du Pont Co.; Kel-F, M. W. Kellogg Co.; and Fluorethane, Bakelite Co. They have been used chiefly in films and simple molded objects rather than in coatings because of their insolubility. However, the recent availability of dispersions of these materials makes possible the application of coatings having outstanding physical and chemical properties. The dispersions may be applied by spray, dip, or flow methods, but the resulting deposit must be fused at temperatures from 675°F to 760°F. The coatings have very poor adhesion and therefore a primer is required on smooth surfaces.

It has not been possible to prepare the monomeric tetrafluoroethylene by direct fluorination of ethylene or acetylene. It has been prepared (3, 5, 6) by pyrolysis of monochlorodifluoromethane at temperatures from 650°C to 800°C according to the following equation:

$$2HCF_2Cl \rightarrow CF_2{=}CF_2 + 2HCl$$

It was not believed that this monomer would form addition polymers since experience indicated that some dissymmetry or the presence of ethylenic hydrogen was necessary. It had been shown that a related monomer, tetrachloroethylene, did not form homopolymers. However, in 1938, Dr. Roy Plunkett of du Pont noted with the keen perception of genius that a pressure cylinder which had contained tetrafluoroethylene gas but from which no more gas could be obtained weighed more than an empty cylinder. He ordered the cylinder cut open and found that it contained the white, solid homopolymer. For this discovery Dr. Plunkett received the John Scott Award from the City of Philadelphia in 1952 (4).

Polytetrafluoroethylene is highly crystalline because of the symmetry of its structure. It has very little thermoplasticity and does not show significant thermal degradation until heated above 750°F. It shows some cold flow or deformation under pressure. It has good flexibility, and its cold brittle point is below −80°C (−112°F). Since it resists all solvents and corrosive agents except hot molten alkali metals it obviously would make a very desirable coating but would be difficult to apply. However, there is no doubt that it will be an important coating material in the future both alone and copolymerized with other monomers (27).

The chlorotrifluoroethylene polymers (Kel-F and Fluorethane) are somewhat less resistant than polytetrafluoroethylene but still considerably more resistant than any other vinyl resins. The molecule is not as symmetrical and the polymer does not show as much crystallinity as that of Teflon. The monomer may be prepared from trichlorotrifluoroethane by dechlorination in boiling ethyl alcohol in the presence of zinc dust as follows:

$$CF_2Cl{-}CFCl_2 + \text{Zinc dust} \rightarrow CF_2{=}CFCl + ZnCl_2$$

The monomer is a gas with a boiling point of −27.9°C and shows less tendency to polymerize than tetrafluoroethylene when stored in pressure cylinders at room temperature. A variety of methods for the polymerization of chlorotrifluoroethylene is discussed by Schildknecht (3).

POLYVINYL CHLORIDE RESINS

Vinyl chloride monomer may be polymerized by the bulk, solvent, suspension, or emulsion method, as described previously. Polyvinyl chloride has a relatively low degree of solubility in organic solvents. Therefore, it is more widely used in calendered film and sheeting and in molded and extruded objects than for the solvent type of coating application. With the introduction of the so-called "paste" or "dispersion" resins, suitable for making organosols and plastisols, the excellent physical and chemical properties of polyvinyl chloride may be incorporated into coatings. The disadvantage of the organosol or plastisol application is that the coating as deposited has no strength. It is a layer of the particles of resin loosely held together by a solvent-type plasticizer or small amount of solvent. This relatively weak coating must be heated to a temperature such that the individual resin particles will fuse together into a continuous film. This temperature is usually between 300° and 350°F. However, the time at this temperature is short since only fusion is required and not a thermosetting or curing process. The dispersion resins are described in detail later in this chapter.

Heat and Light Degradation. Although polyvinyl chloride has many excellent properties it is degraded by heat at temperatures about and above the fusion point, and also by ultraviolet radiation as found in sunshine. This sensitivity to heat and light also applies to the copolymers containing vinyl chloride and to those containing vinylidene chloride. The degradation may be controlled by the incorporation of certain pigments and suitable stabilizers.

A great deal of research has been reported in the literature on the degradation and stabilization of polyvinyl chloride and related materials. Schildknecht (3) lists about 50 patents related to this subject. Brochures are also available from manufacturers of stabilizers such as those listed in Table 2. The polyvinyl chloride resins may vary somewhat in heat and light sensitivity, depending on purity of materials used in resin formulation, conditions of processing, etc. Some resins may have stabilizers incorporated in the resin during manufacture. The types of stabilizers required will be better understood after a brief discussion of the structure of polyvinyl chloride and the degradation products.

Chemical and X-ray analysis have shown polyvinyl chloride to be principally a head-to-tail arrangement of the monomers having a structure as follows:

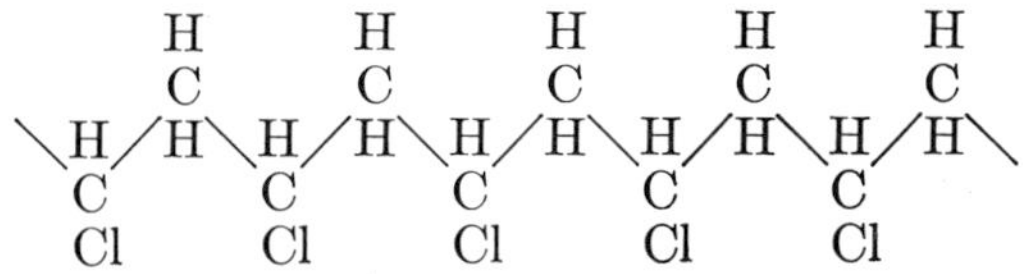

When heated above 212°F polyvinyl chloride will show a series of color changes from cream through orange, brown, and finally black, depending on time and temperature. It has also been shown that hydrochloric acid is a product of the degradation reaction. As a result of the dehydrochlorination the degraded polymer is believed to have the following structure:

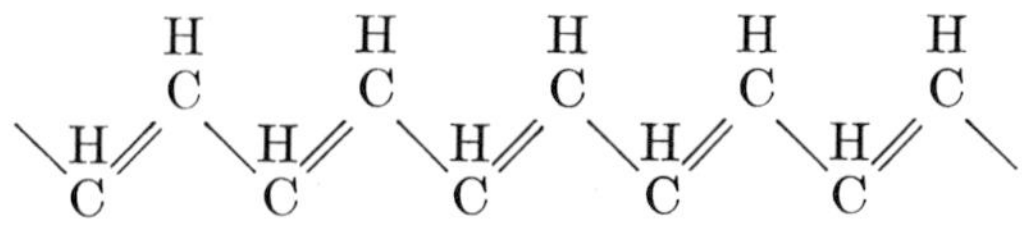

The conjugated diene unsaturation in the above structure has been proved by Hendricks *et al.* (7) by ultraviolet absorption measurements. Oxidation of this structure would produce carbonyl groups along the chain which are chromophores or colored bodies. Hendricks (7) indicated the presence of carbonyl groups by infrared analysis, and they have been reported in heat-degraded polyvinylidene chloride by Boyer (10).

Marvel and Horning (9) showed that the chemical removal of chlorine by reaction with zinc dust was quite different from the above dehydrochlorination; it did not produce a color change, or any unsaturation. They postulated the following structure for this product:

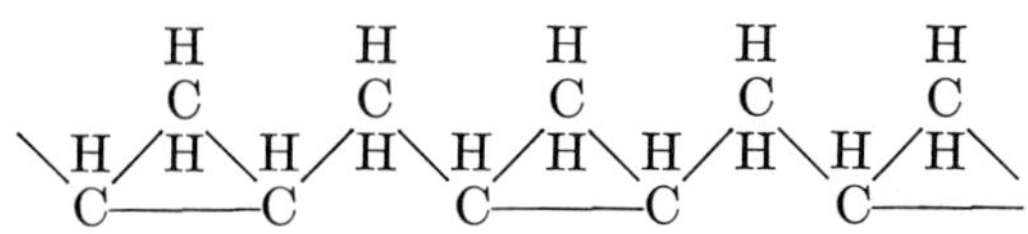

In view of these results it is believed that the degradation of vinyl chloride resins by heat and ultraviolet is the dehydrochlorination reaction as indicated.

Polyvinyl chloride plastics and coatings are subjected to considerable heat during processing, which probably produces some diene unsaturation in the polymer chains. We have seen in Chapter 2 on drying oils that such unsaturation is subject to oxidation by atmospheric oxygen, and that this reaction is accelerated by sunshine. Continued exposure to sunshine not only produces hardening and embrittlement due to oxygen cross-linking between molecules, but it also produces oxidative decomposition due to splitting of the polymer at certain locations where oxygen has been absorbed. The oxidative decomposition of the drying oils can be retarded by the use of pigments which have high absorption for ultraviolet. Similarly polyvinyl chloride may be protected against ultraviolet degradation by certain pigments and other compounds which absorb light in this region, and by the addition of antioxidants. Further evidence of the deleterious effect of heat on light stability is the general experience with films produced from organosols. These films usually show better resistance to sunshine than films produced from solution, probably because both resins and films have had less exposure to heat during processing and film production.

It has been suggested (10) that addition of reactive materials capable of producing a Diels-Alder type of reaction with the diene unsaturation should have a stabilizing effect. Such materials would include the esters of maleic or aconitic acids.

Leistner, Hecker, and Knoepke (32) have disclosed that certain organic compounds, such as organic phosphites, when used with the common metal stabilizers improve the stability to heat and light, and the clarity of clear films of polyvinyl-type resins. They function as chelating compounds by forming complexes with the metallic chlorides produced in the degradation reaction. These complexes are transparent; therefore they do not cause turbidity in clear films. Typical examples are Stabilizers Mark XX and Mark PL of the Argus Chemical Laboratory (33). They do not function alone as stabilizers but must be used in conjunction with regular metallic stabilizers. From this brief discussion it will be seen that the following types of compounds might be useful to stabilize polyvinyl chloride against heat and light:

1. To neutralize HCl and convert it to an insoluble product.
2. To absorb ultraviolet.
3. To act as an antioxidant.
4. To combine with diene unsaturation in a Diels-Alder type of reaction.

A series of such compounds was investigated by Hendricks *et al.* (7, 8) with good results. The fundamental reasons for the yellowing and loss of strength of polystyrene and polyvinylidene chloride during exposure to light were investigated by Matheson and Boyer (28), and improved products were developed from their findings.

Heat and Light Stabilizers. Since there is a large and growing list of heat and light stabilizers which meet the above requirements for protecting polymeric products containing vinyl chloride and vinylidene chloride, it would be impractical to attempt to list all of them here. Some stabilizers are pigments which produce color and opacity; therefore they are not satisfactory for clear films. Stabilizers comprise both the inorganic and organic compounds of lead and sodium, and certain organic compounds of tin, strontium, barium, cadmium, and calcium. Certain epoxy compounds and amines are useful as stabilizers for clear films. A variety of antioxidants also has been suggested to improve heat and light stability. A partial list of these materials is given in Table 2. The use of specific stabilizers is illustrated in the various

Table 2 Heat and Light Stabilizers for Vinyl Chloride Resins

Basic lead carbonate
Basic lead sulfate
Basic silicate white lead
Dibasic lead phosphite
Dibasic lead phthalate
Dibutyl tin dilaurate
Dibutyl tin maleate
Barium ricinoleate
Barium-sodium organic complex (CS-137) *
Strontium organic compounds (SN, V-1-N, V-9) †
Barium-cadmium organic complex (Stabilizer Mark M) ‡
Cadmium organic compound (Stabilizer 21) †
Complex organic compound (Stabilizer E 6B) †
Epox compound (Stabilizer A5) §
Sodium-organophosphate stabilizers ||
Epoxy compound (Stabilizer Mark F) ‡

* National Lead Co.
† Advance Solvents & Chemical Corp.
‡ Argus Chemical Lab.
§ Union Carbide & Carbon Corp.
|| Ferro Chemical Corp.

plastic and coating formulas given in this chapter. It has been found that polyvinyl resin coatings applied to iron, zinc, galvanized iron, and tinplate surfaces are heat-degraded faster than usual because these metals accelerate the reaction. This may be avoided by passivation of the metal surface by standard chemical treatment, or by use of an oil- or resin-type primer.

Commercial Types. The Geon 100 series of polyvinyl chloride resins (B. F. Goodrich Chemical Co.) may be used to illustrate the characteristics and applications of this type of resin. The resins are available as fine white powders having the physical characteristics shown in Table 3.

Table 3 * General Properties of Geon † Resins

Geon Resin	101	101-EP	103-EP	121	126
Specific gravity	1.40	1.40	1.40	1.40	1.40
% Ash content (max.)	0.1	0.1	0.1	0.35	0.35
% Moisture (max.)	0.7	0.7	0.7	0.7	0.7
Specific viscosity ‡	0.52–58	0.52–58	0.43–47	0.57–63	0.35–40

* Courtesy of B. F. Goodrich Chemical Co.
† Trade name, B. F. Goodrich Chemical Co.
‡ Solution in nitrobenzene, special method of B. F. Goodrich Chemical Co.

The high specific gravity or pounds per gallon of these resins is characteristic of materials containing chlorine, as noted previously for chlorinated rubber and chlorinated paraffin wax, and the chlorinated solvents. The specific viscosity of the resins is an indication of the relative molecular weight of the resins. The first three resins are used chiefly for calendered film and molded and extruded objects. The letters EP following the number indicates a faster-mixing or compounding resin. Geon 103-EP is designed chiefly for calendered film and sheeting. Geon 121 is the dispersion type intended for use in plastisols. It may also be used in organosols, but Geon 126 is preferred for this purpose.

Plasticizers. Polyvinyl chloride resins are polymeric materials having excellent toughness but not sufficient flexibility for normal use without plasticizer. The theory and types of plasticizers are discussed in Chapter 10. An interesting comparison of the effects of seven representative types of plasticizers on the physical properties of Geon 101-EP is given in Table 4. The compounds were formulated so as to have equal Durometer hardness, and it will be noted that this required from 50 to 110 parts of plasticizer to 100 parts of resin, depending on the type of plasticizer used.

The value of dioctyl sebacate as a plasticizer for polyvinyl chloride is shown in Table 4. Its good softening action is indicated by the low ratio used. This plasticizer also has low volatility at elevated temperature and excellent retention of flexibility at low temperature, and the results indicate that it imparts good electrical resistance. Dioctyl phthalate also gives good results, but, of course, it is somewhat more

Table 4 * Effect of Plasticizer on Geon 101-EP Resin

Plasticizer	Parts Plasticizer to 100 Parts Resin	Hardness, Durometer A	% Loss, 4 Days at 100°C	Brittle Point, °F	Insulation Resistance †
Dioctyl phthalate	50	85	5.0	−40	Excellent
Tricresyl phosphate	60	85	1.5	−10	Good
Trioctyl phosphate	55	85	10.0	−50	Poor
Dioctyl sebacate	50	85	1.0	−65	Good
Petroleum type ‡	30–30	85	7.0	−30	Poor
Polymeric ester type §	60	85	1.5	−10	Poor
Nitrile rubber type ‖	110	85	0.0	−30	Poor

* Courtesy of B. F. Goodrich Chemical Co.
† Evaluated in compound including clay and basic lead carbonate.
‡ Low-cost extenders in equal amount with dioctyl phthalate; data obtained with Dutrex 20 (Shell Chemical Co.).
§ Paraplex G-50 (Rohm and Haas Co.) for non-migrating use.
‖ Hycar OR-25EP (B. F. Goodrich Chemical Co.) for non-migrating use.

volatile. Frequently, more than one type of plasticizer is used in a polyvinyl chloride compound to obtain best all-round properties. This is illustrated in the formulas which follow shortly.

Sheeting and Film. Thermoplastic polymers may be compounded and formed into relatively thick, rigid sheets or into flexible, continuous sheeting and film. The physical properties are determined by the type of polymer, type and amount of plasticizer, and addition of fillers and pigments. Rigid sheets may be molded, cast, or cut from large blocks of the solid material. Continuous sheeting and film of more flexible compounds may be obtained by extrusion or forcing pastes or solutions through long, slotted openings onto a moving belt. Some films are made by casting the polymer solutions onto a polished belt or rotating drum and stripping the film off after the solvent has evaporated.

Flexible sheeting may also be produced with a calender machine without the use of volatile solvents. This represents a saving in cost and a marked reduction in fire hazard. However, a calender is an expensive machine and requires considerable power to operate. In a simple three-roll calender the compound is fed into the nip between the top and middle rolls. The clearance between these rolls determines the thickness of the sheeting. By proper adjustment of roll temperature the layer of compound adheres to the middle roll. It is removed from this roll after passing between the middle roll and the bottom roll. The sheeting may be obtained as a free film, or it may be deposited on cotton or other fabric by simultaneously passing the fabric between the middle and bottom rolls. Proper adjustment of roll temperatures and pressure will cause the sheeting to adhere to the fabric.

Several types of calenders are available, some of which have more than three rolls.

Calendered Sheeting. An indication of the types of materials used with polyvinyl chloride resin to produce an unsupported blue sheeting suitable for upholstery is given in Formula 1. Of course, considerable

Formula 1 * Calendered Blue Sheeting for Upholstery

	Parts by Weight
Geon 101 resin	100.0
Dioctyl sebacate	15.0
Octyl diphenyl phosphate	15.0
Polymeric plasticizer	7.0
Dioctyl phthalate	17.0
Calcium carbonate	15.0
Basic lead carbonate	3.0
Barium-cadmium laurate	1.0
Stearic acid	0.5
Phthalocyanine blue	2.0

* Courtesy of B. F. Goodrich Chemical Co.

variation may be made to meet performance, color, or cost requirements. Similar compounds may be made with the copolymer vinyl resins described in the following section. The formulation is mixed or compounded in a Banbury mixer or on a two-roll rubber mill to a plastic mass, and is fed to the calender while quite warm so that it will be relatively soft. The mixing time of the compound may be reduced by using the EP-type resins in Table 3.

The four plasticizers in this compound are used to illustrate the properties they impart. Dioctyl sebacate imparts low-temperature flexibility (see Table 4). It also improves the softness or "drape" and the release action from the calender. Octyl diphenyl phosphate imparts flame resistance without decreasing low-temperature flexibility. The polymeric plasticizer may be used as a grinding medium for the pigment and stabilizer to ensure best dispersion of these materials in the compound. A variety of polymeric plasticizers is available as indicated in Chapter 10. Calcium carbonate is a filler added to reduce raw-material cost. It should be kept to a minimum to avoid undue loss of tear strength and abrasion resistance. Basic lead carbonate is a good heat stabilizer. Barium-cadmium laurate is a light stabilizer and also protects the phosphate plasticizer during processing. Stearic acid is a lubricant to facilitate release from the rolls. The phthalocyanine pigment imparts an excellent blue color which is quite stable to heat and light.

Recommendations regarding the use of vinyl chloride resins in extrusion and injection molding are available from the manufacturers. The use of the dispersion-type resins, such as Geon 121 and Geon 126 in Table 3, is similar to that of the copolymer dispersion resins described in the following section. This type of resin has made possible the economical use of vinyl chloride resins in spray, dip, and knife-coating applications.

VINYL CHLORIDE-VINYL ACETATE COPOLYMERS

At the present time the most important vinyl resins for surface coatings are the vinyl chloride-acetate copolymers. They are more soluble than the straight vinyl chloride resins, and they have better compatibility with plasticizers and other film-formers. These copolymers have high vinyl chloride content and therefore retain their sensitivity to degradation by heat and light. They produce extremely strong, tough films having excellent resistance to abrasion and chemicals. They are thermoplastic; therefore they are softened or removed by true solvents and swollen by some non-solvents. Some of these resins may be blended with relatively small percentages of other resins to produce thermosetting films. Some types may be used with alkyd resins, nitrocellulose, etc., to impart increased toughness and chemical resistance to these materials. The copolymers range from slow burning to non-flammable, but the correct plasticizers are required to produce non-flammable coatings.

Commercial Types. The vinyl chloride-acetate copolymers known as "Vinylite* resins" and produced by the Bakelite Co. will serve to illustrate the commercial types. A selected group of the Vinylite copolymer resins is shown in Table 5, together with some of the polyvinyl acetate and polyvinyl butyral resins.

Variations in Composition and Characteristics. The vinyl resins in Table 5 fall into four groups based on composition. The first group contains three of a series based on straight polyvinyl acetate but varying in molecular weight as indicated by their intrinsic viscosities. The second group are hydrolyzed polyvinyl acetate resins and therefore contain some polyvinyl alcohol, depending on the degree of hydrolysis. The third group is the vinyl chloride-acetate copolymers. They show a range of acetate content from 3% to 38%. In addition one resin contains hydroxyl groups from hydrolysis of the acetyl groups, and another is modified by the incorporation of a dibasic acid such as

* Trade mark.

Table 5 ** Vinyl Coating and Adhesive Resins

Product	Approximate Chemical Composition, % by Wt.			Intrinsic Viscosity (Cyclohexanone at 20°C)	Specific Gravity	Solubility Characteristics			Remarks
	VC	VA	Other			Solvents Commonly Used	Concentrations Normally Used %	By Wt. in	Certain members of group, by virtue of reactivity with other materials, possess "limited" thermosetting properties.
Vinyl Acetate Resins									
AYAC	......	100	....	0.11	1.18	Almost all common solvents except aliphatic hydrocarbons (water latent solvent)	80	Acetone	Flows at 200°F, readily dispersed in water.
AYAA	......	100	....	0.39	1.18		55	Acetone	Water-white, odorless, tasteless, nontoxic. Heat-seal readily. Grease resistance and non-support of bacteria useful in food packaging. Block point raised by small amounts waxes or NC. Widely compatible. Hot melts with Bakelite XR-4357. Good heat and light stability of value in certain coatings. AYAT used as adhesive for mfg. of paper hot-drink containers. Resins higher in molecular weight than AYAT have only slightly higher softening point.
AYAT	......	100	....	0.69	1.18		28	Acetone	
Vinyl Alcohol-Acetate Resins									
T24-9	......	91	3.5 *	0.56 †	0.941	Esters, alcohols, alcohol-ethers preferred for thinning	24% (by wt.) in toluene ‡		Oil-and-grease-resistant, heat-sealable coatings. Stronger and higher-impact-resistant bonds than unmodified vinyl acetate resins. Softening point elevated by thermosetting materials, such as Bakelite phenolic resins. Behave similar to vinyl butyral resins in coatings.
MA 28–14	......	86	5.5 *	0.56 †	0.992		28% (by wt.) in methyl acetate ‡		
MA 28–18	......	82	7.0 *	0.56 †	0.992		28% (by wt.) in methyl acetate ‡		
Vinyl Chloride-Acetate Resins									
VYCC *	62	38	....	0.28	1.30	Esters, ether-alcohols, ketones; tolerates hydrocarbons and alcohols	40	*n*-Butyl acetate	Compatible with nitrocellulose, and with other vinyl chloride-acetate resins. Improves resistance to chemicals and outdoor exposure of nitrocellulose finishes.

* Hydroxl (2.6 times greater when calculated as vinyl alcohol).
† Intrinsic viscosity of vinyl acetate resin before modification. (Chemical-composition and intrinsic-viscosity data refer to resin content.)
‡ Available only in solution as indicated.

VYLF	87	13	....	0.24	1.34	Ketones, selected esters. Tolerate aromatic hydrocarbons	35	1:5 Ketone: toluene	Blend with VYHH to increase solids, glass, and build. Lowers flexibility and softening range.
VYHH-1	87	13	....	0.53	1.36		20	1:1 Ketone: toluene	For general coatings which require little or no plasticizer.
VMCH	86	13	1 †	0.53	1.35		20	1:1 Ketone: toluene	Use alone or blend with other vinyl chloride-acetate resins for air-dry and low brake adhesion. Reacts with basic pigments, resins, etc.
VAGH	91	3	2.3 ‡	0.57	1.39	Ketones, selected esters; tolerates hydrocarbons and alcohols	20	1:1 Ketone: toluene	Compatible with wide range of other coating materials. Air-dry adhesion to most coating vehicles.
VYNS-3	90	10	....	0.79	1.36	Ketones	18 32	MEK, atmos. temp. MEK, heated §	Intermediate properties of this group as regards solubility, plasticizer tolerance, heat sealability.
VYCM	91	9	....	1.38	1.37		6 20	MEK, atmos. temp. MEK, heated §	Physical properties similar to VYNW. More soluble. Also used in organosols.
VYNW (formerly VYNW-2)	95	5	....	1.25	1.39	Cyclic ketones, atmos. temps.; methyl ethyl ketone, heated	14 14	Cyclohexanone, atmos. temp. MEK, heated §	Can be highly plasticized to give tough, elastic coating, non-tacky even at 225°F.
VYNV-1	95+	<5	....	1.53	1.39	Disperse (not dissolve) in hydrocarbons and/or plasticizers	35	20:80 Diisobutyl ketone: xylene	For use in organosols. Rapid drying. Fuse at 350°F. Plasticizers permit higher solids.
VYNV-2	95+	<5	....	1.52	1.39		50	Flexol Plasticizer DOP	For use in plastisols (coatings of 100% non-volatile content). Fuse at 350°F.
QYNV	99+	.....	....	1.58	1.40	Disperse in plasticizer	64	Flexol plasticizer DOP	Easy-dispersing grade for use in plastisols. Fuse at 350°F.

* Supplied as 40% solution in *n*-butyl acetate (91% grade).
† 1% interpolymerized dibasic acid (0.7–0.8 carboxyl).
‡ Hydroxyl (2.6 times greater when calculated as vinyl alcohol).
§ Gels on cooling.

Vinyl Butyral Resins

XYHC	......	.....	7 *	0.81	1.12	Alcohols, alcohol plus ketone, esters. Tolerate aromatic hydrocarbons (water latent solvent)	15	Ethanol	Strongly adhesive; excellent resistance to sunlight. Solvent, water resist., and soft. temp. increased by reacting with drying oils, Bakelite phenolic resins, vulcanizing agents, etc. Increase toughness, flexibility, and adhesion of thermosetting materials. Compat. NC, shellac, castor oil, etc.
XYSG	......	.....	7 *	1.16	1.12		10	Ethanol	

* Hydroxyl (2.6 times greater when calculated as vinyl alcohol).
** Courtesy of Bakelite Co., Div. Union Carbide and Carbon Corp.

maleic. The two resins in the fourth group are polyvinyl butyral resins which contain some unreacted hydroxyl groups in their structure.

The difference in molecular weight of the resins in each group is indictated by the difference in the intrinsic viscosities. The effect of composition on solubility of the resins is shown clearly, and should be related to the discussion following Table 1. The good solubility of the polyvinyl acetate resins in almost all common solvents is related to the large acetyl side group and its polar nature. The effect of increasing molecular weight on the concentration of the resin in acetone solution is also shown. It will also be seen that increasing the acetate content in the chloride-acetate group generally widens the range of solvents which can be used. An exception is the resin VAGH; because of its hydroxyl content its solubility is widened. The polyvinyl butyral resins are soluble in alcohols because of their hydroxyl content, and they will tolerate some aromatic hydrocarbons because of the large butyryl side group with a four-carbon chain in it.

The relative toughness and hardness of these resins vary from the soft butyral resins, which require only small amounts of plasticizer, to the hard, high-chloride copolymers which may require as much as equal parts of resin and plasticizer for some applications. Very good oil, grease, and chemical resistance is obtained from the high-chloride copolymer resins, but it should be apparent why the oil and grease resistance obtained from the alcohol-acetate resins is still better. Since the chlorine content is the principal factor in degree of flammability it will be realized that the high-chloride resins are best for nonflammable coatings. On the other hand the resins which do not contain chlorine have excellent light stability and do not require stabilizers. They also have better heat stability in that they do not split off hydrochloric acid, but of course there is a limit to the heat resistance of any organic compound.

Solubility and Compatibility. Since the chloride-acetate copolymers are the ones most widely used in coatings they will now be considered in greater detail. We have seen previously that solubility and compatibility are two important characteristics of coating resins, particularly from economic and application viewpoints. The low solubility of straight polyvinyl chloride resins was greatly improved by copolymerizing of vinyl chloride with vinyl acetate. A good compromise was obtained in the early copolymers by using a molar ratio of 9 chloride to 1 acetate. This ratio corresponds to the 87% chloride, 13% acetate by weight, shown in Table 5 for resins such as VYHH-1. This amount of acetate improved the solubility considerably without much sacrifice of film resistance and hardness. However, it did not increase the compatibility very much. A considerable increase in compatibility with alkyd resins and some oils and varnishes was obtained by incorporation of hydroxyl groups as indicated in resin VAGH. It was necessary to reduce the chloride content considerably more before compatibility with nitrocellulose was obtained, as seen in resin VYCC.

Since the dispersion-type resins VYNV-1, VYNV-2, and QYNV are not applied as solutions their chloride content can be increased with a corresponding increase in hardness and resistance. In addition, their

Table 6 * Quantitative † Solubilities of Vinylite Resin VYHH in Various Solvent-Diluent Mixtures

Solvent-Diluent	Maximum % Resin VYHH Soluble in Solvent-Diluent Mixture Indicated without Thixotropism at 20°C		Maximum % Resin VYHH Soluble in 50% Solvent-50% Diluent Mixture without Gelation at 20°C	Viscosity, centipoises, 15% Resin VYHH in 50% Solvent-50% Diluent Mixture at 20°C
	15% Solvent, 85% Diluent	50% Solvent, 50% Diluent		
	A	*B*	*C*	*D*
Acetone toluene	11.5	18.6	30.0	31
Methyl ethyl ketone-toluene	9.5	17.5	29.8	39
Methyl ethyl ketone-Troluoil	Insol.	10.0	10.0	10 (10% solution)
Methyl *n*-propyl ketone-toluene	10.4	16.6	27.5	50
Methyl isobutyl ketone-toluene (10°C)	6.1	13.8	21.2	82
Methyl isobutyl ketone-toluene	8.3	16.0	22.0	76
Methyl isobutyl ketone-toluene (30°C)	9.0	18.8	29	50
Methyl isobutyl ketone-Troluoil	Insol.	Insol.	Insol.	....
Methyl isobutyl ketone-monochlorotoluene	7.0	11.0	23.0	115
Mesityl oxide-toluene	9.0	17.0	27.8	64
Methyl *n*-butyl ketone-toluene	10.5	16.8	22.8	65
Methyl *n*-amyl ketone-toluene	7.6	16.4	20.0	93
Methyl *n*-amyl ketone-Troluoil	Insol.	Insol.	Insol.	....
Methyl *n*-amyl ketone-xylene	3.0	14.2	20.0	105
Methyl *n*-amyl ketone-Tetralin	11.5 (without gelation)	....	22.5	187
Methyl *n*-hexyl ketone-toluene	6.7	14.5	18.3	120
Cyclohexanone-toluene	11.5	20.5	35.0	80
Methyl cyclohexanone-toluene	17.0 (without gelation)	....	32.5	....
Isophorone-xylene	12.0	19.0	33.2	145
Acetonylacetone-toluene	11.0	16.0	31.0	108
Ethyl acetate (99.3%)-toluene	7.0	14.0	15.8	110
n-Butyl acetate (99.3%)-toluene	Insol.	12.0	13.9	Gel
Nitroethane-toluene	9.8	20.6	27.2	85
1-Nitropropane-toluene	9.8	19.6	27.8	92
Methyl isobutyl ketone-toluene ‡	11.6	19.8	25.5	37

* Courtesy of Bakelite Co., Div. Union Carbide and Carbon Corp.

† Data obtained by interpolation from charts. Slight variations in commercial grades of materials involved will cause corresponding changes in the values.

‡ Data obtained on a 3:2 mixture of resins VYHH and VYLF.

NOTE: It is customary to prepare base solutions, semi-pastes, and finished coatings containing concentrations of resin approximately 2–5% lower than the values given in column *C* above in the thinner mixture indicated. This is, in general, entirely practical because normal room temperatures are higher than the 20°C (68°F) temperature at which these data were obtained. Also a slight amount of thixotropism in the solutions eliminates much of the settling usually encountered in other types of pigmented materials.

intrinsic viscosities indicate that they have the highest molecular weights in the copolymer resin group, which again adds to their film toughness. However, the solvent-non-solvent combinations used with the organosols must be controlled carefully to obtain proper application and film properties. Similarly, a solvent-type plasticizer for these resins must be used for the plastisols.

Fig. 2. Equilibrium Viscosities: Methyl Isobutyl Ketone-Toluene Mixtures

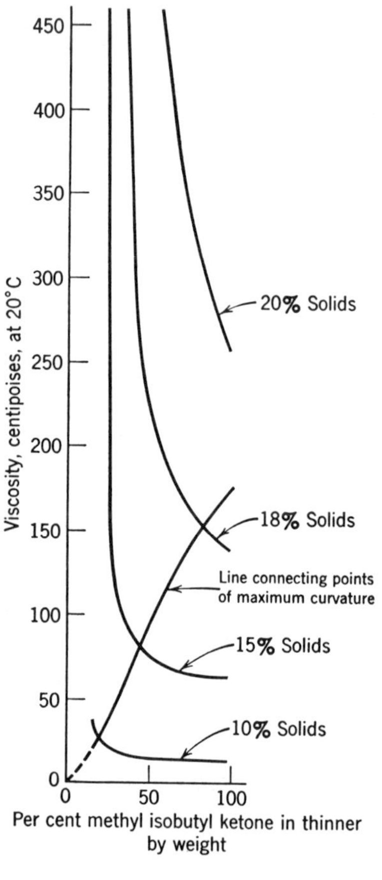

The vinyl copolymer resins exhibit an interesting deviation from most polymeric materials in their solution characteristics. In general, as the concentration of a polymer in a solution is increased the viscosity increases until an extremely viscous mass is obtained. However, when the concentration of a vinyl copolymer resin solution is increased the consistency changes from fluid, to thixotropic, and then to a gel. The concentrations at which these three phases occur depend on the type of true solvent, on the ratio of solvent to diluent in the thinner, and to a lesser extent on the temperature. If the ratio of true solvent to diluent is low the resin will be insoluble at any concentration. Of course, the amount of diluent which can be added to the thinner depends on the relative solvent power of the solvent and the diluent. The maximum concentration of resin VYHH which can be used before thixotropy and gelation occur with thinners composed of 50:50 mixtures of various solvents and diluents is shown in Table 6, columns *B* and *C*. These data are based on diagrams such as those shown in Figs. 2–6.

Figure 2 shows the increase in viscosity of four concentrations of resin VYHH in various ratios of methyl isobutyl ketone and toluene as the thinners. Figure 3 shows the three different consistency phases (fluid, thixotropic, and gel) and the insoluble region, using the same thinners as in Fig. 2 (various ratios of methyl isobutyl ketone and toluene). Figures 4, 5, and 6 show the same solution-consistency phenomena for the following thinner combinations:

Fig. 4, acetone:toluene mixtures.
Fig. 5, methyl ethyl ketone:toluene mixtures.
Fig. 6, methyl ethyl ketone:Troluoil mixtures.

Additional charts are available from the resin manufacturer covering the other solvent combinations, but those shown here illustrate their value in determining the relative solvent power of solvents and diluents, and the consistency characteristics of the solutions. Comparison of Figs. 3 and 4 shows that acetone is a better true solvent than methyl isobutyl ketone. Of course, the much faster evaporation rate of acetone is not shown, and it may be objectionable for some applications. Similarly, Figs. 5 and 6 show that toluene is a better diluent than Troluoil with methyl ethyl ketone as the true solvent.

It should be kept in mind that the data for these charts were obtained on solutions of the resin alone. It may be expected that some variations will occur in commercial coating formulations which contain plasticizers and other modifiers. Since it would be impractical to attempt to cover all possible combinations the above data are useful criteria of the general behavior of this type of resin with solvents.

Two practical advantages may be derived from the three consistency-phase characteristics of vinyl resins. In a dip-coat application the use of a highly volatile thinner will permit dipping in a free-flowing fluid coating which will change to a non-flowing thixotropic coating rapidly after removal from the dip tank because of the loss of solvent. This prevents sagging and increases the thickness of coating which may be applied in each dip operation. The same change in consistency, fluid to thixotropic, may be obtained with a change in temperature. Therefore if the coating were hot it would be fluid but would change to thixotropic when cooled by the object on which it was applied. We shall see that this change in solubility and consistency with temperature may also be utilized to advantage with organosols and plastisols.

The wide range of thinners which may be used with the various vinyl resins is indicated in Table 5. Detailed information regarding these solvent-diluent combinations is available from both resin and solvent manufacturers; therefore it is not deemed practical to include it in this book.

Plasticizers. All the vinyl resins require plasticizers for normal coating use. Frequently more than one plasticizer may be included in a coating to impart required properties as illustrated in Formula 1. It

Fig. 3. Phase Diagram: Methyl Isobutyl Ketone-Toluene Mixtures

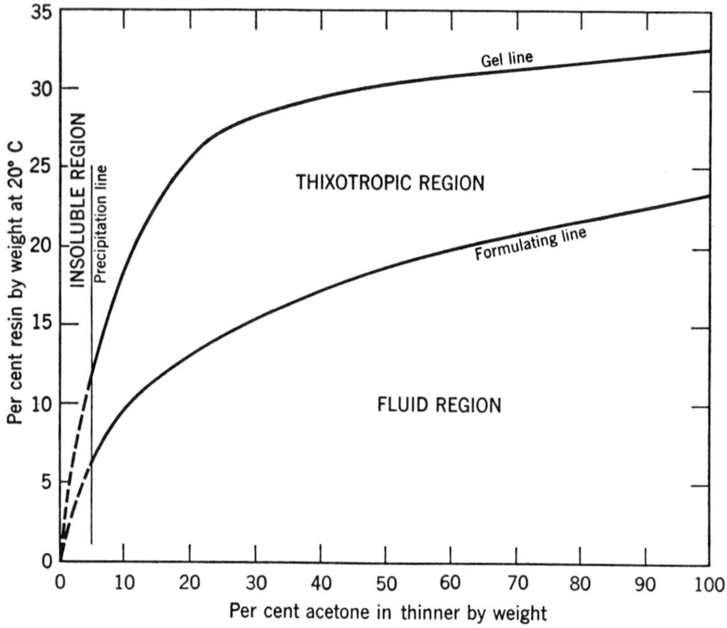

Fig. 4. Phase Diagram: Acetone-Toluene Mixtures

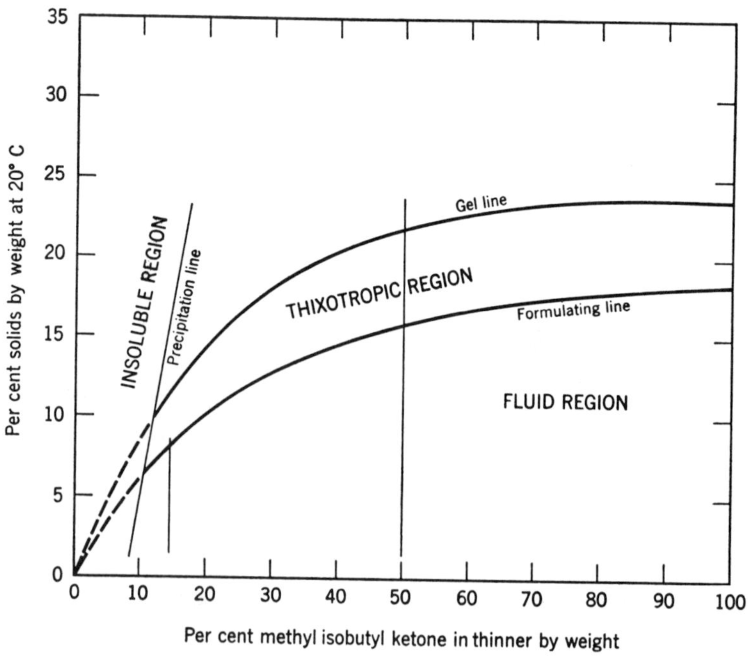

Fig. 5. Phase Diagram: Methyl Ethyl Ketone-Toluene Mixtures

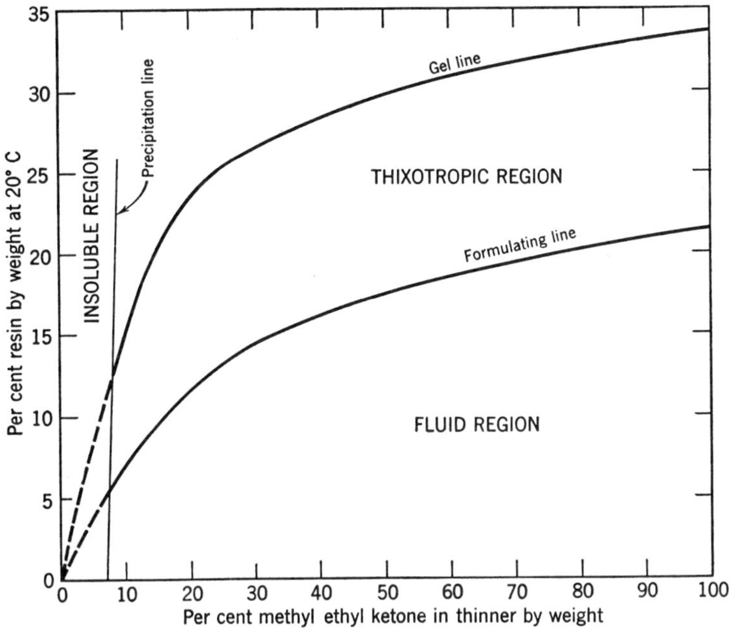

Fig. 6. Phase Diagram: Methyl Ethyl Ketone-"Troluoil" Diluent Mixtures

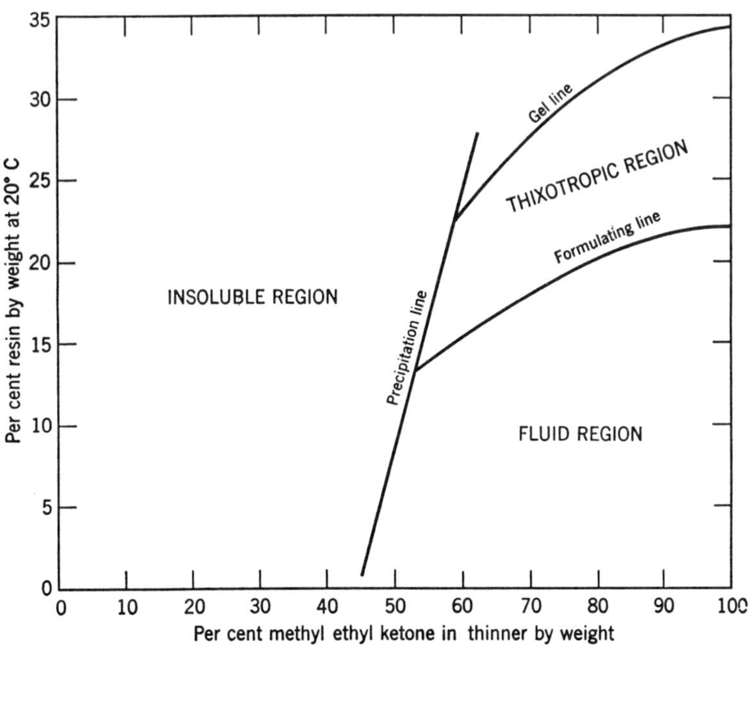

is desirable that the plasticizer exert a solvent action on the vinyl resin. The solvent-type plasticizers show better retention in the coating and therefore better aging qualities. All plasticizers should have very low volatility at normal temperatures, but those used in vinyl coatings should also have low volatility at temperatures of 300–375°F. Many vinyl coatings are heated at this temperature range for short periods to improve adhesion and film properties.

The suitability of the plasticizer for the particular application requirements should be considered carefully by the formulator. For example, in electrical insulation the plasticizer should have good dielectric properties and high moisture resistance. In fabric coatings the softness or drape is important, but the possibility that a particular plasticizer may cause dermatitis must also be considered. Low cost and non-migration are necessary in handbag or novelty coatings; but toxicity, odor, taste, color, and water and gas permeability must be considered in coatings for food packaging. Some coatings are subjected to temperatures of 200–250°F during their normal service life, and the plasticizer must not volatilize, exude, or decompose under such conditions. Other requirements call for retention of flexibility at temperatures as low as −50 to −60°F. Vinyl resin coatings are frequently specified because of their excellent chemical resistance, but the plasticizer in these coatings must also be resistant to the particular chemicals with which the coating will come in contact. The fire retardance of vinyl coatings depends on the chlorine content of the vinyl resin and also on the use of a plasticizer which is non-flammable. The types of plasticizers having the various characteristics noted above are discussed in Chapter 10. Since the vinyl resins have good compatibility with the chemical or monomeric plasticizers and fair compatibility with the polymeric types, little difficulty is experienced in selecting suitable plasticizers.

Reed *et al.* (11, 12, 13) have reported extensive studies on the behavior of plasticizers in vinyl chloride-acetate resins, and other contributors are listed in references 14–19 inclusive. A number of the Paraplex series of resins (Rohm and Haas Co.) are designed especially as polymeric plasticizers for vinyl resins. The polymeric or resinous plasticizers have excellent permanence because they are practically non-volatile even at elevated temperatures. They also impart greater toughness than the chemical plasticizers, but usually they are more expensive. Chemical plasticizers such as dioctyl phthalate, trioctyl phosphate, tricresyl phosphate, and dioctyl sebacate are widely used, but the particular properties of each should be considered for the various applications as indicated above. Some of the chlorinated biphenyls and chlorinated paraffins (see Chapter 9) are used for their electrical properties and moisture resistance and also for their fire-retardant properties. The various ricinoleates are also useful plasticizers for vinyl resin coatings.

Resins. The unmodified vinyl chloride-acetate copolymers such as resins VYLF and VYHH-1 in Table 5 have a very low degree of compatibility with other resinous materials. Consequently, formulations based on these polymers contain practically no other film-modifiers except plasticizers. However, when hydroxyl groups are introduced into the copolymer, as in resin VAGH, the compatibility with other resinous materials is increased considerably. The principal classes of materials compatible with resin VAGH are listed below:

Rosin and its derivatives.
Maleic resins.
Modified phenolic resins.
Pure phenolic resins.
Short- to medium-oil-length varnishes.
Short- to medium-oil alkyd resins.
Urea-formaldehyde resins.
Melamine-formaldehyde resins.

The increased compatibility of VAGH makes it possible to increase the toughness and chemical resistance of varnish and alkyd resin coatings by adding 10–20% of VAGH, based on vehicle solids (see Formula 7); for maximum toughness and chemical resistance VAGH is used with plasticizer only [see Formulas 5(*a*) and 5(*b*)]. The straight coatings are thermoplastic but they may be made thermosetting by addition of 10–20% of heat-reactive phenolic resins, or either urea-formaldehyde or melamine-formaldehyde resins. Without the thermosetting resins the coatings may be air-dried or baked, but a baking schedule is necessary to convert the thermosetting combinations. The thermosetting effect may be due simply to a physical dispersion of the cross-linking resin throughout the film structure, but in some cases there is the possibility of chemical reactivity or condensation of the free hydroxyl groups in the vinyl resin and the cross-linking resin. Of course, the principal value of the thermosetting effect is to minimize softening at higher temperatures, and to increase the resistance of the coating to solvents, but it also improves the surface hardness of the coating.

In order to obtain compatibility with nitrocellulose the acetate content of the copolymer must be increased considerably as shown in resin VYCC. Of course, this reduces its hardness and resistance to solvents, but when VYCC is used instead of alkyd or maleic resins it will add toughness to nitrocellulose coatings. For example, the suggested furniture finishing lacquer given in Formula 2 would produce a much tougher finish than that given by Formula 2, Chapter 11. However, its ease of rubbing and polishing would also be quite different.

Formula 2 * Furniture Finishing Lacquer

	Parts by Weight, Solid Basis
Nitrocellulose, ½ sec	100
Vinylite VYCC	67
Dioctyl phthalate	67

* Courtesy of Bakelite Co., Div. Union Carbide and Carbon Corp.

Modifiers. The vinyl chloride-acetate resins are fairly high-molecular-weight materials. Consequently, they have a strong cohesive force in their films which accounts for their toughness but also detracts from their adhesion to smooth surfaces. A similar condition was discussed in Chapter 11 with respect to nitrocellulose. Coatings based on resins such as VYHH-1 do not have sufficient adhesion when air-dried on metal surfaces, but excellent adhesion may be obtained by baking them at temperatures from 350° to 375°F. To overcome this lack of adhesion in air-dried finishes a modifier was incorporated in the molecular structure of the resin. A reactive dibasic acid, such as maleic, was polymerized with the copolymer as discussed earlier in this chapter. A resin of this type is VMCH, shown in Table 5.

The free carboxyl groups in resin VMCH are highly polar and therefore have a strong affinity for metal surfaces. There is also the possibility of actual chemical reaction between some of the carboxyl groups and the metal. Since carboxyl groups are reactive with basic materials, care must be taken not to use basic pigments or stabilizers in formulations containing resin VMCH. For example, zinc oxide or some of the lead compounds would react with resin VMCH in the container, producing instability on storage and loss of the carboxyl groups required for adhesion. The details regarding pigmentation are given in Volume II.

Table 5 shows that resin VMCH contains only 1% of the dibasic acid. In actual coating applications, resin VMCH is generally blended with resin VYHH-1 rather than being used alone. (See Formula 4.) Since only a very small percentage of the coating is in actual contact with the metal surface the high degree of effectiveness of the carboxyl groups (or increasing the polarity of the coating) in promoting adhesion will be appreciated. It is an indication that the real forces of adhesion operate over a very thin layer at the interface of film and metal. Another point of interest in this connection is the difference in the adhesion of air-dried and baked films of resin VYHH-1. The forces of cohesion in the film while air-drying are greater than the forces of adhesion. Consequently, shrinkage develops from both upper and lower surfaces of the film, and the polar chloride and acetate groups do not approach the metal near enough to develop good adhesion. In the baked films good adhesion does not develop unless the temperature is raised to about the fusion point. This temperature liquefies a layer of the film nearest the metal, thereby permitting orientation of the polar groups and the development of their attractive forces for the metal. The nature of adhesion and the instruments which have been developed in attempts to measure it are discussed in references 21–24 and in Chapter 15.

Applications. The applications of vinyl chloride-acetate resins in surface coatings are discussed in detail in Volume II. However, a few typical applications are included in this chapter, chiefly to indicate the characteristics of the particular resins. These characteristics include the air-drying and baking properties of the resins and their chemical resistance, compatibility, and adhesion. One coating is based on a vinyl butyral resin which is described in a later section. The coatings based on the dispersion resins are also described in a later section.

The effect of the air-drying or baking conditions on the inherent adhesion characteristics of the resins is illustrated in Formulas 3–6

Formula 3 * High-Bake Vinyl Resin Primer

	% by Weight	%	Resin: Plasticizer Ratio
Blue lead	31.3		
Resin VYHH-1	13.0	79.3	100
Dioctyl phthalate	3.4	20.7	26
Methyl isobutyl ketone	21.0	——	
Cyclohexanone	5.2	100.0	
Toluene	20.9		
Xylene	5.2		
	——		
	100.0		

* Courtesy of Bakelite Co.

inclusive, covering a series of metal primers. Formula 3 (resin VYHH-1) has good adhesion only when baked at 350–375°F. Formula 4 (50% resin VYHH-1, 50% VMCH) has good adhesion when either air-dried or baked at low temperatures. Formulas 5(*a*) and 5(*b*) (resin VAGH) may be air-dried, but they give best results when used over Formula 6. Formula 6 is an unusual coating; it illustrates the possibility of using a chemical reaction during the drying process to passivate the metal and develop a strongly adherent organic coating.

This primer, based on resin VYHH-1, does not develop good adhesion unless the coating is baked for a short time at temperatures from 350° to 375°F. The ratio of resin to plasticizer is 100 to 26, and the ratio of pigment to vehicle solids is 1.9 to 1.0.

Formula 4 * Air-Dry or Low-Bake Vinyl Resin Primer

	% by Weight	%	Resin: Plasticizer Ratio
Titanium dioxide	12		
Resin VYHH-1	8 }	84.3	100
Resin VMCH	8 }		
Dioctyl phthalate	3	15.7	18.7
Methyl isobutyl ketone	34.5		
Toluene	34.5	100.0	
	100.0		

* Courtesy of Bakelite Co.

The presence of resin VMCH in Formula 4 develops good adhesion when the primer is either air-dried or baked at temperatures from 200° to 250°F. The ratio of resins to plasticizer is 100 to 18.7, which is

Formulas 5(*a*) and 5(*b*) * Air-Dry Vinyl Resin Primers

(*a*)	% by Weight	%	Resin: Plasticizer Ratio
Red lead	22		
Resin VAGH	15	83.5	100
Tricresyl phosphate	3	16.5	20
Methyl isobutyl ketone	30		
Toluene	30	100.0	
	100		

(*b*)	% by Weight
Basic zinc chromate	7.8
Magnesium silicate	4.8
Resin VAGH	13.9
Methyl isobutyl ketone	44.0
Toluene	29.5
	100.0

* Courtesy of Bakelite Co.

lower than that in Formula 3. The ratio of pigment to vehicle solids is considerably lower, 0.634 to 1.0. The titanium dioxide pigment is not a reactive pigment; therefore it may be used with resin VMCH.

Formulas 5(*a*) and 5(*b*) illustrate two types of corrosion-inhibiting pigments. Both formulations develop good adhesion to metal when they are used over the metal conditioner shown in Formula 6. The ratio of resin to plasticizer is 100 to 20 in 5(*a*), but no plasticizer is used in 5(*b*). There is also a difference in the ratio of pigment to vehicle solids; the ratio in 5(*a*) is 1.22 to 1.0, and in 5(*b*) it is 0.9 to 1.0.

Formula 6 * Air-Dry Vinyl Metal Conditioner

(*a*) Pigment Base	% by Weight
Basic zinc chromate	8.6
Magnesium silicate	1.4
Resin XYHL	9.0
Methyl isobutyl ketone	18.0
Ethyl alcohol	63.0
	100.0
(*b*) Acid Diluent	
Phosphoric acid (85%)	17.2
Water	17.2
Ethyl alcohol	65.6
	100.0

* Courtesy of Bakelite Co.

The pigment base and acid diluent are prepared separately and mixed shortly before application. A very thin coating is applied, 0.3–0.5 mil dry film thickness, and air-dried. The coating has excellent adhesion but is too thin for normal primer use and is referred to as a metal conditioner. Regular primers, such as Formula 5(*a*) or 5(*b*) or other types, are applied over the metal conditioner. Resin XYHL is a polyvinyl butyral type described in a later section of this chapter. It will be noted that no plasticizer is used with it. The ratio of pigment to vehicle solids is 1.11 to 1.0. The correct preparation of the metal surface is important for best results with this coating; other conditions for use of Formula 6 are given in detail in Volume II.

The blue lead pigment in Formula 3 is an excellent heat stabilizer; therefore it protects the resin and keeps heat degradation to a negligible quantity. The baking temperature for Formula 4 is not high enough to require stabilizer. It will be noted that the solvents in Formulas 3, 4, and 5 are ketone-aromatic hydrocarbon mixtures. However, in Formula 6, with the polyvinyl butyral resin, the solvent is a ketone-alcohol mixture. These formulas produce primer coatings having excellent toughness and chemical resistance, but they are somewhat expensive. Formula 7 illustrates the use of the compatible resin VAGH to reduce the drying time and improve the chemical resistance of a corrosion-inhibiting alkyd resin primer.

Formula 7 * Air-Dry Alkyd-Vinyl Resin Primer

	% by Weight	Pounds	Gallons
Red lead	53.6	216.7	3.0
Zinc chromate	10.0	40.7	1.4
Zinc oxide	8.0	35.5	0.7
Graphitic mica	14.7	59.6	2.4
Raw sienna	4.0	16.3	0.6
Magnesium silicate	9.4	37.9	1.6
Aluminum stearate	0.3	1.4	0.2
	100.0		
Resin VAGH	4.7	32.5	2.8
Alkyd resin (65% solids)	42.4	282.0	35.2
Petroleum naphtha	17.1	118.0	16.3
Ethyl butyl ketone	35.0	240.5	35.2
Lead naphthenate (24%)	0.5	3.1	0.3
Cobalt naphthenate (6%)	0.1	0.5	0.1
Manganese naphthenate (6%)	0.1	0.5	0.1
Antiskin agent	0.1	0.9	0.1
	100.0	1083.1	100.0

* Courtesy of Lead Industries Association, *Technical Letter 6*.

The vehicle solids in Formula 7 is a combination of 10% resin VAGH and 90% of a medium-oil-length alkyd resin. The oil in this resin is dehydrated castor, and the phthalic anhydride content is approximately 30%. This primer is sufficiently dry in about 4 hr to apply subsequent coatings, but it would require at least 8 hr, and preferably overnight, without the vinyl resin.

The various architectural and industrial applications for vinyl copolymer resins are given in detail in Volume II. The combination of good toughness and resistance to water and chemicals makes these resins useful in corrosion-inhibiting paint systems for both industrial plants and marine work. They are also used in tank and equipment coatings for breweries, dairies, hospitals, etc. Their use in beer-can linings was referred to previously. Their toughness and flexibility make them suitable for unsupported film and sheeting and also for a wide variety of coatings on textiles, paper, and metal foil.

DISPERSION RESINS

Organosols and Plastisols. The dispersion resins provide a new approach to the application of high polymers in surface coatings through their use in organosols and plastisols. The formulation, manufacture, and use of organosols and plastisols are described in detail in Volume II. A brief outline of the resin characteristics and general procedures is given here.

The dispersion resins are made by emulsion polymerization (30) and were released to the trade first in 1944. Powell (30) reported that 30–35 million pounds of vinyl chloride-type resins were used in dispersion applications in 1950. The dispersion resins are powders of very fine particle size ranging from 0.05 to 1.0 micron in diameter. Dispersions of particles of this size are in the colloidal range; therefore considerable care must be given to the selection of suitable dispersing media. The two types of dispersions which are of particular interest to coating and plastic manufacturers are the organosols and the plastisols. These terms were used first by Powell and Quarles (31) in 1946. The organosols are combinations of dispersion resins, and the plasticizer which may be necessary, dispersed in a mixture of volatile organic liquids which contains both polar and non-polar types. The polar liquid is called the dispersant, and the non-polar liquid is the diluent. Plastisols are the dispersion resins dispersed in suitable liquid plasticizers.

It is the function of the liquids (dispersant-diluent or plasticizer) to maintain the resins in dispersion at low consistency and high resin content for economical application. We have seen previously that the economic disadvantage of high polymers for coatings is the low solid content of their solutions at application viscosity. The organosols and plastisols overcome this disadvantage by replacing solvents with non-solvent dispersing media. However, the coatings which are applied from dispersions are not completely continuous films but are layers of discrete resin particles. The resin particles are adherent because there is a very slight solvent action from the plasticizer or the volatile component, but the film is very weak. However, the film will develop its full strength if it is heated to the fusion point of the resin, which is about 350°F. Only a short time is necessary at this temperature, but heat stabilizers should be added to prevent heat degradation of the resin.

Since volatile components evaporate from dispersions more rapidly than they do from solutions, as discussed in Chapter 6, slower-evaporating liquids are preferred in organosols than are used in solution coatings. For example, the volatile component in the solution-type coatings, Formulas 4 and 5(*a*), is a mixture of equal parts of methyl isobutyl ketone and toluene. In organosols, the methyl isobutyl ketone is frequently replaced with the slower-evaporating diisobutyl ketone, and the toluene is replaced with xylene. In addition the ratios of these solvents and non-solvents are changed considerably; in organosols the methyl isobutyl ketone may range from 5% to 25% and the non-solvent xylene from 75% to 95%. Although xylene is a non-solvent for the dispersion resin it exerts a slight swelling action on it. When it is desirable to reduce this swelling action, part of the xylene may be replaced with an aliphatic hydrocarbon at a further reduction in cost. An indication of the wide range in the composition of the volatiles is given in Table 7. This table also shows the importance of the plasticizer in organosols.

Table 7 * Effect of Plasticizer on Organosol Composition

	Per Cent by Weight		
Organosol	*A*	*B*	*C*
Vinyl resin	33	33	33
Dioctyl phthalate	0	6.6	15.5
Methyl isobutyl ketone	16.7	10.2	1.5
Xylene	30.2	30.1	29.4
Aliphatic hydrocarbon	20.1	20.1	19.6
	100.0	100.0	100.0
Total solids	33%	39.6%	59.5%
Methyl isobutyl ketone in volatile component	25%	17%	3%

* Courtesy of Bakelite Co.

Table 7 shows that a lower-cost volatile component may be used in organosols when a liquid plasticizer is present. The plasticizer has good wetting properties for the resin, and a very slight solvent action on it. The organosols and plastisols are made by dispersing the powdered resin in the liquids in conventional pebble and roller mills. If the initial dispersion is not adequate the consistency will increase on aging because of the weak solvent action of the liquids. Similarly, if the dispersions are stored at temperatures above 120°F, there may be an increase in consistency due to increased solvent action at elevated temperature.

When a coating is applied by conventional methods it must flow or "level out" in order to produce a smooth surface. The flow characteristics of organosols are controlled by the fineness of the ultimate particles of the resin, their degree of dispersion, and their concentration. The degree of dispersion is influenced greatly by the ratio of dispersant to diluent, and the relative polarity of these liquids, as well as the mechanical action in the mill.

In addition to obtaining higher resin content at application consistency and lower-cost volatiles, the organosol technique permits the use of high-molecular-weight resins, and also the 100% polyvinyl chloride type such as vinylite QYNV or Geon 121 and 126. We have seen that higher molecular weight in resins produces films of increased strength. We have also seen that greater resistance to water, solvents, and chemicals is obtained from vinyl resins of high chloride content. It would be impractical to use the high-molecular-weight and high-chloride-content resins in solution coatings because of their low solubility. It will be noted that the intrinsic viscosities (indicating relative molecular weight) of the dispersion resins VYNV-1, VYNV-2, and QYNV in Table 5 are higher than those of any of the other vinyl resins. The methods for preparing pigmented organosols and their applications in coatings will be discussed in Volume II.

Technical bulletins on Pliovic AO (Goodyear Tire and Rubber Co.) suggest the use of surface-active agents to assist in controlling the consistency of organosols. Agents such as the esters of polyethylene glycol or the high-molecular-weight sodium alkyl phosphates are satisfactory. Because of the strong deflocculating action of such agents it may be possible to use only aliphatic hydrocarbons as the volatile components of organosols made with Pliovic AO. It is well known that surface-active agents tend to be highly specific in their action, and that their effects are varied readily by changes in concentration or in other components of the formulation.

Pliovic AO is a unique resin in that it is a copolymer of vinyl chloride and diethyl maleate. This combination provides a degree of internal plasticization so that the resin may be processed at lower temperatures for good working properties, and it requires less plasticizer than polyvinyl chloride for equivalent flexibility.

Plastisols do not contain any volatile thinner; therefore there must be sufficient plasticizer in the formulation to ensure satisfactory dispersion of the resin, pigment, and other ingredients. It will be apparent that the viscosity of the plasticizer will affect the consistency of the plastisol. Some excellent resinous plasticizers are not satisfactory for plastisols because of their high viscosity. Some low-viscosity plasticizers are such good solvents for the resin that excessive consistency would be obtained if they were used alone. They may be blended with plasticizers having less solvent action. Since there is a wide range of plasticizers which may be used in plastisols no difficulty need be experienced in obtaining the desired properties.

There is some tendency for the plastisols to increase in consistency on aging, but this may be kept to a minimum by careful selection of plasticizers of low rate of solvent action on the resin at room temperature. They are frequently ground on a three-roll mill, but the mill should not be allowed to become overheated since the heat will increase the solvent power of the plasticizer and the consistency of the plastisol. Plastisols may be applied by dipping, by knife-coating on paper or textiles, or by a molding operation. In all cases the coating or molding must be fused by heating to at least 325°F to develop proper strength. This requires the incorporation of suitable heat stabilizers as discussed previously.

Further details regarding the formulation, manufacture, and use of plastisols are given in Volume II.

VINYL BUTYRAL RESINS

The raw materials and methods of manufacture of vinyl butyral resins are outlined on page 487. Since this type of resin is produced from polyvinyl alcohol, which is made by hydrolyzing polyvinyl acetate, the cost of the vinyl butyral resins will be higher than that of the copolymer resins. However, the vinyl butyral resins have excellent light and heat stability and may be used where the copolymer resins would not be satisfactory.

Two commercial vinyl butyral resins are listed in Table 5. These resins have the same composition, but they differ in molecular weight as indicated by the difference in their intrinsic viscosities.

Vinyl Butyral Resins XYHL and XYSG

Approximate Composition, % by Weight

Vinyl	54.4
Butyraldehyde	38.3
Acetate	0.3
Hydroxyl	7.0
	100.0

The composition shows that there are some hydroxyl groups as well as butyral groups along the polymer chain. The hydroxyl groups increase the solubility of the resin in alcohols and also its affinity for water. About 5% of water may be used as a latent solvent in combination with alcohols, ketones, or esters. Aromatic hydrocarbons may also be used with alcohols in solvent mixtures. The butyral side group is largely responsible for the affinity of the resin for the aromatic hydrocarbons. The lower-molecular-weight resin Vinylite XYHL is more generally selected for coatings because of its lower viscosity in solution.

For many applications vinyl butyral resins have sufficient flexibility without added plasticizer. However, the resins are compatible with standard chemical and some resinous plasticizers and with castor oil. The vinyl butyral resins are also compatible with some phenolic resins and the urea-formaldehyde and melamine-formaldehyde resins. Such combinations may be "cured" by a short baking schedule, and the resulting coatings will be thermosetting instead of thermoplastic. This increases their resistance to solvents and moisture and produces a harder surface which has less tendency to soften as the temperature is increased. Vinyl butyral resin coatings have very good adhesion to metal, glass, and plastics, and to the more porous surfaces such as wood, paper, or textiles. These coatings do not require a baking schedule to develop adhesion like the unmodified vinyl chloride-vinyl acetate resins.

The vinyl butyral resins have only limited compatibility with other film-forming materials. Their structure and solubility indicate possible compatibility with alcohol-soluble resins. Small amounts of vinyl butyral resins may be added to the alcohol-soluble, heat-reactive phenolic baking resins to increase their flexibility and adhesion without reducing their chemical resistance severely. Conversely, small amounts of the phenolic or the urea-formaldehyde and melamine-formaldehyde resins may be used with the vinyl butyral resins to impart thermosetting properties as mentioned previously. The vinyl butyral resins are compatible with some grades of nitrocellulose, but there would be no distinct advantage in blending these materials since the compatible copolymer VYCC resin has been made available. Blends of vinyl butyral and shellac have been suggested for improved toughness and moisture resistance of the shellac; they also have good sanding property. Vinyl butyral resins could be added to the alcohol-soluble Manila resins to produce shellac substitutes or wood sealers similar to the blends with ethyl cellulose referred to in Chapter 11.

An interesting product resulted from the investigation of the Western Pine Association to develop a knot sealer which would be satisfactory under exterior house paint. The sealer is based on the alcohol-soluble phenolic resin Bakelite BV-9700S and Vinylite XYHL in a ratio of approximately 85:15 on a solid resin basis. The product is known as Knot Sealer WP-578 and is made according to the following formula:

Formula 8 Knot Sealer WP-578

	% by Weight	Solid Basis
Bakelite resin BV-9700S (60% solids)	33.3	85
Vinylite resin XYHL	3.3	15
95% alcohol, denatured	63.4	
	100.0	100

Procedure

Dissolve resin XYHL in the alcohol by thorough agitation, then add resin BV-9700S, and continue agitation until complete mixing is obtained.

Knot Sealer WP-578 may be used directly on the knot and surrounding wood, and the regular exterior house paint applied after the sealer has dried. Since Formula 8 contains only alcohol as the solvent it may be used as a sealer over asphaltic coatings to prevent the asphalt from "bleeding" into subsequent coats of paint. The oils and solvents in paints and enamels dissolve some of the components of asphaltic materials so that they bleed or migrate into the paint coating. Since the asphalt is very dark in color it produces an objectionable discoloration in the paint and frequently retards the drying considerably. The sealer coat is soluble in alcohol, and it is not affected by the oils and solvents in paints and enamels; therefore it prevents the bleeding phenomenon.

The use of vinyl butyral resins in a metal conditioner or "wash primer" for steel and the light metals was discussed on page 513, Formula 6. The two components of this formula must be shipped in separate containers and mixed shortly before application. The mixed material cannot be used if allowed to stand overnight because of reactivity and loss of adhesion. Modifications of this formula are now available which obtain most of the advantages of the two-package "wash primer" in a stable, one-package composition.

Vinyl butyral-plasticizer combinations have been used for special coatings on fabrics. These coatings may be made more resistant to solvents, and less sensitive to softening with heat, if a small percentage of heat-reactive resin is incorporated to impart some thermosetting property. This type of coating may also serve as an adhesive for bonding such widely different materials as cloth, paper, asbestos, cork, wood, metal, glass, and plastics. When this type of adhesive is cured by a short bake it will be more resistant to heat, solvents, and moisture than a thermoplastic adhesive. The wide use of straight vinyl butyral resin as the interlayer for safety glass was referred to previously. Further details regarding the formulation and application of vinyl butyral resin coatings are given in Volume II.

VINYL ACETATE AND ACETATE-ALCOHOL RESINS

The straight vinyl acetate resins and the vinyl acetate-alcohol resins are used much more extensively in adhesives than they are in coatings. Therefore only a brief outline of the types available and their general characteristics will be given. The raw materials for these resins and methods for their preparation were outlined previously.

VINYL ACETATE RESINS

In Table 5 are shown three of a series of five vinyl acetate resins which vary only in molecular weight since their composition is 100% polyvinyl acetate. The relative molecular weights are indicated by the differences in intrinsic viscosities, and the change in per cent solids of acetone solutions as used normally. Other properties of these resins which are affected by change in molecular weight are softening point, tensile strength, and water sensitivity, as indicated in Table 8.

Table 8 * Change in Properties with Molecular Weight of Vinyl Acetate Resins

Vinylite Resin	AYAC	AYAA	AYAT
Softening point, °C	44	66	86.5
Tensile strength, psi	See below †	1500	4200
Water absorption, %			
16 hr at 25°C	2.0	1.6	1.4
144 hr at 25°C	7.3	4.0	3.6
Film condition	Swells, becomes pliable	Swells slightly	Surface attack only; no penetration

* Courtesy of Bakelite Co.
† Film does not break because of ability to cold-flow.

The vinyl acetate resins are soluble in almost all organic solvents except the aliphatic hydrocarbons and anhydrous alcohols (except methanol). However, the lower alcohols become solvents when a small amount of water is added to them. We have seen previously that mixtures of two volatile liquids are often good solvents for resinous materials even though either one alone is a non-solvent. We have also seen that highly soluble polymers tend to retain very strongly a small amount of solvent in their films when air-dried. The retained solvent keeps the film soft and tacky and prolongs the air-drying time considerably. Consequently, films of polyvinyl acetate frequently are heated at temperatures up to 250°F for a few minutes to eliminate retained solvent.

The low softening points for the resins shown in Table 8 indicate that they are heat sensitive and would not be satisfactory in applications which may be heated at temperatures of 65°C (150°F) or higher. These resins are also subject to "cold flow" at normal temperatures. Table 8 shows that the lowest-molecular-weight resin flowed to such an extent when stretched at 25°C that it did not break in the normal manner. Polyvinyl acetate is compatible with the usual ester plasticizers (except DOP) but only low ratios are required, generally not more than 20 of plasticizer to 100 of resin. Castor oil and the linseed-type lacquer oils are incompatible and exude badly on aging. The vinyl acetate resins are incompatible with most alkyd resins and the urea-formaldehyde and melamine-formaldehyde resins. However, they have good compatibility with some of the phenolic resins and with nitrocellulose. The vinyl acetate resins are used in the hot-melt type of adhesives. Some of the low-melting natural and synthetic resins may be added to improve the flow even though they are not completely compatible with polyvinyl acetate.

Polyvinyl acetate adhesives may be applied as solutions in volatile solvents, as latexes with water as the volatile constituent, as hot melts without volatile constituents, or as heat-sealing tape. The adhesive is usually applied to both surfaces being bonded. Because of the solvent-retention characteristic referred to previously a considerable time may be required to develop full bond strength with cold-setting methods. The full bond strength is developed almost immediately in the heat-sealing methods. In these methods the adhesive coating is dried; then the two surfaces are bonded under heat and pressure. The vinyl acetate adhesives are so versatile that they may be formulated to bond almost any combination of materials.

The polyvinyl acetate adhesives and films are extremely stable and show practically no change on aging. This stability and retention of flexibility are quite desirable in coatings for fabrics and leather. However, in order to overcome the softness and heat sensitivity of the surface, a thin coat of nitrocellulose lacquer may be applied as the final finish.

The polyvinyl acetate emulsions or latexes provide a very economical and convenient form in which to apply these resins. They also eliminate the fire hazard connected with the application of solutions in organic solvents, and they have higher solid content at application consistency. The latexes also have much less penetration into porous surfaces, such as paper and textiles, than the solutions. This not only provides better sealing and "hold-out" but also improves the tear strength considerably. When the resin saturates paper or a textile the individual fibers are bonded closely and tear strength is decreased, but only slight loss of tear strength is observed when the resin film is on the surface. The polyvinyl acetate latexes have been used for architectural paints in Europe (20) for some time, and they are now being recommended for wall-sealers and finishes in this country. A description and formulations for these coatings will be given in Volume II.

The latexes are the result of emulsion polymerization and usually contain 1% or less of vinyl acetate monomer. This is toxic if inhaled in high concentrations and may be irritating to the skin. Good ventilation should be provided where the emulsions are being used, and prolonged contact with the skin should be avoided. The latexes also contain a small amount of free acetic acid and are therefore acidic with a *p*H ranging from 4 to 6. If the latex is made basic with ammonia, the addition of small amounts of zinc oxide will stabilize the *p*H. The addition of sodium nitrite has been suggested as a corrosion-inhibitor for use with iron, and sodium bicarbonate with copper or brass. A typical series of latexes is the Elvacet polyvinyl acetate latexes manufactured by the du Pont Co.

VINYL ACETATE-ALCOHOL RESINS

It was stated previously that polyvinyl alcohol is made by hydrolysis of polyvinyl acetate because the monomeric vinyl alcohol is not stable. By controlling the degree of hydrolysis a series of resins may be made ranging from high acetate to high alcohol content. The three vinyl acetate-alcohol resins shown in Table 5 range from 91% vinyl acetate for resin T24-9 to 82% vinyl acetate for resin MA 28-18. It will be noted that these resins are available only as the solutions indicated. The higher polarity of the hydroxyl group in these resins imparts stronger adhesion than can be obtained with the polyvinyl acetate resins. The hydroxyl group also increases compatibility somewhat and permits use of phenolic and urea-formaldehyde or melamine-formaldehyde resins to impart some thermosetting property.

Another series of vinyl acetate-alcohol resins, the Elvanol resins, is shown in Table 9. These range from 43% to 100% hydrolysis of the

Table 9 * Elvanol Polyvinyl Acetate-Alcohol Resins Specifications †

Grades Code No.	Viscosity ‡	% Hydrolysis	*p*H	Max. % Volatiles	Max. % Ash §
50-42	35–45	86–89	6–8	5	1.5
51-05	.....	87.7–89	6–8	5	1.5
52-22	22–28	86–89	6–8	5	1.5
71-30	28–32	99–100	6–8	5	0.75
70-05	4–6	98.5–100	6–8	5	2.0
72-60	55–65	100	6–8	5	0.75

* Courtesy of du Pont Co.

† All specifications on dry basis except volatiles.

‡ Viscosity in centipoises of a 4% water solution at 20°C, determined by means of the Hoeppler falling-ball method.

§ Ash is assumed to be Na_2CO_3 and is calculated as Na_2O.

polyvinyl acetate, and some are available in three ranges of molecular weight. These resins contain less than 2% ash which is assumed to be Na_2CO_3 and is calculated as Na_2O. The resins are hydroscopic and have a volatile content of 5% or less as shipped. Solutions of Elvanol polyvinyl alcohol resins have a *p*H range from 6 to 8.

The code numbers used in designating Elvanol are related both to the degree of hydrolysis and to the molecular weight of the resin. Thus, the first number indicates the degree of hydrolysis. Numbers in the 70's indicate that the polyvinyl alcohol is essentially completely hydrolyzed, and numbers in the 50's indicate that the material is partially hydrolyzed. The second number in the designation indicates the average viscosity of a 4% aqueous solution of the particular Elvanol polyvinyl alcohol resin. Thus, in the code number 72-60, the 72 indicates that the material is fully hydrolyzed, and the 60 indicates that the average viscosity of this material is 60 cp.

The partially hydrolyzed Elvanol resins of the 50 series are soluble in both cold and hot water. The resins in the 70 series will not dissolve in cold water, but if they are dissolved in hot water they will remain in solution on cooling. The completely hydrolyzed Elvanol resins are soluble only in hot water. The Elvanol resins also have a range of solubility in water-ethanol mixtures, and the relationship between degree of hydrolysis and solubility in these mixtures is shown in **Fig. 7.**

Films of Elvanol resins have high tensile strength, ranging from 9000 to 17,000 psi. The films are unaffected by animal and vegetable oils, greases, petroleum hydrocarbons, and many other solvents. Free films may be made by casting the solutions on smooth metal surfaces and subsequently stripping off the dry films. The solutions may also be coated onto paper or textiles to impart grease resistance. The films may be made insoluble in water by adding about 10% of dimethylol urea and catalyst and baking them for 5–10 min at 210–230°F, or ½–1 min over drying cylinders heated with steam at 1–10 lb pressure.

Fig. 7. Relationship between Degree of Hydrolysis and Water-Ethanol Solubility of Elvanol Resins

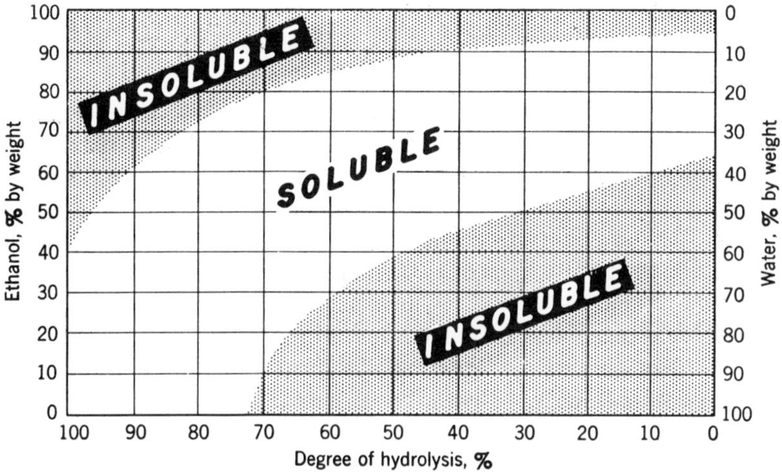

Courtesy of du Pont Co.

Other resins such as trimethylol melamine or water-soluble phenolics may be used as insolubilizers.

The flexibility of Elvanol resin films is increased by addition of plasticizers such as glycerol, the lower glycols, and some amines. However, the flexibility is greatly affected by moisture content and relative humidity. Films cast from Elvanol 72-51 and dried over sulfuric acid may be stretched less than 10% before breaking, but when conditioned at 50% RH it may be stretched to more than 300% without breaking. At 50% RH the film absorbs from 3% to 6% moisture, but it does not become tacky. The water molecules permeate the film and are strongly adsorbed in the micellar structure. This is similar to the action of a solvent-type plasticizer in that it reduces the van der Waals forces, and permits greater movement of the micelles so that the film may be stretched further without breaking. Plasticizers such as glycerol and the glycols have a similar action but are relatively non-volatile and therefore more permanent. The amount of plasticizer required depends somewhat on the relative absorbency of the surface on which the coating is to be applied. Usually from 15% to 20% plasticizer is sufficient for non-absorbent surfaces, but from 50% to 100% may be necessary for surfaces such as paper. We have seen previously that plasticizers reduce the tensile strength of high-polymer films. An unplasticized film of Elvanol 72-51 had a tensile strength of 11,500 psi, which was reduced to 8000 psi with 25% glycerol at 50% RH. However, this film was very flexible since it had an elongation of over 500%.

Some grades of water-dispersible materials, such as starch, dextrin, gelatin, and casein, have limited compatibility with Elvanol resins. These serve as extenders to lower the cost, but they should be checked carefully because of variation in compatibility with various grades.

Elvanol-type resins are used in a variety of molded plastics, free films, and coatings which are required to be resistant to oil, greases, and many solvents. Extruded tubing may be used for aircraft fuel lines or other applications requiring resistance to solvents. The free film may be made into bags or container liners which must be resistant to grease, oils, solvents, etc. Conversely, the bags may be used as water-soluble containers to prevent loss of material or errors in weighing. The bag and its contents are dissolved in a formulation rather than the contents only. Elvanol solutions may be applied to paper to obtain grease-resistant food-packaging materials. The solutions may also be applied to textiles for temporary sizing and removed later in dyeing or other operations. Elvanol resins may be used as emulsifying agents and protective colloids for emulsions which are stable to both acid and alkaline conditions. If added in a sufficient amount they would contribute to the film-forming properties.

The polyvinyl alcohol-type polymers are also manufactured by the Shawinigan Products Corp. and the American Polymer Corp.

VINYLIDENE CHLORIDE-ACRYLONITRILE RESINS

The use of vinylidene chloride for polymeric materials is another interesting example of the value of the copolymerization technique. Polyvinylidene chloride is a very strong, tough material with extremely good water resistance, but it has a high softening point and limited compatibility and solubility. However, by copolymerizing vinylidene chloride with a monomer such as acrylonitrile or vinyl chloride the physical properties are modified so that the copolymer can be used in surface coatings, film and sheeting, and plastics. The copolymers have somewhat less water and chemical resistance than the straight polyvinylidene chloride.

One of the first resins of this type to be marketed was the high vinylidene chloride-low acrylonitrile copolymer known by the trade name of Saran resin, Dow Chemical Co. The extremely low moisture-vapor transmission of films of Saran resins assured them an important place in coatings for paper used in food packaging. Saran F-120 resin is suitable for this purpose. It is applied to paper as a solution in organic solvents, and the coating is baked to remove small amounts of strongly retained solvents. This type of copolymer resin has also been made available in two latex forms which are known as Saran latexes F-122-A15 and F-122-A20. They eliminate the need for organic solvent in the application of these resins with a consequent reduction in cost and fire hazard.

It will be seen from the structures in Table 1 that vinylidene chloride is more symmetrical than vinyl chloride. This permits crystalline areas to develop in the polyvinylidene chloride which impart very high tensile strength and hardness, but difficult solubility characteristics. Copolymerization with acrylonitrile or vinyl chloride practically eliminates the symmetry and crystallinity. This modification reduces somewhat the extreme toughness and chemical resistance, but it improves solubility and compatibility. The vinylidene polymers and copolymers are susceptible to degradation by heat and ultraviolet, similarly to the vinyl chloride copolymers discussed in an earlier section of this chapter. The same types of stabilizers may be used, but they do not always give identical results with vinyl chloride and vinylidene chloride copolymers. These resins are also sensitive to zinc and iron pigments and to iron shipping containers. Reaction with the container may be prevented by giving the container a phosphoric acid wash or by adding a small amount of phosphoric acid to the resin solution.

Commercial Types. The Saran F-120 resin (Dow Chemical Co.) is a commercial high vinylidene chloride-low acrylonitrile copolymer. It is available in two viscosity types, which have the general properties given in Table 10.

The values for tensile strength and elongation indicate good film toughness, but they also indicate the need for plasticizer in flexible coatings for textiles. The very good resistance to both moisture-vapor and toluene-vapor transmission indicates the existence of strong secondary valence forces between the molecules in order to resist these polar and non-polar vapors. The styrene-copolymer films discussed in Chapter 9 would have low moisture-vapor transmission but high

Table 10 * General Properties, Saran F-120 Resin

Properties	Viscosity Types 200 cp	1000 cp
Viscosity range, cp	180–220	1000–1200
Specific gravity at 25/4°C	1.60	
Weight per gallon, lb	13.33	
Refractive index at 25°C	1.580	
Tensile strength, psi	8000–10,000	
Elongation, %	8	
MVT of 1-mil film †	0.15	
TVT ‡	0.26	

* Courtesy of Dow Chemical Co.

† Moisture-vapor transmission in gm/100 sq in./24 hr at 100°F and a relative humidity differential of 90%.

‡ Toluene vapor transmission in gm/100 sq in./24 hr.

toluene-vapor transmission because the styrene copolymer films are more non-polar, and the secondary valence forces are considerably weaker.

Solubility. The solubility of Saran F-120 resin may be expected to be rather limited in view of the above statement regarding the strength of the secondary forces between the molecules. The relatively small size and close packing of the chloride and nitrile groups also contribute to the high resistance to solvents. These groups are sufficiently polar to have strong secondary valence forces, but they are not readily hydrated like the hydroxyl group; therefore the films have good moisture resistance. The absence of hydroxyl, ester, and ether side group suggests that Saran F-120 would not be soluble in alcohol or ester solvents. The most practical true solvent is methyl ethyl ketone (MEK), but others are cyclohexanone, isophorone, and mesityl oxide. A variety of diluents can be used with MEK in limited amounts. The diluents include ketones other than MEK, esters, hydrocarbons, chlorinated hydrocarbons, and alcohols.

Films of Saran F-120 have strong solvent retention similar to that of other polymeric materials. Therefore it is desirable to heat the coatings after application to eliminate the final small percentage of solvent and develop maximum film properties. Coatings on metal may be baked for 1 hr at 250°F, or for less time at somewhat higher temperatures. Coatings on paper may be baked for about 2 min in a tunnel oven with a temperature differential of 95–300°F, followed by a flash bake of 10 sec at 500°F. Coatings having a thickness of less than 0.7 mil may be heated without danger of blistering; thicker coatings require proper adjustment of the heating time and temperature.

Excellent results were obtained (29) with Saran F-120 in air-drying coatings for the interior of gasoline tanks. The Bureau of Ships investigated a series of coating materials to protect the interior of gasoline storage tanks on ships from corrosion. When these tanks are emptied of gasoline they may be filled with sea water for ballast. The tanks are also cleaned occasionally with jets of hot sea water. Under such conditions corrosion of the interior surface of the mild steel tanks can be a serious problem. A report (29) on Saran F-120 stated that when multiple coats are applied to a total thickness of about 8 mils the estimated service will be adequate for 4 years, and that touch-up at 4-year intervals will extend the life expectancy indefinitely. This type of coating is specified in the Bureau of Ships Formula 113/49 and will be described in detail in Volume II under corrosion-inhibiting coatings.

Plasticizers. Since plasticizers generally reduce the moisture and chemical resistance of films it is necessary to use Saran F-120 resin without plasticizer when maximum resistance is required. It may be used alone for many coatings on metal and paper, but from 10% to 30% plasticizer may be necessary for more flexible coatings. The plasticizer should be selected on the basis of film requirements, such as resistance to moisture, solvents, heat, and flammability. A partial list of plasticizers that may be used with Saran F-120 is given in Table 11.

Table 11 Solvent-Type Plasticizers for Saran F-120

Dibutyl phthalate	Tricresyl phosphate
Dioctyl phthalate	Trioctyl phosphate
Ethoxy glycol phthalate	Triphenyl phosphate
Butoxy glycol phthalate	Chloroethyl phosphate
Methyl phthalyl methyl glycolate	Dibutyl sebacate
Butyl phthalyl butyl glycolate	Dioctyl sebacate
	Dibenzyl sebacate

Saran F-120 is used chiefly for its excellent chemical resistance; therefore other resinous materials are not usually blended with it. However, it has limited compatibility with some of the acrylates, phenolics, styrene resins, and rosin esters.

The addition of some of these resins improves the gloss and clarity of the film and facilitates the release of solvent. Coatings based on Saran F-120 are applied on paper, cellophane, textiles, and metal foil used in the packaging industry. They also serve as liners for drums, cement tanks, and chemical-resistant maintenance paints. The coatings may be applied by the brush, dip, roller-coater, and doctor-blade methods, but they do not spray well at normal solids content. The coatings "cobweb" when sprayed instead of atomizing; therefore they do not make smooth films. This phenomenon is used to advantage in packaging instruments, machines, etc., for storage or shipment. The cobwebs adhere to tape applied to the equipment and form a continuous coating around it. A subsequent coating may be applied when maximum protection is desired. The high moisture-resistance property of the Saran coatings prevents corrosion of the equipment during storage or shipment.

VINYLIDENE CHLORIDE-ACRYLONITRILE LATEXES

The two Saran latexes F-122-A15 and F-122-A20 referred to previously are modified with dibutyl phthalate so that they will produce continuous films after application when the emulsion breaks and the resin coagulates. Saran F-122-A15 contains approximately 15% dibutyl phthalate, and Saran F-122-A20 contains 20% by weight on the solids content of the latex. Coatings formulated with these latexes may be applied by any of the conventional methods, including the spray method. The coatings will air-dry satisfactorily, but an improvement in film properties will be obtained if they are baked. The use of the Saran resins in latex form lowers the cost through elimination of organic solvent and removes the fire hazard. However, the films obtained from latex application will be somewhat lower in water and chemical resistance than those obtained from solvent application.

The lower resistance of the films from the latexes is due to the ingredients which must be added to form the emulsion and to produce satisfactory application consistency. The emulsifying agents are usually quite water sensitive, and the thickeners are water-dispersible materials such as the alginates, sodium polyacrylates, and the water-dispersible cellulosic polymers. It is also desirable to use a wetting agent to assist in dispersing the pigment in water before adding the latex. Since most of these materials remain in the film after drying it will be seen that they lower both water and chemical resistance. However, the reduction in film properties is not serious for many applications; therefore the lower cost and absence of fire hazard justify their use. The films from the latexes show considerable discoloration on aging. This feature does not detract from their use in many coatings, but it makes them unsatisfactory for white or pastel coatings. The general properties of the latexes and films are shown in Table 12.

Table 12 * General Properties of Saran Latexes and Films

Properties	F-122-A15	F-122-A20
Total solids, range, %	51.5–53.0	51.5–53.0
Vis. at 25°C, cps (Brookfield)	<50	<50
Average particle size, micron	0.2	0.2
Particle charge	Negative	Negative
Dibutyl phthalate, % in film	14.8–15.8	21.5–23.0
Tensile strength, psi	2000–3000	1500–2500
Elongation, %	150–250	250–350
MVT †	....	Approx. 3.0
Light stability on aging	Discolors	Discolors

* Courtesy of Dow Chemical Co.
† Moisture-vapor transmission in gm/100 sq in./24 hr/2 mil film.

The reduction in tensile strength shown in Table 12 over that shown in Table 11 for the straight resins is due chiefly to the addition of the plasticizer. It will be noted that the films from the latexes are extremely flexible, having an elongation of 150–350%. If the films are baked at elevated temperatures some of the plasticizer may be lost since dibutyl phthalate is somewhat volatile under such conditions. These films may still have sufficient flexibility for the intended application, and they will be stronger and more resistant chemically.

The latexes are shipped at a *p*H range of 6.5–7.5. However, they are stable over a *p*H range of 2–12. They can be made alkaline with ammonia, or acid with monobasic acids such as hydrochloric and nitric. The latexes are not stable in contact with black iron, mild steel, magnesium, or zinc, but are stable with resin-lined equipment or wood, glass, and tile. Of course, all emulsions are sensitive to extremes of temperature, either high or freezing, and also to violent mechanical agitation.

GEON LATEXES

The B. F. Goodrich Chemical Co. produces a series of ten latexes containing "polyvinyl chloride base resins." The composition of these resins is not given in the literature, but the properties of the latexes vary over a wide range, indicating considerable variation in the base resin composition. The latexes are designed for use in coatings for paper and fabrics, and for water-base paints and printing inks.

PLIOVIC LATEX 300

The Goodyear Tire and Rubber Co. produces an aqueous dispersion of a "vinyl chloride copolymer" known as Pliovic latex 300. Since it does not form a continuous film on drying, it is compounded with solvent-type plasticizers or other latexes. It is used in fabric, paper, and leather coatings, special printing inks, etc.

VINYL CHLORIDE-VINYLIDENE CHLORIDE LATEX

A latex based on a copolymer of vinyl chloride and vinylidene chloride is described by Stilbert and Clack (25) and is marketed by

Fig. 8. Physical Properties of Vinyl Chloride-Vinylidene Chloride Copolymers

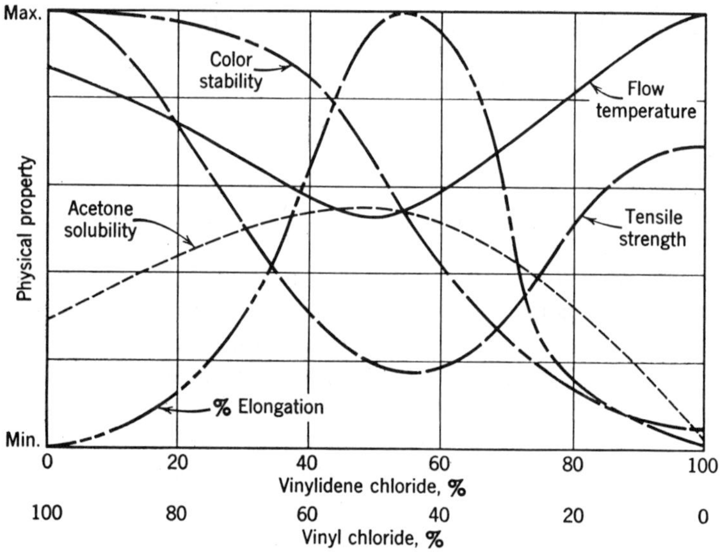

Courtesy of Dow Chemical Co.

Dow Chemical Co. under the trade name of Dow latex 744-B. The effect of blending these monomers on the general properties of the resulting copolymer is shown graphically in Fig. 8. The curves indicate the trend in properties rather than specific values. They show that blends containing 40–60% vinyl chloride have lower tensile strength but higher flexibility and solubility than other combinations or than either homopolymer. They also show that better color stability is obtained with high vinyl chloride content. Since vinyl chloride is a large-volume, lower-cost monomer it is economical to use high proportions of it.

Dow latex 744-B does not form continuous films without modification. This may be accomplished by addition of film-forming latexes, emulsified plasticizers, or liquid plasticizers. The latex may be used in adhesives, paper and textile coatings, and industrial paints. The general properties of the latex and its solid components are given in Table 13.

Table 13 * General Properties of Dow Latex 744-B

Total solids, % by weight	50.0 ± 0.5
Viscosity, cp	20
Average particle size, micron	0.2
*p*H	8.0 ± 0.5
Properties of solid content	
Spec. gravity at 25°C	1.48
Softening point, °C	164
Burning rate	Non-flammable
Heat and light stability	Good
MVT † (5% plasticizer)	Approx. 2.0

* Courtesy of Dow Chemical Co.

† Moisture-vapor transmission in gm/100 sq in./24 hr/1–2 mil film. Value varies with type of plasticizer.

Latex 744-B is compatible with a number of water-dispersible materials and also with certain oleoresinous materials. These include the drying oils, oil-modified alkyd resins, and certain varnishes. The ratio of latex to oleoresinous material determines whether the product will be the usual oil-in-water emulsion or the less widely used water-in-oil type. When the percentage of latex is high the oil-in-water type is produced, and the reverse is obtained when the percentage of latex is low. Either type may be prepared by adding the latex to the oleoresinous material slowly, with good agitation. At first the water-in-oil emulsion is formed, but continued addition of latex inverts the emulsion to the oil-in-water type. In this way vinyl-modified drying oils may be produced which have found use in coated fabrics and printing inks. A blend of 1 volume of latex 744-B and 3 volumes of varnish produces a heavy-bodied, thixotropic, easy-brushing composition for use in semi-gloss or flat paints. Similarly, 1 volume of the latex stirred into 4 volumes of typical metal primers improves film toughness and the bond with subsequent coats and produces easier brushing and less settling of the pigment.

As stated previously, when latex 744-B is to be the principal film-forming component it must be plasticized in order for continuous films to be produced. The latex is compatible with a wide range of plasticizers, but care should be taken that the best plasticizer is selected to meet the special coating requirements of light stability, or resistance to heat, water, solvents, etc.

VINYL CHLORIDE-VINYLIDENE CHLORIDE RESIN

The B. F. Goodrich Chemical Co. produces a solid copolymer of vinyl chloride and high vinylidene chloride for use in organic solvent compositions. The resin is known as Geon 200X20 and is soluble in toluene, xylene, and high aromatic naphthas. These solutions have infinite tolerance for ketones and limited tolerance for esters, higher alcohols, and aliphatic hydrocarbons. The resin has good compatibility with plasticizers and limited compatibility with some other film-forming materials. It is suitable for air-drying coatings, but improved film properties are obtained if the coating can be baked.

It will be apparent from the brief outline of the "vinyl" resins in this chapter that they cover a very wide range of properties. They are still in the development stage so that new types may be on the market before this book appears in print. For example, the vinyl ether polymers have been given considerable study (3, 26) but have not as yet found commercial application as coating resins, although they are being used in pressure-sensitive adhesives. The polyfluorovinyl resins have tremendous possibilities because of their durability and extreme resistance to heat and chemicals, but application difficulties must be overcome first. Lontz and Hapholdt (27) describe the properties of an aqueous colloidal dispersion of the polytetrafluoroethylene (Teflon, du Pont Co.). The dispersion may be converted into thin films by the casting method, with subsequent heat treatment. It may be impregnated into woven fabrics and then laminated under pressure at temperatures above the transition point of 327°C, 625°F. In special formulations it has good adhesion when sprayed on metals.

BIBLIOGRAPHY

1. D. S. Plumb and R. F. Hansen, *Chem. Eng. News, 29,* 52, 5432 (1951).
2. R. G. Heiligmann and E. E. McSweeney, *Ind. Eng. Chem., 44,* 113 (1952).
3. C. E. Schildknecht, *Vinyl and Related Polymers,* John Wiley & Sons, 1952.
4. Staff Report, *Chem. Eng. News, 30,* 2688 (June 30, 1952).
5. J. D. Park *et al., Ind. Eng. Chem., 39,* 354 (1947).
6. Symposium, "Fluorine Compounds," *Ind. Eng. Chem., 39,* 359 (1947).
7. V. W. Fox, J. G. Hendricks, and H. J. Ratti, *Ind. Eng. Chem., 41,* 1774 (1949).
8. J. G. Hendricks, E. L. White, and D. S. Bolley, *Ind. Eng. Chem., 42,* 899 (1950).
9. C. S. Marvel and E. C. Horning, *Organic Chemistry,* p. 701, Volume 1, second ed., H. Gilman, ed., John Wiley & Sons, 1949.
10. R. F. Boyer, *J. Phys. & Colloid Chem., 51,* 80 (1947).
11. M. C. Reed, *Ind. Eng. Chem., 35,* 896 (1943).

12. M. C. Reed and L. Connor, *Ind. Eng. Chem., 40,* 1414 (1948).
13. M. C. Reed and J. Harding, *Ind. Eng. Chem., 41,* 675 (1949).
14. W. A. Woodcock, *Ind. Eng. Chem., 41,* 663 (1949).
15. R. R. Lawrence and E. B. McIntyre, *Ind. Eng. Chem., 41,* 689 (1949).
16. W. J. Sparks and D. W. Young, *Ind. Eng. Chem., 41,* 665 (1949).
17. J. K. Stevenson, L. E. Cheyney, and M. M. Baldwin, *Ind. Eng. Chem., 42,* 2170 (1950).
18. R. J. Meyers *et al., Ind. Eng. Chem., 41,* 709 (1949).
19. W. S. Emerson *et al., Ind. Eng. Chem., 42,* 1431 (1950).
20. P. Whetstone, *Paint, Oil, Chem. Rev., 14,* (July 17, 1952).
21. Saul Moses, *Ind. Eng. Chem., 41,* 2338 (1949).
22. S. Moses and K. R. Witt, *Ind. Eng. Chem., 41,* 2334 (1949).
23. H. A. Gardner, *Physical and Chemical Examination of Paints, Varnishes, and Lacquers,* eleventh ed.
24. A. G. Roberts *et al.,* Preprint Booklet, Division of Paint, Varnish, and Plastics Chemistry, American Chemical Society, Fall, 1952.
25. E. K. Stilbert and H. L. Clack, *Tappi, 34,* No. 8 (1951).
26. C. E. Schildknecht, A. O. Zoss, and C. McKinley, *Ind. Eng. Chem., 39,* 180 (1947).
27. J. F. Lontz and W. B. Hapholdt, Jr., *Ind. Eng. Chem., 44,* 1800 (1952).
28. L. A. Matheson and R. F. Boyer, *Ind. Eng. Chem., 44,* 867 (1952).
29. W. W. Cranmer, Paper presented at 8th Annual Conference, National Association of Corrosion Engineers, Galveston, Texas, March, 1952.
30. G. M. Powell, Paper presented to Division of Paint, Varnish, and Plastics Chemistry, Boston Meeting, April, 1951.
31. G. M. Powell and R. W. Quarles, *Official Digest Federation Paint & Varnish Production Clubs, No. 263,* 696 (1946).
32. W. E. Leistner, A. C. Hecker, and O. H. Knoepke, U. S. Patent 2,564,646 (August, 1951).
33. Argus Chemical Lab., 633 Court St., Brooklyn 31, N. Y., *Tech. Bull., 1.*

Take care to distinguish between knowledge and wisdom. Knowledge is an accumulation of facts; wisdom is the ability to use knowledge effectively.

Chapter 13. ACRYLIC ESTER RESINS

INTRODUCTION

The acrylic ester resins are polymers and copolymers of the esters of acrylic and methacrylic acids.

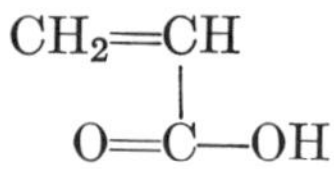
Acrylic acid

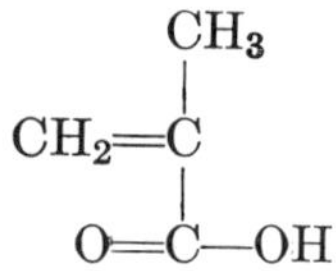
Methacrylic acid

This interesting series of resinous materials ranges in physical properties from soft elastomers having more than 1000% elongation to hard plastics which can be sawed and machined on a lathe. Some members of this series have firm, waxy structures with fairly sharp melting points. All the materials have excellent initial color, and they are stable to light, because they do not discolor on aging or degrade on outdoor exposure. They have very good heat resistance, since they show little or no color change at temperatures up to 350°F. Decomposition occurs at temperatures above 500°F. They are thermoplastic materials and therefore sensitive to some solvents, but they have good resistance to acids, alkalies, water, and alcohol. With the exception of the mineral spirits soluble type, they have good resistance to vegetable and mineral oils and greases. They have low acid number and low reactivity with pigments.

The elastomeric acrylic ester resins are used as base coats in textile, leather, and paper finishing. Some are compatible with the cellulosic polymers and vinyl resins to which they impart improved stability to heat and light. Some of the harder resins are used as the entire film-forming material in heat-resisting white baking enamels. The low acid number and freedom from yellowing on aging make some of them satisfactory for air-drying luminescent finishes, and others are used in fume-resistant finishes. The plastic grades are familiar to everyone under such trade names as Lucite and Plexiglas. The acrylic ester resins are available for coating purposes as solutions in organic solvents and also as oil-in-water type of emulsions or latexes. A series of commercial materials, together with their applications, is described in detail later in this chapter. Some of the methacrylic ester resins are available as solids for use in the plastics industry.

CHEMICAL COMPOSITION AND PHYSICAL PROPERTIES

We have seen in the chapters on rubber resins and vinyl resins that a wide range in physical and chemical properties may be obtained by changing the side group on the monomer when it is used for homopolymers, and also by mixing various monomers before polymerizing to form copolymers. The same situation exists in the acrylic ester resins. Since either acrylic acid or methacrylic acid may be esterified with most of the alcohols, it will be evident that two distinct series of monomers are possible for two corresponding series of homopolymers. When we consider that these monomers may be mixed in various proportions to produce copolymers we realize that a very large group of acrylic ester resins is possible. Commercially, the group is kept as small as possible to maintain inventory at a minimum.

The monomeric units for the polyacrylates and polymethacrylates were shown in Table 1, Chapter 12. It was pointed out that they could be classified as substituted ethylenes, and the relationship of the polyacrylic esters to polyethylene, polystyrene, polyvinyl resins, and various rubbers was indicated. In general, as the molecular weight of any of these polymers is increased, the polymer becomes tougher and somewhat more flexible, but its solution becomes more viscous, or it has less solids at equal viscosity than a solution of a lower-molecular-weight member of the series. It was shown that the nature of the side group determines to a large degree the type of solvent necessary to dissolve the polymer. If the side groups are polar (hydroxyl, ester, etc.) the solvents will be polar (ketones, esters, alcohols, and even water). When the side groups become more non-polar, the solvents required also change to the more non-polar type (ketones, aromatic hydrocarbons, and even aliphatic hydrocarbons).

The water and chemical resistance of the polymers were also shown to be related to the chemical nature of the side group. However, polymers such as polyvinylidene and polytetrafluoroethylene are extremely resistant to solvents and other chemicals because of the symmetry of their structures. This permits close alignment of the polymers to such an extent that they show crystalline patterns by X-ray diffraction. The close packing of the polymers in such arrangements develops the van der Waals forces to a maximum, thereby accounting for the combination of toughness, hardness, and chemical resistance. The acrylic ester resins provide another interesting series of possibilities obtained by varying the nature and position of the side group along the polymer chain.

It is outside the scope of this book to develop the chemical, structural, and physical possibilities in detail, but a few illustrations will serve to point to them. The references given later may be studied for further details. Perhaps the most striking example is the difference in the properties of polymethyl methacrylate and polyethyl acrylate. It will be seen from the monomeric units of these two polymers that they are isomers; they have the same number and types of atoms, but their arrangements are different.

Monomeric Units

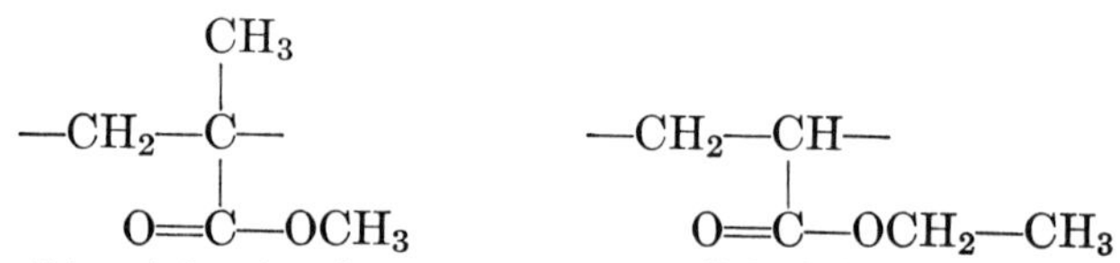

Polymethyl methacrylate — Polyethyl acrylate

The polymethyl methacrylate is a tough, hard plastic which may be sawed or machined, and the polyethyl acrylate is tough, but soft and rubbery, with a high degree of flexibility. The structural formulas for the monomeric units show greater symmetry for the polymethyl methacrylate. This permits closer packing of the polymer chains with the consequent development of greater hardness. The lack of symmetry in the polyethyl acrylate permits the polymer chains to coil more readily, with the consequent flexibility manifested as these chains straighten out when the material is stretched.

The symmetry of the polymethyl methacrylate may be disrupted by increasing the length of hydrocarbon chain on the ester linkage. This point is illustrated very nicely in the series of methacrylic ester polymers marketed by du Pont under the trade name of Hypalon synthetic resins. These resins are white, granular solids which find use in the plastics, rubber, adhesive, and coating industries. The differences in physical properties of the polymers as the polyester is changed from methyl to butyl are shown in Table 1. Note the effect of changing the shape of the side chain in the *n*-butyl and isobutyl polyesters.

The striking differences in physical properties of the polymers shown in Table 1 illustrate the range of possibilities in this field of synthetic resins. Another illustration is the effect on polymer proper-

Table 1 * Effect on Polymer Properties of Change in Length of Side Group

Hypalon † Polymer	Hardness, Rockwell R	Tensile Strength, psi	Elongation, % at break
P-1, polymethyl methacrylate	124	7800	1
P-2, polyethyl methacrylate	116	5400	25
P-4, poly-*n*-butyl methacrylate	−152	500	300
P-5, polyisobutyl methacrylate	112	3600	1
P-6, 50:50 isobutyl *n*-butyl copolymer	88	2100	175

* Courtesy of E. I. du Pont de Nemours & Co.
† Trade name, du Pont.

ties produced by increasing the molecular weight of polyethyl acrylate; this was measured (1) and summarized by the Rohm and Haas Co. (2) as shown in Table 2.

Table 2 * Effect on Polymer Properties of Increasing Molecular Weight of Polyethyl Acrylate

Molecular Weight	Properties
2,200	Viscous, colorless oil. Soluble in benzene. Low viscosity.
7,800	Tough, somewhat fluid mass. Adheres to glass, can be drawn out to threads.
14,500	Tougher. Strong adhesion to glass. Swells in benzene, then dissolves.
41,000	Very tough. Non-flowable.
175,000	Tough, elastic, rubber-like mass. Swells, then dissolves in benzene. High viscosity.
Too high to determine	Very tough. Insoluble in benzene, but swells to 50 times the original volume.

* Courtesy of Rohm and Haas Co.

The usual effect of increasing the length of the side chain of the normal alcohol esters is to soften the polymer and make it more flexible, but branched-chain alcohols have the opposite effect. Compare poly-*n*-butyl with polyisobutyl methacrylate, shown in Table 1. The softening trend continues in the polymethacrylates until the *n*-lauryl polyester is reached. However, the *n*-tetradecyl (14 carbons) and the *n*-hexadecyl (16 carbons) are hard, brittle, and wax-like. The same condition exists in the polyacrylate series, and the transition from flexible polymer to hard wax is at about the same chain length of the side group. This effect is not produced by high molecular weight of the polymers, because measurements have shown their molecular weights to be quite low. It is an interesting case of the side chain forming hard crystallites which actually determine the hardness of the polymer. This condition also affects the temperature at which the polymers become brittle. The data for the brittle points

Fig. 1. Brittle Points of Polymeric *n*-Alkyl Acrylates and Methacrylates

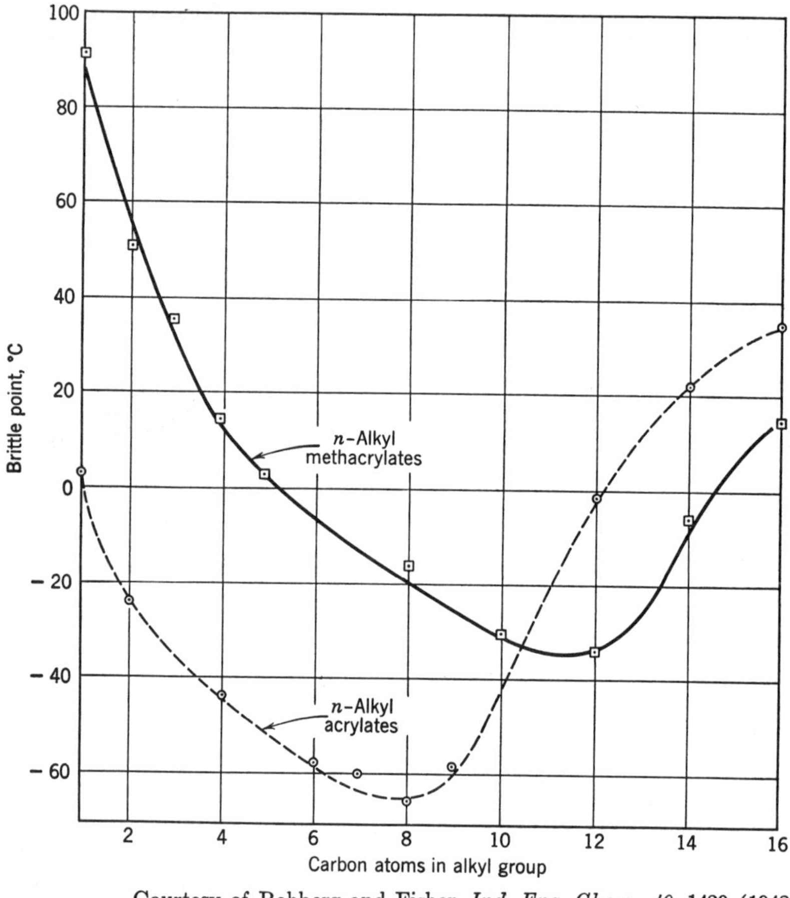

Courtesy of Rehberg and Fisher, *Ind. Eng. Chem.*, *40*, 1429 (1948).

of the two series of polymers were determined by Rehberg and Fisher (3) and are shown graphically in Fig. 1.

It has been shown (4) by X-ray diffraction that the crystallinity of the polymers of the tetradecyl, hexadecyl, and octadecyl acrylates results from the crystallization of the long side chains and is not due to the alignment of the polymer chains. At normal temperatures these materials are hard, waxy solids with melting points corresponding approximately to the brittle points shown in Fig. 1. At temperatures above the melting points the polymers are soft and tacky. It will be realized that this type of crystallinity is quite different from that formed by entire polymer chains grouping together. For example, the polytetrafluoroethylene referred to previously has a softening point of more than 300°C, and it extends over a range of temperatures instead of being sharp.

As may be expected, the solubility of the polyacrylates is similar to that of the polymethacrylates. The latter are somewhat less soluble than the former, and both are affected by the substituent groups on the main polymer chain. In general, both polymers dissolve in esters, ether alcohols, ketones, and aromatic hydrocarbons, and are insoluble in water, alcohols, and aliphatic hydrocarbons. Since they are esters, they might be expected to be sensitive to water, acids, and alkalies. However, the ester groups are well protected by the large bulk of the non-polar molecule, particularly in the polymethacrylate series. Clear films of polymethylacrylate and polyethylacrylate are sensitive to moisture and may develop considerable whiteness when immersed in water. Of course, the higher members of this series are more non-polar, and consequently less sensitive to moisture. Their films have good resistance to 10% solution of sodium hydroxide at room temperature, but the polymethyl and polyethyl acrylates are destroyed after exposure for 15 min at 100°C. Specific characteristics are developed in many commercial polyacrylic resins through copolymerization, etc., to meet the requirements of industry.

MONOMERIC ACRYLIC ESTERS

The monomeric acrylic esters are liquids at normal temperatures with boiling points ranging from about 80°C for methyl acrylate to about 230°C for *n*-octyl acrylate and about 240°C for *n*-lauryl methacrylate. The lower esters are highly flammable, having low flash and fire points; therefore every caution should be taken to prevent fire when handling them. The following monomers are available commercially from the Rohm and Haas Co. and are described in detail in their booklet (2):

Methyl acrylate	*n*-Butyl methacrylate
Ethyl acrylate	*n*-Hexyl methacrylate
Methyl methacrylate	*n*-Octyl methacrylate
Ethyl methacrylate	*n*-Lauryl methacrylate

The du Pont Co. also produces the methyl, ethyl, and *n*-butyl methacrylate monomers.

The monomeric acrylic esters may be prepared by several methods and from different raw materials. The Eastern Regional Research Laboratory of the U. S. Dept. of Agriculture at Philadelphia, Pa., has done a great deal of excellent research on the use of carbohydrates. Fisher *et al.* (5) indicate the production of methyl acrylate from carbohydrates by the following four-step process:

1. Fermentation of carbohydrate →

   lactic acid, CH_3—CH—(OH)—COOH.

2. Esterification with methanol →

   methyl lactate, CH_3—CH—(OH)—COO—CH_3.

3. Reaction with acetic anhydride →

   methyl acetoxy propionate, CH_3—CH—($OCCH_3$)—COO—CH_3.

4. Pyrolysis at about 550°C →

methyl acrylate, CH_2═CH—COO—CH_3, and acetic acid, CH_3COOH.

Methyl acrylate and ethyl acrylate may also be prepared from ethylene cyanohydrin in a single process which includes hydrolysis, dehydration, and esterification according to the following equation:

$$HO{-}CH_2{-}CH_2{-}CN + H_2SO_4 + CH_3OH \rightarrow CH_2{=}CH{-}COOCH_3 + NH_4HSO_4$$

Either acrylic acid or the acrylic esters may be prepared by the so-called Reppe method (6) directly from acetylene, carbon monoxide, and water or alcohol as follows:

$$CH{\equiv}CH + CO + H_2O \rightarrow CH_2{=}CHCOOH \quad \text{acid}$$

$$CH{\equiv}CH + CO + ROH \rightarrow CH_2{=}CHCOOR \quad \text{ester}$$

R = alkyl radical.

Methyl methacrylate may be prepared from acetone and hydrogen cyanide by the following reactions:

$$(CH_3)_2CO + HCN \rightarrow (CH_3)_2{=}C{-}(OH)CN$$

$$(CH_3)_2C(OH)CN + H_2SO_4 + CH_3OH \rightarrow CH_2{=}C(CH_3){-}COOCH_3 + NH_4HSO_4$$

Other methods of preparing the monomeric acrylic esters are described by Schildknecht (7). It has been difficult to obtain good yields of the monomers in the high degree of purity necessary for efficient polymerization. However, considerable work is being done on better production methods. The price of the acrylate and methacrylate ester polymers is higher than that of polystyrene because of higher-cost raw materials, but polymethyl methacrylate is better than polystyrene for some applications. The other acrylic ester polymers have specific uses in the adhesive and coating industries. The monomeric acrylic esters are valuable as chemical intermediates; an excellent description of their reactions is given in the booklet by the Rohm and Haas Co. (2). The acrylic ester polymers may be saponified with sodium hydroxide to produce the sodium polyacrylates, which are useful as thickeners and stabilizing agents in latex and other types of emulsion paints, as referred to previously.

POLYMERIZATION OF ACRYLIC ESTERS

The monomeric acrylic esters may be polymerized by one of the four methods, bulk, solution, suspension, or emulsion, described in Chapter 1, and under styrenated oils (Chapter 2), alkyds (Chapter 7), and vinyl resins (Chapter 12). However, the physical characteristics of the acrylic polymers (hard, or soft and flexible) and the end-use requirements determine which method is used for specific polymers. For example, the bulk method is generally preferred for the hard polymethyl methacrylate, the solution method for the polymers which are sold in solution, and the emulsion method for those sold as latexes.

Bulk Polymerization. In the bulk polymerization of methyl methacrylate it is usually desirable first to partially polymerize the monomer to a syrup having a viscous liquid consistency. This material then is poured into the mold, and the polymerization finished in the presence of a peroxide catalyst. This general method is used for the production of rods, tubes, sheets, and molded objects. The initial polymerization of methyl methacrylate is inhibited by contact with oxygen. The monomer reacts with oxygen to form peroxides, and these act as chain terminators. A detailed description of the reactions involved is given by Schildknecht (7). In the production of small batches of the syrup, the peroxide may be added to the monomer, and the batch heated at temperatures of 80–100°C. For large-scale production it may be necessary to heat the batch without catalyst at first, and then add catalyst in portions until the desired degree of polymerization is obtained. A variety of catalysts may be used, such as benzoyl peroxide, lauroyl peroxide, and *tert*-butyl peroxide, in amounts ranging from 0.02% to 0.50% of catalyst based on the monomer. After the syrup is poured into the mold the polymerization is completed by heating at temperatures ranging from 100° to 120°C for several hours. The time required is affected by the thickness of the object, temperature, catalyst, and purity of the monomer. An excellent description of the various processes and uses for this type of plastic is given by Schildknecht (7).

An interesting phenomenon is the extent of shrinkage of the methacrylic polymers during polymerization. Since the specific gravity of the polymer is greater than that of the monomer, the shrinkage may be calculated from this difference in specific gravity divided by the specific gravity of the monomer:

$$\%\ \text{Shrinkage} = \frac{SG_p - SG_m}{SG_m} \times 100$$

The general subject of shrinkage in addition polymers is discussed by Nichols and Flowers (9), and the values for a series of methacrylic polymers are given by Rohm and Haas Co. (2). It will be noted that the shrinkage decreases as the size of the alcohol substituent increases.

Methacrylate * Polymer	% Shrinkage
Methyl	20.6–21.2
Ethyl	17.8–18.4
n-Propyl	17.1
n-Butyl	14.3–15.5
n-Amyl	13.9
2-Octyl	11.8

* Courtesy of Rohm and Haas Co.

Solution Polymerization. The effect of various solvents on the degree of polymerization attained in the solution method was discussed previously. Solvents such as carbon tetrachloride, which produce free radicals readily, tend to give a lower degree of polymerization than less reactive solvents, such as toluene. Unlike the vinyl polymers the acrylic ester polymers are soluble in the monomers; therefore they do not separate readily when the desired degree of polymerization is reached. The solvent chosen for the solution polymerization is frequently the one in which the polymer is marketed, thereby avoiding the necessity for separation. However, the boiling point of the solvent determines the maximum temperature at which the polymerization may be performed. Usually, the monomer, catalyst, and solvent are heated to initiate the polymerization. The liberated heat of polymerization then carries on the reaction, but final polymerization is done under reflux. The boiling point of the solvent determines the reflux temperature.

Differences in the degree of polymerization may also be obtained by use of different catalysts. The rate of free-radical production of the peroxide catalysts varies with the temperature and type of solvent. A fast rate of free-radical production tends to activate a large number of monomers in a given time and results in a low degree of polymerization. It will be apparent that careful control of all variables in the process is necessary to obtain a uniform product.

Suspension Polymerization. It will be recalled that the polymer is obtained as very small granules or beads in the suspension method. Therefore this method is satisfactory only for polymers which are hard enough to handle in this form. The suspension method has been used for the lower members of the methacrylate series, but the polymers of the higher methacrylates and the acrylates are too soft to be handled satisfactorily. The Hypalon resins described in Table 1 are available in granular form.

Emulsion Polymerization. We have seen in previous chapters that the emulsion polymerization method is very efficient for the production of many types of synthetic polymers. It is particularly economical when the polymer can be used in the latex form. The series of acrylic ester resin latexes known as the Rhoplexes (Rohm and Haas Co.) is described later in this chapter. Usually a water-soluble peroxide such as hydrogen peroxide or ammonium persulfate is used as the catalyst. The emulsifying agent should be chosen with care if the product is to be sold as the latex. Any of the three types of emulsifying agents may be used: the anionic, cationic, or non-ionic. The difference in these types will be discussed later in this chapter. An illustration of the use of the anionic type (soap) is given in Formula 1 in Chapter 9 for the emulsion polymerization of a butadiene-styrene synthetic rubber. Fisher and Mast (8) compared a series of emulsifying agents for the emulsion polymerization of ethyl acrylate. Their standard procedure, as used in the Eastern Regional Research Laboratory, is given in detail in Formula 1. This formula uses the anionic emulsifier (Triton 720, Rohm and Haas Co.) which is described as an alkyl-aryl ether sulfonate.

Formula 1 * Emulsion Polymerization of Ethyl Acrylate

	Parts by Weight
Water	200
Triton † 720 (based on monomer)	0.5–1
Tergitol ‡ paste #4 (based on monomer)	1–2
Monomer (ethyl acrylate)	100
Ammonium persulfate (sufficient to produce polymerization at refluxing temperatures)	0.005

* Fisher and Mast (8).

† Trade name, Rohm and Haas Co.

‡ Trade name, Carbide and Carbon Chemicals Corp.

Procedure

Water and the water soluble ingredients were charged into a round-bottomed, three-necked, ground-glass joint flask fitted with a stirrer, reflux condenser, and thermometer well. The solution was stirred with anchor, paddle, or half-moon-type stirrers at 50–150 rpm. The monomer and any remaining ingredients were then added, and the flask was heated until refluxing occurred. Flasks ranging in size from 200 ml to 12 liters were used and were generally filled to 60% capacity. Heat was applied intermittently, if needed, to induce polymerization. If the polymerization did not start within 10 min after refluxing first occurred, additional ammonium persulfate was added. If excessive quantities of catalyst were required, the monomer was not considered to be of the proper purity. The reaction usually proceeded at a rate sufficient to cause refluxing without external heating for 15–30 min without flooding the condenser. Heat was then applied, and the refluxing temperature raised to 93–95°C for completion. The emulsion was then steam-distilled for 15–30 min. To obtain the polymer, the hot emulsion was run slowly into twice its volume of hot (approximately 90°C), rapidly stirred 3–5% sodium chloride solution. The precipitated polymer in the form of discrete particles was washed with hot water until free of salt and dried. Polymers prepared in this manner were soluble in organic solvents such as toluene, benzene, ethyl acetate, acetone, and dioxane, whereas those made with substantially larger amounts of catalyst were not. If the emulsion itself was desired, it was cooled and poured into storage vessels.

COMMERCIAL ACRYLIC ESTER RESINS

The commercial acrylic ester resins are produced by several manufacturers, including the American Polymer Corp., E. I. du Pont de Nemours & Co., and the Rohm and Haas Co. The series of du Pont methacrylic ester polymers known as the Hypalon synthetic resins was referred to earlier in this chapter. The Rohm and Haas Co. produces a rather complete line of acrylic and methacrylic ester polymers and copolymers for plastics, adhesives, and coatings. The solid resins and organic solvent solutions are known as Acryloid resins, and the resins in emulsion form are known as Rhoplex resin emulsions. These materials will be used in this chapter to illustrate the general composition, properties, and coating applications for the acrylic ester resins. Additional details may be obtained from the Rohm and Haas literature, particularly with respect to new developments in this field. Also, Volume II will contain further information regarding the pigmented acrylic ester resins in architectural and industrial finishes.

We have seen in previous chapters on styrene polymers and vinyl resins that the best all-round resin properties are frequently obtained from copolymers rather than from homopolymers. Although the composition of the Acryloid resins is not given in the literature it may be safe to assume that many of them are copolymers of the various acrylic and methacrylic esters. In fact, it would be a good exercise for the reader to deduce the possible compositions from the known properties of the resins and the theoretical properties which may be expected from polymerization of the various monomers and their combinations. It will be recalled that one of the potent reasons for using the copolymerization technique is the fact that the homopolymers are usually incompatible. If the homopolymers could be blended in various proportions to obtain the desired properties the entire formulation would be much simpler. Table 4 is included specifically to emphasize this feature, although it is recognized that some of the incompatibility shown in this table is due to the different solvents which are used. Table 3 shows the physical properties of a series of Acryloid resins.

Table 3 * Physical Properties of Acryloid † Resins

Acryloid Resin	% Solids, ±1%	Solvent	Spec. Grav.	Pounds per Gallon	Viscosity, cp, at 30°C	Flash Point, °F, closed cup, Tagliabue	General Comparison
A-10	30	Cellosolve acetate	1.03	8.6	710–850	118	Hardest and most solvent-resistant
A-101	40	Methyl ethyl ketone	0.94	7.8	700–1500 ‡	22	Same as A-10 but different solvent
B-72	100	None			....	...	Similar to B-7 but lower molecular weight
B-72	40	Toluol	0.97	8.1	480–640	39	Similar to B-7 but lower in viscosity
B-82	100	None			....	...	Similar to B-72 but lower in cost
B-82	40	Toluol	0.97	8.1	480–640	39	Similar to B-72 but lower in cost
C-10 LV	40	Toluol	0.97	8.1	40–80	39	Softest and most elastic grade
F-10	40	9-MS 1-Amsco F	0.91	7.6	1500	75	Between C-10 and B-72 but soluble in petroleum hydrocarbons

* Courtesy of Rohm and Haas Co.
† Trade name, Rohm and Haas Co.
‡ At 35% solids.

The general comparison in Table 3 indicates the resins which are similar in hardness and general properties. The compatibility data in Table 4 indicate that only the similar resins have good compatibility with each other. The type of solvent in Acryloid F-10 is a deterrent to its compatibility with other members of the series.

Table 4 * Mutual Compatibility of Acryloid † Resins

Other Component	% Acryloid A-10			% Acryloid A-101			% Acryloid B-72			% Acryloid C-10			% Acryloid F-10		
	10	50	90	10	50	90	10	50	90	10	50	90	10	50	90
Acryloid A-10	...	...	...	C	C	C	C	I	C	C	I	I	I	I	I
Acryloid A-101	C	C	C	...	...	...	C	I	C	C	I	I	I	I	I
Acryloid B-72	C	I	C	C	I	C	...	...	...	I	I	I	I	I	I
Acryloid C-10	I	I	C	I	I	C	I	I	I	...	...	...	I	I	I
Acryloid F-10	I	I	I	I	I	I	I	I	I	I	I	I	...	...	...

C = Compatible. I = Incompatible.

* Courtesy of Rohm and Haas Co.
† Trade name, Rohm and Haas Co.

It may be of interest to speculate regarding the possible composition of these resins. It is known that polymethyl methacrylate is a hard resin, and the hardness decreases as the length of the substituent alkyl chain increases up to 12 carbon atoms. The polyacrylates are all soft, elastic materials until the substituent alkyl chain reaches 12 carbon atoms. Since the polybutyl acrylate tends to be slightly tacky at normal temperatures it might be assumed that Acryloid C-10 is the polyethyl acrylate. The description of its properties and uses appears to fit this composition. The hardest resin of the series, Acryloid A-10, might be assumed to be the polymethyl methacrylate, but further consideration indicates that this material would not have sufficient flexibility and adhesion for coating purposes. This suggests a possible copolymer of the methyl methacrylate with one of the other methacrylic esters or an acrylic ester. Obviously further speculation is difficult because of the many possibilities involved. These possibilities concern not only the types of monomers to be used, but also their ratio, type and amount of catalyst, and conditions of polymerization. Another important factor in such a situation would be the relative cost or availability of the monomer to use with the methyl methacrylate. Such problems present an interesting challenge to resin chemists because the best answer, from a commercial viewpoint, is often a compromise of resin properties, cost, and availability.

GENERAL PROPERTIES

The general properties of the acrylic ester resins which justify their use in coatings are excellent stability on aging, water-white original color and good resistance to color change on aging or at baking schedules, good chemical resistance, very low reactivity with pigments. As stated previously they are thermoplastic materials and are therefore susceptible to softening by some solvents and by above normal temperatures.

SOLUBILITY

Table 3 shows that acrylic ester resins can be produced which are soluble in a wide range of solvents. In general, the solubility becomes more critical as the hardness of the resin is increased. It will be seen from Table 3 that the true solvents are esters, ketones, chlorinated hydrocarbons, and aromatic hydrocarbons. Except in special cases, the aliphatic hydrocarbons and the alcohols are non-solvents. However, they may be used in limited proportions as diluents with the true solvents. Also, resins such as Acryloid B-72 may be dissolved in 2B alcohol or isopropyl alcohol for special purposes. The solutions are turbid, but the dry films obtained from such solutions are clear.

COMPATIBILITY

The mutual compatibility of the acrylic ester resins is indicated in Table 4. The resins are compatible with the usual chemical plasticizers such as dibutyl phthalate, dibutyl sebacate, tricresyl phosphate, and chlorinated biphenyls. They are also compatible with nitrocellulose and some of the vinyl resins but have limited compatibility with cellulose acetate. Acryloid resin B-72 is compatible with ethyl cellulose. The acrylic ester resins are not compatible with oil-modified alkyds, oleoresinous varnishes, or drying oils.

SPECIFIC APPLICATIONS

Acrylic ester resins such as Acryloid B-72 and B-82 are the most versatile for coating purposes. This type may be used alone as the entire film-forming component, or in combinations with nitrocellulose, ethyl cellulose, vinyl resins, etc. For certain types of bleached furniture the finish must have little or no color, and it should not darken on aging. A general formulation for this type of finish which indicates the range of the various components is given in Formula 2.

When dissolved in suitable solvents the above combination has a viscosity of about V at 25% solids. It should be thinned to about 14% solids for spray application. This formulation is lower in solids than

Formula 2 * Water-White Wood Finishing Lacquer

	Parts by Weight, Solid Basis
Acryloid B-72	50
Nitrocellulose, RS-½ sec	35
Paraplex G-50 †	15
	100

* Courtesy of Rohm and Haas Co.

† For description see Table 5, Chapter 10.

those shown in Formula 2, Chapter 11, but it has excellent color retention even after prolonged exposure to ultraviolet radiation.

An illustration of the use of Acryloid B-72 with ethyl cellulose as a flexible cloth coating is given in Formula 3.

Formula 3 * Flexible Cloth Coating

	Parts by Weight, Solid Basis
Acryloid B-72	30
Ethyl cellulose	24
Dibutyl sebacate	46
	100

* Courtesy of Rohm and Haas Co.

This formula may be dissolved in toluol or toluol-alcohol combinations. It may be applied to the textile by the knife-coating method or thinned to about 8% solids for dip application. The coating is tack-free after air-drying, or force-drying to remove the solvent. It has good retention of flexibility and is very resistant to hardening and discoloration on aging.

An illustration of the use of Acryloid B-72 with the vinyl resin Vinylite VMCH is given in Formula 4. This formulation produces a hard, high-gloss, white enamel suitable for stove trimmings, hospital equipment, and high-grade household appliances. It may be thinned to about 32% solids for spray application. The coating can be air-dried, but better results are obtained when it is baked for 30 min at 250°F. The coating has very good color retention and resistance to discoloration by heat. It also has good alkali resistance, but it would be sensitive to some solvents.

An illustration of the use of Acryloid B-72 as the entire film-former is given in Formula 5. The low acid number of this resin, together

Formula 4 * White Air-Drying or Baking Enamel

	Pounds	Gallons
Roller mill grind		
Titanium dioxide	139	4.3
Acryloid B-72 (40% solids)	114	14.0
Xylol	16	2.1
Mix with:		
Acryloid B-72 (40% solids)	330	40.7
Vinylite VMCH (20% solids) †	104	14.6
Paraplex G-50 ‡	10	1.2
Methyl isobutyl ketone	80	12.0
Xylol	80	11.1
	873	100.0

Characteristics

	Per Cent
Total solids	40
Volatile	60
Pigment	40
Vehicle solids	60
Composition vehicle solids	
Acryloid B-72	85
Vinylite VMCH	10
Paraplex G-50	5

* Courtesy of Rohm and Haas Co.
† 20% solids in methyl isobutyl ketone; see Table 5, Chapter 12, for resin description.
‡ Resinous plasticizer; see Table 5, Chapter 10.

with its extremely pale color and good color retention, makes it especially suitable for luminescent finishes.

Formula 5 * Luminescent Coating

	Parts by Weight
Phosphorescent pigment †	30
Acryloid B-72 (40% solids)	50
Cellosolve acetate	20
	100

* Courtesy of Rohm and Haas Co.
† New Jersey Zinc Co. of Pa.

The pigment is incorporated into the coating by simple mixing rather than the usual grinding process. Since the phosphorescent pigments do not have good hiding ability it is better to apply two coats of the luminescent paint over a background of similar color. For exterior use it is advisable to apply one coat of clear finish over the luminescent paint. Acryloid B-72 may be thinned to 15–20% solids for the clear finish coat.

The very flexible type of acrylic ester resin, such as Acryloid C-10, may serve as a base coat on fabrics and rubber. Such coatings have excellent adhesion to the base materials and to subsequent finish coats. Combinations of Acryloid C-10 with nitrocellulose may also be used to obtain somewhat faster drying; a typical ratio is shown in Formula 6.

Formula 6 * Fabric Base Coat

	Parts by Weight, Solid Basis
Acryloid C-10	40
Nitrocellulose, RS-½ sec	30
Dibutyl sebacate	30
	100

* Courtesy of Rohm and Haas Co.

Typical nitrocellulose finish coats for fabrics are shown in Formula 5, Chapter 11. A fabric finish coating which is partially thermosetting, and consequently more resistant to solvents, is shown in Formula 7.

Formula 7 * Finish Coat for Fabrics

	Parts by Weight	%, Solid Basis
Acryloid A-101 (40% solids)	112.5	45
Vinylite VAGH † (20% in MEK)	160.0	32
Uformite F-240 ‡ (60% solids)	38.3	23
Methyl ethyl ketone	89.2	
		100
	400.0	

* Courtesy of Rohm and Haas Co.
† See Table 5, Chapter 12.
‡ See Table 2, Chapter 8.

The above finish coat should be baked for at least 30 min at 250°F or for shorter time at 300°F. The urea-formaldehyde resin, Uformite F-240, is thermosetting and produces a cross-linked structure throughout the coating. It may also react with the free hydroxyl groups on the vinylite resin VAGH to impart additional thermosetting property.

The use of Acryloid B-7 with nitrocellulose to produce a chemical-resistant coating for racks and other metal parts to prevent deposition of metals from the electroplating baths is described in Formula 4, Chapter 11. The hardest of this series of acrylic ester resins, Acryloid A-10, may be combined with a small percentage of plasticizer for industrial white baking finishes which do not discolor with heat. These and other Acryloid resin finishes are discussed in detail under industrial finishing systems in Volume II.

ACRYLIC ESTER RESIN EMULSIONS

We have seen in previous sections of this book that the emulsion method of polymerization permits better control of the reaction than either the solution or the bulk method. The polymer may be obtained from the emulsion by coagulating, washing, and drying. It is most economical when the polymer can be applied in the emulsion form, and the polymer coagulated by simple evaporation of the water phase. However, this does not eliminate the emulsifying agents, and other water-soluble materials which would be removed in the washing operation referred to above. Therefore films obtained directly from emulsions are apt to be less resistant to water and chemicals than those obtained from organic solvent solutions.

Typical methods of emulsion polymerization are given in this and other chapters. The stability of emulsions is affected by a number of factors which include the emulsifying agent, the thickening agent or protective colloid, the *p*H, and the conditions of manufacture. The sodium or ammonium soaps are low-cost, efficient emulsifying agents, but they require alkaline conditions. Since the acrylic ester monomers may be saponified to varying degrees under alkaline conditions it is generally preferred to conduct their emulsion polymerization under neutral or slightly acidic conditions. An emulsifying agent, such as Triton 720 in Formula 1, is suitable for this purpose. The work of Fisher and Mast (8) describing a series of emulsifying agents with ethyl acrylate was referred to previously.

Types of Emulsifying Agents. Emulsions made with the non-ionic type of agent have much better tolerance for inorganic salts than those made with the anionic type. It will be recalled that the difference in anionic, cationic, and non-ionic emulsifying agents is related to their behavior in water solution. Both the anionic and cationic agents ionize in water solution, but the non-ionic type does not ionize. The difference between the anionic and cationic agents is related to the direction taken by their ions during electrolysis. If the ion which is effective in producing emulsification travels toward the anode the agent is called anionic. Conversely, if the effective ion moves to the cathode, the agent is known as a cationic agent. Obviously, if the agent does not ionize, there is no characteristic movement of ions, and the agent is called non-ionic. A typical example of each of the three types is given below:

Anionic emulsifying agent: sodium oleate, $(C_{17}H_{33}COO)^{-}Na^{+}$.
Cationic emulsifying agent: lauryl ammonium acetate, $(C_{12}H_{25}NH_3)^{+}(OOCCH_3)^{-}$.
Non-ionic emulsifying agent: glyceryl monolaurate, $C_{11}H_{23}COOC_3H_5(OH)_2$.

In both the anionic and cationic agents the long carbon-chain part of the molecule is the effective part in emulsification. If this were replaced with a very short carbon chain the molecule could not function as an emulsifying agent. It will be noticed that the long carbon-chain ion in soap (sodium oleate) is a negative ion. In electrolysis the negative ion migrates to the anode; therefore it is an anion, and the agent is anionic. The long carbon-chain ion in lauryl ammonium acetate is positive; therefore it migrates to the cathode and is a cation, thereby making the agent cationic. A fairly wide range of the three types of emulsifying agents is available commercially (11, 12). A survey of the literature and procedures for emulsification are given in a paper by the New York Paint & Varnish Production Club (10) and also in Volume II.

Commercial Polyacrylic Ester Emulsions. The series of Rhoplex emulsions (Rohm and Haas Co.) will be used to illustrate the commercial types which are available, together with their film characteristics and fields of application. The acrylic resin components of these emulsions produce films which range from hard and brittle to flexible and non-tacky, and to soft and tacky films. They have the same general resistance to discoloration and hardening on aging as the resins applied from solutions. They are apt to be somewhat less resistant to water and alkali because of the presence of water-sensitive materials in the films, as referred to previously. Some of them may serve as the entire film-forming component of coatings; others may be blended with water-soluble gums, synthetic rubber latexes, etc., to improve flexibility or to increase tack. They are used as base coats and finish coats on paper, fabrics, and metal, and also in adhesives. Table 5 is arranged with the three anionic-type emulsions in the first group and the three non-ionic types in the second group. The emulsions in each group are arranged in decreasing order of hardness.

Stability of Emulsions. One of the important characteristics of an emulsion is its ability to remain stable, or to prevent coagulation of the disperse phase, under a variety of conditions which it may en-

Table 5 * Acrylic Ester Resin Emulsions

Rhoplex † Emulsion	% Solids	Pounds per Gallon	Free Film Properties: Tensile Strength, psi	% Elongation	Relative Hardness
MR	50	8.9	2460	275	Tough, slightly elastic, non-tacky
WC-9	40	8.8	350	710	Soft, elastic, tacky
ER	50	8.9	110	2000	Very flexible, tacky
WN-75	55	9.25			Hard, brittle, non-tacky
WN-80	40	9.0	1870	500–600	Flexible, non-tacky
FRN	40	8.7	290	1200	Flexible, slightly tacky

* Courtesy of Rohm and Haas Co.
† Trade name, Rohm and Haas Co.

counter during storage and use. These conditions include aging for 6 months or more, large variation in temperature, mechanical working as in the grinding of pigment pastes, variation in *p*H, and contact with water containing certain dissolved salts. The stability of an emulsion is determined largely by the type of emulsifying agent used (anionic, cationic, or non-ionic), and the particle size or degree of dispersion of the disperse phase. The average particle size should be not more than 0.2 micron. However, the size distribution is also very important because a relatively small percentage of large-particle material acts as seeds for accelerating coagulation. Emulsions are sensitive to extremes of temperature. At elevated temperatures the continuous film of emulsifying agent around each resin particle becomes thinner and therefore less strong. Also the increased Brownian movement of the particles tends to rupture the film when the particles collide with each other. At freezing temperatures the continuity of the film of emulsifying agent is broken, and it does not return to normal when the emulsion is brought back to room temperature. Mechanical working such as the rapid shearing on a roller mill may rupture the film of emulsifying agent, particularly if finely divided solids are present such as pigments or fillers.

The variation in *p*H may cause reactivity with the emulsifying agent with consequent rupture of the protective film, followed by increase in particle size or perhaps complete coagulation. The various salts of calcium, aluminum, iron, and some other metals which are found in many natural waters also react with some emulsifying agents to produce insoluble compounds which cannot function as emulsifying agents. It is always good practice to use water which has been deionized by distillation or by an ion-exchange process, for the manufacture or dilution of emulsions. The anionic type of emulsifying agents is readily converted to insoluble compounds by salts of calcium, aluminum, etc., which cause instability of the emulsion and coagulation of the disperse phase. The non-ionic type has good tolerance for such salts. Emulsions made with the non-ionic type of agent may also be used over a wide range of *p*H. They may be made alkaline with ammonia, or acid with sulfuric, acetic, or phosphoric acid.

Emulsions vary in their ability to tolerate additions of organic solvents. Usually, the solvents which are miscible with water tend to break emulsions more readily than the non-miscible solvents. It can be seen that if the water-miscible solvents are also solvents for the disperse phase there will be a strong tendency for the resin to go into solution in the organic solvent. It would then be almost impossible to preserve such a mixture as an emulsion because the two phases would be miscible. It will be recalled that the definition of an emulsion is "a stable dispersion of two immiscible liquids."

Usually, the high-polymer emulsions of latexes are very low in viscosity. The addition of thickening agents improves application properties and also the stability. The thickening agents are often referred to as "protective colloids" because of their beneficial effect on stability. These agents include such materials as casein, alginates, polyvinyl alcohol, and sodium polyacrylate. These materials are water dispersible, and since they remain in the film after the emulsion is applied and dried, they tend to reduce its water resistance. It might be assumed that the emulsion could be thickened by raising its solid content. This is not feasible because, in general, the oil-in-water type of emulsion is most stable when the oil phase and the water phase are present in about equal proportions. Since water is the continuous phase, the viscosity does not increase significantly with increase in solid content, but at about 70% solids the emulsion usually inverts to the water-in-oil type. This changes its characteristics considerably so that it is not satisfactory for many purposes. Many emulsions break when thinned to a low solids content because of the tendency to reduce the thickness of the film of emulsifying agent around the oil or resin particles. The thickening agent acts as a protective colloid in such cases and permits greater dilution with retention of stability.

From this brief discussion of some of the factors which affect the stability of emulsions it will be realized that the first three emulsions in Table 5 are less tolerant of many dissolved salts and other addition agents than the second three emulsions because of the type of emulsifying agent they contain. The first three contain the anionic type, and the second three the non-ionic type.

Applications for Rhoplex Emulsions. Table 5 shows that Rhoplex MR is only slightly elastic, whereas the ER is very elastic but tacky. These two products may be blended to produce the degree of flexibility and tackiness required. The degree of tackiness of such blends is shown below:

MR :	ER	Tackiness
100	..	Non-tacky
40	60	Non-tacky
30	70	Slightly tacky
20	80	Tacky
..	100	Tacky

From the above it can be seen that the flexibility of Rhoplex MR can be increased considerably by addition of ER without producing tacky films. Conversely, Rhoplex ER or WC-9 may be added to natural and synthetic rubber latexes to increase tackiness and improve retention of flexibility. These conditions are particularly desirable in the pressure-sensitive and heat-sensitive types of adhesives. This group of Rhoplex emulsions may also be used to impregnate paper and fabrics. Such treatment increases the tear strength of paper tremendously so that it may serve as a base for nitrocellulose or other finishes. The synthetic rubber latexes used for this purpose are discussed in Chapter 9.

The second group of emulsions in Table 5 also has a range in film hardness from hard to soft and tacky. The hard one, Rhoplex WN-75, may be made more flexible by blending with WC-9. It has fair compatibility with the chemical plasticizers, but the resulting films have only fair strengths. It is not compatible with the resinous plasticizers. However, the long-oil architectural type of alkyd resin is compatible and may be stirred into the emulsion directly. Some of the urea-formaldehyde type resins may also be added to Rhoplex WN-75 in this manner. From the above it will be realized that WN-75 may be used alone for chemical-resistant paints for porous surfaces, and it may be blended with flexibilizers for coatings on paper and fabrics.

Rhoplex WN-80 is particularly well suited for paper coating since its films are sufficiently flexible, and they are non-tacky. Usually, WN-80 is thickened by addition of dispersions of sodium polyacrylate, sodium carboxymethyl cellulose, or the other materials mentioned previously. Additions may also be made of water-soluble gums, starch dispersions, or waxes. The compositions are applied to paper by the knife-coating method. Since this method has a tendency to produce foam or bubbles it is usually desirable to add an antifoaming agent. The amount of coating solids which must be applied to obtain completely greaseproof paper depends somewhat on the porosity of the stock. On super-calendered, bleached kraft stock complete greaseproofness may be obtained from a coating weight of 1.5 lb per 1000 sq ft applied in a single operation. On patent board stock, coating weights from 1.9 to 2.4 lb per 1000 sq ft are usually required to greaseproof this material. When only partial greaseproofness is required good films may be obtained from coatings of about 0.9 lb per 1000 sq ft.

Paper coatings may be pigmented with clay to improve the filling property. High concentrations of clay have maximum filling, but the coatings are more pervious to water, grease, etc. A general indication of the ratio of Rhoplex WN-80 solids to clay for resistance to grease and water is given below. The values will vary somewhat with the type of clay and porosity of the stock.

WN-80 Solid	Clay	Coating
45	55	Greaseproof
30	70	Waterproof
15	85	Good water resistance
6	94	Good filling and creasing

The coatings are tough and have good resistance to cracking when creased. They also polish to a high gloss by mechanical rubbing or friction calendering. Lower-cost coatings may be obtained by modification with starch or casein. Combinations of Rhoplex WN-80 with casein have better toughness and water resistance and show less discoloration on aging than straight casein coatings.

Rhoplex FRN produces films which are extremely flexible, but slightly tacky. Therefore they are highly satisfactory for impregnating paper or fabrics which are to be finished with coatings of the cellulosic polymers. Rhoplex FRN may also be knife-coated onto fabrics to produce an anchor coat. Anchor coats provide adhesion to the fabric when it is finished by applying vinyl or other resins at 100% solids on a calender coater machine. This procedure was discussed in Chapter 12.

Rhoplex FRN is compatible with both WN-75 and WN-80 in all proportions and may be added to increase their flexibility. It may be used with many of the synthetic rubber latexes as a permanent, nonmigrating plasticizer since it improves the retention of flexibility on aging. Frequently the films from such blends are somewhat cloudy, but this does not appear to affect their physical properties. Increased stability of the blends may usually be obtained by addition of small amounts of ammonia.

This brief outline of the acrylic ester resins shows that they have many desirable properties as coating materials. They will undoubtedly be used to a greater extent when increased production and new methods of synthesis have been developed to lower their cost. Their unique features are extremely pale color and excellent color retention on aging, and resistance to hardening on aging. They are sensitive to some solvents because their films are thermoplastic, but this disadvantage may be overcome in some cases by blending with a thermosetting resin. They are esters and are therefore susceptible to saponification by alkalies, but this requires rather drastic conditions because the ester groups are somewhat protected. Their extremely wide range of hardness makes the acrylic ester resins useful in plastics, coatings, and adhesives.

An acrylic resin emulsion recently announced by the Rohm and Haas Co., known as Rhoplex AC-33, is designed specifically for architectural water paints. The emulsion has a relatively high viscosity; therefore paints made with it do not require the addition of thickeners such as are used in the emulsion paint formulations given in Chapter 9. This simplifies the manufacture of the paint and should improve package stability, since thickeners tend to change in viscosity on aging. The emulsion is quite resistant to hard water and the soluble salts found in many pigments. This is another advantage in paint manufacture, since normal tap water may be used safely both in making the paint and in its subsequent thinning if necessary for application. Since the paint should be adjusted with ammonia to a *p*H of about 9.4 before packaging, alkali-resistant pigments must be used. The paint is fast drying and may be recoated in 1 or 2 hr; therefore it may not retain a wet edge long enough for good lapping. However, the addition of a small percentage of diethylene glycol increases wet-edge retention without significant effect on drying time or water resistance of the coating. Although the emulsion is compatible with certain oils, alkyd resins, rubber resin latexes, casein, etc., the manufacturer suggests that it be used alone as the complete vehicle for the paint to obtain maximum film properties, stability, and freeze-thaw resistance. The physical characteristics of Rhoplex AC-33 acrylic resin emulsion (Table 6) and a suggested formulation for an interior wall paint (Formula 8) follow.

Table 6 Physical Characteristics * of Rhoplex AC-33 †

Appearance	Milky liquid
Per cent solids	46 ± 0.5
Viscosity at 25°C	65–80 KU
Specific gravity at 25°C	1.04
Pounds per gallon	8.67
*p*H	9.0–9.5

* Courtesy of Rohm and Haas Co.
† Trade name, Rohm and Haas Co.

Formula 8 * Typical Interior Wall Paint Formulation

Pigment Dispersion	Pounds	Gallons
Titanium dioxide	267.0	8.2
Lithopone	76.3	2.1
Mica	51.7	2.3
Silica	81.0	3.7
Tamol † N	6.8	0.46
Diethylene glycol	7.2	0.77
Water	190.0	22.8
Mix with		
Rhoplex AC-33 (46% solids)	516.0	59.7
	1196.0	100.03

Procedure

Dissolve Tamol N in the water, add the diethylene glycol, and then the pigments with good agitation, pass mixture through a Morehouse mill or colloid mill, or grind in a pebble mill from 1–2 hr. Add Rhoplex AC-33 and mix to uniform consistency; then adjust to *p*H of 9.4–9.5 with 28% ammonium hydroxide.

* Courtesy of Rohm and Haas Co.
† Dispersing agent, Rohm and Haas Co.

The above illustrates the type of pigmentation used with this acrylic resin emulsion; it should be compared with the pigmentations given in Chapter 9 for use with rubber resin emulsions. Further details regarding formulation, application, and film properties are given in Volume II.

BIBLIOGRAPHY

1. Staudinger and Trommsdorff, *Ann.*, *502*, 201 (1933).
2. Rohm and Haas Co., *The Monomeric Acrylic Esters.*
3. C. E. Rehberg and C. H. Fisher, *Ind. Eng. Chem.*, *40*, 1429 (1948).
4. Kaufman, Sacher, Alfrey, and Fankuchen, *J. Am. Chem. Soc.*, *70*, 3147 (1948)

5. C. H. Fisher, W. P. Ratchford, and L. T. Smith, *Ind. Eng. Chem., 36,* 229 (1944).
6. Bigelow, *Chem. Eng. News, 1039* (April 14, 1947).
7. C. E. Schildknecht, *Vinyl and Related Polymers,* John Wiley & Sons, 1952.
8. C. H. Fisher and W. C. Mast, *Ind. Eng. Chem., 41,* 790 (1949).
9. F. S. Nichols and R. G. Flowers, *Ind. Eng. Chem., 42,* 292 (1950).
10. New York Paint & Varnish Production Club, Technical Committee 44, H. F. Payne, chairman, *Official Digest Federation Paint & Varnish Production Clubs, No. 302,* 212 (1950).
11. G. M. Sutheim, *Introduction to Emulsions,* Chemical Publishing Co., 1946.
12. C. B. F. Young and K. N. Coons, *Surface Active Agents,* Chemical Publishing Co., 1945.

Be convinced that you are right, then act without being told; that is initiative. There is a continuous demand for those who exercise initiative, and the reward is twofold; it pays financially, and more importantly, it builds strength into character.

Chapter 14. SILICONE RESINS

INTRODUCTION

Silicone resins are based on an inorganic, silicon-oxygen structure which has organic radicals attached to the silicon atoms. The organic radicals may be short-chain hydrocarbons from methyl to amyl groups, ring structures such as the phenyl group, and unsaturated groups such as the vinyl or the allyl group. Also, the ratio of organic groups to silicon atoms may be varied from 1 to 3 organic groups per silicon atom. In addition, silicone products are available in which the silicone is combined chemically with alkyd resins. Silicone resins are a relatively new addition to the raw materials for surface coatings; therefore only a limited number of types are available, although additional types may be expected to be developed.

The distinguishing characteristic of silicone resin coatings is their ability to withstand exposure to higher temperatures than the straight organic resins discussed in previous chapters. Some pigmented silicone finishes have withstood exposure to 500°F for at least 1000 hr without significant change. At about 650°F and higher the organic groups are removed by oxidation, leaving a residue of silica. Silicone finishes have very good color and gloss retention at temperatures of about 450°F, but they show some loss of adhesion. They have excellent exterior durability and resistance to water and chemicals. However, some silicone finishes are quite sensitive to aromatic hydrocarbons.

The exceptional heat stability of silicones is especially valuable when they are used as insulating materials in conjunction with glass cloth, asbestos, and mica. When glass cloth became available to replace cambric, silk, and paper for coated insulating tapes and windings, it was found that glass cloth would withstand higher temperatures than the standard organic insulating varnishes. This spurred the development of silicone resins for coating glass and asbestos cloth and for bonding mica sheets. Silicone resins are also used for impregnating coils and windings for motors and other electrical equipment. The silicone-type insulation makes possible the safe operation of motors and other equipment at temperatures which would destroy the previous organic insulation.

Similarly, silicone rubber has been produced which will withstand operating temperatures which would destroy either natural or synthetic rubbers. It also retains its rubber-like properties over the extreme temperature range of from 500°F down to −120°F. A comparison (9) of silicone rubber with GR-S and natural rubber at very low temperatures showed that GR-S and natural rubber stiffened at −76°F, but silicone rubber stiffened in the range of −112° to −130°F.

The low-molecular-weight silicone polymers with linear structures are oily liquids which can be used for lubrication, hydraulic fluids, antifoam and antifloat agents, and in automobile and furniture polishes. In addition to thermal stability they exhibit a relatively small change in viscosity with change in temperature. Rochow (1) and Wilcock (3) illustrate this difference by the comparison of a silicone oil with a petroleum oil over a temperature range from 210° to −70°F, as shown below:

Change in Viscosity with Temperature

Temperature, °F	Viscosity, centistokes	
	Silicone	Petroleum
210	40	10.8
100	100	100
0	350	11,000
−35	660	230,000
−70	1560	....

The silicone oils may be compounded with silica aerogel, lithium stearate, etc., to produce lubricating greases. These greases have excellent heat stability and relatively small change in consistency with temperature. The silicone oils exert a powerful effect on the surface tension of other oils, resins, paints, and certain aqueous media. As little as 1 part of silicone oil in 100,000 parts of oil or resin acts as an antifoam agent and prevents excessive foaming during heat processing. Also, a very small amount may be used in certain pigmented finishes as an antifloat agent to reduce objectionable color separation. Silicone oils may also be compounded with waxes to produce automobile and furniture polishes. These have exceptional ease of application and produce a high gloss finish with very good water repellence. An interesting use for silicone resins is their application to bakers' bread pans. The silicone coating eliminates the need for greasing the pans after each bread-baking operation. Silicone resins also produce an extremely water-repellent surface on brick and masonry structures. A dilute solution of the silicone may be applied by brushing or spraying, and after about 24 hr it forms a practically invisible coating which prevents moisture absorption into the masonry. This coating has very good exterior durability but should not be used on surfaces below the ground level. It will be apparent that silicone products are unique in some respects. They will no doubt find wider use when their high cost has been reduced by processing and other developments.

STRUCTURE OF SILICONES

The brief outline of the major applications for silicones given in the introduction indicates that silicones are available as oils, rubber, and resins. From the discussions in previous chapters relating molecular structure to physical form it will be realized that there should be three fundamental types of molecular structure in the silicones: (1) low-molecular-weight linear structures (silicone oils); (2) high-molecular-weight linear structures, with little or no cross-linking (silicone rubber); (3) high-molecular-weight, cross-linked structures (silicone resins).

We will discuss briefly the chemistry of these silicone structures, but the reader is referred to the book by E. G. Rochow, *The Chemistry of the Silicones,* second edition, John Wiley & Sons, 1951, for complete details regarding this subject.

It was stated in the introduction that the silicones are based on a silicon-oxygen structure to which various organic radicals are attached. The three fundamental types (oils, rubber, and resins) may be visualized as being built from three unit structures which Rochow (1) has conveniently called *M, D,* and *T* units:

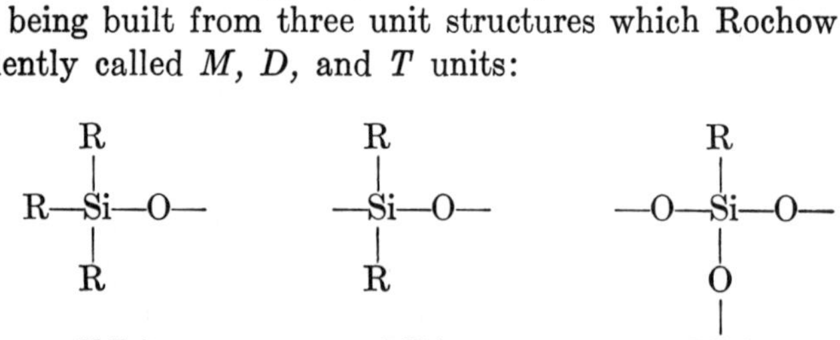

M Unit *D* Unit *T* Unit

It will be seen that the *M* unit is monofunctional, the *D* unit is difunctional, and the *T* unit is trifunctional. From previous discussions of functionality it will be realized that polymers cannot be produced from *M* units. These can be used only for dimers or as end units or chain stoppers. Linear polymers can be produced from the *D* units, with *M* units as the terminating groups. Cross-linked polymers can be produced from the *T* units because of their trifunctionality. Therefore the three general types of silicone materials referred to above can be considered as having the following arrangement of the *M*, *D*, and *T* units:

Silicone oils: $M(D)_xM$, where x is a small number.

Silicone rubber: $M(D)_xM$, where x is a very large number.

Silicone resins: *MDT* combinations.

Four of the many possible arrangements are shown in Fig. 1, page 570. They will be discussed in more detail following a consideration of the chemistry of the units and the chemical bonds they contain. It will be noted that the units contain the silicon-oxygen bond, and the silicon-carbon bond through which the organic radicals are attached to the silicon atoms. The silicon-oxygen bond is extremely heat stable, and it has been found that the silicon-carbon bond is more resistant to oxidation than the carbon-carbon bond on which most organic resins are based. The observing reader may wonder why there are no silicon-silicon bonds, or silicon-hydrogen bonds, corresponding to the carbon-carbon and carbon-hydrogen bonds in organic resins. We shall see that both the silicon-silicon and silicon-hydrogen bonds are readily attacked by oxygen, alkalies, and other chemicals. Therefore these bonds do not usually appear in the structure of silicone materials.

We recognize that the metal silicon is tetravalent from the formula for silica, or silicon dioxide, SiO_2. In this respect silicon is similar to carbon, which forms carbon dioxide, CO_2. However, silica is a hard, crystalline solid with a very high melting point, whereas carbon dioxide is a gas. Therefore, although silicon may replace carbon in many chemical compounds, the resulting silicon compounds may be expected to have very different physical properties from the carbon compounds.

SILICON-HYDROGEN BOND

Carbon and silicon both form a homologous series of hydrides. The simplest member of the carbon series is methane, CH_4, and the simplest in the silicon series is silane, SiH_4. The saturated hydrocarbon series extends from methane to high-molecular-weight polyethylene which may contain more than one thousand carbon atoms in the chain. However, the highest member of the silane series which has been reported (1) is hexasilane, Si_6H_{14}. The higher silanes are unstable; they rearrange to form simpler compounds.

Another marked contrast between the two series of hydrides is their behavior to oxidation. The saturated hydrocarbons do not oxidize readily, but the silanes react with air so rapidly that they ignite. The hydrogens from the silane are liberated, and the end product is the extremely stable compound, SiO_2.

$$SiH_4 + O_2 \rightarrow SiO_2 + 2H_2$$

Another indication of the relative instability of the silicon-hydrogen bond is the reaction of silanes with dilute aqueous alkalies. The silanes hydrolyze to form silica and hydrogen, whereas the hydrocarbons are not affected by such treatment. From these few illustrations it will be realized why the silicon-hydrogen combination does not appear in silicone materials as the carbon-hydrogen combination does in organic materials.

SILICON-SILICON BOND

It was stated in the previous section that attempts to produce silanes higher than the hexasilane, Si_6H_{14}, resulted in a rearrangement with the formation of simpler compounds. This indicates that the silicon-silicon bond is relatively unstable in long chains and may be broken fairly easily. A similar situation exists for the organo-substituted silanes, such as the methyl or phenyl silanes, in which the hydrogen is substituted with methyl or phenyl groups. Even the low-molecular-weight members of these compounds are sensitive to aqueous alkalies and may be split by alkali attack. Therefore it would appear that polymers based on the silicon-silicon bond would be unsatisfactory for coatings even if methods could be found for producing them.

SILICON-CARBON BOND

By comparison with the silicon-hydrogen, or silicon-silicon, bonds the silicon-carbon bond has been found to be quite stable. However, the stability varies somewhat with different structures. For example, the bond $Si{-}CH_3$ is quite resistant to oxidation, but $Si{-}C_6H_{13}$ is not; also, $Si{-}C_2H_5$ is unaffected by water, but $Si{-}C{\equiv}CH$ is broken readily (1). However, the vinyl polysilicone,

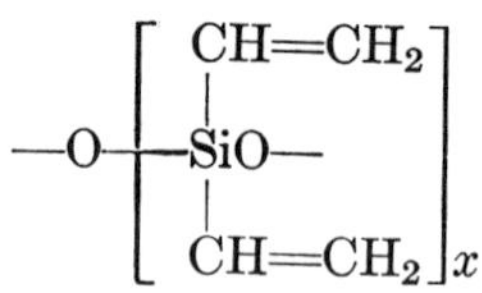

(Sizing GS-1, Linde Air Products Co.) may be applied to glass cloth and baked for 5 min at 530°F to form a strongly adherent coating on the glass fibers, as discussed in detail later and also in references 4–6.

Rochow (1) has described a group of methods which may be used for preparing polymers based on repeating silicon-carbon atoms:

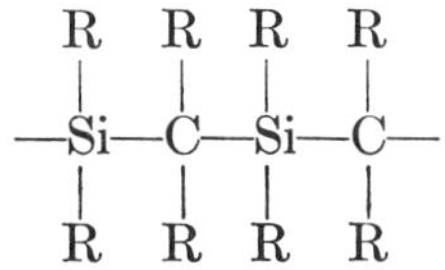

However, although the silicon-carbon bond is relatively stable to heat and is not readily oxidized or hydrolyzed, it is not as stable as the silicon-oxygen bond found in regular silicones.

SILICON-OXYGEN BOND

In view of its heat stability and resistance to water and many chemicals, the silicon-oxygen bond is used as the structure on which the silicone materials are built, as indicated previously. This basic structure has organic radicals attached to the silicon atoms. The arrangement of alternate silicon and oxygen atoms with organic groups on the silicon atoms is variously referred to in the literature as silicone, organosiloxane, and organosilicon oxide. This type of structure will be referred to as silicone in this book.

It will be realized from the discussion of cellulosic and vinyl polymers in previous chapters that considerable variation in properties of the silicones may be made by varying the type of organic groups on the silicon atoms. The most important of these properties are solubility, hardness, and heat and chemical resistance. The important effect of varying the ratio of organic groups to silicon atoms was referred to earlier as the factor which determines whether small molecules, linear polymers, or cross-linked polymers are obtained. These two subjects, type of R and the ratio R/Si, will be taken up in more detail after a brief outline of the two general processes for making the silicones.

PREPARATION OF SILICONES

The various methods for preparing silicones are described in detail by Rochow (1). We shall consider the direct and indirect methods very briefly to point out the types of intermediate materials which are available and the possible variations in the silicones. The four important steps in the preparation of silicones are:

1. Conversion of the metal silicon to a halide, preferably the chloride.
2. Replacement of some chlorine with organic groups.
3. Hydrolysis of remaining chlorine to produce siloxanes.
4. Condensation to silicones with the elimination of water.

These four steps may be visualized by considering four of the many possible products:

$SiCl_4$, silicon tetrachloride.
$(CH_3)_2SiCl_2$, dimethyldichlorosilane.
$(CH_3)_2Si(OH)_2$, dimethylsilanediol.
$[(CH_3)_2SiO]_x$, dimethyl silicone polymer.

The indirect methods start with silicon tetrachloride and a Grignard reagent to produce the organochlorosilanes. The direct methods start with the reaction between the metal silicon and a chlorinated hydrocarbon to produce the organochlorosilanes directly. Somewhat better control of the products is possible with the indirect methods, but the direct methods are more economical.

The reaction between a Grignard reagent and a silicon halide may be illustrated by the following results with silicon tetrachloride. It will be noted that a series of organo-substituted chlorosilanes may be obtained.

$$RMgBr + SiCl_4 \rightarrow RSiCl_3 + MgBrCl$$

$$RMgBr + RSiCl_3 \rightarrow R_2SiCl_2 + MgBrCl$$

$$RMgBr + R_2SiCl_2 \rightarrow R_3SiCl + MgBrCl$$

$$RMgBr + R_3SiCl \rightarrow R_4Si + MgBrCl$$

Under normal conditions the reaction favors the production of the R_2SiCl_2 type. Of course, these organosilanes are converted into silicones by the dual processes of hydrolysis and condensation.

In one of the processes for the direct method the vapors of a chlorinated hydrocarbon are passed through a reaction tube containing the silicon metal in a finely divided state, mixed with a metallic catalyst such as copper. The product is a mixture of compounds, but by careful control of temperature, rate of reaction, type and condition of catalyst, and other factors the predominant compound can be R_2SiCl_2. The product may be separated by distillation into specific components, although for some coating purposes it may be satisfactory as is. The general reaction may be expressed as follows:

$$2RCl + Si \rightarrow R_2SiCl_2$$

ADDITION OF OLEFINS

Another direct method for the production of organochlorosilanes involves the addition of an olefin to a chlorosilane, preferably the trichlorosilane, $HSiCl_3$. A typical example is the addition of 1-octene to trichlorosilane, using diacetyl peroxide as catalyst.

$$C_6H_{13}CH{=}CH_2 + HSiCl_3 \rightarrow C_8H_{17}SiCl_3$$

A series of these addition reactions is described by Rochow (1), and also the free-radical mechanism by which the reaction is believed to take place. Vinyltrichlorosilane may be prepared by this type of reaction, using acetylene as the olefin:

$$CH{\equiv}CH + HSiCl_3 \rightarrow CH_2{=}CH{-}SiCl_3$$

Vinyl chlorosilanes may also be prepared by the direct reaction of elemental silicon and vinyl chloride, although poor yields are obtained (7). A preferred method (7, 8) involves the chlorination of ethyl chlorosilanes followed by a quinoline dehydrochlorination of the chloroethyl groups to vinyl groups.

It will be apparent that a wide variety of organochlorosilanes may be prepared by the direct or indirect method. Some applications have been found for the organochlorosilanes, but more useful coating materials are obtained after hydrolysis and condensation to the silicones, as discussed previously.

SILICONE POLYMERS

The early silicone materials were based on polymers of the methyl silicones and phenyl silicones, and copolymers of these compounds. A variety of products was obtained by varying the ratio of methyl or phenyl groups to silicon atoms. It was found that, in general, the phenyl silicones were harder and more brittle than the methyl silicones, but methyl-phenyl copolymers had better strength and toughness than either of the homopolymers. More recently the substitution of other organic radicals has made available a greater range of silicones. The effect produced by such substitution will be discussed after a brief outline of the types of polymeric structure formed by the silicones.

It was stated previously that the type of polymer produced was related to the *M*, *D*, and *T* units it contained. Since these units vary in the ratio of organic radicals to silicon atoms (R/Si), the polymers may also be related to the R/Si ratio. In general, when the R/Si ratio is greater than 2, linear polymers are formed. Cross-linked polymers are obtained when the R/Si ratio is less than 2. By careful elimination of any *T* units in the monomer it has been possible to produce long-chain polymers of dimethyl silicone for use as silicone rubber. Chains of two thousand or more units $[(CH_3)_2SiO]$ have been pro-

Fig. 1. Relationship of R/Si to Structure

(*a*)

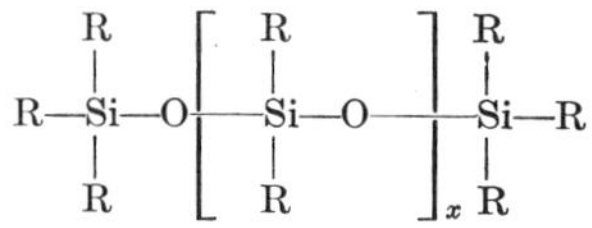

Units: $M[D]_xM$

Ratio: R/Si = 2.66 $x = 1$
2.17 $x = 10$

When x is very large, as in silicone rubber, R/Si = 2.

(*b*)

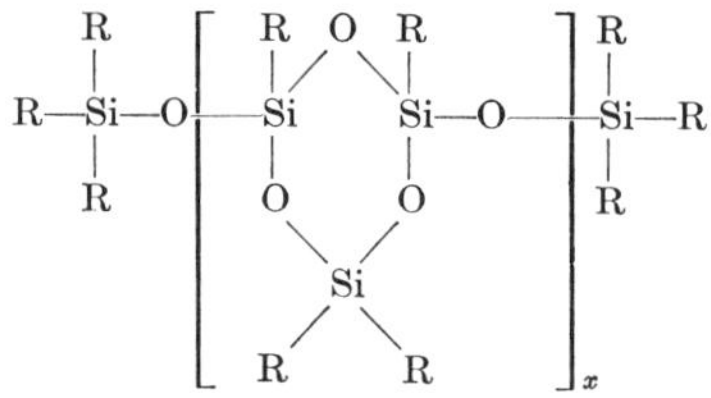

Units: $M[D_1T_2]_xM$

Ratio: R/Si = 2.0 $x = 1$
1.43 $x = 10$

(*c*)

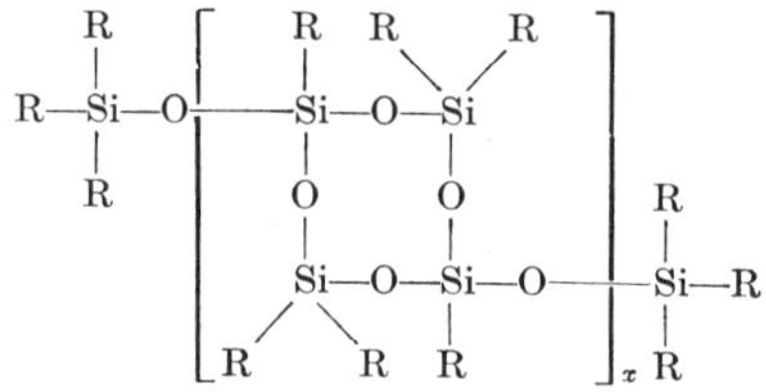

Units: $M[D_2T_2]_xM$

Ratio: R/Si = 2.0 $x = 1$
1.57 $x = 10$

(*d*)

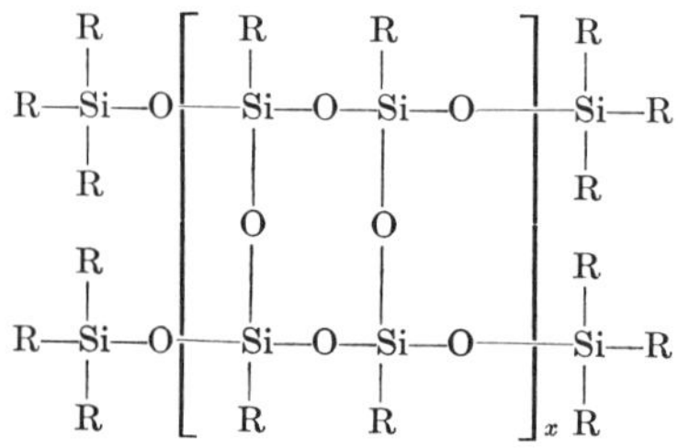

Units: $M_2[T_4]M_2$

Ratio: R/Si = 2.0 $x = 1$
1.18 $x = 10$

duced which form an elastomer similar to rubber. It can be milled on a two-roll mill, and fillers, curing agents, etc., incorporated.

When *T* units are present in the monomer a variety of cross-linked structures is produced. It is believed that these polymers include ring structures based on from three to six silicon-oxygen groups. The silicon-oxygen groups can form stable ring structures up to at least eighteen members because of the large size of the silicon atom and the large valence angle of oxygen. However, the polymers are probably composed of cyclic tetramers to hexamers connected as shown in Fig. 1.

It should be kept in mind that most commercial products would be less symmetrical, or more heterogeneous, than the four of the many possible arrangements shown in Fig. 1.

The structure shown in Fig. 1(*a*) is the linear type which is found in silicone oils and rubbers. When the degree of polymerization is low (x is a small number), liquid, oily products are formed. When x is a large number, flexible, rubbery products are produced. It will be noted that the oils and rubbers are formed from M and D units only. The R/Si ratio is greater than 2 for the oils, and equal to 2 for the high-molecular-weight rubber. Of course, the end groups on the rubber molecules are relatively unimportant because the molecule is so large.

The linear polymers are produced from organodichlorosilanes by hydrolysis, followed with intermolecular condensation:

$$\underset{\text{Organodichlorosilane}}{2R_2SiCl_2} + 3H_2O \rightarrow \underset{\text{Organosilanediol}}{HO\!-\!\overset{\overset{R}{|}}{\underset{\underset{R}{|}}{Si}}\!-\!O\!-\!\overset{\overset{R}{|}}{\underset{\underset{R}{|}}{Si}}\!-\!OH} + 4HCl$$

$$x\left[HO\!-\!\overset{\overset{R}{|}}{\underset{\underset{R}{|}}{Si}}\!-\!O\!-\!\overset{\overset{R}{|}}{\underset{\underset{R}{|}}{Si}}\!-\!OH\right] \rightarrow HO\left[\!-\!\overset{\overset{R}{|}}{\underset{\underset{R}{|}}{Si}}\!-\!O\!-\!\overset{\overset{R}{|}}{\underset{\underset{R}{|}}{Si}}\!-\!\right]_x OH + (x-1)H_2O$$

It will be realized that sufficient organomonochlorosilane must be included to produce M units to react with the terminal hydroxyls on the polymer. It will also be realized that the amount of organomonochlorosilane included will be a factor in determining the chain length of the polymer. The condensation of organosilanediols proceeds much more readily than is usually the case with organic resinous condensations. The application of heat is not generally required at the beginning of the reaction, but both heat and catalyst may be necessary later to effect as nearly complete condensation as possible. Therefore the molecular weight of the polymer will be affected by the ratio of M and D units, the type of catalyst, and the physical conditions of the reaction. Rochow (1) gives considerable details regarding these variables.

It will be seen from structures (*b*) and (*c*) in Fig. 1 that some T units must be included in order to obtain these cross-linked structures. This could be done by using a combination of dichloro- and trichlorosilanes for the hydrolysis and condensation reactions. It will be seen that a co-condensed silicone-alkyd resin could be produced by adding an alkyd, having free hydroxyl groups in its structure, to the silicone before it was condensed completely, then finishing the condensation of the mixture. A chemical combination of silicone and alkyd would be obtained unless the rate of condensation of one resin was very much faster than that of the other. Several commercial products of this type are described later. It will be realized that many types of silicone polymers may be produced by varying the ratio of M, D, and T units and the conditions of the reactions, and by co-condensing other resinous materials with the silicone. A wide range of variation in the physical and chemical properties of the silicone polymers may also be made by changing the type of organic radical attached to the silicon atoms.

VARIATION IN ORGANIC RADICALS

We have seen in previous chapters that side groups on linear or cross-linked polymers affect the physical and chemical properties of the polymers. For example, we recall the difference in solubility of polyethylene, polyvinyl chloride, and polyvinyl alcohol produced by the difference in side groups on these polymers. Another example may be found in the different solubilities of nitrocellulose, ethyl cellulose, and cellulose acetate. These three cellulosic polymers also differ in hardness at equivalent molecular weights because of the difference in configuration of their side groups. Furthermore, they vary in alkali resistance; the ether is more resistant than either of the esters.

It is not always possible to relate specifically the effect of the side groups on one series of polymers to the effect produced in another series. However, we can see that the solubility of the silicones in hydrocarbon solvents is due directly to the hydrocarbon side groups on the silicone polymers. Certainly, a polar silicon-oxygen structure would not be soluble in such non-polar solvents were it not for its non-polar side groups. We could also predict with considerable assurance that the solubility of the silicones in such solvents would increase as the length of the hydrocarbon chain of the side groups increased. Of course, we would recall that the molecular weight of the polymer is also a factor. However, as the length of the carbon chain in the side group increased from methyl to amyl we would not be surprised if the solubility changed from aromatic to aliphatic hydrocarbons. We would also expect that the hardness of the polymer would be decreased with this change, and that its drying or curing time would be increased.

Since the amyl silicone has a higher ratio of carbon to silicone than the methyl silicone we might expect that it would not be as resistant to heat. Rochow (1) reports that when these two silicones are heated in air at about 400°F the amyl silicone cracks and checks after 1 day, whereas the methyl silicone may be satisfactory for 10 years under such conditions. Also, the benzyl silicone is less resistant to oxidation than the phenyl silicone because the aromatic ring of the benzyl groups is connected to the silicon atom by a —CH_2— group. The benzyl silicone behaves more like the butyl or amyl than the phenyl silicone. For example, the benzyl group is not removed by hydrolysis with aqueous acids, but the phenyl group is removed readily under such conditions.

There is a marked difference between the methyl and ethyl silicones and the phenyl silicone. The methyl silicone polymers having a high R/Si ratio are oily liquids, but the corresponding phenyl silicones are hard, brittle solids. The R/Si ratio for methyl silicone must be reduced to about 1.5 before resins are obtained, and these would have a cross-linked structure. Copolymers of methyl or ethyl silicone with phenyl silicone have better strength and toughness than either homopolymer of the alkyl or aryl type. The copolymers also have better dielectric properties and heat resistance than the straight methyl silicone. Rochow (1) gives considerable information regarding these variations. The early silicone materials which appeared on the market were methyl, ethyl, and phenyl silicones and their copolymers, and they are still the most widely used types. Some of the physical characteristics of the methyl and ethyl silicone oils are given in Table 4.

One of the more recent commercial developments is the vinyl silicone materials. Hurd (7, 8) and Wagner *et al.* (5) describe their preparation. A typical laboratory procedure for making vinyl polysilicone is given on page 588. Jellinek (4) and Brown *et al.* (6) describe its use in coatings for glass fibers which are used in polyester laminates. An outline of this application is given on page 591. It was found that vinyl polysilicone could be applied to glass cloth and cured by baking without destroying all the vinyl unsaturation. Subsequent application of another unsaturated coating, followed by a curing schedule, resulted in a copolymerization between the residual vinyl unsaturation and the new coating. In this way a stronger bond was obtained in the laminates because of the chemical bonding instead of the usual van der Waals bonding. Copolymers of methyl and vinyl silicones have been produced, and these cure faster than the straight methyl silicone resins. The introduction of a small percentage of vinyl groups into silicone rubbers also accelerates their cure. Hurd and Roedel (8) discusses a series of copolymers obtained from methyl vinyl silicone, and styrene, acrylonitrile, and both methyl and butyl methacrylates. An increasing number of commercial products are being made available, and the principal ones will be outlined in a later section of this chapter.

RESINOGRAPHY

No discussion of silicones would be complete without reference to the excellent work (1, 18, 19, 20) of the two brothers, T. G. Rochow and E. G. Rochow, and their associates in determining the size of silicone molecules by the technique of resinography. They have defined resinography as the graphic study of resins, through use of the microscope and its adjuncts, in techniques similar to those of metallography and mineralography. Resinography was employed very effectively in a study of the physical structure of fabricated plastics, rubbers, alkyd-styrene copolymers, etc. (18, 20). Such studies yielded information regarding the shape and type of fillers in plastics and the extent to which they were wetted by the resin.

Resinography also indicated the size of the macromolecules of the silicones in silicone rubbers. For this purpose the technique employed was the electron microscopy of fractured surfaces of the silicone rubbers. The rubbers were fractured by cooling them in liquid nitrogen to such low temperatures that they were very brittle. In order to show the irregularities in the fractured surface to best advantage a silica replica was made, which was then "shadowed" to highlight the irregularities. The first step after fracturing the rubber is to coat the surface with a 5% aqueous solution of gelatin. After drying, the gelatin coating is removed and placed in a vacuum evaporator where a layer of silica is evaporated onto it. The gelatin is removed from the silica by washing with water, thereby leaving a positive "print" of the fractured rubber surface in the silica. The print is then shadowed by evaporation and deposition of uranium on it at a small angle. The finished replica is electron-micrographed at magnifications up to 20,000. The micrographs may be enlarged photographically to the size necessary for showing the required details of surface structure. Figures 2 and 3 are typical of the results obtained with this technique.

SIZE OF SILICONE MACROMOLECULES

Figure 2 shows the surface of a silica replica of a fractured white silicone rubber. The white areas are pigment particles embedded in the surface. The elevations represent pigment particles left on or near the surface, and the depressions represent pigment particles removed in the mate to this surface. Figure 3 is the enlargement of a particular area of Fig. 2 chosen to show chiefly the silicone macromolecules between the pigment particles. Rochow (1) calculates the molecular weight of these high polymers by two methods, using the known mag-

Fig. 2. Electron Micrograph of Fracture Surface of White Silicone Rubber (magnification 17,000×)

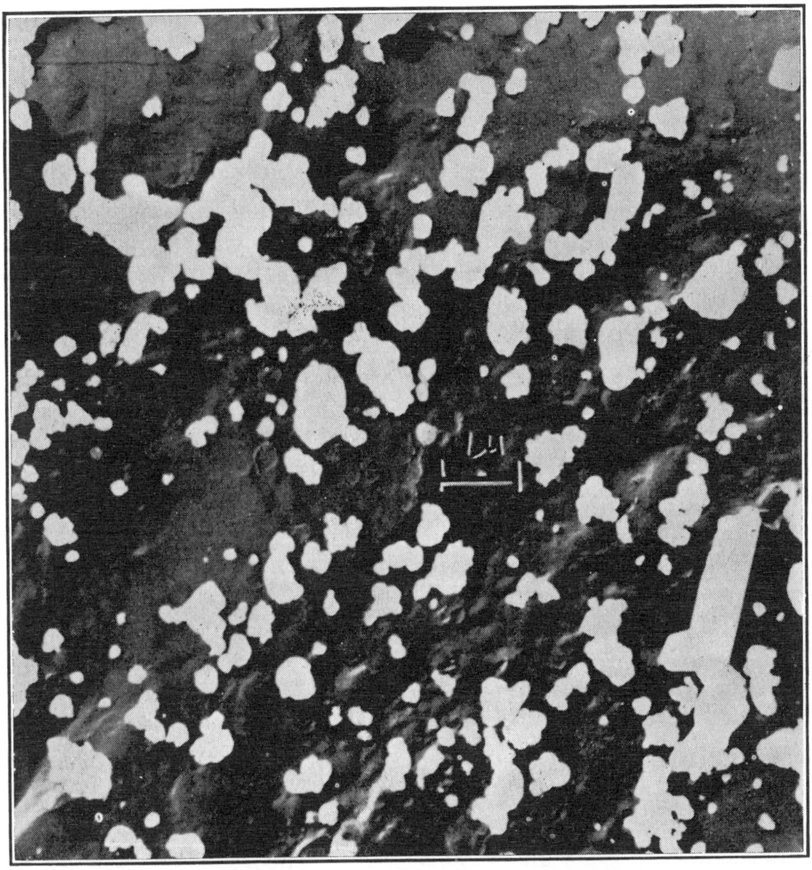

nification to give their size. One method is based on the relationship of size, specific gravity, and Avogadro's number, and the other method on the theoretical molecular size as obtained from molecular models. The first method gave a macromolecular weight of 304,000, and the second method 306,000, certainly a very interesting and important result.

By permission of the authors we are giving the calculations for these results because of their value to students and others interested in this work. Figure 3 is at a magnification of 100,000×; therefore 1 mm corresponds to 100 Å in the original specimen. The largest grains in Fig. 3 appear to be elliptical and are about 0.93 mm by 3.7 mm. The smaller grains are more spherical and are about 0.93–1.9 mm in diameter. For calculation, a sphere of 1.0-mm diameter was taken as generally representative. Since 1.0 mm corresponds to 100 Å in the

Fig. 3. "Close-up" of Fracture Surface of Fig. 2 Showing Silicone Macromolecular Structure (magnification 100,000×)

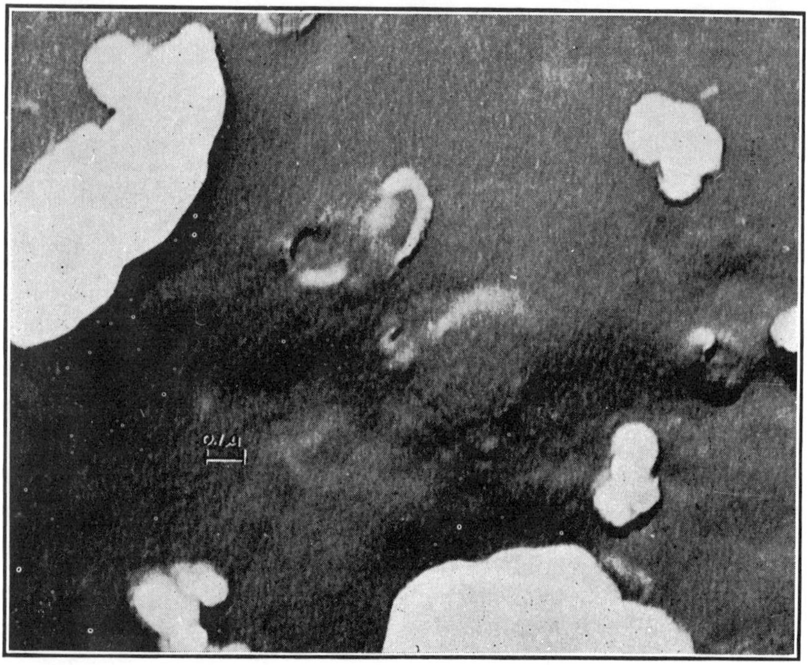

original, the grains would have the volume of a sphere of 100 Å diameter, which is $\frac{4}{3}\pi(50)^3$, or 524,000 Å³. This value was rounded out to 520,000 Å³.

The weight of the grains having a volume of 520,000 Å³ was calculated first from the volume-weight relationship and from Avogadro's number, which is 6.02×10^{23} molecules per gram-mole. The specific gravity was determined experimentally as 0.972. Therefore the macromolecular weight of the grains would be:

$$(520{,}000 \times 10^{-24}\ \text{cm}^3/\text{molecule}) \times (0.972\ \text{gm/cm}^3) \times (6.02 \times 10^{23}\ \text{molecules/mole}) = 304{,}000\ \text{gm/mole}$$

The second method was based on an estimation of the volume of a unit of the macromolecule (CH_3SiO_2), and then division of the total volume (520,000 Å^3) by the volume of the unit to obtain the number of units in the macromolecule. Since the weight of the unit can be calculated from atomic weights, the molecular weight of the macromolecule is obtained by simple multiplication. From molecular models, Norton (21) estimated the dimensions of CH_3SiO_2 to be 3.0 by 7.0 Å in surface area, and 6.0 Å in depth, giving a volume of 126 Å^3. Therefore the number of units in a volume of 520,000 Å is 4120. The molecular weight of CH_3SiO_2 calculated from atomic weights is 74.1. Therefore the molecular weight of the macromolecule is 4120×74.1, or 306,000. This value checks very closely with the 304,000 obtained by the first method of calculation. The reader is referred to the original references for further details regarding this interesting and valuable technique for obtaining the size and weight of macromolecules from visual observations and well-established calculations.

COMMERCIAL SILICONE MATERIALS

At present there are four manufacturers of the raw materials for silicone resins and the silicone oils, rubber, and resins. These are Dow Corning Corp., General Electric Co., Linde Air Products Co., and Plaskon Division, Libbey-Owens Ford Glass Co.

Since the silicone resin field is relatively recent, new developments are being added at a steady rate. It is realized that some of the products of the above manufacturers listed on the following pages may have become obsolete, or have been replaced by improvements, before this book is printed. However, the products listed here cover the major applications for the silicones, and the reader should consult with the manufacturers for possible changes or new products.

The products listed include some organochlorosilanes, and organoethoxysilanes from which silicone oils and resins may be made by hydrolysis and condensation. There are also the methyl, ethyl, and phenyl silicone oils and resins which find wide use in the electrical insulation industry and the coating industry. Some of these are used as antifoam and antiflooding agents in the surface coating industry, and also in polishes for furniture and automobiles. Included also are silicone oil emulsions which are used in the rubber and plastics industries as mold-release agents. These emulsions may find use in the coatings industry at some future time. Several of the silicone products are recommended for clear, water-repellent coatings on masonry; these include a water-soluble sodium methyl siliconate. This material may be used as a coating or included in the masonry mixture; it also serves as a water repellent for paper and textiles.

In order to extend the use of silicones in the coating industry products have been developed which may be cold-blended with alkyds, and also products which result from co-condensing alkyds and silicones. It will be realized that in most cases the actual composition of the commercial silicone products is not available.

Silicone resins are used in both clear and pigmented finishes. Some of the applications for clear coatings will be described in this section. However, the types of pigments and pigmented coatings will be discussed in Volume II.

Table 1 * Silicone Products

Silicon Resin #	Solids, % ± 1	Solvent	Viscosity at 25°C Centi-poises	Viscosity at 25°C Gardner-Holdt	Color, Max., Gardner	Pounds per Gallon
		Products for Coating Materials				
801	60	Xylene	100–300	D–L	2	8.75
802	50	Xylene	100–200	D–H	2	8.4
804	60	Toluene	20–40	A3–A	2	8.75
XR-129G	90	Xylene	5–50	.....	Light straw	
XR-379	50	Xylene Cyclohexanone	50–100	.. .	10	8.4
XR-398	50	Xylene Cyclohexanone	100–1000	.....	5–7	8.9
		Products for Insulating Materials				
935	50	Xylene	80–150	.....	Light straw	8.4
996	50	Xylene	100–200	.....	Dark brown	8.4
993	50	Toluene Xylene	100–200	.....	Light straw	8.4
993	70	Toluene Xylene	500–1000	.....	Light straw	9.0
2103	60	Toluene	60–125	.....	Light straw	9.0
2104	60	Toluene	20–40	.....	Light straw	8.75

Silastic stocks, pastes, and tapes. See page 581.
DC 200 silicone fluids. See description on page 583.

* Courtesy of Dow Corning Corp.

Dow Corning Silicone Resin 801. This resin may be used for white and colored heat-resistant coatings having good exterior durability. It is suggested particularly for aluminum coatings on boilers, stacks, mufflers, etc. Coatings without catalyst require 1–4 hr at 400°F for good film properties. Addition of 0.2% cobalt or zinc drier as metal to resin solids reduces the baking schedule to 1 hr at 400°F. Catalyzed coatings are stable for at least 6 months. Longer baking time develops maximum resistance, but the films are sensitive to chlorinated and aromatic hydrocarbons. Resin 801 is soluble in a variety of organic solvents, including hydrocarbons having KB values greater than 50. It is compatible with chlorinated biphenyls, ester gum, ethyl cellulose, and non-reactive phenolics.

Silicone Resin 802. Resin 802 is somewhat more flexible than Resin 801 but requires a longer and higher baking schedule. It forms a tack-free film after 1–3 hr at 480°F, but it should be baked an additional 3–16 hr to develop maximum film properties. When catalyzed with 0.1–0.2% cobalt or zinc the baking schedule may be reduced to 3 hr at 400°F. Blends may be made with Resin 801 if desired. Baked films have good heat resistance; they do not yellow on aging, have good color and gloss retention up to 500°F, and are resistant to checking until 3000–5000 hr at 480°F. Resin 802 has about the same solubility and compatibility as Resin 801.

Silicone Resin 804. Resin 804 will air-dry to a tack-free film. It will bake for 1 hr at 480°F to produce a harder, less flexible film than Resin 802. Driers are not usually recommended, but shorter baking time may be obtained from use of 0.05% zinc. The catalyzed finishes have poor package stability. Resin 804 has solubility characteristics similar to those of the other resins in this series, but it is somewhat different in compatibility. It is completely compatible with linseed and tung oils, non-reactive phenolics, melamine-formaldehyde resins, methacrylates, and ethyl cellulose. It is used where faster baking schedules are required in both coatings and insulation.

Silicone Resin XR-129G. This product is intended as a water repellent for application to exterior above-ground masonry surfaces. Rather than being a resin in the true sense of the word, it is a relatively low-molecular-weight alkoxy silicone which hydrolyzes after application to the masonry. This results in the substitution of hydroxyls for the alkoxyls, which hydroxyls then condense to siloxane linkages. In this manner, a silicone resin is formed within the masonry, thereby producing a silicone coating in the capillaries.

It is suggested that XR-129G be thinned with a solvent having a KB value not less than 75, a flash point minimum of 85°F, and a boiling range of 250–350°F. Water should be avoided since it forms an insoluble precipitate through hydrolysis of the alkoxy radicals. It should be thinned to 3–5% and may be applied by brushing or spraying. When spray application is employed, it is essential that a low pressure be used since high-pressure spraying will produce a dangerous fog of solvent. The masonry surface should be cleaned and free from moisture. One gallon of diluted material covers from 100 to 250 sq ft, depending on the porosity of the masonry. Since the resin formed merely coats the inside surfaces of the capillaries and does not fill them, it does not prevent the passage of air or of water under a hydrostatic head. Accordingly, this type of material allows the drying of brick which is so coated but prevents effective use as a water barrier on surfaces below the ground level. The coating does not change the appearance of the masonry, but it successfully prevents the unsightly efflorescence so often seen on brick. This treatment develops water repellency within 24 hr, but it requires about 2 weeks to reach maximum repellency. It has good durability and may be expected to last at least 5 years.

Silicone Resin XR-379. This resin has been described (10) as a copolymer of silicone and alkyd containing approximately 75% silicone and 25% alkyd on a solid resin basis. It is completely compatible with non-oxidizing alkyds, phenolics, urea-formaldehyde resins, chlorinated biphenyls, coumarone-indene resins, ester gum, nitrocellulose, and ethyl cellulose. It is partially compatible with oxidizing alkyds, melamine-formaldehyde resins, vinyl chloride-acetate, and vinyl butyrate. It is soluble in many common organic solvents and mixtures with aromatic hydrocarbons. The wide range of compatibility of Resin XR-379 permits its addition to many finishes to improve their heat resistance, and color retention at high temperatures.

Finishes made with XR-379 alone have improved adhesion in comparison with straight silicone resin finishes. In white enamels XR-379 has good color retention at temperatures up to 450°F and good exterior durability. It may be used with a range of colored pigments, but these should be tested for suitable compatibility and heat resistance.

Silicone Resin XR-398. Resin XR-398 is also a silicone-alkyd copolymer containing the reverse ratio of silicone and alkyd to Resin XR-379, i.e. 25% silicone and 75% alkyd. It has compatibility and solubility similar to those of XR-379, but it is not as heat resistant. However, finishes made with it have good color retention at temperatures up to 400°F. Such finishes also have good adhesion because of the high alkyd content.

Silicone Insulation. The silicone insulation resins meet the AIEE requirements for Class H materials. Equipment with this type of insulation can operate about 100°C higher than that with Class B materials, and has a longer life. Silicone varnishes are used to impregnate glass and asbestos fabrics and to bond together mica splittings and other finely divided particles. They are also used to impregnate coils, motors, transformers, and other electrical equipment. The fabrics are often coated by drawing them through a bath of the varnish, removing the excess by squeeze rolls, and drying the coating by passage up a tower drier. Several coats may be applied continuously by multiple arrangement of baths and tower driers. The coils, etc., may be dipped, followed by careful baking to avoid air holes from bubbles. They may also be placed in a chamber from which the air is evacuated and the varnish then drawn in. The resins have a wide range of hardness and flexibility to meet various requirements. They also vary in the temperature required for proper cure.

Silicone Resin 935. Resin 935 is suggested for coating glass cloth and sleeving, bonding mica, etc. It should be baked in the tower drier for 20–30 min at 500–550°F. The films are quite flexible and retain this property at temperatures down to −65°F.

Silicone Resin 996. This resin is used for impregnating coils and also for coating glass and asbestos cloth. Care must be taken to drain, air-dry, and then bake the impregnated coating to ensure proper cure without air holes, etc. The equipment is frequently heated before dipping to eliminate air and to facilitate impregnation. Several typical schedules are given on page 582 to illustrate the procedure. The properly cured films are hard and tough and have good heat resistance and electrical properties.

Silicone Resin 993. This resin is especially designed for impregnating coils, etc., and has better heat stability than Resin 996. However, it requires a higher temperature for curing: 480°F instead of 400°F for Resin 996. The films are more flexible than those from Resin 996, and they have good resistance to humid and corrosive atmospheres.

Silicone Resin 2103. This resin is a partially condensed silicone resin which condenses further on baking to a hard, rigid laminate with glass or asbestos cloth and other fillers. After saturating the cloth, it is air-dried to remove the solvent and then baked. The resin melts about 158°F, and the condensation reaction begins when the resin is heated to about 300°F and produces bubbles from the elimination of water. The bubbles decrease about 365°F, and the resin thickens about 465°F. Further heating to 480°F sets the resin to the infusible stage. Catalysts such as triethanolamine may be added to reduce the curing time and temperature.

Silicone Resin 2104. This resin is another laminating resin similar to Resin 2103. The curing schedule depends on the pressure used, and the amount of catalyst XY5 supplied with the resin to meet specific requirements. It is a thermosetting resin and cures to an insoluble, infusible product.

Silastic Stocks, Pastes, and Tapes. The Silastic materials are based on silicone rubber compounded with various fillers. The Silastic

Curing Schedules

Standard Industrial Motors Rewound with Class H Materials and Impregnated with Dow Corning 996

A. Stators and Wound Rotors.

1. Prebake at 300–400°F for 3–4 hr.
2. Cool to 160–170°F.
3. Immerse for 10–20 min or until all bubbling stops.
4. Drain and air-dry for 3 hr.
5. Bake at 175°F for 6–8 hr.

For succeeding impregnations:

6. Immerse for 3–4 min.
7. Drain and air-dry for 3 hr.
8. Bake at 175°F for 6–8 hr.

For final impregnation:

9. Immerse for 3–4 min.
10. Drain and air-dry for 3 hr.
11. Bake at 175°F for 6–8 hr.
12. Bake at 300°F for 12–16 hr.

B. Armature and Field Coils.

1. Prebake at 300–400°F for 3–4 hr.
2. Cool to 160–170°F.
3. Immerse for 10–20 min or until all bubbling stops.
4. Drain and air-dry for 1–3 hr.
5. Bake at 175°F for 2–4 hr.
6. Bake at 300°F for 4–6 hr.
7. Immerse for 3–5 min.
8. Drain and air-dry 1–3 hr.
9. Bake at 175°F for 2–4 hr.
10. Bake at 300°F for 4–6 hr.

Note: Time required for the silicone-insulated unit to reach temperatures given above must be added to the baking times. Baking times within the ranges given will depend upon the size and complexity of the motor or coil.

Industrial Motors, Traction Motors, and Magnet Coils Designed for Class H Rating

A. Stators and Wound Rotors.

1. Prebake at 300–400°F for 3–4 hr.
2. Cool to 160–170°F.
3. Vacuum-impregnate or immerse for 10–20 min or until all bubbling stops.
4. Drain and air-dry for 3 hr.
5. Bake at 175°F for 6–8 hr.

For succeeding impregnations:

6. Immerse for 3–5 min.
7. Drain and air-dry for 3 hr.
8. Bake at 175°F for 6–8 hr.

For final impregnation:

9. Immerse for 3–5 min.
10. Drain and air-dry for 3 hr.
11. Bake at 175°F for 6–8 hr.
12. Bake at 300°F for 4–6 hr.
13. Bake at 400°F for 4–6 hr.

B. Armature, Field, and Magnet Coils.

1. Prebake at 300–400°F for 3–4 hr.
2. Cool to 160–170°F.
3. Vacuum-impregnate or immerse for 10–20 min or until all bubbling stops.
4. Drain and air-dry for 1–3 hr.
5. Bake at 175°F for 2–4 hr.
6. Bake at 300°F for 3–6 hr.

For succeeding impregnations:

7. Immerse for 3–5 min.
8. Drain and air-dry for 1–3 hr.
9. Bake at 175°F for 2–4 hr.
10. Bake at 300°F for 2–6 hr.

For final impregnation:

11. Immerse for 3–5 min.
12. Drain and air-dry for 1–3 hr.
13. Bake at 175°F for 2–4 hr.
14. Bake at 400°F for 4–6 hr.

stocks can be fabricated into a variety of shapes by molding, extruding, laminating, calendering, and sheeting; or they can be dispersed in a solvent and applied by spraying, dipping, or brushing. The Silastic pastes are produced in a range of consistencies, without volatile solvent, for coating, laminating, calking and sealing, and embedding electrical coils. The Silastic tapes are used for winding around coils. The tapes are only partially vulcanized materials which are further vulcanized after application, thereby solidifying the tape with the elimination of voids at the tape edges. Detailed information regarding the types and specific applications for the Silastic materials is available from the manufacturer.

Silicone Fluids, DC 200 Series. These silicone fluids are available in five viscosities ranging from 0.65 to 250,000 centistokes at 25°C. They are used as hydraulic fluids, lubricators, mold-release agents, etc., and also find application in the coating industry as antifoam and antifloat agents, and in polishes. The percentage of liquid silicones to use

Table 2 * Silicon Chemicals

		Properties of Pure Compound					Grades Available	
Name and Type	Formula	Molecular Weight	Boiling Point, °C †	Specific Gravity at 25/25°C	Refractive Index at 25°C	Pounds per Gallon	Purified, %	Technical, %
Chlorosilanes, $RSiCl_3$								
Methyltrichlorosilane	CH_3SiCl_3	149.5	66.4	1.270	1.4085	10.57	99	95
Ethyltrichlorosilane	$C_2H_5SiCl_3$	163.5	99.5	1.236	1.4257	10.28	98	95
Dodecyltrichlorosilane	$C_{12}H_{25}SiCl_3$	303.7	288	1.026	1.4521	8.54	95	90
Hexadecyltrichlorosilane	$C_{16}H_{33}SiCl_3$	359.9	269 ‡	0.996	1.4568	8.29	95	90
Octadecyltrichlorosilane	$C_{18}H_{37}SiCl_3$	387.9	291 ‡	0.984	1.4583	8.19	95	90
Phenyltrichlorosilane	$C_6H_5SiCl_3$	211.5	201	1.321	1.5240	10.99	98	93
Chlorosilanes, R_2SiCl_2								
Dimethyldichlorosilane	$(CH_3)_2SiCl_2$	129.0	70.3	1.073	1.4023	8.93	99	97
Diethyldichlorosilane	$(C_2H_5)_2SiCl_2$	157.1	130.4	1.053	1.4309	8.76	98	93
Diphenyldichlorosilane	$(C_6H_5)_2SiCl_2$	253.2	304	1.218	1.5773	10.13	95	90
Chlorosilanes, R_3SiCl								
Trimethylchlorosilane	$(CH_3)_3SiCl$	108.6	57.0	0.854	1.3893	7.11	99	95
Triphenylchlorosilane	$(C_6H_5)_3SiCl$	294.8	378	Solid §	......	.....	95	90
Ethoxysilanes								
Methyltriethoxysilane	$CH_3Si(OC_2H_5)_3$	178.3	143.5	0.885	1.3820	7.36	99	95
Dimethyldiethoxysilane	$(CH_3)_2Si(OC_2H_5)_2$	148.2	113.5	0.837	1.3775	6.96	99	95
Trimethylethoxysilane	$(CH_3)_3SiOC_2H_5$	118.2	75.7	0.757	1.3743	6.30	99	97
Tetraethoxysilane	$(C_2H_5O)_4Si$	208.3	168.1	0.936	1.3832	7.79	99	95
Siloxanes								
Hexamethyldisiloxane	$(CH_3)_3SiOSi(CH_3)_3$	162.2	99.5	0.751	1.3750	6.24	99	97
Silanes								
Tetramethylsilane	$(CH_3)_4Si$	88.2	26.9	0.632	1.3556	5.26	99	95
Tetracresoxysilane	$(C_7H_7O)_4Si$	456.6	425	1.118	1.5574	9.30	98	‖
			Melting Point, °C					
Silanols								
Diphenylsilanediol	$(C_6H_5)_2Si(OH)_2$	216.3	160	Solid	......	.....	99	97
Triphenylsilanol	$(C_6H_5)_3SiOH$	276.4	151	Solid	......	.....	98	90

* Courtesy of Dow Corning Corp.
† Boiling point at 760 mm of mercury.
‡ Boiling point at 100 mm of mercury.
§ Melting point of triphenylchlorosilane, 96°C
‖ 75% distillable tetracresoxysilane; balance, cresyl polysilicates.

for antifoam and antifloat purposes is quite small and somewhat critical. For example, it has been reported (12) that 1 part of silicone per million of hot oil is adequate to prevent foaming. When a considerable excess is used there is a tendency for films of the oil or resin to form pinholes or small craters in their surfaces. A similar situation exists when these fluids are used as antifloat agents; an excess may make the condition worse (10, 11, 12). This phenomenon will be discussed in Volume II on pigmented finishes, but a detailed report by the New York Paint & Varnish Production Club is available (11). This report states that the methyl silicones are more effective antifloat agents than the ethyl, diethyl, or phenyl silicones.

A special DC Anti-Foam A is available for use in both aqueous and non-aqueous systems.

An interesting application for silicone fluids is in automobile polishes. This is disclosed in U. S. Patent 2,523,281, issued September 26, 1950, to Chester C. Currie, and assigned to the Dow Corning Corp. Further details regarding these polishes and the cleaner-polish emulsions are given on page 589.

Table 2 shows a comprehensive list of silicon chemicals produced by the Dow Corning Corp. These are used in the manufacture of silicone oils, rubber, and resins as described previously and also on page 588.

Table 3 shows some of the silicone products manufactured by the General Electric Co.

General Electric Silicone Resin SR-32. This resin is specially designed for coating applications on glass cloth where flexibility, heat life, and electrical properties are important. Insulating tapes made from silicone-impregnated glass cloth have found wide utility in the manufacture of certain electrical equipment. A typical curve schedule is 1 hr at 250°C.

Silicone Resin SR-61. SR-61 is a relatively hard resin, having good heat and chemical resistance and exterior durability. It is compatible with the Aroclors and Acryloid F-10 and has limited compatibility with drying oils, ethyl cellulose, and Resimene 877. It is not compatible with alkyds, varnishes, urea- and most melamine-formaldehyde resins. SR-61 may be used with many pigments to produce enamels which withstand heat tests at 500°F for 1000 hr without significant change. Zinc octoate is the preferred catalyst to increase film hardness and maintain good package stability. Zinc naphthenate may be used, but it gives slightly darker color. Iron octoate or naphthenate is also satisfactory except for dark color. The usual amount is 0.5% zinc as metal to resin solids.

Table 3 * Silicone Products, General Electric Co.

Silicone Product	Solids, %	Solvent	Viscosity at 25°C, cp	Pounds per Gallon
Resins				
SR-32	60	Toluene	150–225	8.7
SR-61	60	Pet. spirits		8.45
SR-82	60	Xylene	25–40	8.8
SR-98	50	Xylene	75–150	8.3
Elastomers				
SE-76	100	Gum polymer		
SE-450	100	Rubber compound		
SE-550	100	Low-temperature rubber compound		
SE-972	100	Wire-covering rubber compound		
SS-07	100	Rubber-coating paste		
SS-15	100	Rubber-coating paste		
Fluids				
SF-96	100	Water-white, non-volatile liquids of various viscosities		
SF-69	100	Antifloat fluid		
Emulsions				
SM-61	35	Water emulsion of silicone fluid		
Chemicals				
SC-50	20	Sodium methylsiliconate solution		
SC-01		Trimethylchlorosilane		
SC-02		Dimethyldichlorosilane		
SC-03		Methyltrichlorosilane		
Specialties				
SS-24		Antifoam		
SS-91		Bouncing putty		

* Courtesy of General Electric Co.. Silicone Products Dept.

SR-61 coatings should be baked to develop maximum properties; when catalyst is used the following schedules are suggested:

20–30 min at 500°F
50–60 min at 400°F
80–90 min at 350°F
120–180 min at 300°F

Details regarding the colored pigments to use with SR-61 are given in Volume II. A typical formulation for high-heat-resistant aluminum paint is given in Formula 1.

Formula 1 High-Heat-Resistant Aluminum Paint

	Per Cent by Weight
Standard aluminum paste	22.5
GE SR-61 (60% solids)	31.2
Aroclor (50% solids)	37.4
Zinc octoate (8%)	1.2
Petroleum spirits	7.7
	100.0

The above paint may be brushed or sprayed on boiler stacks, etc., which are subjected to temperatures as high as 1000°F. Under such conditions the Aroclor volatilizes from the film, and the silicone slowly decomposes, leaving the aluminum as a durable coating on the metal.

Silicone Resin SR-82. This resin was designed to have good compatibility with alkyds, urea- and melamine-formaldehyde resins, chlorinated biphenyls, and ethyl cellulose. Combinations with alkyds not only lower the cost, but also produce improved adhesion and resistance to certain solvents. Such combinations do not have the extreme heat resistance of straight silicone finishes, but they have been shown to be superior to alkyd-melamine combinations. The SR-82 offers many possibilities for up-grading the standard baking finishes, particularly with respect to heat and water resistance.

Silicone Resin SR-98. This resin is suitable for impregnation and bonding applications, particularly in electrical equipment. It has outstanding mechanical strength at elevated temperatures. A typical cure schedule is 1–2 hr at 250°C. Longer cures are required for thick, complicated structures.

Silicone Gums and Elastomers. SE-76 silicone gum is a clear, colorless polymer which can be compounded into a wide variety of silicone rubber compounds. Reinforcing fillers and vulcanizing catalysts are mixed with SE-76 on conventional rubber milling equipment.

Silicone rubber compounds may be molded, extruded, calendered, and bonded by techniques similar to those used with natural and synthetic rubber compounds.

Temperature resistance to 500°F, flexibility to -120°F, excellent compression-set characteristics, and good chemical inertness have contributed to a wide range of applications in the aircraft, automotive, and electrical industries.

Silicone Fluids SF-96 and SF-69. SF-96 fluids are widely used in furniture and automobile polish preparations. Polishes containing these fluids are usually characterized by high gloss, long life, and easy rub-out. The fluids also find application as capacitor and transformer dielectrics, as damping fluids, and as mold-release agents.

SF-69 is used in small quantities as an antifloat agent in paints.

Silicone Emulsions. SM-61 and similar emulsions of silicone oil in water are diluted and used as mold-release agents in the rubber industry and the shell-molding foundry process.

Sodium Methyl Siliconate, SC-50. This material is water soluble and is believed to have structure *A* in water solution and structure *B* after drying and dehydrating.

$$(A)\quad \left[CH_3{-}\overset{\displaystyle OH}{\underset{\displaystyle OH}{\overset{|}{\underset{|}{Si}}}}{-}O{-} \right]^- Na^+ \qquad (B)\quad CH_3SiO_2Na$$

The SC-50 is available in approximately 50% solution in water which is miscible with water, methanol, and ethanol in all proportions. However, the methanol solutions of SC-50 will gel on aging, whereas the ethanol solutions are stable indefinitely. The SC-50 is strongly basic, having a *p*H of 13. Addition of acids or acidic salts will precipitate methyl siliconic acid. The SC-50 reacts with soluble calcium salts to produce the insoluble calcium methyl siliconate. Dilute solutions of SC-50 react with salts of metals such as iron or copper to produce soluble metallomethyl siliconates.

The SC-50 has possibilities as a water-repellent agent for masonry, textiles, and paper. It may also be used in mercerizing and xanthate solutions to add water repellence to fibers. The copper methyl siliconate imparts mildew resistance as well as water repellence to textiles.

Silicone Specialties: SS-24 and SS-91. Silicone SS-24 is used as an antifoam agent in aqueous and non-aqueous media. The addition of very small amounts reduces the foam which develops in processing alkyd resins, phenolic resins, etc. The reduction of foam frequently permits increased loading in the kettle and consequent greater production per unit of apparatus. Some paints develop considerable foam or air bubbles during brush application which may produce irregularities in the surface when the bubbles break. Addition of the antifoam corrects this condition. The minimum amount should be determined for each requirement to avoid using an excess which may cause difficulties. Silicone SS-24 is not suitable for addition to food products.

SS-91 bouncing putty is an unusual silicone material which resists sudden shocks (bounces) and yet is putty-like and subject to slow cold-flow.

Table 4 shows some of the silicone products, as manufactured by the Linde Air Products Co., and their properties.

Applications for Products in Table 4. The organo-substituted chlorosilanes and ethoxysilanes may be hydrolyzed and condensed to form silicone polymers, as described previously. Both homopolymers and copolymers may be produced, depending on requirements. A typical laboratory procedure for making a vinyl silicone polymer is given by Wagner *et al.* (5).

Table 4 Silicone Products, Linde Air Products Co.

Silicone Product	Molecular Weight	Boiling Range, °C	Flash Point, °F	Density at 20°C
Chlorosilane monomers				
Diethyldichlorosilane	157.1	126–130	70	1.05
Ethyltrichlorosilane	163.5	97–98	80	1.24
Amyltrichlorosilane	205.6	160–164	155	1.15
Phenyltrichlorosilane	211.5	197–201	195	1.32
Vinyltrichlorosilane	161.5	88–91	70	1.28
Ethoxysilane monomers				
Ethyltriethoxysilane	192.3	155–160	130	0.895
Amyltriethoxysilane	234.4	195–210	185	0.889
Vinyltriethoxysilane	190.0	157–159	130	0.90

	Vis. at 25°C, centistokes	Flash Point, °F	Spec. Gravity	Refractive Index
Silicone oils				
Ethyl silicone oil L-41	1000	430	0.982	1.447
Ethyl silicone oil L-41	5000	430	0.985	1.449
Methyl silicone oil L-45	200	550	0.972	1.403
Methyl silicone oil L-45	350	550	0.972	1.403
Methyl silicone oil L-45	500	550	0.972	1.403

Silicone oil emulsions LE-81 and LE-85. See page 591.
Linde sizing GS-1 for glass fabrics. See page 591.

Formula 2 Vinyl Silicone Polymer

(*A*)	Ether	300 cc
	Ice	2000 gm
(*B*)	Vinyltrichlorosilane	250 cc
	Ether	250 cc
Conc. NH_4OH		3 cc

Procedure

Place *A* in a 4-liter beaker, and add *B* rapidly with agitation. Allow to stand for 10 min, separate aqueous layer, and extract with ether. Wash ether extracts with water, shake with 3 cc conc. NH_4OH, wash again with water, and dry over anhydrous calcium sulfate. Remove about 90% of the ether by distillation not above 45°C. Remove residual ether under reduced pressure and below 60°C. The final vinyl polysilicone is a brittle solid with a softening point of 87°C.

Variation in the amount of NH_4OH gave polymers of different softening point and hydroxyl content. Increasing the NH_4OH produced harder polymers, having lower hydroxyl content.

Wagner (5) also gives results obtained by using a series of amines as catalysts instead of the NH_4OH. It was found (5) that vinylpolysiloxanes and vinyltriethoxysilane were readily polymerized through the vinyl group with peroxide catalysts and heat. A variety of products was obtained, ranging from viscous liquids to brittle solids. Wagner (5) found that vinyltrichlorosilane did not react with maleic anhydride or ethyl maleate, but vinyltriethoxysilane gave addition products, when peroxides were used as catalysts. The products were hard, brittle resins, having molecular weights up to 3000, and they were soluble in ether, dioxane, benzene, and toluene.

It will be apparent that the chlorosilane and ethoxysilane monomers in Table 4 may be hydrolyzed and condensed alone, or with other reactive compounds, to produce a variety of oils and resinous materials.

Linde Silicone Oils. The Linde ethyl and methyl silicone oils, L-41 and L-45 respectively, are linear polymers of the $M(D)_xM$ type described previously. They are used as special lubricants, mold-release agents, antifoam and antifloat agents, and water repellents, and in automobile and furniture polishes.

The polishes have been quite successful because they produce a high gloss, with a minimum of rubbing during application, and they are very water repellent. Best results are obtained on automobiles if the surface is cleaned with an emulsion-type cleaner-polish before applying the solvent-type polish. However, a combination cleaner and polish may be made by using higher silicone content. The solvent-type polishes are combinations of silicone oils and waxes dispersed in a solvent such as mineral spirits. The medium- to high-viscosity silicone oils are generally preferred. A wide variety of paraffin and natural and synthetic waxes may be used since many of them are compatible with the silicone oils and solvent. The general range of components and procedure for making silicone-wax polishes are illustrated in Formula 3.

The emulsion-type cleaner-polishes contain a mild abrasive such as fine-particle-size calcium carbonate or clay in addition to the silicone oil and wax. They must also contain a satisfactory emulsifier which will maintain a stable emulsion. A great variety of emulsifiers is available. Frequently, best results are obtained with the type which is formed during emulsification. This type is illustrated in Formula 4, using the oleic acid-morpholine combination as the emulsifier.

Here, again, selection may be made from a variety of waxes and abrasives to obtain specific results.

Formula 3 Silicone-Wax Polishes, Solvent Type

	Per Cent by Weight
Silicone oil	1.5–3.0
Wax	2.0–3.0
Mineral spirits	94–95

Procedure

Melt the silicone oil and wax in a small amount of the mineral spirits, add remainder of mineral spirits, and maintain temperature about 140°F until a clear solution is obtained. Avoid open flames because of fire hazard. Cool rapidly with vigorous agitation so as to obtain a fine dispersion of the wax in the silicone-mineral spirits combination. Further details regarding specific silicones and waxes are available from the manufacturers.

Formula 4 * Silicone Cleaner-Polish, Emulsion Type

		Per Cent by Weight
(*A*)	Linde silicone oil L-45 (350 centistokes)	2.0
	Crown Wax #23 †	1.6
	Oleic acid	2.0
	Mineral spirits	36.6
(*B*)	Snowfloss ‡	9.8
	Franklin Clay §	2.4
	Morpholine	0.9
	Water	44.7
		100.0

Procedure

Melt the silicone oil, wax, and oleic acid in a small amount of mineral spirits, add balance of mineral spirits, and maintain temperature about 140°F until a clear solution is obtained. In a separate container make a slurry of the Snowfloss and clay in the water-morpholine mixture, and heat to 140°F. Add *A* to *B* with vigorous agitation, and continue stirring until emulsion is cool.

* Courtesy of Linde Air Products Co.

† Petrolite Co., 30 Broad Street, New York City.

‡ Johns-Manville Corp., 22 East 40th St., New York City.

§ United Clay Mines Corp., 109 Oakland St., Trenton, N. J.

To meet the need for a single-application cleaner-polish compound the percentage of silicone oil may be increased to 4% or higher. However, a somewhat higher gloss would be obtained from a double application of a solvent-type silicone-wax polish over the emulsion-type cleaner.

Some difficulty has been experienced in refinishing automobiles which have been polished with a silicone-wax polish. The refinishing enamels or lacquers show poor adhesion and tend to "crawl," producing an uneven surface, even when the old finish has been cleaned with naphtha, toluene, or carbon tetrachloride. It appears to be quite difficult to remove the last traces of silicone compounds from the surface. An answer to this problem was suggested by Raynolds (12). He proposed adding 1% of a 1% solution of SF-69 in toluol to the refinishing enamel or lacquer. He found that this produced excellent adhesion to the old finish coat.

Silicone Oil Emulsions. The silicone oil emulsions LE-81 and LE-85 are oil-in-water type emulsions of ethyl silicone and methyl silicone oils respectively. They contain 35% by weight of silicone oil and have a consistency about equal to that of water. They are resistant to repeated cycles of freezing and thawing and will withstand several days at 125°F without "creaming" or separating. They can be diluted to 3% oil and retain excellent stability at room temperature, or for several days at 175°F.

The diluted emulsions are used as mold-release agents for rubber and plastics, and may be applied to the mold by brushing, spraying, or dipping. The emulsions are not satisfactory as antifoam or antifloat agents.

Linde Sizing GS-1. Linde sizing GS-1 is a 25% solution in toluene of a vinyl polysilicone resin. This resin has a vinyl group, $CH_2{=}CH{-}$, attached to each silicon atom in the polymer. The sizing has given outstanding results (4, 6) when applied to glass cloth used for glass cloth-polyester laminates. For application to the glass cloth GS-1 sizing is thinned with toluene to about 0.5% resin content, and triethanolamine added in the ratio of 1% based on the resin. The glass cloth is immersed in this solution, and the excess removed by squeeze rolls or other means. The coated cloth is then baked for 5–10 min at 530°F to cure the resin. About 0.25–0.3% of resin based on the glass is applied. Brown *et al.* (6) found that an optimum amount was 0.26%, which corresponded to about 30 layers of polymer molecules on the glass surface. The glass cloth-polyester laminates have very high structural strength combined with light weight and have found extensive use in military and industrial applications. They are made by laminating under relatively low heat and pressure a number of layers of glass cloth saturated with a polyester resin. If the bond between the glass cloth and the polyester resin is not strong, or is weakened by some means such as immersion in water, the strength of laminate will be reduced correspondingly.

It was found that application of silicone resin to the glass cloth improved its adhesion to the laminating resin. This was believed to be due to a strong van der Waals attraction, or perhaps actual chemical bonding between the —Si—O— groups and the glass fiber surface, and also the natural affinity of the organic radicals on the silicone resin for the polyester resin. Thus the silicone resin acted as a strong tie between the glass cloth and the polyester resin in the laminate. Brown *et al.* (6) showed that when ethyl silicone resin and other standard sizings were applied to glass cloth, in comparison with GS-1 sizing, there was a considerable loss in strength of all the laminates after they were immersed in water except those sized with GS-1 sizing. It is believed that the water permeated the twenty or thirty molecular layers of sizing material, thus weakening the bond between them. This did not happen with GS-1 sizing because the molecular layers were chemically bonded together through polymerization of the vinyl groups. It is also believed that the residual vinyl unsaturation in the layer adjacent to the laminating resin formed a copolymer with the unsaturation in the laminating resin. Therefore, in this system, all the molecular layers between the glass cloth and the laminating resin would be chemically bonded together, and also chemically bonded at the interfaces between them and the glass cloth and laminating resin. Since chemical bonds are stronger than the van der Waals forces the laminates made with the vinyl silicone sizing had greater initial dry strength, and a lower loss of strength in the water-immersion test. Values obtained (4, 6) showed that the laminates with the vinyl silicone sizing were actually stronger when wet than the older commercial standard laminates were when they were dry.

Table 5 shows the physical properties of three silicone resins manufactured by Plaskon Division, Libbey-Owens Ford Glass Co.

Table 5 Silicone Products, Plaskon Division, Libbey-Owens Ford Glass Co.

Silicon Resin	Solids, %	Solvent	Viscosity, Gardner-Holdt	Color, Gardner 1933	Pounds per Gallon
ST-856	55	Xylene	U–X	2–4	8.75
ST-873	60	Min. spirits	T–W	5–9	8.0
ST-881	60	Xylene	A–E	8–10	8.7

The three silicone products listed in Table 5 are co-condensed silicone-alkyd resin combinations. As discussed earlier in this chapter it is theoretically possible to co-condense a siloxane having free hydroxyl groups with another resin, also having free hydroxyls, to obtain a chemically combined product. Such products should have better drying or curing rates and better film properties than physical blends of silicone and the other resin. Since alkyd resins may be readily prepared with free hydroxyls, and alkyds also have the flexibility and adhesion which are lacking in the silicones, the silicone-alkyd combination is a logical development. Silicone combinations with nonoxidizing alkyds would require baking, but in combinaton with oxidizing alkyds the product may be air-dried. Of course, air-drying resins may also be baked to reduce the drying time or to develop greater resistance of the films to solvents and chemicals.

Plaskon Silicone Resin ST-856. Resin ST-856 is a baking resin for industrial finishes requiring good heat resistance and exterior durability. It is suitable for finishes on space heaters, stove parts, engines, hospital equipment, exterior metal signs, etc. The films have good adhesion in one-coat finishes on metal. The suggested baking schedule for pigmented finishes is 400°F for 30 min. However, these finishes may be baked as low as 325°F for 100 min or as high as 450°F for 8 min. Pigmented finishes which were properly formulated and applied maintained high gloss and film integrity after exposure at 450°F for more than 300 hr.

Silicone Resin ST-873. Resin ST-873 is an air-drying product which may be used in coatings for outdoor exposure on wood or metal surfaces. The higher cost of the silicone-alkyd resins over straight alkyd resins for such applications may be partly offset by reduced maintenance expenses due to greater durability and improved appearance on aging. The gloss retention of this type of resin is reported to be considerably better than that of the long-oil-type alkyds normally used for such applications. Enamels made with Resin ST-873 dry in 4–6 hr and are hard overnight. Lead and manganese driers may be used if suitable antioxidants are also included, but such additions should be checked for adequate package stability. The air-drying properties are derived chiefly from the oxidizing alkyd component.

Silicone Resin ST-881. Resin ST-881 may be air-dried or baked, but heating is necessary to develop resistance to solvents. The resin is intended specifically for high-heat-resistant aluminum coatings on boilers, stacks, mufflers, etc. Good exterior durability is obtained on metal surfaces subjected to a wide range of temperatures up to 1500°F. For coatings which are not subjected to high temperatures but are required to have good solvent resistance, a baking schedule at 400°F for 30 min is suggested. This schedule may be varied from 325°F for 100 min to 450°F for 8 min.

ETHYL SILICATE

Ethyl silicate has been used for a long time for adhesives, for impregnating porous materials, for weatherproofing stone, for lacquers which adhere to glass, and as a source of finely divided amorphous silica. It was among the first organic silicon compounds to be used commercially, but it should not be confused with the silicone materials. Its properties and uses are described by Cogan and Setterstrom (13), and reprints of this article are available from Carbide and Carbon Chemicals Co. The article describes three commercial types which are available:

1. Tetraethyl orthosilicate, $(C_2H_5O)_4Si$.
2. Condensed ethyl silicate, a mixture of tetraethyl orthosilicate and polysilicates.
3. Ethyl silicate #40, a mixture of ethyl polysilicates containing 40% silica.

Typical Molecule * of Ethyl Silicate #40

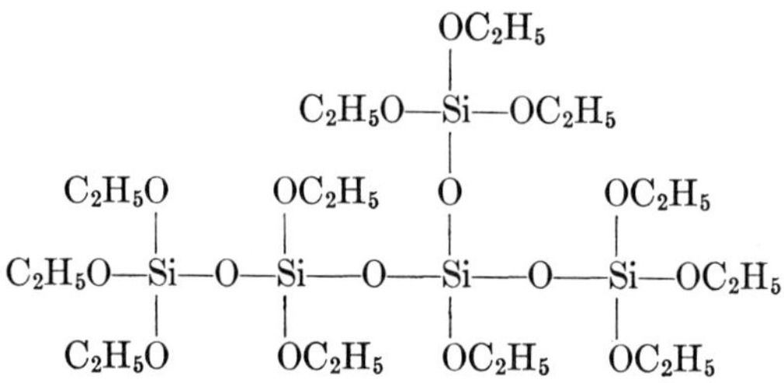

* Courtesy of Cogan and Setterstrom (13).

These ethyl silicate products are made by reacting silicon tetrachloride with ethyl alcohol under conditions which prevent excessive hydrolysis and polymerization. The ethyl silicates hydrolyze slowly in the presence of moisture to produce silicic acids, which undergo simultaneous dehydration and polymerization to give amorphous silica as the final product. Ethyl silicate is not soluble in water, but may be dissolved in ethanol, isopropanol, or water mixtures of these alcohols. The hydrolysis of ethyl silicate is very slow in neutral solutions, but it may be catalyzed by acids or bases such as hydrochloric acid or ammonia. In this process the material passes through a sticky, colloidal stage, and then dries to a hard, vitreous material which is insoluble in water and resistant to high temperatures. For this reason it is used as a mold binder in which metals are cast, and as a bonding agent for refractory crucibles, electrical insulators, and other ceramic-like products.

Solutions of ethyl silicate can be prepared which will penetrate porous materials and then react to deposit adhesive silica, which acts as a bonding and protective agent. It has been used for this purpose on artistic and architectural stonework. There is some shrinkage on drying, so that complete waterproofing is not obtained by this process, but it does exert a weatherproofing action. Satisfactory films of ethyl silicate alone have not been produced because of brittleness and lack of adhesion. Pigmented ethyl silicate films have better continuity and adhesion, particularly if a laminar pigment such as mica is used. Such paint coatings have good heat and chemical resistance and have been investigated for fire-retardant paints. Combinations of ethyl silicate and various polymers, such as nitrocellulose, vinyl resins, and urea-formaldehyde resins, have been used for strongly adherent coatings on glass. A variety of other uses for the ethyl silicates is given by Cogan and Setterstrom in reference 13. These authors state that the development of a satisfactory plasticizer for the ethyl silicates would increase their applications considerably.

TITANIUM ESTERS

The development of titanium esters for use in protective coatings is analogous in some respect to the development of silicon esters and the silicones. Titanium belongs to the same group in the periodic system of the elements as carbon and silicon; therefore it should be possible to produce organotitanium compounds. An incentive for such investigation was provided by the large quantity of titanium tetrachloride which became available in Australia after World War II. During the war this compound was used to produce smoke screens. In Australia, the Division of Industrial Chemistry of the Commonwealth Scientific and Industrial Research Organization, together with the Defense Research Laboratories, issued a report entitled "Preparation and Utilization of Esters of Titanium" (14). A review of the position of these esters as of January, 1951, is given by Winter (15–17). At the dedication of the expansion of the du Pont Experimental Station at Wilmington, Del., May, 1951, it was disclosed that the development of titanium esters for high-heat-resistant paints was under way. It was indicated that there were problems in obtaining satisfactory adhesion of the paints to metal surfaces, and in developing their resistance to attack by salt water.

Winter (15) reports that the butyl ester has received most attention because of performance and economic advantages. It may be prepared by several methods, one of which is to react titanium tetrachloride with butyl alcohol in the presence of ammonia:

$$TiCl_4 + {}_4C_4H_9OH \xrightarrow{NH_3} Ti(OC_4H_9)_4 + NH_4Cl$$

Both monomeric and polymeric esters may be prepared. The monomeric ester is hydrolyzed readily by atmospheric moisture, but the polymer is much more stable. Hydrolysis is followed by condensation which, when carried to the maximum, results in the production of amorphous titanium dioxide, TiO_2. In some respects this procedure is similar to that described for ethyl silicate in the preceding section. Neither butyl titanate nor ethyl silicate is a true film-former since its coating becomes brittle and does not adhere to smooth surfaces. However, when these materials are suitably pigmented, coatings are formed on metals and porous surfaces. Aluminum paints were formulated (14) with butyl titanate and aluminum powder for use on boilers, smoke stacks, etc., which had excellent heat resistance and adhesion to steel. The original paints accelerated the corrosion of steel under humid conditions. Investigation showed that this was due in part to residual combined chlorine in the ester as an impurity. However, even with carefully purified esters which contained only traces of chlorine the corrosion was not eliminated entirely. Paints were then made with butyl titanate and standard anticorrosive pigments. Better results were obtained with the polymeric butyl titanate than with the monomeric material. The paints containing red lead and zinc chromate were powdery and lacked adhesion. Paints containing red iron oxide, titanium dioxide, and chromium oxide did not give protection against corrosion. However, paints containing zinc dust gave good protection under humid conditions and were very heat-resistant. Later work showed that metallic zinc leaf gave better results than zinc dust because thicker coats could be applied with the zinc leaf. The tendency of the zinc-dust paint to settle hard in the container was overcome by addition of a small amount of high-viscosity-grade ethyl cellulose. A typical formulation given by Winter (15) is shown in Formula 5.

A suggested system for a heat-resistant, protective coating on steel consists of two coats of butyl titanate zinc-dust paint, followed by two coats of butyl titanate aluminum paint. Sand blasting is considered to be the best method for preparing the steel surface for painting. The paints may be applied by brushing or spraying, but

Formula 5 * Butyl Titanate, Zinc-Dust Paint

	Per Cent by Weight
Zinc dust	56.9
Butyl titanate (polymer)	20.4
Ethyl cellulose (high vis.)	0.6
Mineral turpentine	22.1
	100.0

* Courtesy of G. J. Winter (15).

care should be taken to apply thin coats. Each coat should air-dry for 15–20 min and then should be heated to 600–900°F. These coatings will withstand exposure to 1300°F for very long periods. They may be applied directly to steel at about 600°F if the mineral turpentine is replaced with a light mineral oil.

Some interesting possibilities for butyl titanate were reported by Kraitzer *et al* (14). He found that butyl titanate could be incorporated into a drying oil by reacting it with the monoglycerides of linseed oil. It could also be incorporated into alkyd resins by reacting drying oil monoglycerides and phthalic anhydride, then adding butyl titanate, and finishing the reaction. A titanated castor oil was prepared by the ester-interchange method. This was compatible with nitrocellulose and was suggested as a possible plasticizer for lacquers. It is apparent that the titanium esters offer some interesting possibilities in coatings, if their economic and performance characteristics justify continued development.

BIBLIOGRAPHY

1. E. G. Rochow, *Chemistry of the Silicones,* second ed., John Wiley & Sons, 1951.
2. G. S. Cook and D. W. Kennedy, *Official Digest Federation Paint & Varnish Production Clubs, No. 321,* 629 (October, 1951).
3. Wilcock, *Gen. Elec. Rev., 49,* No. 11, 14 (1946).
4. M. H. Jellinek, Preprint of 7th Annual Technical Session, Society of Plastics Industry, Inc., April 9–11, 1952.
5. G. H. Wagner, D. L. Bailey, A. N. Pines, and M. L. Dunham, Paper presented at Atlantic City Meeting, American Chemical Society, Sept. 14–19, 1952.
6. D. B. Brown, S. Camialo, and M. H. Jellinek, Paper presented at Atlantic City Meeting, American Chemical Society, Sept. 14–19, 1952.
7. D. T. Hurd, *J. Am. Chem. Soc., 67,* 1813 (1945).
8. D. T. Hurd and G. F. Roedel, *Ind. Eng. Chem., 11,* 2078 (1948).
9. K. E. Polmenteer, P. C. Servais, and G. M. Konkle, *Ind. Eng. Chem., 44,* 1576 (1952).

10. I. W. Hutchison, *Am. Paint J.*, p. 80, Feb. 5, 1951; p. 62, Feb. 12, 1951.
11. New York Paint & Varnish Production Club, *Official Digest Federation Paint & Varnish Production Clubs, No. 323,* 867 (December, 1951).
12. J. W. Raynolds, *Official Digest Federation Paint & Varnish Production Clubs, No. 310,* 896 (November, 1950).
13. H. D. Cogan and C. A. Setterstrom, *Ind. Eng. Chem., 39,* 1364 (1947).
14. I. Kraitzer, K. McTaggart, and G. Winter, *J. Oil & Colour Chemists' Assoc., 31,* No. 340 (October, 1948).
15. G. Winter, *J. Oil & Colour Chemists' Assoc., 34,* No. 367 (January, 1951).
16. G. Winter, *Paint, Oil, Chem. Rev., 12* (June 21, 1951).
17. G. Winter, *Can. Paint & Varnish Mag., 14* (December, 1951).
18. T. G. Rochow and F. G. Rowe, *Anal. Chem., 21,* 461 (1949).
19. T. G. Rochow and E. G. Rochow, *Science,* No. 2881, 3, 271 (1950).
20. T. G. Rochow and R. L. Gilbert, in *Protective and Decorative Coatings,* J. J. Mattiello, ed., Volume V, Chapter 5, John Wiley & Sons, 1946.
21. F. J. Norton, *Gen. Elec. Rev., 47,* No. 8, 6 (1944).

Set the apertures of the mind wide open to admit the beauties of nature, the noble thoughts of mankind, and the promptings of the spirit, but close them to the trivial and the sordid, for we build our lives according to that which we take in, and we reflect that which we admit.

Chapter 15. TEST METHODS

INTRODUCTION

There are three general purposes for using testing procedures in the coating industry, namely:

1. To determine whether incoming raw materials meet established specifications.
2. To determine whether the coating material is being processed properly during manufacture.
3. To determine whether the finished coating material meets the established specifications and application requirements.

A great many testing procedures have been developed to meet the needs of these three general purposes. At present, there is a considerable lack of uniformity in the procedures used by various companies and organizations. Attempts are being made to establish standard test methods for the coatings industry, and it may be expected that eventually the industry will have its own booklet on test methods. The Federation of Paint & Varnish Production Clubs has made progress in this direction. It has established a number of Federation Standard Methods of Test to date, and it has worked with the American Society for Testing Materials (ASTM) to establish joint Federation-ASTM Standard Methods of Test. These Standard Methods may be found in the *Official Digest* of the Federation, and in the ASTM Standards for Paint, Varnish, Lacquer, and Related Products under Committee D-1.

The American Oil Chemists' Society (AOCS) has set up a very comprehensive series of test methods, available in loose-leaf booklet form, for testing raw materials, finished oils and fats, and the by-products of the industry. The book by Gardner and Sward, *Physical and Chemical Examination of Paints, Varnishes, Lacquers, and Colors,* published by the Henry A. Gardner Laboratory, Inc., Bethesda, Md., is frequently referred to as the "bible" of the industry because of its extensive coverage of testing procedures. The federal and other governmental and military agencies have established specifications and testing procedures to cover the wide variety of coating materials purchased by them. Many of these tests are included in Federal Specification TT-P-141b, 15 January 1949. Copies of Federal Specifications and their cost may be obtained from the Superintendent of Documents, Government Printing Office, Washington 5, D. C. Volume V of *Protective and Decorative Coatings,* J. J. Mattiello, editor, John Wiley & Sons, 1946, is devoted to the analysis and testing methods for oils, resins, metal finishes, and some pigments. A review of analyses and test procedures for coatings is given by Rochow and Stafford, *Anal. Chem., 22,* 206 (February, 1950). This review appears periodically. Many raw-material suppliers and large paint companies have developed special test procedures to assure uniformity in their products and compliance with specifications.

In view of the wide variety of test procedures and their availability from many sources, it would be impractical to describe all of them in detail in this book. However, it was believed advisable to outline the tests for various materials and to indicate their purpose and significance. The discussion in this chapter will be restricted to tests for oils, resins, high polymers, etc., covered in this book. The tests for pigments and pigmented coatings will be given in Volume II. The tests are arranged alphabetically in Table 1 for ease of location. However, they are discussed in groups which are related by procedure, materials, or other factors.

It is extremely important when testing commercial products to be sure that the sample to be tested is completely representative of the batch of material. Good sampling techniques should be established because, if the sample does not represent the batch, all test results will be valueless. It is also important that clean sample containers be used to avoid contamination of the samples. Tightly closed containers are always essential for hot samples which contain volatile solvents. When testing the application and film properties of clear coatings, several factors must be considered carefully. The wood or metal surface to which the coating is to be applied must be thoroughly cleaned and prepared according to a standard prescribed method. The material must be applied carefully so that a uniform coating is deposited on the surface to be coated. The actual thickness of the dried coating should be determined by one of the several methods which are available. Since each of a series of materials under test may not give the same thickness per coat even when applied in

Table 1 Alphabetical List of Test Methods

PAGE		PAGE	
648	Abrasion Resistance	614	Iodine Number
624	Acetone Number	646	Kauri Reduction Test
624	Acetyl Value	645	Mandrel Test
607	Acid Number	602	Non-Volatile Content
649	Adhesion	654	Permeability to Water
656	Alcohol Resistance	643	Print Test
655	Alkali Resistance	611	Refractive Index
631	Application Methods	622	Saponification Number
613	Break Test	638	Skinning
647	Cold-Check Test	655	Soap Resistance
607	Color	626	Softening Point
617	Diene Number	602	Solid Content
640	Drying Time	608	Specific Gravity
657	Electrical Resistance	617	Spectroscopic Analysis
652	Exterior Durability	644	Tensile Strength
634	Film Thickness	634	Thickness Measurement
644	Flexibility	616	Thiocyanogen Number
639	Gasproofness	625	Titer Test
611	Gel Test	623	Unsaponifiable Matter
642	Hardness	603	Viscosity
617	Hexabromide Test	602	Volatile Content
624	Hydroxyl Number	653	Water Resistance
647	Impact Resistance	608	Weight per Gallon

the same manner, it is necessary to measure the final film thickness instead of assuming that all coatings in the series will be the same thickness because an equal number of coats of each material were applied. The results from tests may be affected greatly by the temperature and relative humidity under which the tests are made. Most large manufacturers maintain constant-temperature and humidity rooms in which the tests are performed. The standard conditions in the paint industry for these rooms are 77°F or 25°C as the temperature, and 50% relative humidity. When such rooms are not available the prevailing temperature and humidity should be reported with the test results.

TESTS FOR VARNISHES AND RESIN SOLUTIONS

It may be well to start this discussion of test methods with the group of tests usually applied to varnishes and resin solutions to determine their compliance with standard physical specifications. The tests in this group are:

% Solids.
Viscosity.
Color.
Acid number.
Pounds per gallon.

% Solids, or Non-Volatile Content

References

ASTM Designation D154-47.
Federal Specification TT-P-141b, Method 404.1.

Most varnishes and synthetic resins are marketed as solutions in a volatile solvent. It is important for cost and formulation purposes that the solid content of these solutions remain uniform from batch to batch. The solid content may be determined by weighing a small quantity into a shallow dish and heating it in an oven to drive off the volatile solvent. Subsequent weighing will give the weight of solid material in the original sample, from which the percentage may be calculated.

There are several precautions which must be taken in order to obtain uniform results with this apparently simple test. The weight of the sample taken for test should be related to the area of the shallow dish so that the layer will not be too thick. Thick layers tend to dry on the surface, or "skin over," thereby trapping some of the volatile material. Some methods include a piece of stiff wire in the shallow dish which can be used to break a skin if it forms. Other methods suggest a layer of clean sand in the bottom of the dish to provide a greater surface area, which will give a thinner film with less tendency to entrap the solvent. The most widely used "dish" is the metal cover of a friction-type, ½-pint paint can. Of course, glass petri dishes could be used but these would require cleaning, whereas the can covers are discarded after use.

The temperature of drying in most methods is 105°C ± 2°, or 221°F ± 4°, and the time of test is 3 hr. If the sample contains a high percentage of oxidizable material it was found (5) that a sufficient weight of oxygen was absorbed under these conditions to affect the results. This applied particularly to long-oil alkyd resins during World War II when the phthalic anhydride content was held within very strict limits. The increase in weight from the absorbed oxygen gave a high value for the solid content, resulting in an unduly low value for the phthalic anhydride when calculated on this basis. To overcome this condition it was recommended that the temperature be maintained at 105°C, but the time reduced to 2 hr instead of 3 hr, and the drying be done in a vacuum oven instead of the normal air-circulating oven.

The weight of the sample of varnishes or alkyds should be between 1 and 2 gm, and the sample should be weighed by difference from a weighing bottle. In the determination of the solid content of urea- or melamine-formaldehyde resins, the film thickness and the drying time and temperature are very critical. These resins condense when heated; therefore, if the temperature is too high, or the heating time too long, there will be a loss of solids through the condensation reaction. The film thickness is also a factor in this situation. As a result, specific conditions for the test have been established by the resin manufacturers. These include special aluminum dishes having perfectly flat bottoms, and these should be maintained in a level position in the drying oven. The sample should be weighed into the dish from a weighing bottle and should contain 0.5 ± 0.05 gm of resin solids. The sample in the dish should be diluted with 5 cc of xylene, then placed in an air-circulating oven in a level position, and maintained at 105°C for 2 hr. Subsequent cooling and weighing give the solid content.

Viscosity

References

ASTM Designation D154-47.
ASTM Designation D301-48.
Federal Specification TT-P-141b, Method 427.1.

There is an important relationship between solid content and viscosity of varnishes and resin solutions. If the varnish or resin has been overpolymerized during processing, the viscosity will be above normal when thinned to the standard solid content. Conversely, the solid content will be below normal if the solution is thinned to the standard viscosity. The combination of the two tests serves to check this situation. Since varnishes, oils, and resin solutions are Newtonian liquids, it is correct to use the term "viscosity" for their rate of flow under specified conditions. We shall see in Volume II that the term "consistency" is preferred for pigmented coatings, since they are not Newtonian in their flow behavior, but show various degrees of plasticity, thixotropy, etc.

The viscosity of clear coatings is usually expressed in poises or centipoises, or in the letters of the standard tubes of the Gardner bubble-tube method. The conversion of values from poises to Gardner tubes, and vice versa, will be found in the following table.

Viscosity Conversion Table

Standard Poise	Gardner-Holdt Tubes	Seconds, #4 Ford Cup	Seconds, Gardner-Holdt	Lithographic Ink Standards	
				Mattiello	Gardner
0.50	A	20			
0.65	B	26			
0.85	C	34			
1.00	D	40	1.46		
1.25	E	46	1.83		
1.40	F	51	2.05		
1.65	G	57	2.42		
1.80	G–H	60	2.64		000
2.00	H	65	2.93	00000	
2.25	I	75	3.30		
2.50	J	85	3.67		
2.75	K	96	4.03		
3.00	L	108	4.40	00000	
3.20	M	117	4.70		
3.40	N	123	5.00		
3.70	O	127	5.40		
4.00	P	131	5.80		
4.35	Q	137	6.40	000	
4.70	R	144	6.90		
4.80	R+	147	7.03		00
5.00	S	154	7.30		
5.50	T	166	8.10		
6.27	U		9.20		
8.00	U–V		11.60		0
8.84	V		13.00		
10.70	W		15.70	00	
12.90	X		18.90		
14.40	X+		21.10	0	1
17.60	Y		25.80		
22.70	Z		33.30		
23.50	Z+		35.00		2
27.00	Z1		39.60	1	
34.00	Z2		49.85	2	
36.20	Z2+		53.10		
46.30	Z3		67.90		
62.00	Z4		91.00	3	4
63.40	Z4+		93.00		
98.50	Z5		144.50		
120.00	Z5+		176.41		5
148.00	Z6		217.10	5	
200.00				6	6

Occasionally, the viscosity is expressed in stokes or centistokes, instead of poises. This value corrects for the difference in densities of the samples being tested. When the viscosity is expressed in stokes it is known as the kinematic viscosity (see ASTM Designation D44). The viscosity in stokes may be obtained by dividing the viscosity in poises by the density of the sample. Since viscosity is a logarithmic function of the temperature, it will be realized that extreme care must be taken to control the temperature of the sample being tested. The standard temperature at which the majority of samples is measured is 25°C. Higher temperatures may be used for viscous oils which have no volatile content.

There are four general methods for measuring the viscosity of clear liquids:

1. Bubble-tube method.
2. Falling-ball method.
3. Efflux cup method.
4. Torsional method.

The bubble-tube method is the one most widely used for oils, varnishes, and resin solutions. The standard tubes and various accessories are described by Gardner (1). The sample is placed in a special glass tube having a length of 112 mm and a diameter of 10.75 mm. The tube is filled to a designated mark near the top, and a cork inserted to another designated mark. Therefore the air space between the top of the liquid and the cork will be uniform in all tests and will control the size of the bubble. The tube containing the sample is brought to standard temperature, 25°C, and then inverted so that the bubble travels from one end of the tube to the other. The rate of rise of the bubble is a measure of the viscosity of the liquid. The viscosity may be expressed as the time in seconds for the bubble to travel from one end of the tube to the other. The rate of rise of the bubble also may be compared with that of the bubbles in several tubes containing liquids of known viscosities, and marked with designating letters. The viscosity of the sample may then be expressed as the letter on the standard tube having the nearest rate of rise. The viscosity in such letters may be converted into poises from the values in the table on page 604. The bubble-tube method has an accuracy of about 10%, provided the temperature is controlled carefully. This accuracy is sufficient for general production and laboratory work, and the method is very simple to operate.

The falling-ball method is used in the lacquer industry for measuring solutions of nitrocellulose (see ASTM Designation D301-48). The nitrocellulose solution of specified concentration is placed in a glass cylinder 14 in. in length and 1 in. in diameter. The cylinder is marked about 2 in. from each end to provide a distance of 10 in. between the marks. A 3/8-in. steel ball weighing 2.035 gm is allowed to fall through the solution in the tube, and the time in seconds to fall between the two marks is taken as the viscosity. The temperature must be controlled carefully to obtain reproducible results. The viscosities of various nitrocellulose solutions obtained by this method are given in Chapter 11.

A variety of efflux-type viscometers is described by Gardner and Sward (1). These instruments consist of variously shaped cups which have an orifice in the bottom. The cup is filled with the material to be tested and then emptied by allowing the material to run through the orifice. The time for efflux is taken as a measure of the viscosity. The Ford Cup is the one most widely used in the coatings industry, although some reports give viscosity in terms of the Parlin Cup and the ASTM Cup. The Zahn Cup is designed so that it may be dipped into a batch of hot varnish or oil, thereby measuring the viscosity during processing. The weak point in the efflux-type method is the lack of good temperature control. The bubble-tube method is fully as accurate and just as simple to operate. However, the efflux-type viscometer, such as the Ford Cup, is very useful for coatings having low pigment concentration and is discussed in more detail in Volume II.

The torsional-type viscometer consists of a cup in which a disc or a cylinder is suspended on a wire. The viscosity of the liquid placed in the cup is indicated by a device which measures the angle through which the wire is twisted when the cup or the disc is rotated. The three principal instruments of this type are the MacMichael, the Stormer, and the Green (Interchemical) Rotational viscometers. The MacMichael may be used for measuring the viscosity of heavy-bodied oils because it has better temperature control, particularly at elevated temperatures, than the Gardner tubes. The Stormer and the Green viscometers are used chiefly on pigmented materials and are discussed in Volume II. There are two other instruments which indicate viscosity by measuring the resistance the fluid offers to a disc rotated in it. These are the Brookfield Synchro-Lectric and the De Vilbiss Electro-Viscometer. They may be immersed in the liquid in the original container and the viscosity read directly or obtained from conversion charts. The disc in the Brookfield instrument may be rotated at four speeds, and a range of disc sizes is available. By varying the speed of rotation and the size of the disc, a wide range of viscosities may be measured with this instrument. The Brookfield and De Vilbiss instruments are illustrated in Gardner and Sward (1).

Color

Reference

ASTM Designation D154-47.

The most convenient and widely used method for measuring the color of oils, varnishes, and resin solutions is by comparison of the sample with the so-called "Gardner 1933 Standards." These standards consist of various mixtures of ferric chloride solution, cobalt chloride solution, and hydrochloric acid. The mixtures are graduated to produce a set of 18 standard colors, which are sealed into glass tubes. Since these tubes are the same size as those used for the bubble-type viscosity tests described previously, a direct comparison for color may be made on the sample taken for viscosity measurement. The standard color tubes are arranged in a simple holder which has a space between adjacent tubes to insert the sample tube. In this way the sample tube may be moved along the set until its color is seen to be the same as that of one of the standards. The color of the sample is then designated by the number on the Gardner 1933 Standards tube which it matches. If the sample has a color between that of two standards, it is reported as 4–5, 10–11, etc. It is recognized that commercial varnishes and resin solutions require a color range or tolerance in their specifications. Therefore, the specifications may list the color as 4–8, 7–10, etc., indicating that shipments of a particular product may vary in color from the number 4 standard to the number 8 standard, etc. The compositions of the standard solutions are given in ASTM D154-47 and also by Gardner and Sward (1). It should be remembered that, whereas the color of the oil, varnish, or resin is important, it is not necessarily related to the color of the film of these materials when they are dried and aged. The original color of two films may be identical, but one may discolor much more than the other as it ages.

Acid Number. The acid number of an oil or varnish is the number of milligrams of potassium hydroxide required to neutralize the free acids in 1 gm of these materials under the conditions of the test. The acid number of varnishes or resin solutions should be expressed on the basis of the solid components, not on the solution. The acid number is an indication of the relative reactivity of the oil or varnish with basic pigments used in paints and enamels. Usually, the acidity is the result of free fatty acids present in the oil. However, in varnishes, alkyd resins, etc., some of the acidity may be due to resin acids, phthalic acid, and other carboxylic materials. The various acids may not all behave in the same manner with basic pigments as do the fatty acids from oil.

Determination of the acid number is made by titrating a mixture of sample and solvent with a carefully standardized potassium or sodium hydroxide solution free from carbonates, using phenolphthalein as indicator, and a pink coloration as the end point. Some analysts prefer an alcoholic potassium hydroxide solution instead of the aqueous solution, but either may be used. A satisfactory sodium hydroxide solution may be prepared by dissolving 200 gm NaOH in 200 mm of water, allowing it to settle in a stoppered bottle for a few days, decanting the clear liquid into a clean bottle, adding barium hydroxide solution until no further precipitation occurs, and then allowing the solution to settle again until clear. The clear solution should be drawn off and diluted with about 10 liters of freshly boiled distilled water. This should be kept in a bottle having a large soda lime tube in the stopper. The normality of this solution should be determined by titration against a primary standard such as potassium acid phthalate (Bureau of Standards #84a) or benzoic acid (Bureau of Standards #39f), using phenolphthalein as the indicator. The strength of the NaOH solution should be about 0.25 *N*.

Prepare a solvent mixture of equal parts of 95% ethyl alcohol and reagent grade benzene, and titrate it with dilute alkali solution until it shows a pale pink color with phenolphthalein as indicator.

Weigh from 5 to 10 gm of sample into a 300-cc Erlenmeyer flask and add 50 cc of the solvent mixture. Add the indicator, and titrate rapidly with the sodium hydroxide solution until a pink color is obtained which persists for several minutes. The acid number is calculated from the following equation:

$$\text{Acid number} = \frac{\text{cc std. alkali} \times N \times 56.11}{\text{wt. of sample, gm (solid basis)}}$$

Several modifications of this method are described by Gardner and Sward (1). In the ASTM D154-47 method for varnishes, the sample is refluxed before titration in the presence of a measured quantity of 0.1 *N* oxalic acid to decompose any soaps which may be present in it.

Specific Gravity, Pounds per Gallon. The specific gravity of a material is the ratio of the weight of a volume of it to the weight of the same volume of water, both measured in air, and at a stated temperature. When both are measured at the same temperature the value for the specific gravity should be followed by an expression such as 25°/25° or 20°/20°, thereby indicating the temperature. When they are measured at different temperatures the expression should so indicate: 25°/4°, etc.; the temperature below the line is that of the water. Since water weighs 8.33 lb per gal, we can obtain the pounds per gallon of oil, varnish, etc., by multiplying 8.33 by the specific gravity of the material. We can also calculate the inverse of this, namely, the number of gallons obtained from 1 lb, or 100 lb, of oil, varnish, solvent, pigment, etc. The latter value is known as the bulking value. It is important to keep this in mind because many materials in the paint industry are purchased by the pound and sold by the gallon. If two materials are competitive in every respect, and one of them has a higher bulking value than the other, obviously it would be advantageous to purchase the one with the higher bulking value.

The specific gravity of relatively fluid materials may be obtained with a hydrometer. Better accuracy is possible with a balance of the Westphal type. However, for best results a pycnometer should be used which has been calibrated carefully against water. The usual pycnometer is a pear-shaped bottle, with a ground-glass stopper; the better ones have a thermometer in the stopper. Pycnometers are also available having a cylindrical shape and wide mouth, for use with very viscous materials. Extreme care must be taken to eliminate all air bubbles from the sample in the pycnometer before weighing it. The pycnometer should be filled at the temperature specified, although temperature corrections may be made with materials for which the coefficient of expansion is known. Since the specific gravity is the ratio of the weights at equal volumes, the calculation for the sample is made simply by dividing the weight of it which the pycnometer holds, by the predetermined weight of the water in the pycnometer.

The pounds per gallon may be obtained quite readily, and with fair accuracy, by means of a covered cup which contains exactly 83.3 gm of water when full. This type of cup is described by Gardner (1). Simple calculation will show that the weight in grams of material in the cup, divided by 10, will be the weight in pounds per gallon of the material in the cup. Also, the weight in grams of the material in the cup, multiplied by 0.012, will be the specific gravity of the material. This type of cup is widely used for determining the pounds per gallon of architectural and industrial paints. The determination of the specific gravity of pigments requires a special technique which will be discussed in Volume II.

TESTS FOR OILS AND FATTY ACIDS

The identifying characteristics of oils and fatty acids usually include the following:

Viscosity.
Color.
Specific gravity.
Refractive index.
Gel test.
Break test.
Iodine number.
Reagents and solutions.
Diene number.
Spectrophotometric analysis.
Acid number.
Saponification number.
Unsaponifiable matter.
Acetyl number.
Hydroxyl number.
Acetone number.
Titer test.

It will be seen that some of these characteristics are the same as those discussed in the previous section on varnishes and resin solutions. The tests for these characteristics will not be repeated unless special modifications of them are necessary. *Test Methods and Analyses* of the AOCS covers tests for oils and fatty acids in great detail. The section on drying oils and oil seeds in Gardner and Sward (1) also contains a good discussion of the test methods and the various modifications which have been investigated. The work of Hilditch (6) is a valuable fundamental study of the chemical constitution of oils and fats, and Ralston (7) has written an exhaustive treatise on the fatty acids and their derivatives. The two books referred to are valuable contributions to the literature of oils and fats.

Viscosity, Color, and Specific Gravity. The viscosity of oils is usually measured by the Gardner bubble-tube method described in the previous section. The standard tubes cover the viscosity range of A (0.5 poise) to T (5.5 poises), the "heavy-body" standards from U (6.27 poises) to Z6 (148 poises), and the "rubber series" from Z7 (388 poises) to Z10 (1066 poises). Since the rate of rise of the bubble is quite slow in the highly viscous oils, some people prefer to use a method such as the MacMichael, which can be operated at elevated temperatures. The color of oils is measured satisfactorily by the Gardner 1933 Standards. The specific gravity of high-viscosity oils is usually measured with a wide-mouth pycnometer because of the difficulty of eliminating air bubbles. In addition, the pycnometer may be filled at higher temperature than normal, so that the oil is more fluid and the air bubbles are released more easily. By application of the proper temperature correction to the value obtained, the specific gravity may then be expressed at normal temperature. The correction for temperatures between 15° and 30°C is 0.00064 per degree. This amount multiplied by the number of degrees above 15° should be added to the value obtained at the higher temperature.

Refractive Index. The refractive index of a material is a measure of the extent to which a beam of light is bent, or refracted, when it passes from the air through the material. The numerical value of the RI is the ratio of the speed of light in a vacuum to the speed in the material. This value is related to the chemical composition of the oil and its degree of unsaturation, and it is used primarily to indicate changes in these characteristics. However, it is also affected by any oxidation of the oil or by changes in its degree of polymerization. The changes in the refractive index of linseed oil as it is polymerized are given in Tables 6 and 7 in Chapter 2. In view of these changes that occur the refractive index is not used as a criterion for commercial oils as much as it was some years ago. The refractive index is a very simple test to make, provided a good refractometer is available. We shall see in Volume II that the difference between the refractive indexes of oils and resins, and that of pigments, is related to the hiding power of paints and other pigmented coatings.

Gel Test. The heat-gel test is frequently referred to as the Browne Heat Test and is included in ASTM Designation D555-47. A number of variations are described by Gardner and Sward (1). The test is a useful indication of the adulteration of tung oil with other oils. A distinguishing characteristic of tung oil is its rapid gelation at temperatures above 525°F; it also forms a special type of gel. The gel is dry and crumbly and may be cut with a knife like bread, whereas other oils tend to be sticky in the gel form. The rate of gelation is affected by the quantity of oil and the temperature; therefore standard conditions must be provided in order to obtain reproducible results. The apparatus developed by the ASTM is illustrated in Fig. 1.

The general procedure for making the heat-gel test is as follows. Place some cottonseed oil in the 1000-ml tall-form glass beaker to the height indicated. Place the thermometer in the heating bath so that the bottom will be level with the bottoms of the test tubes when they are in place. Heat the bath slowly to 560°F (293°C); then place the two tubes containing the sample and the standard in the bath. Note

Fig. 1. Heat-Gel Test Apparatus (ASTM Designation D557-47)

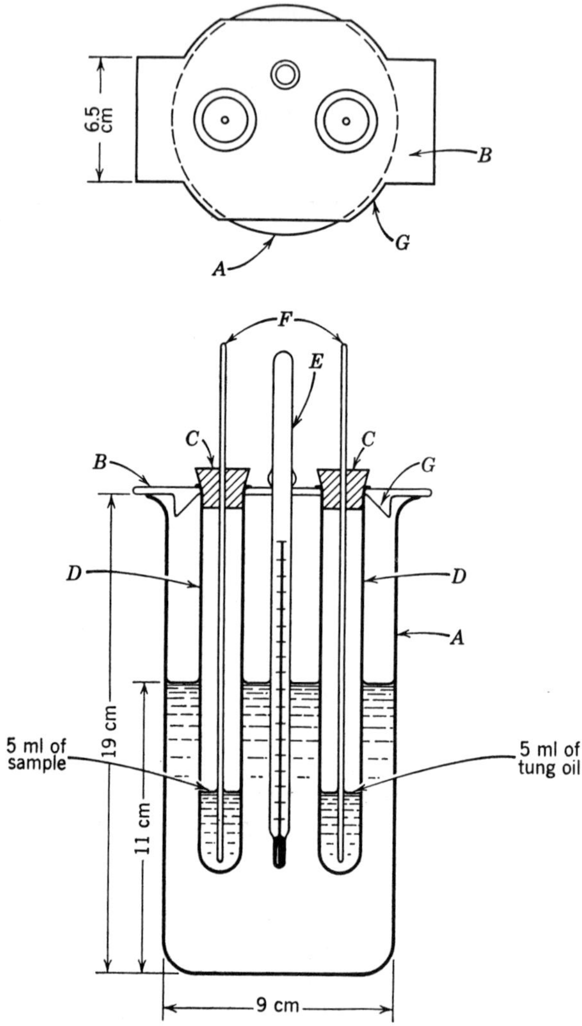

A = 1000 ml tall-form beaker, 9 cm in diameter by 19 cm in height.

B = support plate about 6.5 cm in width, made of Monel metal, aluminum, or stainless steel.

C = corks.

D = test tubes 15 cm by 16 ml.

E = thermometer, range 240°–310°C.

F = glass rods 3 ml in diameter.

G = guide to prevent cover from slipping.

the time, remove the source of heat for about 45 sec, and then reapply the heat source. The temperature of the bath should have dropped to 540°F (282.2°C) before 2 min of elapsed time, and it should be maintained at 540°F. After the samples have been in the bath for 9 min, the rods should be raised at intervals of 15 sec. The gel point is taken as the time required for the oil to become so stiff that the entire tube may be lifted by the rod. The end point should be quite sharp, since the oil changes from fluid to a stiff gel in a few seconds. A good-quality tung oil should gel in 12 min in this test. However, it is desirable to include the known standard oil for purposes of direct comparison. Oiticica oil may also be tested in this apparatus, but it requires about 22 min to gel.

Five milliliters of oil are used as the amount to test, and the test tubes are marked to indicate this quantity. A perforated cork is placed in each tube so that the 3-mm glass rod may be moved freely in it. The thermometer should comply with the requirements of ASTM 32C-44 tung oil type, having a range of 240–310°C. The test does not require that a stirrer be placed in the heating bath, but a screen around it assists in maintaining the temperature uniform.

A great many modified and processed oils have been placed on the market in recent years. One of the principal objectives for such oils is to reduce the heating time of varnishes made with them. No specific test method has been established for evaluating these oils because they vary considerably in the rate of bodying and the temperature at which they should be bodied. However, a satisfactory procedure is to heat from 500 to 1000 gm of oil in an insulated, stainless steel beaker, having a good source of heat. The rate of heating to the top temperature should approximate that of commercial batches of varnish. The top temperature may be from 550° to 585° or 600°F, depending on the type of oil. A sample should be taken when the top temperature is reached, and at 15- to 30-min intervals, depending on the expected rate of bodying. The oil should be taken to the gel stage, and the character of the gel noted. The samples should be tested for viscosity and color, and also acid number if necessary. The rate of change of viscosity with time is usually plotted on semi-log paper for ready comparison. This general procedure also may be applied to the evaluation of a series of varnish resins with a specific oil.

Break Test. The "break" in oils is described in Chapter 2 as the small percentage, about 0.3%, of sterols, phosphatides, and carbohydrates which "break" or separate as a mucilaginous sludge when the oil is heated to varnish-making temperatures. Many attempts have been made (1) to reproduce this effect with a small sample in the laboratory without complete success. It appears that simply heating the small sample to the temperature required, or to higher temperatures, does not give reproducible results, particularly with some oils. The method given below from Federal Specification TT-P-141b, Method 516.1, has been used with some success.

The general procedure is to weigh 25 gm of oil into a 100-ml beaker, add 3 drops of CP HCl (specific gravity 1.19), and stir thoroughly. Place the beaker and contents on a wire gauze, supported on a tripod. Suspend a thermometer in the sample in such a way that the bulb is in the center, and apply heat so that the temperature rises at the rate of 133–144°F (74–80°C) per minute. Do not disturb the sample while it is being heated. Continue heating to 550°F (288°C); then remove flame and allow to cool to room temperature. When cold, stir in about 50 ml of carbon tetrachloride, to dissolve the oil. Filter the sample through a weighed Gooch crucible, containing the usual asbestos mat, by use of suction. Carefully wash all break from the beaker with a stream of carbon tetrachloride, using a rubber "policeman" if necessary. Wash the filter and break with not less than 100 ml of carbon tetrachloride; then dry the crucible and contents at 221° ± 4°F (105° ± 2°C) to constant weight. From the increase in weight of the crucible, and the weight of sample used, the percentage of break may be calculated.

Iodine Number. The iodine number represents the grams of iodine absorbed by 100 gm of oil, which is the per cent iodine absorbed. Since iodine is absorbed at the double bonds, the iodine number should be an indication of the degree of unsaturation of the oil. This is quite satisfactory for comparison of oils such as linseed, safflower, soybean, and others which have non-conjugated unsaturation. However, a comparison of the iodine numbers of linseed and tung oils, when determined by the same method, does not reveal the relative unsaturation in these oils. This is due to the unsaturation in tung oil being of the conjugated type. The conjugation reduces the amount of iodine absorbed to about one-half to two-thirds of the theoretical amount. As a result of this difficulty, a great many variations in the method have been reported in the literature, and complete agreement on this controversial subject has not been reached. However, the Wijs method with its modification for tung oil (larger sample for test) is the one most widely used at present. The method as given below is taken from ASTM Designation D555-47 and is very similar to the method of Federal Specification TT-P-141b, Method 506.1. The Wijs method is rather complicated and time consuming, but attempts to simplify it have not been entirely successful. The reagents and solutions for the Wijs method must be prepared and stored carefully, and the procedure must be followed exactly to obtain uniform results. The method requires that the oil be dissolved in chloroform, followed by addition of the Wijs solution, and storing for 1 hr in a dark place at a specified temperature. Then, potassium iodide solution and water are added, followed by titration wtih sodium thiosulfate solution, using starch solution as the indicator. The end point is indicated by the disappearance of the blue color of excess iodine.

Reagents and Solutions. Acetic Acid: Glacial acetic acid (99%) having a melting point of 14.8°C, and free from reducing impurities. If these requirements are not complied with, the results of iodine-number determination will be erratic. Determine the melting point, and test for reducing impurities as follows.

Melting-Point Determination: Fill a 6-in. test tube about two-thirds full of the acetic acid and insert into the acid an accurate thermometer (ASTM Designation E1-49) by means of a cork stopper fitting the test tube. The amount of acid should be at least double the quantity required to cover the bulb of the thermometer when the bottom of the latter is ½ in. from the bottom of a test tube. Suspend this tube in a larger test tube by means of a cork. Cool the acid by immersing the assembly in ice water until the temperature is 10°C; then withdraw the assembly from the ice water and stir the acid rather vigorously for a few moments, thus causing the super-cooled liquid to crystallize partially and give a mixture of liquid and solid acid. Take a thermometer reading every 15 sec and consider as the true melting point that temperature at which the reading remains constant for at least 2 min.

Test for Reducing Impurities: Dilute 2 ml of the acetic acid with 10 ml of distilled water and 0.1 ml of 0.1 N $KMnO_4$ solution and maintain at 24° ± 3°C. At the end of 2 hr the pink color should not be discharged.

Wijs Iodine Monochloride Solution: Dissolve 13 gm of iodine in a liter of the acetic acid, using gentle heat if necessary, determining the strength by titration with thiosulfate. Set aside 50–100 ml of the solution, and introduce dry chlorine gas into the remainder until the characteristic color change occurs, and the halogen content has been nearly doubled as ascertained by titration. If the halogen content has been more than doubled, reduce it by adding the requisite quantity of the iodine-acetic acid solution. A slight excess of iodine does no harm, but an excess of chlorine should be avoided.

Chloroform: USP grade, acetone-free.

Standard Sodium Thiosulfate Solution: Dissolve pure sodium thiosulfate in distilled water that has been well boiled to free it from carbon dioxide, so that 24.83 gm of $Na_2S_2O_3 \cdot 5H_2O$ will be present in 1000 ml of the solution. Allow this solution to stand for a few hours; then standardize with pure resublimed iodine or potassium biiodate. This solution will be approximately 0.1 N, and it is best to leave it as it is after determining its exact iodine value, rather than to attempt to adjust it to exactly 0.1 N strength. Preserve in a stock bottle with a guard tube filled with soda lime.

Starch Solution: Make a paste of 0.2 gm of soluble starch (potato) in cold water and pour into 100 ml of boiling water, boil for 5 min, cool, and bottle.

Potassium Iodide Solution: Prepare fresh by dissolving 10 gm of KI, free from iodate, in 90 ml of distilled water.

Procedure. Place a small quantity of the oil to be tested in a weighing burette or small beaker. Weigh accurately. Transfer, by dropping, from 0.11 to 0.13 gm of oil (0.16–0.19 gm for tung oil) to a 500-ml bottle having a well-ground glass stopper, or an Erlenmeyer flask having a specially flanged neck for the iodine test. Reweigh the burette or beaker to determine the amount of sample used. Add 10 ml of chloroform. Whirl the bottle to dissolve the sample. Add 10 ml of chloroform to each of two empty bottles like that used for the sample. Add to each bottle 25 ml of the Wijs solution, and let stand with occasional shaking for 1 hr in a dark place at a temperature of 21–23°C. Add 10 ml of the potassium iodide solution (10%) and 100 ml of water. Titrate with 0.1 N sodium thiosulfate, using starch as an indicator. The titrations on the two blank tests should agree within 0.1 ml. From the difference between the average for the blank titrations and the titration on the sample, and the iodine value of the thiosulfate solution, calculate the iodine number of the sample tested. The iodine number is centigrams of iodine for 1 gm of oil, or grams of iodine for 100 gm of oil, which is equivalent to the per cent iodine absorbed.

In view of the difficulties involved in the determination of unsaturation by the iodine absorption procedure, other methods have been tried. By far the most satisfactory method is the spectrophotometric procedure, but this requires rather expensive apparatus. Another method is the complete hydrogenation of the oil, but the procedure is cumbersome, and the results are not believed reliable for heat-treated oils. Another attempt was made by comparing thiocyanogen $(SCN)_2$ absorption with iodine absorption. It is believed that thiocyanogen adds selectively to different degrees of unsaturation. For example, it adds to the double bond in the oleic radical, to one of the two double bonds in the linoleic radical, and to two of the three double bonds in the linolenic radical. By comparing the iodine and cyanogen additions, and setting up a set of simultaneous equations, the proportions of the various saturated and unsaturated fatty acid radicals could be calculated. The procedure and method of calculation are given in detail by Bradley and Kropa (8) and Gardner and Sward (1). The procedure is time consuming and is not used much since spectrophotometric analysis became available. The hexabromide method is in the same category. It depends on the selective absorption of bromine by the different fatty acids, but the reliability of the method for quantitative purposes is doubtful. The method is described in references 1 and 8.

Diene Number (MAV). The diene number, otherwise known as the maleic anhydride value MAV, is used to indicate the amount of conjugated unsaturation in drying oils and fatty acids. It is based on the well-known 1:4 addition (Diels-Alder reaction) of maleic anhydride to conjugated unsaturation, as found in tung, oiticica, and dehydrated castor oils. However, it is shown in Chapter 2 that maleic anhydride also adds to the non-conjugated unsaturation in linseed, soybean, and other oils. In view of this situation it is necessary to establish a carefully standardized procedure in order to obtain reliable results. The method originated with Kaufmann and Baltes (9, 10), and a number of modifications have been proposed (11–13). Details of the method are available in references 1 and 8. In view of the uncertainties of the method it is generally preferred to use spectrophotometric analysis whenever possible.

Spectrophotometric Analysis. Spectrophotometric analysis is based on the fact that organic compounds selectively absorb light of certain wave lengths. These wave lengths include the infrared, the visible, and the ultraviolet regions. The long waves of the infrared are absorbed by molecular groups, such as the hydroxyl, carboxyl, and phenyl. Absorption in the visible is indicated by the characteristic color of various compounds which we see. Absorption of the short waves in the ultraviolet region indicates atomic configurations, such as carbon-carbon unsaturation. Therefore selective absorption in the ultraviolet may be used to determine the type and amount of unsaturation in oils, fatty acids, and their esters.

We shall see in Volume II that spectrophotometers, operating in the visible region, are used to measure the color of transparent liquids and films, and opaque coatings on various substrates. It is well known that the color of transparent or opaque objects which we see is the selected wave lengths of white light which the objects transmit or reflect to us. The other wave lengths of white light are absorbed by the objects. For example, if an object absorbs the blue and green, and transmits or reflects the red, it will appear to us as a red object. Of course, if it absorbs all the visible wave lengths, it will not transmit or reflect any color, and therefore it will appear to be black. When selective absorption occurs in the infrared and ultraviolet regions it is not visible to the eye as color, but it may be photographed on certain films, or re-

Fig. 2. Relationship between Solution Properties and Extinction Coefficient, also between Wave Length and Wave Number

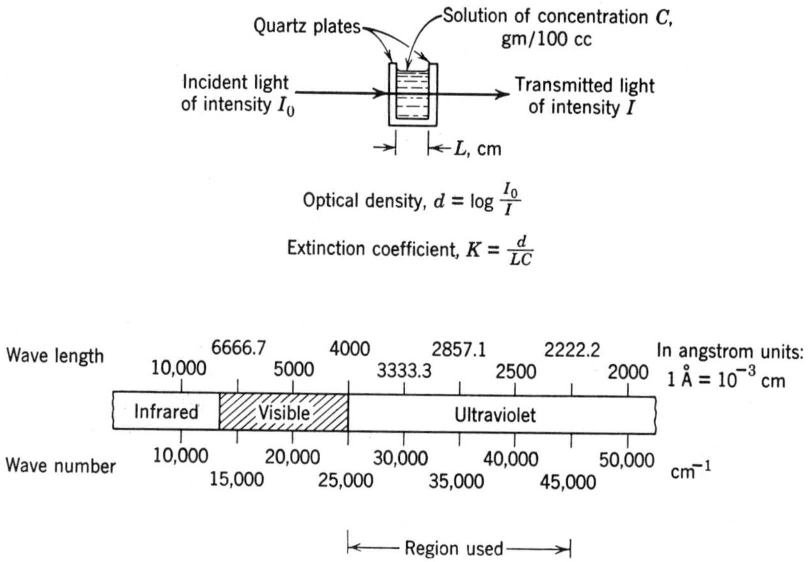

Courtesy of J. P. Kass, in *Protective and Decorative Coatings*, J. J. Mattiello, ed., Vol. IV.

corded on charts by spectrophotometers. The charts used for the visible region have the per cent transmittance or reflectance as the ordinate, and the wave lengths of visible light as the abscissa. However, for ultraviolet absorption it is more convenient mathematically to plot the results in terms which are not so readily visualized. These terms are the logarithm of the extinction coefficient for the ordinate, and the wave number $\times 10^{-3}$ (cm^{-1}) for the abscissa. These terms involve the optical density of a solution of the sample, the concentration of the sample in the solution, the size of the cell in which it is being tested, and the wave length of the light which it absorbs. The relationship between the wave length in angstrom units and the wave number in centimeters $^{-1}$ is as follows:

$$\text{Wave length (Å)} = \frac{10^8}{\text{Wave number (cm}^{-1})}$$

The factors involved in the extinction coefficient are shown in Fig. 2. An indication of the results which may be obtained from the spectrophotometric method is given in Figs. 3 and 4. These are taken from the excellent chapter, "Ultraviolet Absorption Studies of Drying Oils,"

Fig. 3. Absorption Spectra of Pure Fatty Acids

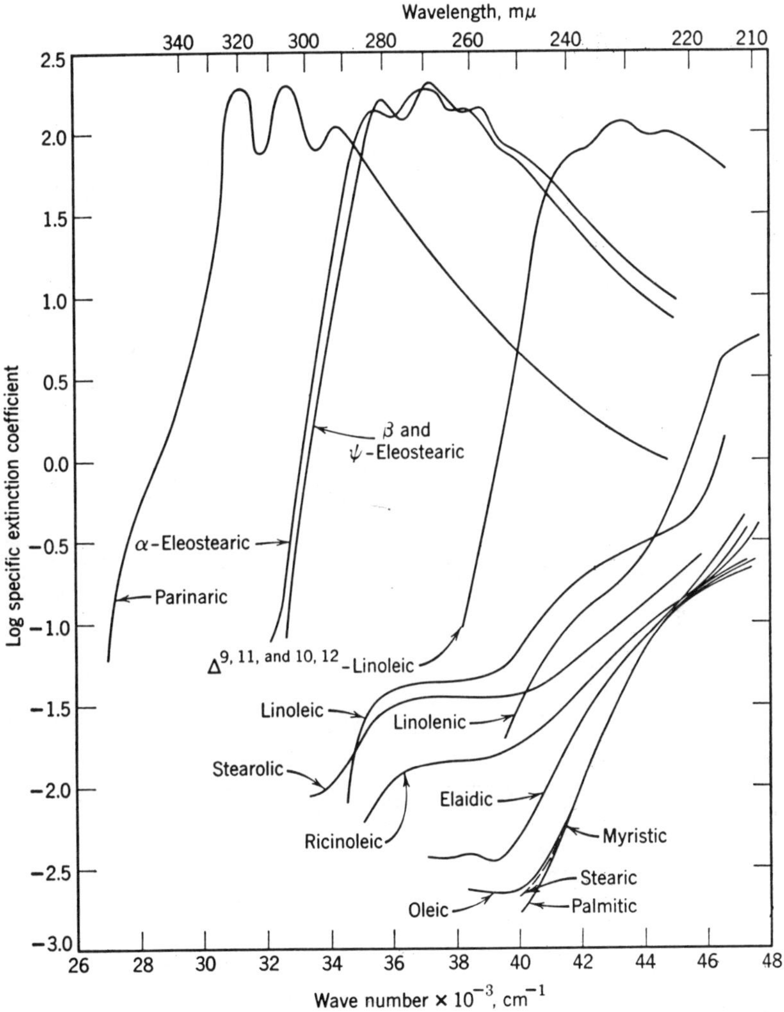

Courtesy of J. P. Kass, in *Protective and Decorative Coatings,* J. J. Mattiello, ed., Vol. IV.

by J. Peter Kass (14) in Volume IV of *Protective and Decorative Coatings*, J. J. Mattiello, ed., John Wiley & Sons, 1944.

Figure 3 shows the ultraviolet absorption spectra of a series of fatty acids. This series includes saturated and unsaturated acids and con-

Fig. 4. Absorption Spectra of Drying, Semi-drying, and Non-drying Oils

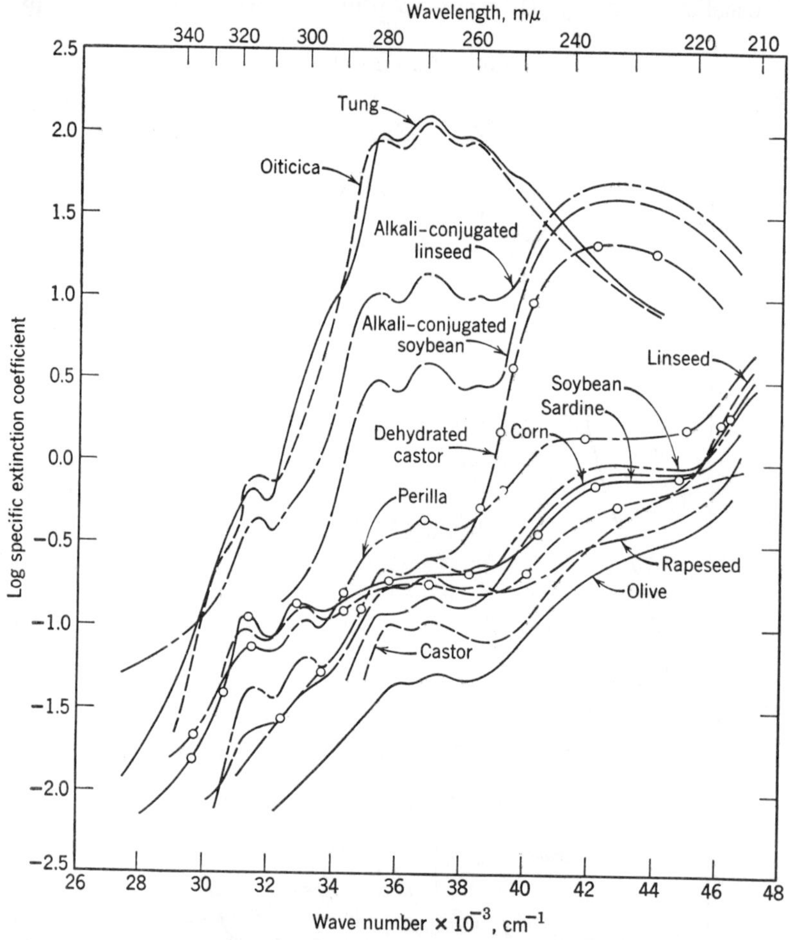

Courtesy of J. P. Kass, in *Protective and Decorative Coatings*, J. J. Mattiello, ed., Vol. IV.

jugated and non-conjugated unsaturation. The formulas for these acids are given in Table 2 for direct comparison with the curves in Fig. 3.

Comparison of the curves in Fig. 3 shows that they fall into two general groups: those having high absorption coefficients, and those

Table 2 Saturated and Unsaturated Fatty Acids

Acid	Formula
Myristic	$CH_3(CH2)_{12}COOH$
Palmitic	$CH_3(CH2)_{14}COOH$
Stearic	$CH_3(CH2)_{16}COOH$
Oleic *	$CH_3(CH2)_7CH{=}CH(CH2)_7COOH$
Ricinoleic	$CH_3(CH2)_5CH(OH)CH_2CH{=}CH(CH2)_7COOH$
Stearolic	$CH_3(CH2)_7C{\equiv}C(CH2)_7COOH$
Linoleic	$CH_3(CH2)_4CH{=}CHCH_2CH{=}CH(CH2)_7COOH$
$\Delta^{9,11}$-Linoleic	$CH_3(CH2)_5CH{=}CHCH{=}CH(CH2)_7COOH$
$\Delta^{10,12}$-Linoleic	$CH_3(CH2)_4CH{=}CHCH{=}CHCH_2(CH2)_7COOH$
Linolenic	$CH_3CH_2CH{=}CHCH_2CH{=}CHCH_2CH{=}CH(CH2)_7COOH$
Eleostearic	$CH_3(CH2)_3CH{=}CHCH{=}CHCH{=}CH(CH2)_7COOH$
Parinaric	$CH_3CH_2CH{=}CHCH{=}CHCH{=}CHCH{=}CH(CH2)_7COOH$

* Isomeric with elaidic.

having low absorption coefficients. Comparison of the formulas shows that the acids in the high-absorption group have conjugated unsaturation, and those in the low-absorption group either have no unsaturation or have the non-conjugated type. This information alone would be of distinct value since it distinguishes between the two types of unsaturation. However, it will be seen that Fig. 3 also distinguishes among the dienoic, trienoic, and tetraenoic acids. The conjugated acids having 2, 3, and 4 double bonds have absorption maxima at different wave lengths. The dienoic acids have a maximum absorption at about 230 mμ, the trienoic acids at about 270 mμ, and the tetraenoic acid at 380mμ. Kass (14) gives further details regarding the precise differences between the absorption characteristics of the acids in Fig. 3. However, it will be quite evident that the spectrophotometric method is a very valuable tool for distinguishing between conjugated and non-conjugated unsaturation, and also for indicating the number of double bonds in a fatty acid radical.

Another excellent illustration is provided in Fig. 4, which shows the absorption spectra of the oils instead of the fatty acids. Again, there are two general groups of curves. The oils in the low-absorption group either have no unsaturation, or have saturation of the non-conjugated type. The curves in this group are not sufficiently different, or characteristic of specific oils, to justify using them for identifying specific oils in this group. However, the oils in the high-absorption group have a characteristic amount and type of absorption. Comparison of the absorption maximum for tung oil in Fig. 4 and eleostearic acid in Fig. 3 shows both at 270 mμ. The tung oil shows no peaks at either 230 mμ for dienoic acids, or at 308 mμ for tetraenoic acids, as may be expected from its known composition. However, the oil has a maximum specific extinction coefficient of 140 (log K = 2.146), whereas the value for eleostearic acid is 189 (log K = 2.276). Since these values are proportional to the concentration, the percentage of eleostearic acid in the sample of tung oil is (140/189) $\times$ 100, which is equal to 74%.

If we consider the curve for dehydrated castor oil we see that the maximum is about 230 mμ, which corresponds to the conjugated diene as may be expected. There is a very weak absorption at 270 mμ, indicating only traces of conjugated triene. The specific absorption coefficient at 230 mμ is 24 (log K = 1.38), and the value for $\Delta^{9,11}$-linoleic acid in Fig. 3 is 115. Therefore the per cent of this acid in the oil is 24/115 = 22%, which is about normal.

The alkali conjugated linseed and soybean oils are referred to in Chapter 2 under Isomerized Oils. Since linseed oil contains both double and triple bonds we would expect the isomerization to produce both types in the conjugated form. These are shown clearly in the curve in Fig. 4, and the calculations indicate 42.7% dienoic and 6.6% trienoic acids. The curve for the isomerized soybean oil indicates about 40% conjugated dienoic and 2% conjugated trienoic acids, which are average values for this type of oil.

The early work on the spectrophotometric method was done by Kass and Burr (15, 16) in connection with their investigation of alkali isomerization. Bradley and Richardson (17, 18) also reported on the value of this method for the identification of conjugated unsaturation in oils and fatty acids. The early measurements were made on the quartz-prism spectrograph with a Hilger Spekker photometer, but a simpler instrument is now available in the Beckman Quartz Photoelectric Spectrophotometer. It is manufactured by the National Technical Laboratories, South Pasadena, Calif. Descriptive literature is available from this company regarding spectrophotometers for analysis in the infrared, visible, and ultraviolet regions of the spectrum. The analyses require only a very small amount of sample, they are quite rapid, and they may be made readily by a trained laboratory assistant. The fundamentals of operation and construction of the spectrophotometer are described by Cary and Beckman (19).

Acid Number of Oils. The acid number of oils is determined in the same manner as that for varnishes, described in the previous section. It is expressed as the number of milligrams of potassium hydroxide required to neutralize the free acid in 1 gm of oil.

Saponification Number. The saponification number is the number of milligrams of potassium hydroxide required to saponify 1 gm of sample. The saponification number is determined by refluxing a mixture of oil and alcoholic potassium hydroxide for 30–60 min and then titrating the excess alkali with acid. This procedure includes neutralizing the free acid in the oil as well as saponifying the oil. To obtain the ester number of the oil it is necessary to subtract the acid number from the saponification number. The saponification equivalent is defined as the milligrams of sample which is saponified by 1 mole (56.11 gm) of potassium hydroxide.

Procedure. Weigh about 2 gm of the sample in a 300-ml Erlenmeyer flask, and add 25 ml of 0.5 *N* alcoholic potassium hydroxide solution. Reflux the mixture gently for 30–60 min. Cool to room temperature, add phenolphthalein as indicator, and titrate the excess alkali with 0.5 *N* sulfuric acid. Run a blank on the potassium hydroxide solution, and subtract the number of milliliters used for the blank from the number used for the sample. The saponification number is then calculated from the following equation:

$$\text{Sap. number} = \frac{56.11 \times 0.5\,N \times \text{difference (ml)}}{\text{Weight of sample (gm)}}$$

The saponification equivalent is obtained as follows:

$$\text{Sap. equivalent} = \frac{56.11}{\text{Sap. number}}$$

Unsaponifiable Matter. Oils and fats may contain small amounts of substances which are not saponified by caustic alkalies but are soluble in normal solvents for oils and fats. These substances include higher aliphatic alcohols, sterols, pigments, and certain hydrocarbons. The unsaponifiable matter is expressed as the per cent by weight of the sample and may be determined by the following method.

Procedure. Weigh 8–10 gm of the oil or clear vehicle of known solids content. Transfer to a 250-ml long-neck flask. Add 5 ml of a 50% aqueous solution of sodium hydroxide, and 50 ml of 95% ethyl alcohol. Put a condenser loop inside the neck of the flask and boil for 2 hr. Occasionally agitate the flask to break up the liquid, but do not project the liquid onto the side of the flask. At the end of 2 hr remove the condenser and allow the liquid to boil down to about 25 ml.

Transfer to a 500-ml glass-stoppered separatory funnel, rinsing with water. Dilute with water to 250 ml. Add 100 ml of redistilled ether. Stopper and shake for 1 min. Let stand until the two layers separate sharp and clear. Draw all but one or two drops of the aqueous layer into a second 500-ml separatory funnel and repeat the process, using 60 ml of ether. After thorough separation, draw off the aqueous solution into a 400-ml beaker, then the ether solution into the first separatory funnel, rising down with a little water. Return the aqueous solution to the second separatory funnel and shake out again with 60 ml of ether in a similar manner, draw off and discard the aqueous solution, and rinse the ether solution into the first separatory funnel with a little water.

Carefully swirl and tumble the combined ether extracts and water rinsings as necessary to obtain complete extraction but avoid emulsion formation; after the water and ether layers have clearly separated, draw off the water. Wash the ether solution with 25-ml portions of water until all the alkali has been removed as determined with phenolphthalein indicator.

Swirl the separatory funnel so as to bring the last drops of water down to the stopcock and draw off until the ether solution just fills the bore of the stopcock. Wipe out the stem of the separatory funnel with a bit of cotton on a wire. Draw the ether solution (portionwise if necessary) into a 250-ml flask and distill off. While still hot drain the flask into a small weighed beaker, rinsing with a little ether. Evaporate this ether, cool the beaker, and weigh. (The unsaponifiable oil from adulterated drying oils may be volatile and as a consequence may evaporate on long heating. Therefore heat the beaker on a warm plate, occasionally blowing out with a current of dry air. Discontinue heating as soon as the odor of ether is gone.) Report the unsaponifiable matter as percentage by weight of solids.

Acetyl and Hydroxyl Numbers. There has been considerable controversy over the value of these two procedures for determining the free hydroxyl groups in oils and resins. The hydroxyl number is a later development and is generally preferred to the acetyl number. The acetyl number is the number of milligrams of potassium hydroxide required to neutralize the acetic acid obtained by saponifying 1 gm of the acetylated sample. The procedure involves acetylating the sample with an excess of acetic anhydride, and then measuring the difference in saponification values of the acetylated material and the original sample. The hydroxyl number is obtained by measuring the excess of acetic anhydride which remains after reacting the sample with standard acetic anhydride in pyridine, and correcting the value for the acidity of the sample.

Acetone Number. The acetone number of polymerized drying oils is the number of grams of acetone which must be stirred into the oil to produce sufficient turbidity to obscure ordinary newsprint. The unbodied and low polymerized oils are freely miscible with acetone, but highly polymerized oil is immiscible and develops turbidity when acetone is mixed with it. The acetone number may be used as a control test to indicate the relative uniformity of polymer size in bodied oils. The test is made by adding acetone from a burette to a weighed sample of the oil in a beaker with constant stirring until sufficient turbidity develops to obscure the newsprint beneath the beaker. Subsequent weighing of the beaker indicates the weight of acetone added. When the rate of heating of bodied oils is not uniform from batch to batch, or when the top temperature is not the same, the batches may be brought to the same viscosity by either reducing or increasing the heating time. However, the polymer size distribution will not be uniform among such batches, and the variation may be indicated by the acetone-number method. For example, if a batch of oil has been overpolymerized the viscosity may be adjusted to specification by addition of a percentage of low-viscosity oil. This blended oil will have a higher acetone number than the standard oil because of its overpolymerized component. A similar test may also be made with blown oils, using mineral spirits, acetone, or other solvents to indicate differences in degree of oxidation during processing.

Titer Test. The titer test is made to determine the temperature of solidification of fatty acids, fats, etc. Since industrial materials may be handled more readily in the liquid condition, it is important to know the temperature of solidification, particularly if this temperature is near normal room temperature. There is a considerable difference in the titer of saturated and non-conjugated unsaturated fatty acids. There is also a large difference between the conjugated and non-conjugated trienoic acids, linolenic and eleostearic. The values in the literature sometimes are given as titer and sometimes as the melting point. The values vary somewhat because of the difficulty of obtaining pure fatty acids. The relative melting points of a few representative acids are given below:

Fatty Acid	Melting Point, °C
Lauric	44
Myristic	54
Palmitic	63
Stearic	69
Oleic	13
Linoleic	−6
Linolenic	−11
α-Eleostearic	48
β-Eleostearic	71

TESTS FOR VARNISH RESINS

A great deal of useful information on the characteristics of varnish resins may be found in *Coating and Ink Resins* by Dr. William Krumbhaar, Reinhold Publishing Corp., 1947. Varnish resins usually are hard, brittle materials, and they include the natural resins, rosin derivatives, terpene, coumarone, and petroleum resins, and pure synthetics such as the 100% phenolic resins. The characteristics of varnish resins which are of general interest are the following:

Softening point.
Acid number.
Saponification number.
Color.
Solubility in oils and solvents.
Bodying rate with oils.
Reactivity with oils and ester gum.

Considerable information on the analysis of resins is given by Gardner and Sward (1) and Stafford and Williams (20)

Softening Point. One of the characteristics of varnish resins, as described in Chapter 3, is that they are "mixtures of mutually soluble materials." Therefore they soften slowly when sufficient heat is applied instead of melting sharply as a single chemical compound does. For this reason it is not satisfactory to make a melting-point determination on varnish resins by the capillary-tube method. Many methods have been developed to determine the softening point of resins, and most of them are described by Gardner and Sward (1). The most widely used method is the ball and ring method (ASTM Designation E-28-42T). The mercury method is preferred by some resin manufacturers, and others have developed specific methods for their particular use.

In the ball and ring method, a steel ball is placed on a layer of resin held in a ring, and the assembly heated in a bath until the resin softens sufficiently to permit the ball to fall through it to the bottom of the bath. The temperature of the bath when the ball reaches the bottom is taken as the softening point. The dimensions of the ball, ring, bath, etc., are given in the ASTM test method. Care must be taken in the preparation of the sample for testing. The resin is melted carefully and then poured into the ring; air bubbles or foam must be avoided to obtain good results. Water is used as the bath for resins which soften below 80°C, and glycerol for resins which soften above 80°C. These liquids have little or no effect on the resin at the softening point, but the Navy Department developed a modification of the method to avoid any possibility of such effects. In this modification the ball and ring assembly is heated in a test tube immersed in a glycerol bath, thus avoiding direct contact of the resin with the glycerol. Consistent results can be obtained with the ball and ring method when the exact procedure is followed carefully.

There are also several modifications of the mercury method. A rather simple procedure is to place a known weight (about 3 gm) of resin in a ⅝-in. by 6-in. test tube with a thermometer. The resin is melted carefully and then allowed to cool and solidify. It is covered with 50 gm of mercury, and the test tube and contents are placed in a glycerol bath. The bath is heated at a uniform rate, and the softening point is taken as the temperature at which the first penetration of resin occurs through the mercury to its surface.

Further information may be obtained by these two methods in addition to the softening point, by observing the character of the resin from the time of initial softening to the end point. Some resins melt quite sharply, whereas others are slow melting and have a stringy consistency. The latter condition usually indicates that the sample has a higher molecular weight or it may be nearer a gel stage. Of course, the softening-point determination is not made on heat-reactive resins, such as the heat-reactive phenolics, because these resins are not heat stable at their softening points.

The softening point is an indication of the relative hardness of resins. The hardness of the resin directly affects the drying time and film hardness of oleoresinous and spirit varnishes. In general, as the hardness of a resin increases it becomes more difficult to dissolve in either oils or volatile solvents. Harder resins usually have better solvent release than softer resins, as discussed in Chapter 3.

Acid Number. Varnish resins may be considered to have low acid numbers if they range from 4 to 8, medium acid numbers from 8 to 12, and high acidity from 12 upward. High acid number may be detrimental if the resin is to be used in varnishes which will be pigmented with reactive pigments. However, high acidity tends to retard the gelation of tung oil and other oils containing conjugated unsaturation, thereby giving better control of the varnish during processing at the top temperature. Since the type of acid may be different among the various varnish resins a single procedure for the acid number may not give complete results in all cases. The general method is to dissolve the resin in a suitable solvent, usually a benzene-ethyl alcohol mixture, and then proceed as indicated for acid number of varnishes on page 607. Several modifications of this general procedure are described by Gardner and Sward (1). The results should be expressed on the basis of resin solids, not on the basis of resin solution.

Saponification Number. The saponification number is obtained on the solution of the resin as used for the acid number. The same general procedure is used as described for varnishes. However, some resins are more difficult to saponify than others; therefore the time of refluxing should be increased in such cases. The results should be expressed as the milligrams of potassium hydroxide required to saponify 1 gm of resin solids, and not on the basis of the resin solution.

Color. The color of resins is usually determined on a solution of the resin, rather than on the solid resin. The exception is the rosin cube standards set up by the Dept. of Agriculture and modified in 1936. The original cubes were made of rosin, but ⅞-in. glass cube standards are now available from the Dept. of Agriculture on deposit of $100.00. Secondary standards of actual rosin cubes are also available from the Supervising Inspector of Naval Stores, Savannah, Ga. An attempt was made by Gardner (1) to correlate the rosin cube standards with 50% rosin solutions in toluol measured by the Gardner color standards. The results were not very satisfactory, but the values obtained are given in Table 3. This table also gives the original names corresponding to the letters used to designate the colors of the rosin cube standards.

Table 3 Color Designation of Rosin Standards

Rosin Cube Standards		Gardner 1933 Color Standards
Letter	Name	50% Rosin in Toluol
X	Extra	6–7
WW	Water White	6–7
WG	Window Glass	7–8
N	Nancy	8–9
M	Mary	9–10
K	Kate	10–11
I	Isaac	10–11
H	Harry	11
G	George	12–13
F	Frank	14–15
E	Edward	16–17
D	Dolly	>18

As stated above the general procedure for obtaining the color of varnish resins is to dissolve the resin in a suitable solvent and compare the color of the solution with the Gardner 1933 Standards described on page 607. When reporting the color by this method care should be taken to state the solvent used and the resin concentration.

Solubility in Oils and Solvents. Some varnish resins are used only for cooking with oils for oleoresinous varnishes, others are dissolved in solvents for spirit varnishes and lacquers, and some are satisfactory for both uses. Many varnish resins are soluble in the various drying oils by simply heating the mixture of resin and oil to the temperature at which the resin is melted completely. Some resins require higher temperatures and longer times to become solubilized in the oil. Other resins may be solubilized in unbodied oils but not in bodied oils. Finally, some resins become insoluble in the oil if the temperature is raised too rapidly. These various types of varnish resins are described in Chapter 3, and the necessary precautions are given regarding their use. A simple test method may be established for evaluating these types of resins by heating a specified quantity of resin and oil under controlled conditions in comparison with a standard resin of known performance. After heating at the specified time and temperature, the sample should be thinned with the required solvent, and the solution compared with the standard for clarity, color, and viscosity. The solutions should be set aside and examined periodically for change in clarity or viscosity, which would indicate stability or change in solubility on aging.

For solubility in solvents the resin should be dissolved by cold cutting or mild heating as required, in the necessary solvents, to the specified resin concentration. The solution should be compared with that of a standard resin for viscosity, color, and clarity. It is general practice to consider that the resin has lower solubility in the solvent if the solution has higher viscosity than that of the standard resin. Partial insolubility is indicated by turbidity in the solution.

If the resin is intended for use in nitrocellulose lacquers or other compositions, its compatibility with these materials should be determined. This is done by mixing the resin solution with the other ingredients in the required proportions and noting whether turbidity occurs in the mixture. If the mixture remains clear it should be poured on a clean glass plate and allowed to dry in a vertical position, under controlled air and temperature conditions. If the film is clear and free from haziness the ingredients of the composition are compatible. It frequently happens that the ingredients are compatible in solution but not in the dry film; therefore both solution and dry-film compatibility tests must be made. It may also happen that the resin is compatible in certain ratios with the other ingredients but not in all ratios. Therefore the compatibility tests are often made in 90:10, 50:50, and

10:90 ratios to determine the general range. Further ratios may be tested if it is considered necessary.

Bodying Rate with Oils. Some of the maleic, penta, and phenolic resins are designed to reduce the time of cooking with oils in varnish making. Since there are many of these resins on the market it may be desired to set up a standard procedure for evaluating them. This may be done quite readily by heating the necessary ratio of resin and oil in a stainless steel beaker, under carefully controlled conditions of temperature increase, and taking samples periodically for viscosity measurement. The sample may be taken to gelation if necessary to determine the rapidity of gelation near the gel stage. The rate of bodying is indicated by plotting the viscosity increase against time at a specified heating schedule. This procedure is similar to that discussed on page 611 for studying the bodying rate of oils.

Reactivity with Oils and Ester Gum. Varnish resins may be classified into two general groups as being non-heat-reactive and heat-reactive with other resins and oils. The non-heat-reactive type simply dissolves in the other resins and oils during varnish making, and its effect on the hardness, drying time, or chemical resistance of the films is generally additive only. However, the heat-reactive type of resin produces a greater effect on these properties than may be expected from the percentage of it in the mixture. The nature of this reaction is somewhat controversial and is discussed in Chapter 3.

The heat-reactive resins are frequently combined with relatively soft resins such as ester gum to raise the softening point. Resins of this type may be evaluated by heating various mixtures of the two resins under controlled time and temperature conditions and then determining the softening point of the resulting composite resin. It may also be desirable to determine the change in solubility of the resin as the percentage of heat-reactive resin is increased. The reactivity of the resin with oils is more difficult to evaluate. Some heat-reactive resins foam considerably while being cooked with the oil, but this may be due simply to the elimination of water as the resin condenses further under the heat treatment. Similarly, an increase in the bodying rate of the oil is not necessarily an indication of actual reactivity with the oil. Of course, the actual structure of the resin-oil complex can be determined, as discussed in Chapter 3, but this involves a difficult and time-consuming analytical procedure. Probably the simplest indication of resin reactivity with the oil is the large increase in water and chemical resistance and exterior durability imparted to the dry varnish film by relatively low ratios of resin. Varnishes of this type are described in Chapter 4.

METHODS OF APPLICATION

It has been well established that the appearance and useful life of a coating may be seriously impaired by faulty application. There are several factors which must be considered for the successful application of coatings. The surface of the object to be coated must be clean and properly prepared. The coating material must have the correct consistency and the proper rate of evaporation of its solvent for the particular method of application being used. It must then be applied uniformly and within the required film-thickness range. Finally, the coated object must be air-dried or baked under carefully controlled conditions of temperature, and in air which is relatively free from dust particles. The most widely used methods of application are brushing, spraying, dipping, and knife coating. A great deal of effort has been expended in the development of laboratory procedures and apparatus for applying coatings of uniform thickness by these various methods. The results are described in considerable detail by Gardner and Sward (1).

Brush Application. Two important considerations in brush application are ease of brushing, or "brushability," and degree of leveling of the coating after it is brushed on. The several pieces of apparatus for quantitatively measuring brushability are illustrated and described by Gardner and Sward. However, good ratings for the relative brushability of a series of coating materials can be obtained from hand brushing by a skilled operator. Brush application is used chiefly for architectural paints. A painter can cover a greater surface per day if the paint brushes easily. Easy brushability is obtained from the use of low-bodied oils or resins in the vehicle. A slow-evaporating solvent makes a paint easier to brush than a fast solvent. Also, a paint with a low pigment content brushes easier than one with a high pigment content. A proper balance must be obtained among these various factors in the formulation of architectural paints.

The flow or leveling of a paint is important because it affects the appearance and gloss. Paints made with raw oils or low-bodied oils and resins usually have poor leveling, whereas paints with high-bodied oils and resins have good leveling. It will be realized that the first type will be easy brushing, but the second type will be hard brushing. It is well known that exterior house paints, made with a large percentage of raw oil in the vehicle, are easy brushing, but they flow so poorly that the brush marks may be seen quite readily in the dry paint film. On the other hand, high-gloss enamels are much more difficult to brush on, but they flow out to smooth, level films. Further discussion of the various pigmented coatings will be given in Volume II. The principal clear coatings which are applied by brushing are varnishes for floor, furniture, marine work, etc. Varnishes flow out readily because they are made from bodied oils and resins and are usually thinned with the relatively slow-evaporating mineral spirits. Attempts to make a brushing type of nitrocellulose lacquers have not been entirely successful because of the very high consistency of the solid portion of the vehicle and the relatively fast rate of evaporation of the solvents. When very slow-evaporating solvents are used the brushability is improved, but the rate of drying is retarded to such an extent that the lacquer loses its significant property of extremely fast drying.

Spray Application. Application of coatings by the spray method is simple and quite rapid, and this procedure is widely used for industrial coatings. There are no brush-mark problems with spray application, but a comparable problem arises known as "orange peel," which is a particular type of surface roughness. It is produced by a combination of fast-evaporating solvent, above-normal viscosity of the coating material, and too great a distance between the spray gun and the object being coated. Obviously, orange peel may be eliminated by adjusting these three factors correctly.

When it is necessary to evaluate a series of coatings which must be applied by the spray method, it is difficult to apply them all at the same film thickness by manual spraying. To overcome this difficulty, the Bell Telephone Laboratories developed an automatic spraying machine which is described in detail in ASTM Designation D823-47T. The apparatus consists of a regular spray gun which can be moved mechanically along a spray table on which test panels are placed. By adjusting the consistency of the coating, the rate of travel of the gun, and the distance of the gun from the panels, any number of panels may be spray-coated so as to have the same thickness of coating on all panels. This apparatus is particularly useful when a large number of panels must be coated for subsequent testing.

Dip Application. Dip application is used whenever possible for industrial coatings because it is the most economical method, and all parts of the object, inside and outside, are coated. However, in order to maintain a reasonable speed on production lines the objects must be withdrawn from the dip tank too fast to permit application of coatings having uniform thickness on all parts. The lack of uniformity in thickness is not sufficient to be of importance for many industrial objects, but it would seriously affect the results obtained on panels prepared for test purposes. Therefore, a laboratory dip-coating apparatus was developed (21) to prepare coated panels for subsequent testing. The apparatus shown in Fig. 5 is known as the Fisher-Payne

Fig. 5. Fisher-Payne Dip-Coater

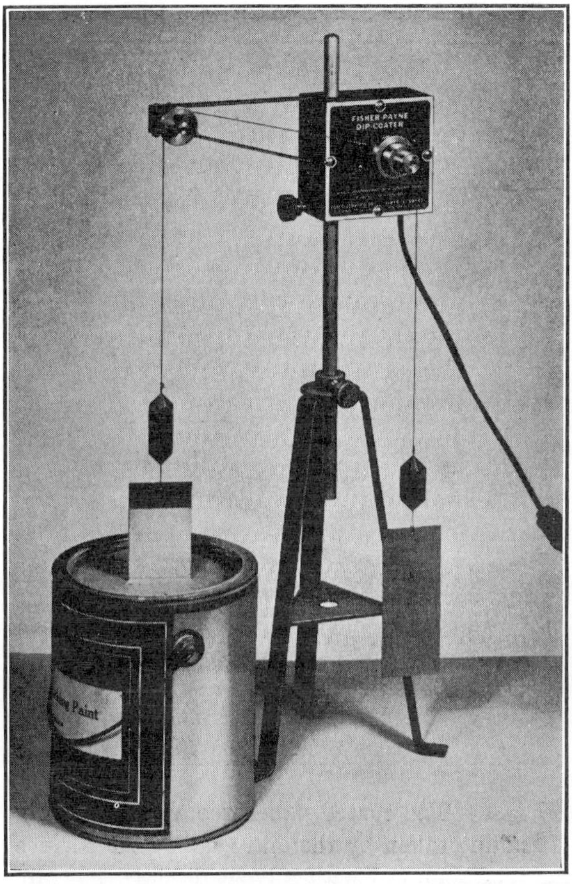

Courtesy of Fisher Scientific Co., Pittsburgh, Pa.

Dip-Coater and is described in ASTM Designation D823-47T. The Dip-Coater may be obtained from the Fisher Scientific Co., or its affiliate, the Eimer and Amend Co.

The Dip-Coater may be used to apply most coatings, but it does not produce uniform films with heavily pigmented or thixotropic materials, since these do not flow well. In the original work with this apparatus, Payne (21) investigated the relationship between viscosity of the coating material and rate of withdrawal of the panel from the dip tank. Various coating materials at different viscosities were applied to steel panels 12 in. long and 3 in. wide, at withdrawal rates ranging from 2 to 12 in. per min. After the coatings were dry, the thickness of the film was measured at 2-in. intervals along the panels

Fig. 6. Film-Thickness Measurement with Dial Gage and Clamp

as shown in Fig. 6. The coated panel was clamped firmly to the base plate, and a reading taken by the dial gage on the surface of the film. The film was then removed with solvent directly beneath the stem of the gage, and a second reading was made. The difference between the two dial readings was taken as the film thickness. Typical results obtained with the Dip-Coater at rates of withdrawal from 2 to 12 in. per min are shown in Fig. 7.

Figure 7 shows that the film thickness was not uniform over the entire panel when the withdrawal rates were greater than 4 in. per min. Therefore the Dip-Coater was designed to have withdrawal rates of 2, 3, and 4 in. per min. Practically any desired film thickness may

Fig. 7. Variation of Film Thickness with Viscosity of Coating Material and Rate of Withdrawal

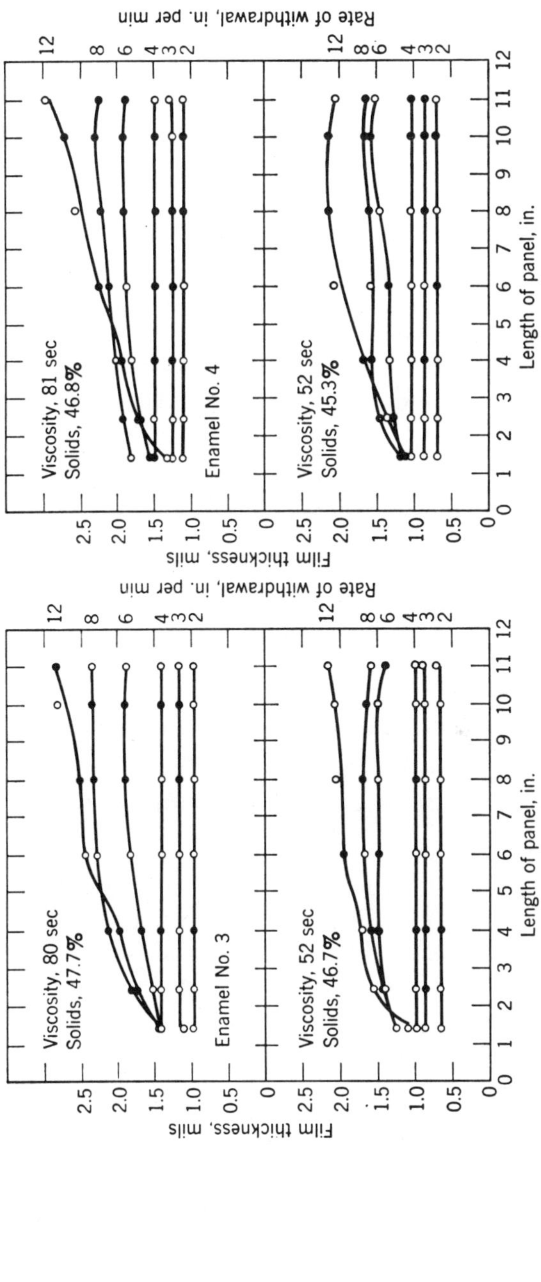

H. F. Payne, *Ind. Eng. Chem.*, *15*, 48 (1943).

be applied at one of these rates with suitable adjustment of the viscosity. The apparatus may be operated by a laboratory assistant to produce uniformly coated panels for test purposes.

Knife-Coat Application. Knife-coat or "doctor-blade" application is widely used for coating paper and textiles, as described in previous chapters. Modifications of this method have been made for laboratory application of coatings to panels for test purposes. Several types of these applicators are described by Gardner and Sward (1). One of the most widely used is the Bird Film Applicator or Bradley Blade (23). It is a rectangular piece of steel which is machined so that it has a beveled undercut to contain the coating material, and a specified clearance which allows a film of the coating to be applied as the blade is drawn down the panel. Blades are available in 2-, 3-, and 6-in. lengths having clearances of 0.5, 1.5, 3.0, and 6 mils. The wet-film thickness is not precisely the same as the clearance because of the drag of the blade on the coating. The dry-film thickness depends on the solid content in the coating material, so that the thickness must be measured directly if the exact dimension is required. An applicator of this type, but having an adjustable blade so that the clearance may be varied, is available from the Henry A. Gardner Laboratory, Inc. Close examination of a coating applied with such a blade shows slight irregularities in film thickness because it is not possible to manually draw the blade down the panel without a slight irregularity in movement. An automatic applicator was developed by Dunn and Baier (22) in which the blade slides down an inclined plane. The apparatus is described by Gardner and Sward (1) and in ASTM Designation D823-47T.

FILM-THICKNESS MEASUREMENT

Instruments for measuring wet-film thickness immediately after the coating is applied, and also dry-film thickness, are described by Gardner and Sward (1). Those for determining wet-film thickness include the Pfund Gage and the Interchemical Gage. The Pfund Gage is based on forcing a convex lens through the paint film and then measuring the diameter of the paint area on the lens. The wet-film thickness (t) is obtained from the relationship between the diameter (D) of the paint area and the radius (R) of curvature of the lens, as follows: $t = D^2/16R$. The Interchemical Gage consists of an eccentric center wheel supported by two concentric wheels which are graduated to show the eccentricity of the center wheel. The measurement is made by placing the gage on the wet film so that the eccentric wheel is above the film and then rotating the gage through one half revolution. The film thickness is read directly from the graduation on the concentric wheel which corresponds to the place where the paint first makes contact with the eccentric wheel.

Devices for determining dry-film thickness include instruments for direct measurement, such as the dial gage shown in Fig. 6, and for indirect measurement by electrical and magnetic devices. When direct measurements are made with a dial gage, care should be taken to measure the panel thickness at exactly the same spot at which the over-all panel-film measurement was made, because a panel varies somewhat in thickness over its area. The indirect instruments are particularly useful for measuring the thickness of films on coated objects as they come off the production line. The instrument is held against the coated object, and the film thickness is read without having to remove or damage the coating. This type of instrument includes the Magne-Gage (American Instrument Co.) and the Enamel Thickness Gage (General Electric Co.). The Gardner thickness gage is designed to force a needle through the paint film and measure the point of contact with the metal base electrically. Direct measurement is obtained from the difference in points of contact of the surface of the film and the metal base. By using a suitable attachment the Interchemical wet-film gage may be used to measure dry-film thickness. This method requires that two strips of film be removed so that the concentric wheels can be rotated on the metal base. The point of contact of the eccentric wheel and the dry-film surface can be located by noting when a light source is shut off visually by the contact. Then the thickness of the film may be read directly from the graduations on the concentric wheels.

It has been well established that the thickness of organic coatings directly affects their service performance. There is an optimum thickness range for most coatings; films which are either below or above this range do not provide maximum service. Illustrations of this statement will be given in Volume II in connection with architectural and industrial finishes.

PERFORMANCE TESTS

Performance tests are made on coating materials and their films to determine their suitability for application, drying schedule, and service requirements. Only the tests which apply to clear coatings will be considered in this section; the tests for pigmented coatings will be discussed in Volume II. The principal materials used for air-drying clear coatings are varnishes and nitrocellulose lacquers. The comparison of varnishes with alkyds for air-drying properties (Chapter 4, page 205) showed that alkyds have only fair through-dry when air-dried. Combinations of alkyds and amino resins have excellent color and hardness when utilized as clear baking finishes. Clear coatings are used for floor varnishes, marine or spar varnishes, furniture finishes, food-can linings, gloss overprint coatings, insulating varnishes, and paper coatings. The tests which will be discussed in this section are:

Skinning.
Gasproof.
Drying schedule.
Hardness.
Flexibility.
Cold-check.
Impact resistance.
Abrasion resistance.
Adhesion.
Exterior durability.
Water resistance.
Permeability to water.
Soap and alkali resistance.
Alcohol resistance.
Electrical resistance.

Skinning Test. Coating materials should be designed so that a "skin" does not form on their surface in full containers. However, many fast-drying coatings tend to form a skin in partially filled containers because of oxidation from the enclosed air. The skin may range in consistency from a soft gelatinous layer to a firm "tight skin." The latter may be removed fairly readily with a spatula, and broken pieces of a tight skin may be eliminated by passing the coating material through a strainer. However, the soft skins are extremely troublesome. They cannot be removed readily with a spatula, and the broken particles slide through the interstices of all regular strainers. They also resist removal by centrifuging the material because they are not of sufficient difference in gravity to be thrown out. If they remain in the coating material they appear as irregularities in the surface of the finish, and constitute one of the forms of "seeding" referred to in Chapter 5.

The skinning test for varnishes in ASTM Designation D154-47 requires that 6 fluid oz of sample be placed in a wide-mouthed, 8-oz glass jar of 2-in. diameter and 4½-in. height, and the cover screwed on tightly. The jar is then inverted, stored in a dark place, and examined periodically for skin formation. The dark place is necessary because light through the glass would accelerate the oxidation and skin formation. Coating materials are stored normally in opaque containers, but skin formation could not be observed without removing the lid, thereby admitting more oxygen.

Gasproof Test. Coatings which contain conjugated oils such as tung, oiticica, and dehydrated castor may form frosted films if dried under certain conditions. The degree of frosting may range from incipient crystal formation, or "crow's footing," to a completely frosted or crystalline appearance. The frosted condition may be caused by air-drying the coating in a cold drafty place, or baking it in a "foul oven," that is, an oven in which there are products of combustion from the burners due to poor ventilation or lack of air circulation.

The gasproofness of a coating may be determined by drying it in a closed container in which there are products of the incomplete combustion of a kerosene lamp or a gas burner. Federal Specification TT-P-141b describes both types; Method 416.1 uses a kerosene lamp in a glass bell jar, and Method 416.2 uses a gas burner with a yellow flame in a metal cabinet with glass sides and door. In general, the gas burner with a yellow flame is a more severe test and is to be preferred. Gardner and Sward (1) describe a similar cabinet used in the laboratory of the Scientific Section. The dimensions of the cabinet are not critical, but certain precautions must be observed when making the test. The first is that the varnish film should be air-dried no more than 15 min before placing it in the gas-test cabinet. If the film has dried to the extent that it is set, it may not frost when placed in the cabinet. The burner with the yellow flame should be placed in the cabinet 10–15 min before the test panels. However, the temperature in the cabinet should be no more than 120°F, and preferably lower. If the oven is too hot the varnish film dries before the frosting can develop. Cabinets of various sizes have been used up to about 2 ft square. It is convenient to have the door and sides of glass so that the development of frosting can be seen and timed. The top should have means to allow the escape of some of the fumes. The yellow flame of the burner should be about 3 in. from the bottom of the cabinet in a central position. The varnish to be tested is poured on tin panels and allowed to drain in a nearly vertical position for 10–15 min. The panel is then placed in the cabinet under the conditions noted above. It may be maintained in either a horizontal or near vertical position, and checked periodically for possible development of frosting. If frosting has not developed in 3–4 hr, the film is usually dry enough to prevent it.

Drying Schedules. Air-drying tests should be performed at the standard 77°F and 50% relative humidity whenever possible. When these conditions are not available, the prevailing temperature and humidity should be mentioned in the report. The coatings may be applied by any of the methods discussed previously, but varnishes are usually poured on tin panels and allowed to drain in a nearly vertical position. The following stages of air drying have been recognized, although not all of them are always reported: dust free, set-to-touch, tack free, dry hard, print free. The definitions of these terms are found in Federal Specification TT-P-141b. This specification is given in detail because it describes the empirical finger-touch method which is widely used, despite the development of various instruments for determining the several stages of drying. Some of the instruments are discussed after the specification.

Method 406.1
15 JANUARY 1949

DRYING TIME (VARNISHES, LACQUERS, AND ENAMELS)

1. PROCEDURE

1.1 Draw down a film of the material on thoroughly cleaned plate glass to a wet-film thickness of approximately 0.0015 inch (unless otherwise specified), place the panel in a horizontal position and allow to dry under standard laboratory air-drying conditions (see sec. 200). The film is tested at points not less than 1.0 centimeter (½ inch) from the edges of the film. It shall be considered to have set to touch when a light touch of the finger shows a tacky condition but none of the coating adheres to the finger. The material shall be considered to have dried hard when the maximum pressure that can be exerted between the thumb and finger (without twisting) does not move the film nor leave a mark which remains noticeable after the film is lightly polished with a soft cloth. It shall be considered to be free from after-tack when, on the test for the dry hard condition (exerted for a maximum of 2 seconds), the film shows no pull on the thumb.

NOTE 1. In the event of a disagreement on the results on the test for after-tack, the umpire method outlined in 1.2 shall be used.

1.2 Umpire method. Place on the film to be tested for after-tack a piece of "No. 1 sulfate gumming kraft" paper measuring 2 by 3 inches. Place a 2-inch-diameter cylinder weight, sufficient to produce a pressure of 2 pounds per square inch on the paper (6.28 pounds), and allow to remain for 5 seconds. Then turn upside down and note whether the paper drops off of its own accord within 10 seconds. To pass the after-tack requirement, the paper should drop off. The paper shall meet the following requirements:

Weight (24 by 36, 500), pounds	90 ± 5
Thickness, inch	$0.0078 \pm 5\%$
Bursting strength, not less than, points	90
Smoothness, Bekk tester, either side, seconds	25 ± 10
Water resistance, dry indicator method, seconds	150 ± 25

All tests to be made on paper conditioned in accordance with the standard method of the Technical Association of the Pulp and Paper Industry.

NOTE 2. Where specification requirements for drying time are stated in terms other than those used above, the definitions in paragraph 2 shall apply. The test films shall be prepared by spraying steel panels to the thicknesses specified in method 204.1.

2. DEFINITIONS

2.1 Dust-free. A film shall be considered dust-free when the finger, without pressure, can be lightly run over the surface of the paint without picking up any paint on the finger. The surface shall be reasonably hard and nonfluid. While the surface may be sticky, no pronounced marks should be left by the finger.

2.2 Tack-free. A film shall be considered tack-free when the finger with a slight pressure will not leave a mark. The surface shall not be sticky.

2.3 Dry through. A film shall be considered dry through when it cannot be distorted or removed by the following test: Place the panel in a horizontal position at a height such that when the thumb is placed on the film the arm of the operator is in a straight line from wrist to shoulder. Bear downward on the film with the thumb, exerting the maximum pressure of the arm, at the same time turning the thumb through an angle of 90° in the plane of the film. Examine for loosening, detachment, wrinkling, or other evidence of distortion of the film.

2.4 Full hardness. A film shall be considered to have reached full hardness when it is very difficult to remove with the fingernail and shows excellent resistance to a knife edge.

2.5 Dry for recoating. A film shall be considered dry for recoating when a second coat can be applied without the development of any film irregularities, such as lifting or loss of adhesion of the first coat, and the drying time of the second coat does not exceed the maximum specified (if any) for the first coat.

Several of the instruments for measuring drying time are based on the fact that powdered materials will adhere to a film while it is drying, but will not adhere after the film has dried. These instruments are designed to deposit the powdered material on the coated panel continuously or periodically over a time period sufficient to include time of application and final dry condition. A variation of this design is to have a steel ball or weighted pin press against the film during drying and to note the condition of the path in the film.

The Sanderson Drying Time Meter (1, 24) deposits a track of sand on a revolving disc on which the coating has been applied. The time at which the sand no longer adheres to the coating can be calculated from the rpm of the instrument and the number of spirals of sand on the disc. The Gardner Circular Drying Time Recorder (1) uses a weighted pin traveling in a circular path to leave a record in the film. Since the pin starts in contact with the wet film some of the coating will adhere to it, which leaves an intermittent series of marks in the film. These marks continue until the film has dried sufficiently so that the pin makes only a smooth track on the surface of the coating.

Dust-Free Tester. The time required for a coating to become dust-free so that it will no longer pick up dust from the atmosphere is important for both architectural and industrial coatings. Therefore a technical committee of the New York Paint & Varnish Production Club developed a Dust-Free Tester (1, 25). It is designed so that a small rectangular layer of powdered material can be deposited manually on the film at intervals during the drying period. After the film has dried, it is blown with a light stream of air and wiped with a soft cloth. The point at which no powder adheres to the film can be determined and is considered the dust-free time. A comparison of various powders, such as fabric flock, walnut shell flour, and colloidal calcium carbonate (Multifex), showed very little difference in results. Therefore, the Multifex (Diamond Alkali Co.) was chosen because of its general availability in paint plants. The instrument was compared with the finger method, and its dust-free time was slightly longer. It is believed that this measurement of the dust-free time is more accurate than can be obtained with the finger method.

Tack Tester. The Atlas Tack Tester (Atlas Powder Co.) is a very simple device to indicate the tack-free time. It is a rectangular strip of metal, bent at an angle of somewhat more than 90°, to give a short arm 1 sq in. in area and a longer arm. A piece of aluminum foil is wrapped around the short arm and then placed in contact with the coating, and a weight is placed on it. The weight is removed after a specified time, usually 5 sec, and if the coating is tack-free the tester will fall over because the long arm overbalances the short arm by an amount equal to 5 gm. The device is particularly useful for fast-drying coatings such as lacquers, but some modification in conditions may be required for slow-drying varnishes and enamels.

Hardness Test. The Sward Rocker (1) is the most widely used instrument for measuring the hardness of organic coatings. The Rocker operates on the principle of the pendulum. Two flat, chromium-plated bronze rings, 4 in. in diameter and 2 in. apart, are weighted so that, when oscillated, they will slowly come to rest. The rate at which the oscillations are slowed down depends on the hardness of the surface on which the Rocker is oscillating. Only a limited number of the oscillations are used in making a hardness rating, and these are determined by a bubble-tube arrangement on the Rocker. The instrument is standardized against plate glass so that the Sward ratings express the hardness of coatings in terms of the percentage of the hardness of plate glass.

The Sward Rocker is a very simple and useful instrument. It is so simple to operate that it is often used without due care being given to the conditions necessary to obtain reproducible results with it. The coating to be tested should be applied to a flat panel, the panel should be in a leveled position during the test, and it should be held rigidly, not simply laid on a bench top. In addition, the hardness of coatings is affected by temperature and humidity; therefore standard conditions should be used if possible. When all these factors are given proper consideration, reproducible results will be obtained. The Rocker gives the best results in the hardness range of 20–45 and is not as sensitive above or below. Air-drying varnishes and enamels range in hardness from 4 to 12 after drying overnight, and may require a week to become as hard as 16–20. Baking finishes are rarely below 20, and the majority are between 30 and 45. The very hard alkyd-amino resin finishes frequently range from 50 to 55. These will be discussed in more detail in Volume II.

Print-Free Test. Since most manufactured articles are wrapped and packaged as soon as possible after the finish has been applied, it is necessary to know whether or not the wrapping material will stick to the finish and mar its appearance. No standard method has been established for determining print-free time because of the wide variation in packaging materials and weight of the articles. The general procedure is to test the finish after air-drying or baking for the specified time by placing a piece of fabric on its surface, applying a weight to the fabric, maintaining the assembly at room temperature or at 110–120°F for a stated time, then removing the weight and fabric, and noting the appearance of the finish. The fabric may be flannelette, burlap, a hard twill, or a coarse paper. The lint from the flannelette adheres if the finish is not print-free. A hard twill cloth is an excellent test fabric since it does not require the finish to be tacky enough to remove lint, but will leave a definite impress of its twill pattern in the finish if it is not print-free. The weight varies from 1 to 10 psi, with 2 psi being a good compromise. The test temperature should be above normal room temperature because many articles shipped in freight cars during the summer reach 120°F. A good method is to prepare two test panels, place fabric and weights on each panel equivalent to 2 and 5 psi, place one panel in an oven at 120°F, and maintain the other under room conditions. Examine the panel in the oven each hour during the day and after overnight. Examine the panel in the room each 24 hr for 1 week. Many satisfactory finishes will show a slight marring with the 5-lb weight at 120°F overnight, or at room temperature for 1 week. The majority of satisfactory finishes will resist marring with the 2-lb weight under these conditions. The print-free test is used chiefly on articles such as furniture and wooden toys which have been finished at air-drying or force-drying schedules. There should be no difficulty in producing baking finishes for metal goods which are print-free under normal shipping and handling.

Flexibility. The flexibility requirements for coatings vary with the substrate on which the coating is applied and the particular use of the coating. Coatings on rigid substrates, such as brick or plaster walls or metal castings for machinery, require only sufficient flexibility to withstand the stress of expansion and contraction with temperature change and possible impact by extraneous objects. Wood may be considered a rigid substrate, but it expands to a greater extent across the grain than it does with the grain under conditions of increasing humidity. Wood is also subject to considerable stress with sudden and extreme temperature drop as used in the cold-check test. Coatings on metal sheets, such as automobile bodies and tin containers, are subject to considerable flexing during normal use; therefore they require a fair degree of flexibility combined with good adhesion and toughness. The extreme case in this direction is coatings on upholstery fabrics for furniture and automobiles. These coatings require a high degree of flexibility and maximum toughness. However, coatings for shower curtains and draperies are required to be "soft" or to have a good "hand" in conjunction with good tensile strength. Coatings on paper vary considerably with the use; most of them require a good degree of flexibility, but some are designed to stiffen and strengthen the paper.

The flexibility of coatings for fairly rigid substrates is usually measured by bending a coated metal panel over a mandrel, and noting the size of the mandrel at which cracking begins. The coatings for nonrigid substrates are evaluated as free films by a combination of an elongation test and a tensile strength test. These tests, and the modulus of elasticity, are described in Chapter 10, together with the values obtained for typical coating materials and plasticizers. A series of stress-strain curves is shown in Chapter 10, Fig. 2, as obtained on the Scott Tester IP-4 (26). Gardner and Sward (1) describe smaller instruments for tensile strength-elongation determinations.

Two types of mandrels are used in the flexibility test: a set consisting of nine mandrels ranging from ⅛ in. to 1 in. in diameter, and the conical mandrel. The mandrel set and suitable holder are available from the Henry A. Gardner Laboratory, Inc. (1). In operation, the coated panel is bent over the mandrel of specified size and examined immediately for possible cracks in the coating. The conical mandrel is described in ASTM Designation D522-41, and is made by Henry Zuhr, 187 Lafayette Street, New York City. The mandrel is a cone-shaped piece of steel 8 in. long, with diameters of 1½ in. at the base and ⅛ in. at the apex. A drawbar with handle arrangement enables the operator to bend the test panel around the mandrel. The panel is 4½ in. in width with a maximum length of 7½ in. and maximum thickness of $\frac{1}{32}$ in. The coated panel is held in a clamp on the instrument with the coated side away from the mandrel. The coating is covered with two sheets of kraft paper lubricated with talc to protect it from the movement of the drawbar. The drawbar is then moved through an angle of 180° in 15 sec, and the coating examined immediately for cracks. The end point is marked and measured where the first continuous crack starts from the small end of the mandrel. The drawbar is then returned to starting position and the panel removed. The results may be expressed as the per cent elongation at which cracking occurs, from the known relationship between diameter of mandrel at point of failure and the combined thickness of the panel and coating. The relationship was developed by Schuh and Theuerer (27), and the method for plotting it on charts so that the per cent elongation may be read directly is given in ASTM Designation D522-41. The mandrel test is simple and is widely used in the industry. It must be recognized that the results are affected by prevailing humidity, and particularly by the temperature. Therefore standard conditions, 77°F and 50% relative humidity, should be used unless other conditions are specified. When it is necessary to determine flexibility at low temperatures, the panel and assembly are held at the required temperature in a cold chamber and bent as described above.

The "knife-curl" test is an empirical test of value in experienced hands, but the results cannot be translated into numerical terms for reports. The coating is cut with the pointed end of a penknife, and the extent of curl is noted. Brittle coatings chip under such action, coatings with medium flexibility curl slightly and then the curl crumbles, but the curl from flexible coatings remains continuous. The test is given in Federal Specification TT-P-141b as Method 617.1. The relative toughness and adhesion of coatings may be obtained simultaneously. A modification of this test is discussed later under adhesion tests.

The kauri reduction test was developed to indicate the flexibility and toughness of varnishes. The kauri reduction values are given in Chapter 4 for some of the varnish formulations. The test requires the addition to the varnish of a 33.3% run kauri gum solution in turpentine until a dried film of the mixture cracks when bent over a ⅛-in. mandrel. Details of the test are given by Gardner and Sward (1) and in Federal Specification TT-P-141b, Method 415.1, and ASTM Designation D154-47. The test was developed many years ago when varnishes were made chiefly with natural resins and were slower drying than those in vogue at present. It gives fair results with present-day varnishes provided the details are followed closely, and the run kauri solution is completely compatible with the varnish. However, no provision is made for measuring the film thickness of the mixture being tested. This may vary considerably, depending on the ratio of run kauri solution to varnish, and the initial viscosity of the varnish.

The run kauri solution is made by heating kauri gum to about 600°F in a distillation flask until 25% has been distilled off. The product is dissolved in freshly distilled turpentine to produce a 33.3% solution. The run kauri solution is mixed with not less than 10 gm of varnish in a specified proportion, and the mixture flowed on a tin panel and maintained in a nearly vertical position at 70–90°F for 1 hr. The panel is baked for 5 hr at 203–212°F in a horizontal position and then removed from the oven and cooled at 77°F for 30 min. The panel is then bent, coated side up, over a ⅛-in. mandrel and examined for cracks at the bend. To pass the test no cracks must appear in the bent area between points ¼ in. from each side of the panel. Varnishes are rated as being able to pass a stated per cent kauri reduction without cracking. The per cent is calculated on the rather confusing combination of solid content of the varnish but total solution of the run kauri. For example, if a varnish having 50% solid content passes a 100% kauri reduction the following mixture will be made:

10 gm varnish (solid content, 5 gm)
5 gm run kauri solution (solid content, 1.66 gm)

This mixture, dried and tested as above, should not crack at the bend.

Cold-Check Test. The cold-check test is designed to indicate the flexibility, or resistance to checking, of clear finishes on wood when subjected to a sudden temperature change. The original test and cabinet were designed by Wayne C. Norris (29). A suitable cabinet is now available commercially from the American Instrument Co. The phenomenon of cold-checking is discussed in Chapter 11 under Furniture Finishes. The finish to be tested is applied to a wood panel and dried in its regular schedule. The type of panel most widely used is 12 × 6 in., 5-ply walnut or mahogany veneer, but other types may be substituted if desired. The finishing system is applied to the surface of the panel, but the back and edges should also be coated to prevent uneven absorption of moisture. The coated panels are placed on a rack and maintained at 120°F for 1 hr. The rack is then removed from the oven and placed in the cold-check cabinet as rapidly as possible. The cabinet is automatically controlled to produce a temperature of −5°F, and the rack and panels are removed after 1 hr. The panels should be examined immediately after removal from the cold cabinet for typical cross-grain cold checks. A cold-check cycle consists of 1 hr at 120°F and 1 hr at 5°F. A poor finish fails in 1 cycle, and a good finish at 10 cycles, but many high-quality finishes do not cold-check at 25 cycles. The test is usually not run for more than 25 cycles.

Impact Resistance. The impact-resistance test measures a combination of toughness and adhesion of coatings. A variety of instruments has been developed for measuring impact resistance (1), most of them based on a falling weight striking the panel from a known height. The Bell Laboratory Impact Tester (28) consists of a mechanically rotated hammer arranged so that it strikes the panel with a glancing blow. The rotation can be varied from 400 rpm to 1900 rpm and the panel moved so that no two blows strike the same spot. The machine shows clearly the type of failure exhibited by various coatings under impact, and the energy required to produce the failure can be calculated from the weight of the hammer and the known rpm.

The falling-weight apparatus is simple to construct; therefore many paint laboratories have made instruments to meet their specific requirements. Two instruments are available from Gardner (1): the Hart Impact Tester and the Variable Impact Tester. A hammer in the Hart Tester may be adjusted to strike the panel from a variety of angles. In the Variable Impact Tester a weight is dropped on the panel from different heights corresponding to impacts ranging from 2 to 28 in.-lb at intervals of 2 in.-lb. The weight falls on the back of the panel so that the coated side is bulged out to an extent determined by the shape of the weight and the height of fall. The height is adjusted until the coating fails, and failure is indicated by cracks radiating from the point of impact. An indication of the relative adhesion of a coating may be obtained by noting the ease with which it can be removed with a penknife from the bulged-out area. Two coatings may crack to about the same extent, but one may have much better adhesion than the other. This also applies to coatings which do not crack under the specified impact conditions.

Abrasion Resistance. Abrasion resistance is a measure of the toughness of a coating, or its resistance to wear. Extremely hard coatings usually have less resistance to wear than softer, rubbery coatings. Gardner and Sward (1) describe the various methods which have been developed, but the three in general use are the falling-sand method, the air-impinged abrasive method, and the rotating-wheel abrader. The falling-sand apparatus is the simplest to construct and is described by Gardner (1) and ASTM Designation D968-48T. The essential elements of the apparatus are a tube through which the sand can fall freely from a funnel at the top. The bottom of the tube is a specified distance from a panel holder which is inclined at an angle of 45°. Below the panel holder is a box or other container in which the sand may be collected. The procedure is to pour 5 liters of 20- to 30-mesh Ottawa sand in the funnel, and remove a disc so that the sand falls through the tube onto the test panel in 52–59 sec. This is repeated until a circular area of the coating 4 mm in diameter is worn away by the sand. The abrasion resistance is expressed as the liters of sand used to remove the 4-mm area of film. The abrasion coefficient is the liters of sand required per mil of film, and is obtained by dividing the volume of sand used by the film thickness in mils. This simple test gives reproducible results when directions are followed carefully.

The apparatus described in ASTM Designation D658-44 uses Carborundum as the abrasive; it is impinged onto the test panel by a stream of compressed air at the rate of 44 ± 1 gm per min. The test panel is placed at an angle of 45° to the horizontal, with the end of the nozzle touching the panel. The abrasive is impinged onto the panel until an area of 2-mm diameter is worn away. The abrasion coefficient is expressed as the grams of abrasive to wear through 1-mil thickness of film under the conditions of the test.

The Taber Abraser is typical of the rotating abrader wheel type of instrument. The coated panel revolves in a horizontal position, and two weighted abrasive wheels are rotated by friction against the panel. The wheels revolve in opposite directions, one sliding radially outward and the other sliding toward the inside of the wear track. Wheels having different degrees of abrasiveness are used, depending on the type of coating under test. The wheels should be cleaned or refaced with emery paper or a refacing stone after each 1000 revolutions or less. The test panel rotates at 50–70 rpm, and the actual revolutions are recorded on a counter. The metal test panels are 4 in. square, having a ¼-in. hole in the center to fit over a pin on the instrument. The area abraded by the wheels is approximately 10 sq cm.

It might appear that the end point could be taken as the number of revolutions required to wear through the coating to the metal panel. However, this is not satisfactory because the irregularities in the panel and the coating do not permit a uniform wearing through over the test area. Therefore the weight of coating lost in a given number of revolutions is often used as the abrasion rating. Since pigmented coatings may vary considerably in specific gravity, the weight loss is not a good indication of the thickness which has been worn away. If the weight loss is converted to volume loss by the relationship between specific gravity and weight, the ratings are more satisfactory. There is a tendency for very hard coatings to chip after they have been worn thin, causing high weight loss. Conversely, coatings which soften under the heat of abrasion tend to pick up abrasive from the wheels, causing a low weight loss. However, good results may be obtained with the majority of coatings.

The results obtained with the falling-sand type of instrument may not check with those from the rotating-wheel type. Furthermore, the results obtained with either instrument may not correspond exactly to the ratings obtained from an actual wear test. This discrepancy may be related to the fact that the abrasion in the instruments is almost continuous, whereas in wear tests such as people walking up a set of stairs or along a passageway on the coating the abrasion is intermittent.

Adhesion. There are two major factors which affect the adhesion of coatings to the surface of various substrates: one is the roughness of the surface, the other is the affinity of the coating for the surface. The first is called mechanical adhesion, and the second specific adhesion. If the surface of the substrate is sufficiently rough practically all coating materials will adhere to it. However, some coatings may be peeled fairly readily from smooth surfaces. Invariably, these coatings are high-molecular-weight materials such as nitrocellulose lacquers, or amino resins which are polymerized during the baking operation.

We have seen in previous chapters that high polymers have great cohesive strength; their cohesion is greater than their adhesion to smooth metal. This would indicate that specific adhesion is a function of the actual molecular surface of the coating in contact with the metal. Improved adhesion of high polymers may be obtained in two ways: one is to add relatively large proportions of low-molecular-weight resins and plasticizers which reduce the cohesive strength, and the other is to add relatively small proportions of a highly polar material, such as the maleic acid in vinyl polymers, which increases the specific adhesion through change in degree of polarity. Both of these methods for increasing adhesion suggest that specific adhesion is a force which operates at the specific interface of coating and substrate. It is most likely that this force does not extend into the coating for more than a few layers of molecules. Therefore any attempt to measure adhesion by the force required to cut away the coating from the metal must apply this force at the specific interface precisely. This would appear to be a practical impossibility because of the relative roughness of even very smooth metal in the dimensions under consideration.

The methods which have been developed for measuring adhesion include the empirical scratching with a knife, measurement of the force required to push away the coating with a mechanically operated knife, and determination of the force required to throw the coating off the surface by very rapid vibration or by very high-speed centrifugal action. From the above discussion of adhesion it would appear that true adhesion can be measured only by the vibration or the centrifugal method. However, information of much practical value can be obtained from the various knife methods, although it does not seem correct to call the property measured adhesion. Consideration of the factors involved indicates that it is a combination of film toughness, adhesion, and condition of the substrate.

Gardner and Sward (1) discuss the many variations which have been developed in methods for testing adhesion. Soller *et al.* (31) discuss the principles and practice of the ultracentrifugal method and compare it with the Interchemical Adherometer (1, 32, 33), the Arco Microknife (34, 35), and the ASTM tensile method (40), Designation D897-49. Moses and Witt (36, 37) describe the evalution of adhesion by the ultrasonic vibration method. Several attempts (38, 39, 40) have been made to measure adhesion by drying the coating on a metal base, then cementing it to a strip of wood or metal, and subsequently pulling the assembly apart in a tensile-strength machine. Two inherent difficulties in this method are (*a*) to find a satisfactory adhesive which will not affect the dried coating in any way, and (*b*) to pull the strip of coating from the metal completely so that the results will be uniform in all tests. The wide variation in sensitivity, hardness, flexibility, and adhesion of coatings precludes a simple solution to these difficulties.

The ultrasonic vibration method (36, 37) developed at the Naval Research Laboratory produces longitudinal ultrasonic vibrations in a metal cylinder on which the coating has been applied. The coating on the free end of the cylinder separates from the metal when the force due to acceleration exceeds the adhesion at the coating-metal interface. The ultracentrifugal method was developed at the University of Cincinnati by Malloy *et al.* (31) and is being investigated by the Bureau of Aeronautics of the U. S. Navy with the possibility that it may be used to determine the suitability of coatings for planes which travel at supersonic speed. In this method the coating is applied as a spot on a ¼-in.-diameter steel ball. The ball is used as the rotor in an electromagnetic type of ultracentrifuge, in which the rotor rotates at speeds from 35,000 to 60,000 rps, in a vacuum, without mechanical support. The speed of the rotor at which the paint spot is thrown off can be obtained from the frequency of an oscillator attached to the instrument. The adhesion in psi can be obtained from this speed, the dimensions of the ball, the weight and specific gravity of the paint, and the equations given by Malloy *et al.* (31). It would appear that this method has possibilities for determining the specific adhesion of coatings, provided all the coating leaves the metal.

The Adherometer has been used satisfactorily with many standard coatings, but it does not give good results with very hard coatings or very tough coatings. The test is made on a 2 × 6 in. steel panel which has been coated except for a ½-in. strip at one end of the panel. The coated panel is placed on a movable table in a horizontal position. The knife, which is placed on the bare strip at the end of the panel against the coating, is 4 mm wide and weighted to keep it in contact with the metal. Since the knife is made of ivory it is not hard enough to cut into the steel. The knife is attached to a weighted pendulum connected to a dial gage. When the table and panel are moved the force which the knife exerts to remove the coating deflects the pendulum to an amount which may be read on the gage. This force is a composite of four separate forces: (1) the force to overcome the friction of the knife against the metal; (2) the force to cut through the coating on each side of the knife; (3) the force to overcome the toughness of the coating as it is being pushed off; (4) the force to overcome the adhesion of the coating to the metal. The first three can be determined separately and then subtracted from the total to give the force of adhesion. The force of friction is determined simply by running the knife over the same track after the coating has been removed, and noting the dial reading. The force required to cut the coating on each side of the knife is determined by removing another strip adjacent to the first one. Since only one side of the knife cuts the coating in this case, the difference in over-all force for the first and second strips must be that amount required to make one cut through the coating. The force to overcome the toughness of the coating is directly proportional to its thickness. Therefore a series of panels is prepared, having coatings of increasing thickness, and the values obtained are plotted on a curve and extrapolated to zero thickness. The remaining value is considered to be the force of adhesion. A value for the specific toughness of the coating per mil may be obtained from the curve. It will be realized that a very hard coating may chip and produce a jagged path instead of a clean cut, and a very tough extensible coating may pile up in front of the knife and tear away in a jerky manner. Both these conditions would make it difficult to obtain a uniform reading of the force on the dial.

The Arco Microknife instrument (34, 35) has a cone-shaped diamond for a cutting tool, and it is arranged so that parallel cuts through the coating can be made at spacings of 1–28 mils apart. When the cuts are made sufficiently close together, the strip of coating between them will be loosened from the metal. The distance between the centers of the cuts is an indication of the relative adhesion of coatings. As can be seen, it would be difficult to obtain clean cuts with hard, brittle coatings or highly extensible tough coatings. Similar tests have been made with other instruments, but making a cross-hatch pattern instead of parallel lines. This general type of test gives good comparative results for a series of coatings when run under carefully controlled conditions. The results do not determine true adhesion but a combination of toughness and adhesion.

Exterior Durability. Exterior durability tests will be considered in detail in connection with house paints, corrosion-resistance paint, etc., in Volume II. Spar varnishes are used as clear coatings on wood for pleasure craft and some architectural applications. These are usually tested for exterior durability by applying two or three coats on mahogany or maple panels and exposing the panels under several types of climatic conditions. One of the major difficulties in this test is the impossibility of obtaining absolutely uniform test panels. No two panels are exactly alike even if cut from the same board. The panels should be protected on the back and edges with at least two coats of aluminum paint. It is advisable to put on the final coat of aluminum paint after the last coat of varnish has dried so that it may be brought over the edge and onto the face of the panel for about ¼ in. This helps to prevent edge creeping of moisture under the varnish film which would cause premature failure.

When a comprehensive test is required, the finished panels can be exposed at several locations, such as a northern urban location, at the seashore, in Florida, and in a southern dry location. Sunshine and moisture are the two major factors in the deterioration of coatings; therefore the Florida test usually gives the fastest results. It has been well established that results from accelerated tests, such as the Florida exposure or a Weatherometer Machine (41), must be used with considerable judgment because they may not be proportional to the results obtained under less severe conditions.

Occasionally, it may be necessary to expose clear finishes on steel panels. Great care must be taken to clean these panels completely, and not to handle them with bare hands after cleaning. Perspiration and other material from the hands will accelerate rusting under the finish, and this rusting would not be uniform on all panels being handled. The hands may be covered with clean cotton gloves, and the clean panels handled by the edges only.

Water Resistance. The resistance of varnish films to immersion in cold water is usually determined by the method given in ASTM Designation D154-47. This requires flowing the varnish on a tin panel, and after it has dried in a vertical position in the laboratory for 48 hr, placing the upper end of the panel in cold water. The panel is examined periodically for at least 18 hr for the development of a whitened appearance. After a stated period of immersion (18, 24, or 48 hr), the panel is removed, and the degree of whitening, if any, is noted. In some varnish films the whitened condition remains permanently; in others it may disappear almost immediately. The time required for whitening to disappear should be recorded. About 48 hr after the immersion test the relative toughness and adhesion of the part of the film which was immersed should be compared with those of the section which was not immersed. This is usually done with the thumbnail or penknife.

If sufficient water penetrates the film and is distributed throughout it, there is enough difference in the refractive indexes of the water and the film material to diffract the light so that it becomes white and opaque. When the film is removed from the water bath the water evaporates out of the film so that it mav become clear again. However, if the spaces in the film occupied by the water do not collapse or return to normal, they will be filled with air. Since there is a greater difference in refraction of air and film material than of water and film, it will be seen that fewer spaces would be required for whiteness to persist.

Many varnishes are used in the tin decorating industry for food containers in which food is processed after the can is decorated. These coatings are usually baked as discussed in Chapter 4. The baked films may be tested for resistance to processing by immersion in boiling water for 10–30 min, and noting any development of whitening and blistering and the disappearance of these effects after removal from the water. However, some specifications may require that the coated panel be placed in a steam chest under the same conditions used in processing the food. The boiling-water test is faster than the cold-water test, but it may not give a true comparison for a series of varnishes which are to be used only at normal temperatures. The tests for resistance to high humidity and salt fog will be discussed in Volume II.

Permeability to Water. The rate at which water permeates a film is of particular interest for coatings on packaging materials and for paints designed to act as water barriers. Several modifications of the permeability-cup method for testing have been developed. They are reviewed in an article by E. J. Dunn (42) in connection with the permeability of paints for wall construction. Standard Method 9–39 of the Federation of Paint & Varnish Production Clubs uses the Payne Permeability Cup, manufactured by the Fisher Scientific Co. (43). The development and use of this cup method are described by Payne (44–47) and by the New York Paint & Varnish Production Club (48). The Payne Cup is a shallow metal cup having a wide flange and a separate ring of the same dimensions as the flange. The film to be tested is placed between the flange and the ring and held in position with clamps. The opening in the ring has an area of 10 sq cm. Usually, water is placed in the cup before the film is put in position, and the assembly maintained in a desiccator at a specified temperature. The cup is weighed periodically to determine the loss, which corresponds to the weight of water which has permeated through the 10-sq-cm area of film in the time period. This condition provides practically 100% relative humidity on one side of the film, and zero humidity on the other side. This can be expressed as the vapor-pressure differential at the particular temperature used. Since many paint laboratories do not have a constant-temperature room maintained at 77°F, it may be desirable to place the desiccator in an oven held at 90° or 100°F so that cooling will not be required in the summertime. The permeability increases rapidly with temperature; therefore temperatures higher than 120°F should not be used unless specifically required. A desiccant such as calcium chloride may be used in the cup instead of water, and the cup maintained in a cabinet with controlled humidity conditions. In this way a range of vapor-pressure differentials across the film may be obtained, depending on the humidity in the cabinet. Under these conditions the cups will increase in weight instead of losing weight.

The Payne Cup method is applicable to testing a wide variety of free films or coatings on various porous substrates. In some cases, the film may be too brittle, or too easily stretched as a free film to handle satisfactorily. These coatings may be applied to special paper (48) or untreated Cellophane as a support. Coatings which require baking to cure them may be supported on glass fiber cloth (46). Ordinarily, the film has sufficient compressibility to act as an efficient gasket between the flange on the cup and the ring to prevent leakage of moisture edgewise. If extremely hard films are being tested, or coatings on thin wooden panels, it is desirable to place a thin gasket of rubber or polyethylene on each side of the test specimen. In all cases the coated side of the specimen should be placed toward the water in the cup. The mechanism of moisture permeation has been discussed by Payne (49) and will be treated in considerable detail in Volume II in connection with corrosion-resistant paint systems. However, it should be realized from previous discussions in this volume that permeability is related to the relative polarity of film-formers and the cohesive forces which bind the micelles in the film.

Soap and Alkali Resistance. The alkali-resistance test is frequently required for varnishes, but the soap-resistance test is used more often for pigmented finishes and will be described in more detail in Volume II. It involves immersing the coated panels in a hot soap solution, and noting the degree of blistering, softening, and loss of gloss after stated time intervals. The concentration of soap is usually 0.5–2%, and the test temperature 160°F. The alkali test is made with solutions of sodium hydroxide ranging from 0.5% to 10% concentration, with the usual being 2% and 5%. Except for extreme cases, the test is made at room temperature. Good results are obtained by placing a drop of the alkali solution on the coated panel, covering it with a watch glass, and washing it off after a stated period. There is a tendency for the drop of alkali solution to spread to different degrees on various types of coatings. This may be minimized by covering the drop with a cover made by cutting off the end of a viscosity bubble tube about ½ in. from the bottom. When the size of the drop is not too large the solution will be held by capillarity to this cover, and also prevented from spreading and evaporating. When comparing a series of coatings for alkali resistance a drop of the solution may be placed on each panel at 15- or 30-min intervals for 2–6 hr, depending on the expected resistance; then all the drops are washed off under running water. The extent to which the various coatings are affected at the different time intervals can be seen readily and given a numerical rating. A plot of the ratings against time shows readily the relative alkali resistance of the series.

ASTM Designation D154-47 specifies that the varnish be applied to a 1×6 in. test tube by dipping the tube into the varnish, immediately inverting the tube, and allowing the varnish to dry for 48 hr. The coated tube is then placed in a 3% solution of sodium hydroxide and examined at various time intervals. The type of failure should be reported as whitening, blistering, extent of removal, etc., after the coated tube has been removed from the test solution, washed under running water, and allowed to air-dry for 30 min. As may be expected, there is considerable difference in the effects produced by the soap test and the alkali test; therefore the relative resistance of a series of coatings in one test may be different from that in the other test.

Alcohol Resistance. Finishes for certain types of furniture, bar tops, etc., are required to be resistant to alcohol-water mixtures. The test is usually made by finishing a wooden panel according to the schedule employed and then placing one or two drops of an equal part mixture of alcohol and water on the finish. The mixture is covered with a small watch glass, or the viscosity-tube bottoms described in the alkali-resistance test, until it has evaporated entirely. The finish is then examined for softening, blistering, or impression of the glass in the surface. This test simulates the condition which exists when a drinking glass is in contact with a wetted surface and allowed to remain overnight. This is not the condition which exists when an alcohol-water mixture is wiped from the surface shortly after being spilled on it. A test to simulate this condition may be made by placing a series of drops on the surface of the finish at 5-min intervals, and covering each drop with a viscosity-tube bottom. Five minutes after the last drop is applied the covers are removed from all drops, and the panel is wiped with one rapid movement of a clean cotton cloth. The surface of the finish will be removed, dulled, or not affected, depending on the extent to which it was softened. If the alcohol-water mixture is replaced with alcohol-ethyl acetate or other solvents used in nail polishes, the test can be used to indicate nail-polish resistance.

Electrical Resistance. Electrical-resistance tests for insulating varnishes are used only by a specialized section of the coatings industry; therefore they will not be included in this book. ASTM Designation D115-48T describes some of the methods for testing electrical insulation. The types of varnished fabric and paper are described in a brochure by D. O. Woodbury, published by the National Electrical Manufacturers Assoc., 155 East 44th Street, New York 17, New York. Chatfield and Wredden (50) describe the type of cloth, methods of manufacture, and properties of varnished cloths for electrical insulation. They do not include the silicone materials.

BIBLIOGRAPHY

1. H. Gardner and G. Sward, *Physical and Chemical Examination of Paints, Varnishes, Lacquers and Colors,* Henry A. Gardner Laboratory, Bethesda, Md.
2. American Society for Testing Materials, Committee D-1, "Standards on Paint, Varnish, Lacquer, and Related Products," 1916 Race St., Philadelphia 3, Pa.
3. American Oil Chemists' Society, *Test Methods and Analyses,* 35 E. Wacker Drive, Chicago 1, Ill.
4. Federal Specifications, Supt. of Documents, Government Printing Office, Washington 5, D. C.
5. N. Y. Paint & Varnish Production Club, *Official Digest Federation Paint & Varnish Production Clubs, No. 241,* 559 (December, 1944).
6. T. P. Hilditch, *The Chemical Constitution of Natural Fats,* John Wiley & Sons, 1947.
7. A. W. Ralston, *Fatty Acids and Their Derivatives,* John Wiley & Sons, 1948.
8. T. F. Bradley and E. L. Kropa, in *Protective and Decorative Coatings,* J. J. Mattiello, ed., Volume V, Chapter 2, John Wiley & Sons, 1946.
9. H. P. Kaufmann and J. Baltes, *Fette u. Seifen, 43,* 93 (1936).
10. H. P. Kaufmann, *Ber.,* 70B, 900 (1937).
11. B. A. Ellis and R. A. Jones, *Analyst, 61,* 812 (1936).
12. R. S. McKinney, N. J. Halbrook, and W. G. Rose, *Oil & Soap, 19,* 141 (1943).
13. G. W. Priest and J. D. von Mikusch, *Ind. Eng. Chem., 32,* 1314 (1940).
14. J. P. Kass, in *Protective and Decorative Coatings,* J. J. Mattiello, ed., Vol. IV, Chapter 12, John Wiley & Sons, 1944.
15. J. P. Kass and G. O. Burr, *J. Am. Chem. Soc., 61,* 3292 (1939).
16. J. P. Kass and G. O. Burr, *J. Am. Chem. Soc., 62,* 1796 (1940).
17. T. F. Bradley and D. Richardson, *Ind. Eng. Chem., 32,* 963 (1940).
18. T. F. Bradley and D. Richardson, *Ind. Eng. Chem., 34,* 237 (1942).
19. H. H. Cary and A. O. Beckman, *J. Optical Soc. Am., 31,* No. 11, p. 682 (1941).
20. R. W. Stafford and E. F. Williams, in *Protective and Decorative Coatings,* J. J. Mattiello, ed., Volume V, Chapter 1, John Wiley & Sons, 1946.
21. H. F. Payne, *Ind. Eng. Chem., 15,* 48 (1943).
22. E. J. Dunn and C. H. Baier, *Ind. Eng. Chem., Anal. Ed., 13,* 427 (1941).
23. Bird & Son, East Walpole, Mass.

24. *Proc. ASTM, 25,* Part II, 407 (1925); *26,* Part II, 556 (1926).
25. N. Y. Paint & Varnish Production Club, *Official Digest Federation Paint & Varnish Production Clubs, No. 286,* 836 (1948).
26. Scott Testers, Inc., Providence, R. I., *Bull. Scott Serigraph,* IP-4.
27. Schuh and Theuerer, *Ind. Eng. Chem., Anal. Ed., 9,* 9 (1937).
28. H. G. Arlt, *Bell Lab. Record, 15,* 108 (December, 1936).
29. W. C. Norris, *Proc. ASTM, 37,* Part II, 478 (1937).
30. Taber Instrument Corp., North Tonawanda, N. Y.
31. A. M. Malloy, W. Soller, and A. G. Roberts, Preprint Booklet, Division of Paint, Varnish, and Plastics Chemistry, American Chemical Society, September, 1952.
32. H. Green and T. P. Lamattina, *Ind. Eng. Chem., Anal. Ed., 20,* 523 (1948).
33. C. J. Rolle and T. L. Dietrich, *Ind. Eng. Chem., Anal. Ed., 21,* 996 (1949).
34. Arco Co., *Ind. Eng. Chem., Anal. Ed., 16,* 144 (1944).
35. Arco Co., *Chem. Inds., 54,* 82 (January, 1944).
36. Saul Moses, *Ind. Eng. Chem., 41,* 2338 (1949).
37. S. Moses and K. R. Witt, *Ind. Eng. Chem., 41,* 2334 (1949).
38. R. P. Courtney and H. F. Wakefield, *Ind. Eng. Chem., Anal. Ed., 6,* 470 (1934).
39. New York Paint & Varnish Production Club, *Official Digest Federation Paint & Varnish Production Clubs, No. 130* (1939).
40. ASTM Designation D897-49T.
41. Atlas Electric Devices, Chicago, Ill.
42. E. J. Dunn, *Official Digest Federation Paint & Varnish Production Clubs, No. 329,* 365 (June, 1952).
43. Fisher Scientific Co., 717 Forbes St., Pittsburgh 19, Pa.
44. H. F. Payne, *Official Digest Federation Paint & Varnish Production Clubs, No. 159,* 297 (1936).
45. H. F. Payne, *Ind. Eng. Chem., 29,* 893 (1937).
46. H. F. Payne, *Ind. Eng. Chem., 11,* 453 (1939).
47. H. F. Payne, *Ind Eng. Chem., 32,* 737 (1940).
48. New York Paint & Varnish Production Club, *Official Digest Federation Paint & Varnish Production Clubs,* Convention Issue, 1937.
49. H. F. Payne, in *Protective and Decorative Coatings,* J. J. Mattiello, ed., Volume IV, Chapter 7, John Wiley & Sons, 1944.
50. H. W. Chatfield and J. H. Wredden, *Varnished Cloths,* Chemical Publishing Co., 1947.

INDEX